Monographs on Endocrinology

Volume 29

Jürg Müller

Regulation of Aldosterone Biosynthesis

Physiological and Clinical Aspects

With 55 Figures and 13 Tables

Springer-Verlag
Berlin Heidelberg New York
London Paris Tokyo

Prof. Dr. med. Jürg Müller

Steroid Laboratory,
Dept. of Medicine, University Hospital
CH-8091 Zurich/Switzerland

This volume is the second, revised edition of Vol. 5 published within the same series

ISBN-13:978-3-642-83122-5 e-ISBN-13:978-3-642-83120-1
DOI: 10.1007/978-3-642-83120-1

Library of Congress Cataloging-in-Publication Data
Müller, Jürg, 1931-
Regulation of aldosterone biosynthesis / Jürg Müller. - 2nd rev. ed.
p. cm. - (Monographs on endocrinology; v. 29)
Bibliography: p. Includes index.
ISBN-13:978-3-642-83122-5(U.S.)
1. Aldosterone - Synthesis - Regulation. I. Title. II. Series.
[DNLM: 1. Aldosterone - biosynthesis. W1 MO57 v. 29 / WK 755 M958r]
QP188.A28M84 1988 599'.01927 - dc 19
DNLM/DLC 87-28520 for Library of Congress CIP

Typesetting: Georg Appl, Wemding.

2127/3020-543210

To John P. Coghlan
in friendship and admiration

Preface to the Second Edition

Sixteen years is a long time, not only in human life but also in the rapid history of contemporary endocrinology. Since the publication of the first edition of this monograph, numerous new lines of research and discoveries have greatly contributed to our knowledge of the physiological and pathological regulation of aldosterone biosynthesis in man and animals. The first reports about a sensitive radioimmunoassay for plasma aldosterone and about a preparation of dispersed zona glomerulosa cells were published in 1970 (Mayes et al. 1970; Haning et al. 1970). These two developments alone turned out to have a tremendous impact on research in aldosterone physiology (for reviews see Coghlan et al. 1979b; J.F.Tait et al. 1980b). In 1971, atrial natriuretic peptides, somatostatin, and the precursor molecule of ACTH had not yet been discovered. Angiotensin antagonists and converting-enzyme inhibitors were not yet available. The clinical syndrome of hyporeninemic hypoaldosteronism was unknown. The possible roles of prostaglandins and dopamine in the control of aldosterone production had not been considered. Cyclic AMP was then the only substance with a clearly established second-messenger function.

The following is an attempt to update the first edition of this monograph by including the most important developments since 1971, but without altering the basic concept and general structure of the book. It is still a heavily biased, one-man's overview of a vast, complex, and rapidly growing area of physiological research. Although the volume of the text and the number of references have been substantially increased, the revised monograph is far from being complete. It is intended to be a road map rather than an encyclopedia.

Zurich, September 1987 — Jürg Müller

Acknowledgements

This book is dedicated to John P. Coghlan, Melbourne, whose own research has greatly contributed to our knowledge of the physiological regulation of aldosterone biosynthesis and who has been to me a constant source of inspiration and encouragement since 1959, when we met in the laboratory of Dr. Ralph E. Peterson. In the winter (or rather Australian summer) of 1973/74, I had the privilege of working at the Howard Florey Institute of Experimental Physiology and Medicine. Many ideas expressed in this book are based on extensive discussions I had then and thereafter with John Coghlan and his colleagues, among them Bruce Scoggins, John McDougall, Marelyn Wintour, and Derek Denton.

All my own research cited in this review has been carried out in the Steroid Laboratory, Metabolic Unit, Department of Medicine, University Hospital of Zurich. It has been generously and continuously supported by the Government of the Canton of Zürich, with substantial contributions under Research Grants 2579, 3853, 4742.3, 3.325.70, 3.018.73, 3.704-0.76, 3.804-0.79, 3.929-0.82, and 3.882-0.85 from the Swiss National Foundation for Scientific Research. I am most grateful for the outstanding contributions made by my inspired, imaginative, and hard-working research assistants Klaus Baumann, Bernhard Trost, Jan Komor, Philippe Haldy, Conradin Meuli, and Markus Lauber and for the excellent technical assistance by Daniella Vogelsang, Catharina Modéer, Gisela Möhren, Lotti Berchtold, Elsbeth Läuffer, Margrit Wellauer, Lilian Frei, Athena Reichardt, Idda Rutz, Birgit Hauri, Eva-Grethe Lund, Lilian Hofstetter, Patricia Canlas, Heidi Seiler, Cornelia Henzi, Carmen Matt, and Sonia Maggetti. Some of the studies were performed in collaboration with Dr. W. Joe Weick, Dr. Walter H. Ziegler, Dr. Rolf Huber, the late Prof. Franz Gross, Dr. Urs Hunziker, and Dr. Dorette B. Brunner. Sincere thanks for timeless support and for giving a wide berth to my always complex but often unsuccessful research ventures are due to my past and present superiors Prof. E. R. Froesch, Prof. A. Labhart, the late Prof. P. H. Rossier, and Prof. P. Frick. I am still very much indebted to Dr. Ralph E. Peterson, who initiated my investigations on the regulation of aldosterone biosynthesis and who was my major teacher in steroid biochemistry. This edition of the monograph would have never been completed without the untiring efforts, patience, and proficiency of Mrs. Rosmarie Meister and Mrs. Martha Salman, two of the best secretaries I have had the pleasure to work with. I am also grateful for the skillful professional work of Mr. I. Glitsch (artist) and Mr. S. Bernasconi (photography). The constructive critical advice given by Dr. Alessandro Capponi and his valuable help in editing Sects. 4.1.5, 4.3, and 4.14 are much appreciated.

Jürg Müller

Preface to the First Edition

Most of our knowledge of the physiological control of aldosterone secretion is based on animal experiments and clinical studies which were carried out in the 1950s and early 1960s by a large number of inspired, ingenious, and meticulous researchers. Their work has been excellently reviewed by – among others – A. F. Muller (1963), Blair-West et al. (1963), Laragh and Kelly (1964), Ganong et al. (1966), Mulrow (1966), J. O. Davis (1967), and Gross (1967). According to the majority of these investigators, aldosterone secretion is primarily regulated by the renin-angiotensin system, with plasma sodium and potassium levels and pituitary secretion of ACTH playing important secondary roles. During the past 6 years, this hypothesis has been generally accepted and only occasionally challenged.

The following is an attempt to take – from the perspective of a relatively simple in vitro model – a new look at the efferent axis of an apparently very complex control system maintaining adequate aldosterone production in the mammalian organism. My views are based mainly on a series of experiments which I have performed in order to study more closely the interactions of adrenocortical tissue with substances capable of directly influencing aldosterone biosynthesis. Since all these studies were carried out in vitro and with rat adrenal tissue only, the information obtained by means of this particular experimental model will be collated with the findings of other investigators who have used a different experimental approach to the study of aldosterone biosynthesis and its regulation in the rat as well as in other animal species and in man.

Contents

1 Zona Glomerulosa of the Adrenal Cortex: Source of Aldosterone

The adrenal cortex is the only organ capable of producing aldosterone in man and other mammalian species. In adrenalectomized patients, sheep, and dogs, as well as in some (but not all) patients with Addison's disease, aldosterone plasma concentrations and secretion rates or the urinary excretion of aldosterone metabolites have been found to be either unmeasurably low or within the range of the reagent blank of the analytical method used (Kliman and Peterson 1960; Bojesen and Degn 1961; Kowarski et al. 1964; Brodie et al. 1967; Coghlan and Scoggins 1967a; J.J. Brown et al. 1968a; Schaison et al., 1970; Scholer et al. 1972). Three cases of ectopic secretion of the hormone by ovarian tumors have been reported. Two androblastomas (Sertoli cell tumor, Ehrlich et al. 1962; malignant sex cord tumor, Jackson et al. 1986) and a malignant arrhenoblastoma (Todesco et al. 1975) were the sources of excessive aldosterone secretion causing elevated blood pressure and potassium loss in a 9-year-old girl, in a 60-year-old woman, and in a 31-year-old woman respectively.

According to a preliminary communication by Antreassian et al. (1972), homogenates of rat or pork kidneys converted tritiated 18-hydroxycorticosterone to aldosterone to a similar extent as did a homogenate of rat adrenals. The conversions were very low, and the experiment needs confirmation. It points to the interesting possibility that some of the 18-hydroxycorticosterone secreted by the adrenal cortex as a by-product of aldosterone biosynthesis might function as a prehormone (Baird et al. 1968) of aldosterone in certain peripheral tissues. However, high amounts of 18-hydroxycorticosterone given as infusions or bolus injections to normal human subjects did not alter the plasma aldosterone concentration and did not induce any measurable change in the rectal potential difference or in the urinary sodium/potassium ratio characteristic of mineralocorticoid activity (Huston et al. 1981). Thus, conversion to aldosterone must have been negligible. Moreover, in adrenalectomized rats, 18-hydroxycorticosterone did not influence sodium and potassium excretion but had a distinctive effect on acid-base balance not shared by aldosterone (Damasco et al. 1979, 1983).

Based on indirect, purely morphological evidence, the concept of a functional zonation of the mammalian adrenal cortex was proposed many years before the discovery of aldosterone. In a review on the pituitary-adrenocortical relationship, Swann (1940) suggested that the "salt and water hormone" was secreted by the glomerular layer of the adrenal cortex, because hypophysectomy led to degeneration of the inner layers only and hardly impaired the vital "salt and water function" of the adrenal cortex. Additional evidence for the production of mineralocorticoids by the zona glomerulosa was found by Greep and Deane (1947) and Deane et al. (1948). Potassium loading or sodium restriction resulted in morphological and histochemical changes in the zona glomerulosa of the rat adrenal cortex, which indicated increased secretory activity. Administration of deoxycorticosterone acetate or potassium restriction caused cytological alterations which were suggestive of inactivity of the zona glomerulosa. In adrenals from hypertensive patients treated with dietary sodium restriction, Peschel and Race (1954) found an increased thickness and a decreased lipid content of the zona glomerulosa. During sodium restriction of rats, increases in zona glomerulosa thickness were associated with an increased capacity of the adrenals to produce aldosterone in vitro, according to Eisenstein and Hartroft (1957). Similarly, the first indications of a close functional relationship between the juxtaglomerular apparatus of the kidney and the secretory activity of the zona glomerulosa of the adrenal cortex were found in morphological studies many years before renin or aldosterone could be measured (Deane and Masson 1951; Hartroft and Hartroft 1955).

Physical spearation of the zona glomerulosa from the zona fasciculata and reticularis in excised rat adrenals was achieved by "decapsulation" by Giroud et al. (1956a). Decapsulated adrenals produced only negligible amounts of aldosterone, whereas the "capsules" produced the same quantities of aldosterone as the complete adrenals. Histological examination indicated that the capsular portion of the adrenal gland contained the whole zona glomerulosa and approximately 20% of the zona fasciculata. Pure zona glomerulosa cells can be isolated from collagenase-treated capsular rat adrenal tissue either by unit gravity sedimentation (J.F. Tait et al. 1972; Braley and Williams 1980) or by gel filtration chromatography (B.C. Williams et al. 1979). Evidence for a zona glomerulosa origin of aldosterone in another species, i.e., cattle, was obtained by Ayres et al. (1956), who incubated selected slices of ox adrenal cortex consisting either mainly of zona glomerulosa or mainly of zona fasciculata (Fig. 1). Per unit weight, "zona glomeru-

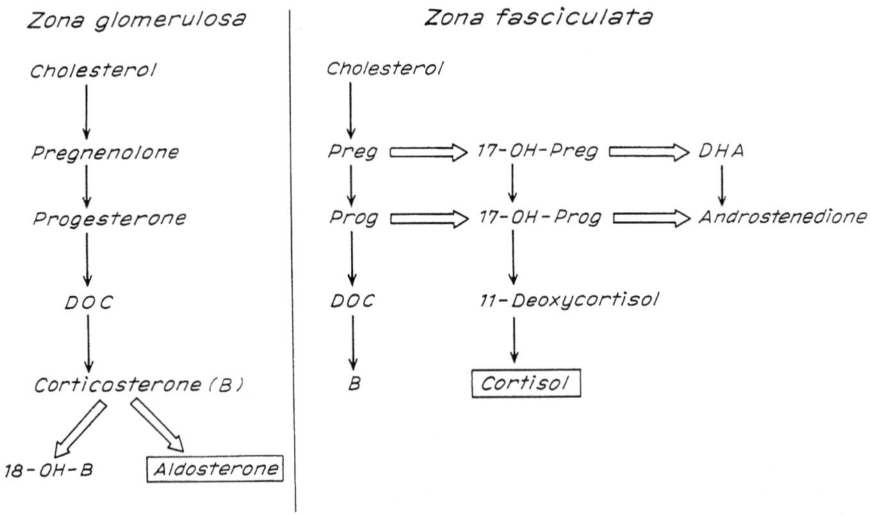

Fig. 1. Zone differences in steroid hormone biosynthesis in the mammalian adrenal cortex. Biosynthetic steps occurring exclusively or mainly in the zona glomerulosa and in the zona fasciculata are shown by *thick open arrows*

losa slices" produced 500% more aldosterone and 40% more corticosterone, but 60% less cortisol than "zona fasciculata slices". According to Peytremann et al. (1973), cells prepared from the inner layers of the calf adrenal cortex, consisting mainly of zona fasciculata, contained ten times less aldosterone than cells prepared from the zona glomerulosa. Most likely, the trace amounts of aldosterone apparently produced by "zona fasciculata cells" were due either to tissue contamination or to methodological limitations. No measurable aldosterone production by decapsulated portions of rat adrenals was observed by Baniukiewicz et al. (1968) or in our laboratory (J. Müller 1970b), not even when tissue from sodium-deficient rats was maximally stimulated by ACTH. Marusic and Mulrow (1967a) found no definite conversion of ^{14}C-labeled corticosterone to aldosterone by beef adrenal zona fasciculata-reticularis homogenates or mitochondria, although these tissue preparations converted measurable amounts of the precursor steroid to 18-hydroxycorticosterone. Thus, we may assume that, at least in these two animal species, all the aldosterone derives from the zona glomerulosa. However, it is difficult to judge from the available evidence whether cortisol is produced exclusively by cells of the inner zones or whether some normal adrenocortical cells can produce both cortisol and aldosterone. Human "aldosterone-producing tumors", i.e., adrenocortical adenomas surgically removed from patients with primary aldosteronism, have generally been found to contain not only high quantities of aldosterone and corticosterone, but also considerable amounts of cortisol (Neher 1958; Louis and Conn 1961; Biglieri et al. 1963; Kaplan 1967). Aldosterone-producing tissue converted added cortisol to 18-hydroxycortisol and to 18-oxocortisol (Ulick et al. 1983). High amounts of these two steroids were found in the urine of patients with either an aldosterone-producing adenoma or glucocorticoid-suppressible aldosteronism (Chu and Ulick 1982; Ulick and Chu 1982; Gomez-Sanchez et al. 1984). Accordingly, at least in these two disorders, aldosterone and cortisol might be produced by the same cells. Although this does not seem to be the case in normal man, small but definite amounts of 18-hydroxycortisol and 18-oxocortisol were measured in the plasma and urine of normal human subjects (Corrie et al. 1985; Gomez-Sanchez et al. 1984). According to Corrie et al. (1985), plasma 18-hydroxycortisol was approximately 100 times lower than plasma cortisol but eight times higher than plasma aldosterone. Plasma 18-hydroxycortisol paralleled plasma cortisol in its diurnal variation and its suppressibility by dexamethasone. It was raised by sodium restriction but not by an angiotensin-II infusion. Most likely, 18-hydroxycortisol is produced by the zona glomerulosa from circulating cortisol during a second passage through the adrenal cortex. According to Vinson et al. (1978a), capsular portions of rat adrenals produced cortisol and testosterone from endogenous precursors and converted labeled progesterone to 17-hydroxylated steroids such as cortisol, 11-deoxycortisol, 17α-hydroxyprogesterone, and testosterone to the same extent as the decapsulated portions did. Nevertheless, these investigators concluded from indirect evidence that 17α-hydroxylation probably did not occur in the zona glomerulosa. Moreover, it is still uncertain whether the rat adrenal cortex produces any cortisol at all except from added 17α-hydroxysteroids (Bush 1953; Péron 1960; Brownell et al. 1963; Laplante et al. 1964; R. B. Young and Sweat 1967; K. Baumann and Müller 1972b). In our laboratory, we have observed no measurable cortisol production by rat adrenal tissue from either endogenous precursors or added progesterone.

In nonmammalian vertebrates the adrenal does not show the typical zonation of the mammalian gland. However, Varma (1977) has presented indirect ultrastructural evidence which indicates that in bullfrog adrenal tissue aldosterone and corticosterone may be produced by different cells. Tissue slices of the subcapsular zone of duck adrenals produced relatively more aldosterone than tissue slices from the inner zones did during superfusion with ACTH (Klingbeil 1985b). Only subcapsular tissue responded with an increased aldosterone output to superfusion with angiotensin II.

2 Pathway of Aldosterone Biosynthesis

2.1 Classical Pathway and Alternative Pathways

The biosynthetic reactions required for the production of aldosterone from cholesterol are shown in their generally accepted sequence in Fig. 2. Since the exact nature of the two final steps, i.e., the ones involved in the conversion of corticosterone to aldosterone, is still uncertain (Neher 1979), the less presumptive terms "methyl oxidation 1" and "methyl oxidation 2" as proposed by Ulick (1976) are preferable to the more conventional terms "18-hydroxylation" and "18-hydroxydehydrogenation". In particular, the conversion of 18-hydroxycorticosterone to aldosterone appears to be a reaction more complex than the simple dehydrogenation of a hydroxyl group (see p. 9).

Several other pathways have been suggested by Wettstein (1961) among others, who incubated tritiated progesterone with bovine adrenal homogenates and found that it was incorporated into 18-oxoprogesterone, 11β-, 18-dihydroxyprogesterone, and 21-deoxyaldosterone, three possible intermediates of aldosterone production by alternative routes. Added 18-hydroxy-11-deoxycorticosterone was converted to aldosterone by amphibian and mammalian adrenal tissue, according to Kahnt and Neher (1965), Nicolis and Ulick (1965), De Nicola and Birmingham (1970), Melby et al. (1971), Grekin et al. (1973), Fattah et al. (1977), and J. Müller (1980a). Aldosterone was formed from labeled 18-hydroxyprogesterone in bullfrog adrenal slices (Nicolis and Ulick 1965). 11-Dehydrocorticosterone-4-^{14}C was converted to aldosterone by rabbit adrenal tissue (Fazekas and Kokai 1967) as well as by homogenates of monkey and frog adrenals (Sharma 1970). Addition of substrate amounts of 11β-hydroxyprogesterone to bovine adrenal slices (Stachenko and Giroud 1964) and of 21-deoxyaldosterone, 11-dehydroaldosterone, or 21-hydroxypregnenolone to a bovine adrenal homogenate (Kahnt and Neher 1965) led to an increase in aldosterone output. Thus, the reactions listed in Fig. 2 can theoretically occur in a number of alternative sequences. During the past 25 years, most studies on aldosterone biosynthesis from endogenous precursors have yielded results which are at least compatible with the assumption that most if not all of the hormone derives from the "classical" sequence cholesterol $\rightarrow$ pregnenolone $\rightarrow$ progesterone $\rightarrow$deoxycorticosterone $\rightarrow$corticosterone $\rightarrow$18-hydroxycorticosterone $\rightarrow$aldosterone. The extent of aldosterone production by other pathways remains to be investigated. However, the available evidence does not exclude the possibility that the activation of an alternative biosynthetic pathway could play an important role in the physiological control of aldosterone production (Blair-West et al. 1970b, 1971d).

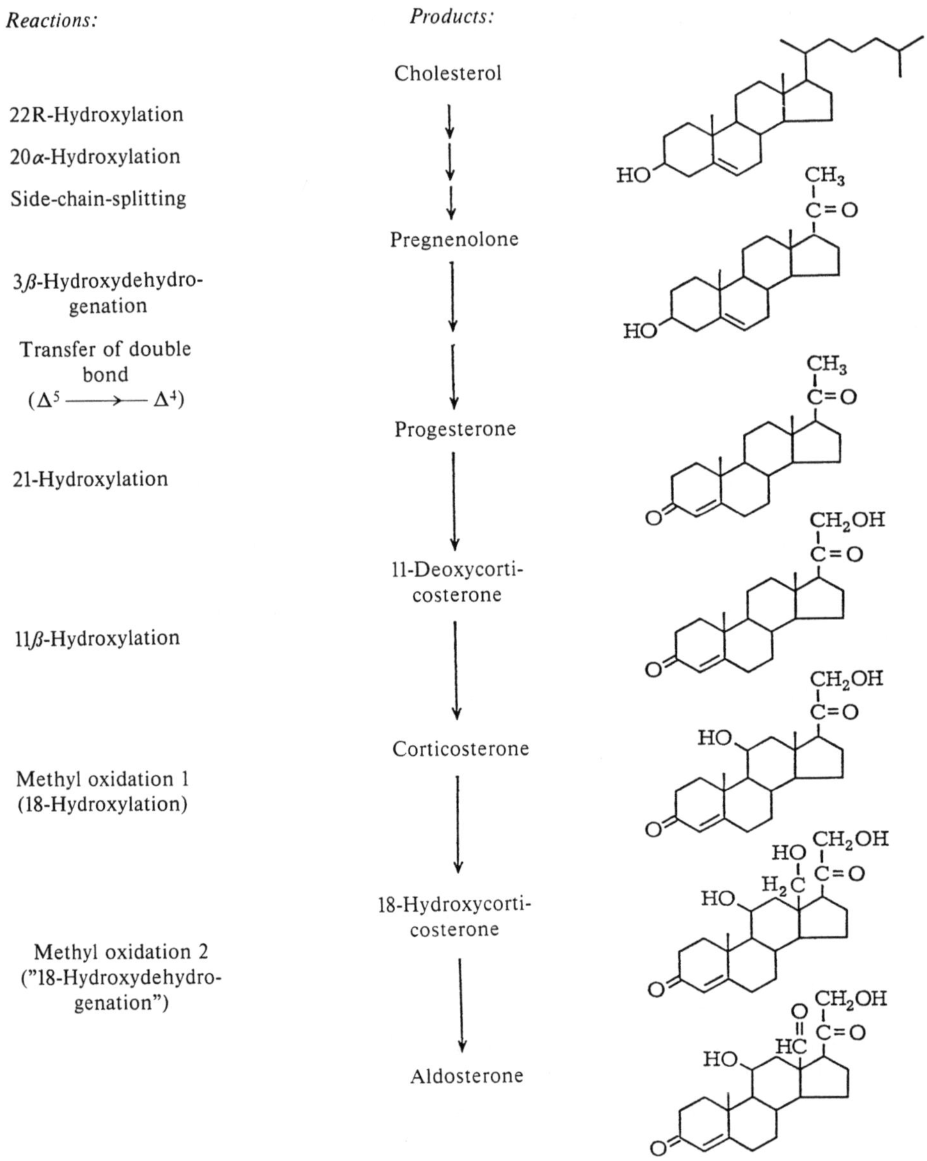

Fig. 2. Major pathway of aldosterone biosynthesis

2.2 Corticosterone as an Intermediate Product

Labeled corticosterone is converted to aldosterone in high yield when incubated with adrenal tissue of man (Mulrow and Cohn 1959), ox (Ayres et al. 1975), dog (W. W. Davis et al. 1968), rat (J. Müller 1966), duck (Donaldson et al. 1965), and bullfrog (Nicolis and Ulick 1965). Based on the final specific radioactivities of corticosterone and aldosterone isolated from bovine adrenal capsule strippings

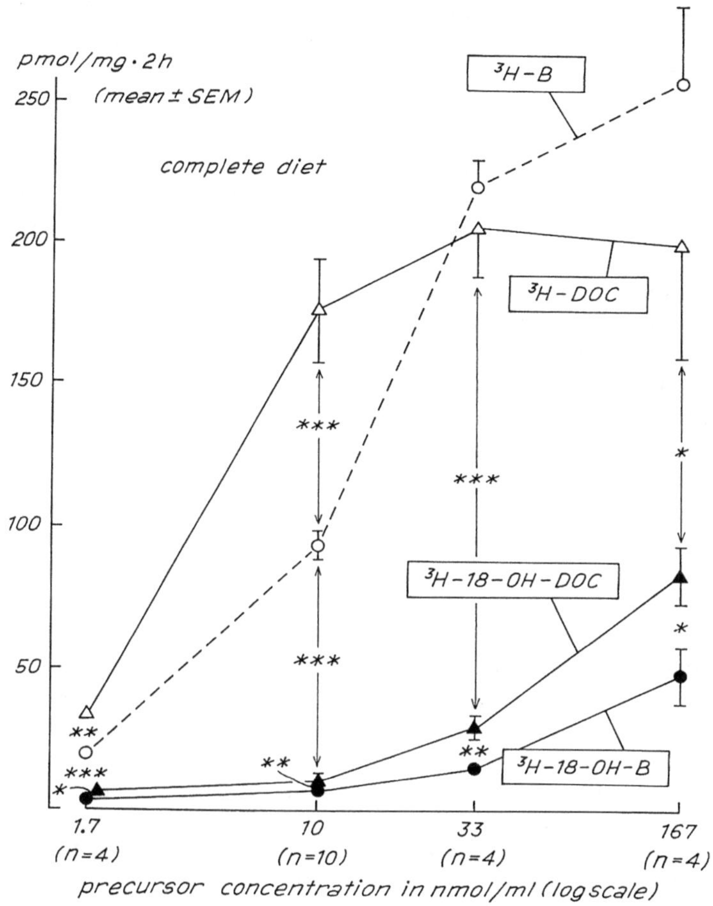

Fig. 3. Conversion to aldosterone of tritiated deoxycorticosterone *(DOC)*, corticosterone *(B)*, 18-hydroxy-11-deoxycorticosterone *(18-OH-DOC)*, and 18-hydroxycorticosterone *(18-OH-B)* by capsular adrenals of rats kept on a complete diet. *Asterisks* denote the significance of the differences between mean values on adjacent curves as calculated by *t*-tests: *$P<0.05$, **$P<0.01$, ***$P<0.001$. (From J. Müller 1980a)

("zona glomerulosa") which had been incubated with trace amounts of ^{14}C-labeled corticosterone, Ayres et al. (1960) calculated that at least 50% of the aldosterone produced was derived from corticosterone. When labeled corticosterone was added in substrate amounts, at least 92% of the aldosterone was produced from this precursor. The results of more recent experiments carried out in our laboratory (J. Müller 1980a) are consistent with these older findings. When capsular portions of rat adrenals were incubated with low, nonsaturating precursor concentrations, labeled deoxycorticosterone was converted to aldosterone 1.5–2 times more efficiently than labeled corticosterone (Fig. 3). Thus, up to 50% of the aldosterone derived from deoxycorticosterone appeared to have proceeded from a pathway not leading through corticosterone. At higher, near-saturating or saturating concentrations, corticosterone was converted to aldosterone at least as efficiently as deoxycorticosterone, indicating that all of the aldosterone derived from deoxycorticosterone could have been formed via corticosterone. Slaga and Krum (1973) incubated rat and rabbit adrenal tissue with trace amounts of radioactive pregnenolone and progesterone. According to the order of decreasing specific radioactivities of various conversion products, corticosterone, 11-dehydrocorticosterone, and 18-hydroxycorticosterone were possible precursors of aldosterone in the rabbit adrenal, whereas corticosterone, 18-hydroxycorticosterone, and 18-hydroxy-11-deoxycorticosterone were possible precursors of aldosterone in the rat adrenal.

In the mammalian adrenal cortex, the zona fasciculata is the major source of corticosterone. In the rat species, corticosterone is the main glucocorticoid (Bush 1953); most of the corticosterone originates from the inner zones of the adrenal cortex and is produced under ACTH control (Baniukiewicz et al. 1968). In man, the secretion rate and plasma level of corticosterone (Peterson 1957, 1959) follow closely those of cortisol, which is presumably produced only by the zona fasciculata-reticularis. According to Bledsoe et al. (1966), approximately 90% of the corticosterone production of normal man is ACTH dependent. When endogenous ACTH secretion was suppressed by dexamethasone, corticosterone secretion was altered in the same direction as aldosterone secretion in response to changes in sodium balance. Elevated secretion rates of corticosterone were observed in some patients with primary aldosteronism but not in patients with secondary aldosteronism due to renovascular or malignant hypertension (Biglieri et al. 1968). Corticosterone is produced by pure rat adrenal zona glomerulosa cell preparations (J. F. Tait et al. 1972; B. C. Williams et al. 1979). In impure zona glomerulosa cell or tissue preparations, the corticosterone output from endogenous precursors increases in response to added stimulators such as angiotensin II, potassium ions, and serotonin, which act exclusively or preferentially on the zona glomerulosa (Kaplan and Bartter 1961; Kaplan 1965; Baniuskiewicz et al. 1968; S. A. S. Tait et al. 1970; J. Müller 1970b; Haning et al. 1970; McDougall et al. 1976).

2.3 18-Hydroxycorticosterone: Intermediate Product or By-product?

It is generally assumed that the conversion of corticosterone to aldosterone is a two-step reaction in which 18-hydroxycorticosterone is an obligatory intermediate product. This assumption is based mainly on analogy and indirect evidence; the direct evidence is rather conflicting. Tritiated 18-hydroxycorticosterone was converted to aldosterone in high yield by slices of the adrenal cortex and an adrenal tumor surgically removed from a patient with primary aldosteronism (Pasqualini et al. 1964). A considerable rate of conversion of tritiated 18-hydroxycorticosterone to aldosterone by sheep adrenal mitochondria was observed by Raman et al. (1966). According to Fattah et al. (1977), tritiated 18-hydroxycorticosterone (added in trace amounts) was incorporated into aldosterone by incubated capsular portions of rat adrenals to approximately the same extent as labeled corticosterone, deoxycorticosterone, and progesterone. However, a majority of investigators found added 18-hydroxycorticosterone to be a rather poor precursor of aldosterone. Incubation of radioactive 18-hydroxycorticosterone with bullfrog adrenal homogenate or mitochondria did not result in detectable synthesis of aldosterone (Psychoyos et al. 1966); under the same experimental conditions corticosterone was converted in good yield to 18-hydroxycorticosterone and aldosterone. Bovine, frog, and human adrenal tissue slices or homogenates converted 18-hydroxycorticosterone to aldosterone (Sandor and Lanthier 1963; Nicolis and Ulick 1965; Kahnt and Neher 1965), but the yields of aldosterone were substantially (3–80 times) larger when corticosterone was incubated under the same conditons. Similarly, the rate of conversion of deoxycorticosterone and progesterone to aldosterone was markedly higher than the rate of conversion of the 18-hydroxy derivatives of these two steroids. In our own laboratory, labeled corticosterone and deoxycorticosterone were converted to aldosterone 9–25 times more efficiently than their respective 18-hydroxy derivatives when added in low substrate amounts to incubated capsular portions of rat adrenals (J. Müller 1980a; Fig. 3). At higher concentrations the convertibility of the two 18-hydroxysteroids increased, but it did not reach the convertibility of the 18-deoxysteroids.

It was previously suggested by Nicolis and Ulick (1965) that the low rate of conversion of exogenous 18-hydroxysteroids could be due to the fact that in aqueous solution these compounds existed mainly as 18,20-cyclic hemiketals and were more resistant to enzymatic dehydrogenation in this form than in the open α-ketol form. At that time, these authors thought that a dehydrogenase was responsible for the conversion of 18-hydroxycorticosterone to aldosterone, a view also held by Greengard et al. (1967). However, according to more recent evidence (Marusic et al. 1973; Rapp and Dahl 1976; Aupetit et al. 1977b, 1979, 1983; I. Kojima et al.

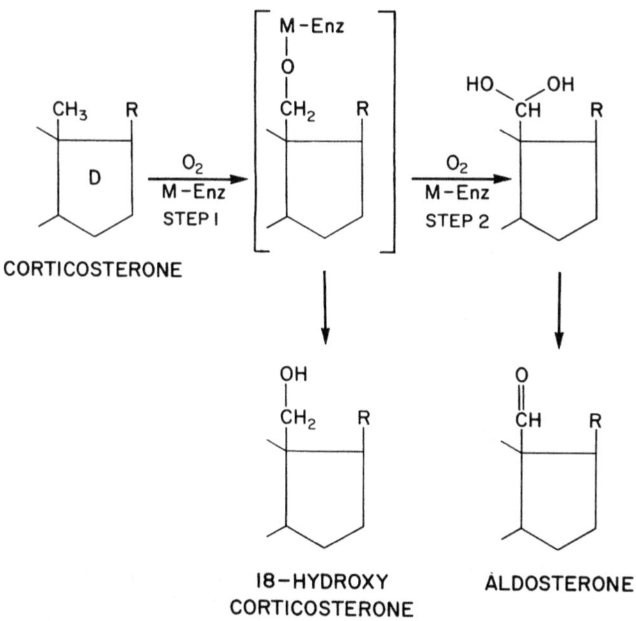

Fig. 4. Suggested structure of hypothetical intermediate and mechanism for the two-step, mixed-function oxidation of corticosterone to aldosterone. *M-Enz*, metalloenzyme; *R*, $-CO-CH_2OH$. (From Ulick 1976, with permission)

1982, 1984a), the enzyme involved in the ultimate step of aldosterone biosynthesis is a mixed-function oxidase rather than a dehydrogenase. Accordingly, the conversion of 18-hydroxycorticosterone to aldosterone may not be a direct dehydrogenation, but may more likely consist of a second hydroxylation at C-18, leading to an unstable intermediate product which spontaneously decomposes into aldosterone and water (Rapp and Dahl 1976; Ulick 1976; Fig. 4). In an extension of this hypothesis, Ulick (1976) has suggested that the true intermediary product of the two corticosterone methyl oxidations might not be free 18-hydroxycorticosterone but rather a reactive complex of the mono-oxygenated steroid and the enzyme which has formed it. This reactive derivative might then be either further oxidized to aldosterone (by the same or a second enzyme) or converted by means of a spontaneous side reaction to 18-hydroxycorticosterone. This hypothesis could explain why the free 18-hydroxycorticosterone appearing in the incubation medium or extracellular fluid is a relatively stable end product (Sandor et al. 1972) and why exogenous 18-hydroxycorticosterone is only poorly converted to aldosterone. An alternative to the above-stated two-hydroxylation hypothesis has been elaborated mainly on the basis of theoretical considerations by Conradin Meuli in the course of his research in our laboratory (1979–1984) and is shown in Fig. 5. While working with molecular models of steroids, Meuli observed that there would be hardly

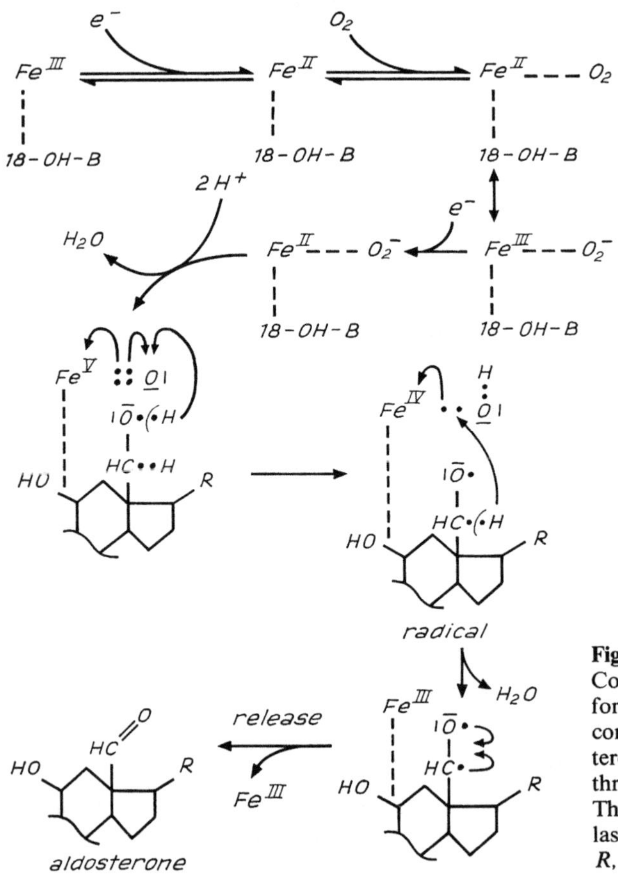

Fig. 5. Unpublished hypothesis by Conradin Meuli about a mechanism for the conversion of 18-hydroxycorticosterone *(18-OH-B)* to aldosterone by cytochrome P-450 through the formation of a radical. The participating electrons for the last three reactions are shown. *R*, $-CO-CH_2OH$.

sufficient space for accommodating two hydroxyl groups at C-18. He therefore suggested that in the second oxidation a radical might be formed, with the oxygen at C-18 carrying the free electron. This radical might decompose by the transfer of a hydrogen atom to the hydroxyl group bound to the central iron of the cytochrome P-450, thus yielding aldosterone and water as a reduction product. Such a chain of events would be consistent with the recognized mode of action of liver microsomal cytochrome P-450 in hydroxylations and in the conversion of ethanol to acetaldehyde (R. E. White and Coon 1980; Macdonald et al. 1982; Morgan et al. 1982).

Perhaps the variable convertibility to aldosterone of added 18-hydroxylated steroids observed in different studies was partly due to the form in which the steroids were added. Both 18-hydroxycorticosterone and 18-hydroxy-11-deoxycorticosterone are known to exist in two spontaneously interconvertible forms (Dominguez 1965; Aragones et al. 1978), i.e., a so-called M form (more polar), which, according to A.K. Roy et al. (1976), is a cyclic hemiketal, and an L form (less polar), which appears to be a mixed ketal, formed by reaction with an alcoholic solvent in the presence of acid impurities. Damasco and Lantos (1975) observed that the L form of 18-hydroxycorticosterone was more efficiently converted to aldosterone than the M form by rat adrenal tissue. Neher (1979) has suggested that the "L form" of 18-hydroxycorticosterone might be identical with 18-deoxyaldosterone, i.e., a product of a nonenzymic dehydration of 18-hydroxycorticosterone (Schmidlin and Wettstein 1961). This compound was converted to aldosterone twice as efficiently as 18-hydroxycorticosterone by bullfrog adrenal tissue (Ulick 1973). Excellent conversions to aldosterone (10%) were observed by Cozza et al. (1985), both when acid-treated 18-hydroxycorticosterone (80% 18-deoxyaldosterone) was incubated with quartered rat adrenals at pH 4.8 and when 18-hydroxycorticosterone (M form) enclosed in rat adrenal liposomes was incubated without enzymes at pH 3.3. In conversion experiments with radioactive precursors and trapping of intermediates with unlabeled steroids, an operative enzymatic pathway from corticosterone to aldosterone via 18-hydroxycorticosterone and 18-deoxyaldosterone optimal at pH 4.8 was disclosed by these investigators. Aupetit et al. (1977a) found that the 18-hydroxycorticosterone produced by duck adrenal mitochondria during incubation with tritiated corticosterone was not further converted to aldosterone, in contrast to added synthetic 18-hydroxycorticosterone. The convertibility of the biosynthetically obtained steroid could be restored by extraction and purification.

The assumption that the conversion of corticosterone to aldosterone is a two-step reaction is strongly supported by two different abnormal patterns of steroid secretion found in children with renal salt loss due to an inborn defect in the terminal part of the aldosterone biosynthetic pathway. In the patients described by Visser and Cost (1964) and by Degenhart et al. (1966), a decreased aldosterone secretion was associated with a decreased secretion of 18-hydroxycorticosterone and an increased secretion of corticosterone (corticosterone methyl oxidase defect type 1; Ulick 1976). By contrast, in the cases reported by Ulick et al. (1964), David et al. (1968), Rappaport et al. (1968), Touitou et al. (1970), and Milla et al. (1977) only the secretion of aldosterone was decreased, while both 18-hydroxycorticosterone and corticosterone secretions were abnormally elevated (corticosterone methyl oxidase defect type 2).

The conversion of 18-hydroxycorticosterone to aldosterone appears to be the only biosynthetic step which is practically limited to the zona glomerulosa. 18-Hydroxycorticosterone can be produced by the zona fasciculata as well as by the zona glomerulosa of bovine and rat adrenals (Sheppard et al. 1963; Marusic and Mulrow 1967a; Baniukiewicz et al. 1968). Decapsulated portions of rat adrenals converted both tritiated corticosterone and 18-hydroxy-11-deoxycorticosterone (18-OH-DOC) to 18-hydroxycorticosterone, but to a substantially lesser extent than the capsular portions (J. Müller 1980b). Sonicated mitochondria from bovine adrenal cortex (without zone separation) produced 18-hydroxycorticosterone from deoxycorticosterone mainly via 18-OH-DOC when the amount of added precursor was limited (Kim et al. 1983). High concentrations of deoxycorticosterone inhibited the conversion of 18-OH-DOC to 18-hydroxycorticosterone. In man, 18-hydroxycorticosterone appears to be mainly a by-product of aldosterone biosynthesis. However, in patients with hypoaldosteronism following surgical removal of an aldosterone-producing adenoma, most of the circulating 18-hydroxycorticosterone was apparently secreted by the zona fasciculata (Kater and Biglieri 1982). In 17α-hydroxylase deficiency, excessive amounts of 18-hydroxycorticosterone are secreted by the zona fasciculata, whereas aldosterone production is suppressed (Kater et al. 1982). The secretion rate of 18-hydroxycorticosterone was found to be 60%–200% greater than the secretion rate of aldosterone in normal human beings (Ulick et al. 1964; Ulick and Vetter 1965). The ratio of these two steroids in the plasma and that of their major metabolites, i.e., 18-hydroxytetrahydro-11-dehydrocorticosterone:tetrahydroaldosterone, in the urine vary between 1 and 3 (V.I. Martin et al. 1975; Ulick 1976; Mason et al. 1977; Kater et al. 1985b). The secretion rates and plasma concentrations of both steroids were proportionally increased in patients with primary aldosteronism or secondary aldosteronism due to cirrhosis of the liver, congestive heart failure, nephrotic syndrome, Bartter's syndrome, salt-losing nephropathies, or unilateral renal artery occlusion (Ulick et al. 1964; Kater et al. 1985b). Equal fractional increases in the production of both steroids were induced in normal subjects by sodium depletion or treatment with ACTH. Parallel increases in the plasma concentrations of aldosterone and 18-hydroxycorticosterone were observed in response to angiotensin-II infusions in normal, sodium-replete subjects, while the concentrations of other corticosteroids remained unchanged (Mason et al. 1976). The plasma concentration of 18-hydroxycorticosterone in normal human beings was lower when they were in the recumbent than in the upright position (V.I. Martin et al. 1975). Sodium depletion resulted in an increase in plasma 18-hydroxycorticosterone more extensive than the simultaneous increase in plasma aldosterone; however, in sodium-depleted subjects angiotensin II elicited a greater fractional increase in plasma aldosterone than in plasma 18-hydroxycorticosterone (Mason et al. 1977).

2.4 Alternative Pathway Through 18-Hydroxy-11-deoxycorticosterone?

18-Hydroxy-11-deoxycorticosterone is one of the major secretory products of rat adrenals (Birmingham and Ward 1961) and is synthesized in relatively small amounts by the human adrenal cortex (De Nicola and Birmingham 1968). Most of this steroid derives from the zona fasciculata-reticularis (Sheppard et al. 1963; Stachenko et al. 1964; Baniukiewicz et al. 1968), and its biosynthesis appears to be predominantly under ACTH control (Cortés et al. 1963; Melby et al. 1972; G. H. Williams et al. 1976a; Tuck et al. 1977; Tan and Mulrow 1978).

However, small amounts of 18-OH-DOC are also produced by the capsular portions of rat adrenals from endogenous as well as from added precursors (Sheppard et al. 1963; R. Lucis et al. 1965; Baniukiewicz et al. 1968). According to Baniukiewicz et al. (1968), the decay curves of 18-OH-DOC output by superfused capsular adrenals of hypophysectomized rats indicated that this steroid was produced by the zona glomerulosa rather than by contaminating zona fasciculata tissue. The 18-OH-DOC output of incubated rat adrenal zona glomerulosa (impure) cells increased in response to two specific stimulators of aldosterone biosynthesis, i.e., angiotensin II added in vitro or a high potassium concentration in the medium (Moore et al. 1978). These stimulators had no effect on the 18-OH-DOC output of zona fasciculata cells. Moreover, it has been observed by several investigators, although not consistently, that the plasma concentration or the secretion rate of 18-OH-DOC in man was not only influenced by ACTH, but also by alterations in sodium intake, change in posture, angiotensin-II infusion, and treatment with a mineralocorticoid or spironolactone (Melby et al. 1972; G. H. Williams et al. 1976a; Tuck et al. 1977; Abshagen et al. 1978). An increased 18-OH-DOC secretion rate was also found in patients with primary aldosteronism (Melby et al. 1972).

As mentioned above (see page 5), 18-OH-DOC is a possible precursor of aldosterone. However, the relative importance of alternative biosynthetic pathways leading through 18-OH-DOC to aldosterone is difficult to evaluate. The observed low conversion of added 18-OH-DOC to aldosterone is not conclusive. Since 18-hydroxycorticosterone is generally assumed to be a true intermediate product of aldosterone biosynthesis, although not necessarily in the form in which it is secreted (see p. 9), in spite of its apparent low convertibility to aldosterone there is little reason to exclude 18-OH-DOC from a similar role. The only evidence against an operative pathway through 18-OH-DOC to aldosterone has been presented by Vecsei et al. (1968). These investigators added substrate amounts of various unlabeled steroids to quartered rat adrenals incubated with radioactively labeled progesterone and deoxycorticosterone and found that corticosterone and 18-hydroxycorticosterone but not 18-OH-DOC inhibited the incorporation of radioactivity into aldosterone. On the other hand, at low precursor concentrations, significantly more tritiated deoxycorticosterone than tritiated corticosterone was converted to aldosterone by capsular adrenals of normally fed and sodium-restricted rats (J. Müller, 1980a; Fig. 3). Similarly, rat adrenal zona glomerulosa cells converted more deoxycorticosterone than corticosterone to aldosterone at low precursor concentrations (Aguilera and Catt 1979). Under these experimental conditions, up to 50% of the aldosterone formed from deoxycorticosterone may have proceeded from a pathway not leading through corticosterone and thus, by exclusion, through 18-OH-DOC (Fig. 6). 18-OH-DOC has been found to be converted in good yield to 18-hydroxycorticosterone by human adrenal tissue and capsular portions of rat adrenals (Grekin et al. 1973; Fattah et al. 1977; J. Müller 1980b). This confirms the existence of the biosynthetic pathway 2 shown in Fig. 6. Indirect and inconclusive evidence for the existence of a second alternative pathway lead-

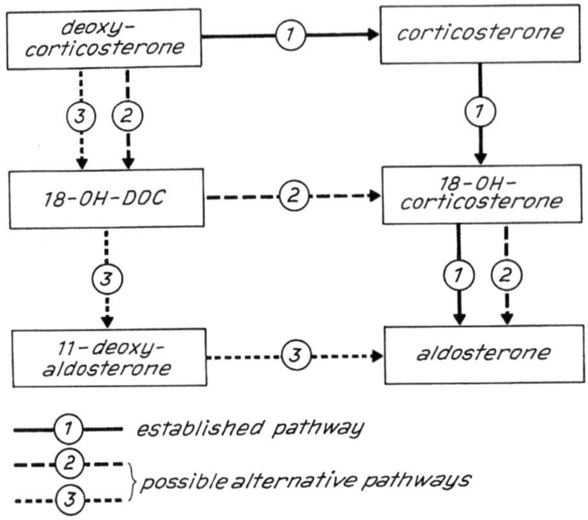

Fig. 6. Biosynthetic pathways from deoxycorticosterone to aldosterone. (From J. Müller 1980a)

ing via 18-OH-DOC and 11-deoxyaldosterone to aldosterone, i.e., pathway 3 (Fig. 6), may be taken from the observation that under most experimental conditions tritiated 18-OH-DOC was converted to aldosterone twice as efficiently as tritiated 18-hydroxycorticosterone by capsular rat adrenals (J. Müller 1980a). By itself, this could indicate that up to 50% of the aldosterone derived from 18-OH-DOC was formed by a pathway leading not through 18-hydroxycorticosterone but rather through 11-deoxyaldosterone. However, 11-deoxyaldosterone has as yet not been established as a natural product of the adrenal cortex, and it is not known whether and to which extent it would be converted to aldosterone. A possible biological precursor of this steroid, 18-oxoprogesterone, was incubated with bullfrog adrenal tissue as well as with slices of a human aldosterone-producing adenoma by Ulick (1973). It yielded only traces of aldosterone and was mainly converted to 11-deoxycorticosterone-18-oic acid.

2.5 Deoxycorticosterone Secretion

Deoxycorticosterone is the immediate precursor of corticosterone and of 18-OH-DOC and is theoretically also of dual origin in the adrenal cortex. The capsular and the decapsulated portions of adrenal glands of sodium- and potassium-replete rats produced approximately the same amount of this compound under basal conditions of incubation (Vinson and Whitehouse 1969; Müller 1970b). In intact man, the plasma levels and secretion rate of this steroid closely parallel of those of cortisol, indicating that most of the deoxycorticosterone is produced by the zona fasciculata under the control of pituitary ACTH (R. D. Brown and Strott 1971; Schambelan and Biglieri 1972; Oddie et al. 1972; Tan and Mulrow 1975). According to R. D. Brown et al. (1972a), the plasma deoxycorticosterone did not increase in response to angiotensin II or potassium administration in dexamethasone-treated human beings, but only in subjects receiving both dexamethasone and metyrapone. However, in hypopituitary patients or in normal subjects treated with dexamethasone, the plasma deoxycorticosterone significantly increased in response to sodium restriction (Schambelan and Biglieri 1972) and to an angiotensin-II infusion (Tan and Mulrow 1975).

Pathologically increased deoxycorticosterone secretion rates are characteristic of congenital deficiencies of 11β-hydroxylation (Eberlein and Bongiovanni 1955) or 17α-hydroxylation (Biglieri et al. 1966b). In patients with these forms of congenital adrenal hyperplasia, the production of deoxycorticosterone appears to be mainly under ACTH control. Among patients with mineralocorticoid hypertension without a primary adrenal lesion, increased secretion of deoxycorticosterone was observed in cases with increased ACTH secretion, but not in cases with increased activity of the renin-angiotensin system (Biglieri et al. 1968). The secretion rate of deoxycorticosterone was found to be normal or decreased in most patients with secondary aldosteronism associated with edema (Crane and Harris 1970). In late stages of normal human pregnancy (after 23-24 weeks of gestation), maternal plasma deoxycorticosterone concentrations were found to be substantially elevated (R. D. Brown et al. 1972b). Even higher levels of deoxycorticosterone were found in fetal umbilical mixed-cord plasma. Most of this deoxycorticosterone is probably produced by extra-adrenal 21-hydroxylation of circulating progesterone in maternal and fetal nonendocrine organs such as the kidneys (Winkel et al. 1980b, 1981; Casey and MacDonald 1982).

In hypophysectomized dogs, the secretion of deoxycorticosterone was stimulated by an angiotensin-II infusion, but not by sodium depletion or potassium infusion (Taylor et al. 1972).

3 Aldosterone Biosynthesis by Cell-Free Systems

Aldosterone is synthesized by homogenized adrenocortical tissue from endogenous and from exogenous precursors. According to Kahnt and Neher (1965), the endogenous precursor of aldosterone production by bovine adrenocortical tissue homogenate is most likely free cholesterol. After centrifugal fractionation of sheep (Raman et al. 1966), bullfrog (Psychoyos et al. 1966) and rat (Marusic and Mulrow 1967b) adrenocortical homogenates, only the mitochondrial fractions were found to actively convert corticosterone to 18-hydroxycorticosterone and aldosterone. Aupetit et al. (1977b) demonstrated that the conversion of added 18-hydroxycorticosterone to aldosterone was also restricted to the mitochondrial fraction of duck adrenocortical tissue rather than being localized in both the microsomal and the mitochondrial fraction as had been previously reported by Aupetit (1972). In sheep adrenal mitochondria, the site of both reactions was found to be the internal membrane (Antreassian et al. 1979). Attempts by Raman et al. (1966) and by Sandor et al. (1972) to bring the mitochondrial enzyme system involved in the two final steps of aldosterone biosynthesis into a soluble form were unsuccessful. But Rapp and Dahl (1976) succeeded in obtaining, by ultrasonification from rat capsular adrenal mitochondria, cytochrome P-450 particles capable of catalyzing the conversion of corticosterone to 18-hydroxycorticosterone and aldosterone. The conversion of tritiated corticosterone to aldosterone and 18-hydroxycorticosterone by a purified bovine adrenocortical cytochrome P-$450_{11\beta}$ preparation – reconstituted to an active enzyme system by the addition of adrenodoxin and adrenodoxin reductase – was reported by Wada et al. (1984, 1985). The conversion to aldosterone and – to a lesser extent – the conversion to 18-hydroxycorticosterone were substantially enhanced by the addition of a lipidic extract of bovine adrenocortical zona glomerulosa mitochondria (Wada et al. 1984) as well as by a number of different neutral or acid phospholipids added separately or in mixtures (Ohnishi et al. 1984).

The conversion of corticosterone to aldosterone by isolated mitochondria requires the presence of NADPH, which can be provided directly or by the addition of NADP plus either a citric acid cycle intermediate such as fumarate or malate (Tallan et al. 1967) or a NADPH-generating system such as glucose-6-phosphate and glucose-6-phosphate dehydrogenase. Studies on the coenzyme requirement for the conversion of exogenous 18-hydroxycortiosterone to aldosterone have yielded conflicting results. Kahnt and Neher (1965) observed that exogenous 18-hydroxycorticosterone was converted to aldosterone at the same rate as exogenous corticosterone by bovine adrenal homogenates when incubated with NAD or NADP but without fumarate, although the production of aldosterone from endogenous precursors was markedly decreased under these conditions.

However, according to Raman et al. (1966), NADPH or an NADPH-generating system was necessary for the conversion of added 18-hydroxycorticosterone to aldosterone; in the presence of NAD or NADP, sheep adrenal mitochondria converted 18-hydroxycorticosterone no longer to aldosterone but rather to 18-hydroxy-11-dehydrocorticosterone. This is consistent with the finding by Marusic et al. (1973b) that bovine adrenal mitochondrial conversion to aldosterone of 18-hydroxycorticosterone as well as of corticosterone required the presence of NADPH and that it did not occur in the presence of FAD, NAD, NADP, or NADH.

Greengard et al. (1967) presented the following evidence for the participation of cytochrome P-450 in the conversion of corticosterone to aldosterone by bullfrog adrenal mitochondria: appearance of a characteristic absorption spectrum upon reduction and addition of carbon monoxide, inhibition of aldosterone biosynthesis by carbon monoxide, and reversal of this inhibition by light with a maximum effect at a wave length of 450 nm. Cytochrome P-450 is a component of mixed-function oxidases, which require both molecular oxygen and NADPH, and has been found to be involved in a number of steroid and drug hydroxylation reactions (Omura et al. 1965). The observation by Marusic et al. (1973b) and by I. Kojima et al. (1982, 1984a) that both oxygen and NADPH are needed for the conversion of exogenous 18-hydroxycorticosterone to aldosterone indicates that not only the 18-hydroxylation of corticosterone (as assumed by Greengard et al. 1967) but also the final step of aldosterone biosynthesis, i.e., methyl oxidation 2, involves a mixed-function oxidase. Consistent with this assumption is the observation by Sandor and Lanthier (1963) and by Kahnt and Neher (1965) that the conversion of 18-hydroxycorticosterone to aldosterone is irreversible. Moreover, the conversion of 18-hydroxycorticosterone to aldosterone by sonicated mitochondria from bovine adrenal zona glomerulosa was almost completely inhibited in an atmosphere of 100% CO in the presence of NAD^+ and was significantly reduced in an atmosphere of 90% CO and 10% O_2 in the presence of NADPH (I. Kojima et al. 1982, 1984a). The conversion of 18-hydroxycorticosterone to aldosterone by duck adrenal mitochondria was dependent on energy metabolism and required reducing power (Aupetit et al. 1980, 1983, 1984). The final step of aldosterone biosynthesis was also inhibited by several drugs known to affect cytochrome P-450, such as metyrapone, Su-8000, Su-10603, amphenone B, and spironolactone, according to I. Kojima et al. (1982, 1984a). However, according to Aupetit et al. (1979), the conversion of 18-hydroxycorticosterone to aldosterone was inhibited by three different spirolactones but not by metyrapone.

Two distinct cytochrome P-450 proteins have been separated from bovine adrenocortical mitochondria, one of them, P-450_{SCC}, catalyzing the cholesterol side-chain cleavage, the other one, P-$450_{11\beta}$, catalyzing 11β- and 18-hydroxylations of deoxycorticosterone as well 11β- and 18-hydroxylations of androst-4-ene-3,17-dione (Jefcoate et al. 1970; Suhara et al. 1978). According to a substantial amount of evidence, one single enzyme system appears to be responsible for both 11β-hydroxylation and 18-hydroxylation of deoxycorticosterone in bovine and rat adrenocortical mitochondria (Bjoerkhem and Karlmar 1975; Rapp and Dahl 1976; Watanuki et al. 1978; Sato et al. 1978). In contrast, the available evidence allows no definite conclusions about the number of different cytochromes P-450 involved in the three-step conversion of deoxycorticosterone to aldosterone in zona glomerulosa mitochondria. In rats susceptible to salt-induced hypertension, Rapp and Dahl (1976) observed a decrease of Warburg's partition constant, not only for the cytochrome P-450 catalyzing the conversions of deoxycorticosterone to corticosterone and 18-OH-DOC, but also for the capsular adrenal cytochrome P-450 involved in the conversion of corticosterone to 18-hydroxycorticosterone. On the other hand, there was no difference between rats susceptible and rats resistent to salt-induced hypertension in the Warburg partition constant for the conversion of corticosterone to aldosterone. These observations indicated that the same cytochrome P-450 was responsible for both 18- and 11β-hydroxylations in the zona glomerulosa as well as in the inner zones but that the conversion of 18-hydroxycorticosterone to aldosterone was catalyzed by a different cytochrome P-450. This hypothesis was only partially supported by Wada et al. (1984, 1985), who found that cytochrome P-450$_{11\beta}$- purified to electrophoretic homogeneity from bovine adrenocortical mitochondria (without zone separation) - converted tritiated corticosterone to 18-hydroxycorticosterone and aldosterone and tritiated 18-hydroxycorticosterone to aldosterone, provided that a lipid extract of adrenocortical mitochondria was added to the reconstituted enzyme system. These investigators concluded that 18-hydroxycorticosterone was a substrate of cytochrome P-450$_{11\beta}$ and that aldosterone was produced from corticosterone via 18-hydroxycorticosterone by cytochrome P-450$_{11\beta}$. Yanagibashi et al. (1986), who purified bovine and porcine adrenocortical 11β/18-hydroxylase, also came to the conclusion that one single enzyme, i.e., cytochrome P-450$_{11\beta}$, catalyzed 11β- and 18-hydroxylations in both the zona glomerulosa and the zona fasciculata and converted 18-hydroxycorticosterone to aldosterone in the zona glomerulosa. These investigators observed a more active aldosterone synthesis in intact mitochondria of the zona glomerulosa than in intact mitochondria of the zona fasciculata. However, partially purified extracts of mitochondria from the two zones were equally active in catalyzing all three steps in the conversion of deoxycorticosterone to aldosterone. No activity corresponding to any of the three steps was found in an extramitochondrial fraction or a mitochondrial fraction not containing 11β/-18-hydroxylase.

There are several lines of evidence against the hypothesis that the same cytochrome P-450$_{11\beta}$ is responsible for both 11β-hydroxylation and 18-hydroxylation in the zona glomerulosa and in the inner zones as well as for the two-step conversion of corticosterone to aldosterone. First, aldosterone is uniquely and 18-hydroxycorticosterone preferably produced by the zona glomerulosa of the mammalian adrenal cortex, whereas 11β-hydroxylation occurs to equal extents in the inner and in the outer zones. Second, there appears to be a fundamental zone difference in the substrate specificities of both "18-hydroxylase" and "11β-hydroxylase". According to Björkhem and Karlmar (1975), bovine and rat adrenocortical mitochondria (presumably mainly originating from the zona fasciculata) converted deoxycorticosterone much more efficiently than corticosterone to its respective 18-hydroxyderivative. We observed that in the decapsulated portions of rat adrenals deoxycorticosterone was converted to 18-hydroxy steroids 20 times more abundantly than corticosterone, whereas in the capsular portions of the same glands both precursors were 18-hydroxylated to approximately the same extent (K. Baumann and Müller 1972b). Such a zone difference was not found in the bovine adrenal cortex by Karlmar (1982). He measured approximately the same ratio of 18-hydroxylation of deoxycorticosterone over 18-hydroxylation of corticosterone in mitochondria from the outer layer (46) and in mitochondria from the inner layer (58). However, Brode (1965) reported that mitochondria from the zona glomerulosa of bovine adrenals converted progesterone to 11β-hydroxyprogesterone more efficiently than mitochondria from the inner zones, whereas there was no zonal difference in the 11β-hydroxylation of deoxycorticosterone. In decapsulated rat adrenal glands the "11β-hydroxylase" appeared to have a much higher affinity for deoxycorticosterone and 11-deoxycortisol than for 18-OH-DOC; in contrast, 18-OH-DOC, deoxycorticosterone, and 11-deoxycortisol were converted to 11β-hydroxysteroids to similar extents in capsular adrenal tissue (J. Müller 1980b). Third, substantial alterations in the activity of the enzyme(s) involved in the conversions of corticosterone to 18-hydroxycorticosterone and aldosterone experimentally induced in animals by various manipulations were not accompanied by parallel and commensurate alterations in zona glomerulosa 11β-hydroxylase activity. Thus, sodium restriction of rats led to an increased conversion of corticosterone to 18-hydroxycorticosterone and aldosterone but not of deoxycorticosterone to corticosterone, and resulted in an enhanced binding to capsular mitochondrial cytochrome P-450 of corticosterone but not of deoxycorticosterone (Marusic and Mulrow 1967b; R.E. Kramer et al. 1979). In reverse, potassium depletion or mineralocorticoid treatment of rats had a much greater suppressive effect on the conversions of corticosterone to 18-hydroxycorticosterone and aldosterone than on zona glomerulosa 11β-hydroxylase activity (K. Baumann and Müller 1972a, 1972b; Hunziker and Müller 1977). Prolonged infusion of angiotensin II to intact rats stimulated the conversion of corticosterone to aldosterone but not the conversion of deoxycorticosterone to corticosterone by zona glomerulosa mitochondria (Aguilera et al. 1980b). In the same study, hypophysectomy of rats lowered the zona glomerulosa mitochondrial conversion of deoxycorticosterone to corticosterone while raising the conversion of corticosterone to aldosterone. Prolonged treatment with ACTH stimulated the capsular adrenal conversion of deoxycorticosterone to corticosterone and 18-OH-DOC while suppressing the conversion of corticosterone to aldosterone and 18-hydroxycorticosterone (J. Müller 1978). All these observations are strongly suggestive of a separate cytochrome P-450 involved in the final steps of aldosterone biosynthesis. Hornsby and Crivello (1983b) have proposed the designation "cytochrome P-450_{CMO}" (for *c*orticosterone *m*ethyl *o*xidase) for a still-hypothetical mitochondrial enzyme catalyzing the two-step conversion of corticosterone to aldosterone. This enzyme would be located exclusively in the zona glomerulosa of the mammalian adrenal cortex. Its biosynthesis would be induced by sodium restriction, long-term angiotensin-II infusion, or potassium intake and would be repressed by potassium restriction, sodium loading, ACTH treatment, or mineralocorticoid administration. Cytochrome P-450_{CMO} might be closely related to cytochrome P-$450_{11\beta}$ and have a similar molecular weight. Alterations in sodium or potassium intake had very similar effects on the following conversions by rat capsular adrenal tissue: corticosterone to 18-hydroxycorticosterone, 18-OH-DOC to 18-hydroxycorticosterone, and 18-hydroxycorticosterone to aldosterone, while not affecting the conversions of deoxycorticosterone to corticosterone and to 18-OH-DOC (J. Müller 1980a, 1980b; Fig. 7). These findings would be at least compatible with the assumption of a single enzyme (i.e., cytochrome P-450_{CMO}) catalyzing both corticosterone methyl

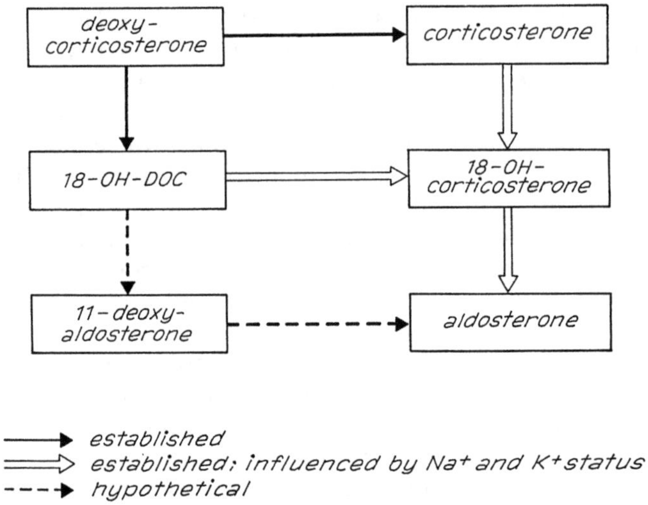

Fig. 7. Influence of sodium and potassium intake on single steps in the biosynthetic pathways from deoxycorticosterone to aldosterone

oxidations as well as most of the 11β-hydroxylation of 18-OH-DOC in the zona glomerulosa cell. Cytochrome P-450$_{CMO}$ might be identical with a 49-K mitochondrial protein which was induced in rats by potassium intake or sodium restriction and which occurred only in the zona glomerulosa (Meuli and Müller 1983b, 1984). On the other hand, cytochrome P-450$_{11\beta}$ might be present in all the zones of the adrenal cortex. This enzyme would be relatively independent of the sodium and potassium status. It would catalyze the conversions of deoxycorticosterone to corticosterone and to 18-OH-DOC in all the zones as well as the conversion of 18-OH-DOC to 18-hydroxycorticosterone in the inner zones.

Some conflicting observations on the requirement for metal ions of the mitochondrial conversions of corticosterone to 18-hydroxycorticosterone and aldosterone by different investigators may have been due to species differences. Optimal conversions by duck adrenal mitochondria required the presence of all Na^+, K^+, Mg^{2+}, and Ca^{2+}; among them, calcium ions appeared to be the most important ones (Sandor et al. 1972). Conversions by sheep (Raman et al. 1966) and bullfrog (Psychoyos et al. 1966; Sharma et al. 1967) adrenal mitochondria depended considerably on the presence of bivalent metal cations in the incubation medium. Magnesium ions appeared to be optimal for both reactions. When Mg^{2+} was replaced by Ca^{2+}, the rate of conversion of corticosterone to 18-hydroxycorticosterone remained equal, but the conversion of corticosterone to aldosterone was markedly decreased (Psychoyos et al. 1966). Both reactions were inhibited by sodium ions (in concentrations above 100 mmol/l) and stimulated by potassium ions (in concentrations above 65 mmol/l), according to Sharma et al. (1967). The conversion of corticosterone to aldosterone by rat adrenal mitochondria was less dependent on cations; it was only 30% lower in Tris chloride buffer (without any metal cations) than in Krebs-Ringer bicarbonate buffer (Marusic and Mulrow 1967b). The absence or presence of sodium or potassium ions in the incubation medium did not affect the conversion of corticosterone to aldosterone by rat

(Marusic and Mulrow 1967b) or dog (Marusic et al. 1973a) adrenocortical mitochondria.

The addition of ACTH or angiotensin II to the incubation medium stimulated neither aldosterone biosynthesis from endogenous precursors or tritiated cholesterol by bovine adrenal homogenates (Kahnt and Neher 1965) nor the conversion of corticosterone to aldosterone by dog adrenal mitochondria (Marusic et al. 1973a).

4 Substances Directly Influencing Aldosterone Biosynthesis in Short-term Incubation or Perfusion Experiments

4.1 Angiotensins

4.1.1 Angiotensin II

Angiotensin II[1] is widely considered to be the physiologically most important regulator of aldosterone secretion. This characterization is debatable. However, angiotensin II has generally been found to be a potent stimulator of aldosterone output in man as well as in several mammalian and submammalian animal species. The peptide exerts this activity by (a) a direct action upon the aldosterone-producing cell and (b) by enhancing aldosterone biosynthesis rather than by stimulating the release of preformed steroid. There is ample direct experimental evidence for a specific interaction of angiotensin II with isolated zona glomerulosa cells, which will be summarized on the following pages. On the other hand, the assumption that the peptide stimulates the biosynthesis of aldosterone was originally based mainly on indirect evidence. Thus, the increments in aldosterone output of isolated dog or sheep adrenals induced by small doses of angiotensin II (Ganong et al. 1962; J.O. Davis 1962; Blair-West et al. 1962a) were much larger than the total aldosterone content of normal adrenal glands[2]. In the rat, systemically infused angiotensin led not only to a significant increase in aldosterone secretion but also to a sixfold increase in the adrenal content of aldosterone (Dufau and Kliman 1968a).

The aldosterone-stimulating activity of angiotensin II was first demonstrated in normal man (Laragh et al. 1960a; Biron et al. 1961). Intravenous infusion of the peptide over periods of several hours was followed by increased aldosterone secretion or urinary excretion rates. Later, two- to fivefold increases in the plasma

---

[1] The synthetic analogue val^5-angiotensin II-asp$^1\beta$-amide (asn^1,val^5-angiotensin II = Hypertensin, Ciba-Geigy Ltd., Basel) was used in most in vivo studies as well as in earlier in vitro investigations referred to in this chapter. In more recent adrenocortical cell incubation experiments, the free acids ile^5-angiotensin II or val^5-angiotensin II have been preferentially used. Asp^1,ile^5-angiotensin II induced a larger increase in aldosterone production from rat adrenal zona glomerulosa cells than did asn^1,val^5-angiotensin II (Mendelsohn 1980). This could have been due to a more rapid degradation of the amide form than of the free-acid form of the peptide. In man, as well, the pressor and steroidogenic activities of asp^1,ile^5-angiotensin II were found to be significantly greater than those of asn^1,val^5-angiotensin II (Kono et al. 1985a).

[2] A mean content of 0.22 μg aldosterone per gland was found in five sheep adrenals (Coghlan and Blair-West 1967). The mean aldosterone secretion rate in sodium-replete sheep of 0.54 μg/h increased severalfold over periods of hours during infusions of angiotensin II (Blair-West et al. 1962a).

aldosterone concentration were induced by intravenous angiotensin II infusions, according to - among others - Horton (1969), Michelakis and Horton (1970), Mendelsohn et al. (1972), Scholer et al. (1973), and G.H. Williams et al. (1977). Usually, pressor doses of the peptide were used. However, subpressor doses of 1-2 ng/kg/min were also clearly effective in several instances. Responses could be observed within 10 min after starting the infusion.

In nephrectomized-hypophysectomized dogs, injections or infusions of angiotensin II into the arterial supply of isolated adrenal glands led within minutes to a distinct elevation of aldosterone, corticosterone, and 17-hydroxycorticosteroid secretion (J.O. Davis 1962; Ganong et al. 1962). This steroidogenic effect was obtained by small doses of the peptide (0.5 μg injected or 42 ng/min infused) which were ineffective then given systemically. In intact sodium-replete sheep with adrenal autotransplants to the neck, infusions of angiotensin II into the adrenal artery at rates of 0.1-0.5 μg/h elicited marked increases in aldosterone secretion without a significant effect on cortisol and corticosterone outputs (Blair-West et al. 1962, 1970b). In sodium-replete dexamethasone-treated rabbits, angiotensin II infusions significantly stimulated the secretion of aldosterone and corticosterone and did not affect deoxycorticosterone secretion (Braverman and Davis 1973).

According to early studies, angiotensin II appeared to have little if any aldosterone-stimulating activity in the rat, in contrast to its marked effects in man, dogs, and sheep. Neither angiotensin II nor rat renin affected the aldosterone output by intact or hypophysectomized rats (Eilers and Peterson 1964; Marieb and Mulrow 1965). However, very high doses of the peptide, in excess of those needed to raise blood pressure - 3.6 μg/min (Dufau and Kliman 1965a), 1 μg/min (Cade and Perenich 1965) - stimulated aldosterone secretion[3] in intact rats. Perhaps the apparent lack of aldosterone-stimulating activity of angiotensin II was partially due to the experimental procedures involved in these studies, in which aldosterone secretion was measured in the adrenal venous blood of anesthetized rats. On the one hand, experimental stress may have led to increased secretion of endogenous renin and ACTH and thus to an elevated aldosterone secretion under baseline conditions. On the other hand, anesthesia may have attenuated the adrenals' response to exogenous angiotensin II (Coleman et al. 1974). Aldosterone output rose in response to angiotensin II in sodium-replete (Marieb and Mulrow, 1965) and sodium-deficient (Kinson and Singer 1968) hypophysectomized-nephrectomized rats. In conscious, intact, sodium-replete rats, small doses of angiotensin II (1-5 ng/min) consistently raised peripheral plasma aldosterone concentrations within 20 min (W.B. Campbell et al. 1974; Coleman et al. 1974). Bojesen (1966) had previously succeeded in raising peripheral plasma aldosterone concentrations by a 2-h infusion of angiotensin II (10 ng/min). In a freshwater teleost fish *(Catostomus macrocheilus)*, injections of angiotensin II or of a fish renin preparation did not influence plasma aldosterone concentration (Reinking 1983). In bullfrogs,

[3] In intact rats, the steroidogenic effect of systemically administered angiotensin II may be partially due to ACTH release, unless ACTH secretion is blocked by glucocorticoids or morphine (van der Wal and de Wied 1968; Dufau and Kliman 1968b). An ACTH-stimulating effect of systemically infused angiotensin II was also observed in normal human subjects (Rayyis et al. 1971).

angiotensin II infusions raised blood pressure but failed to affect the adrenal secretion of aldosterone and corticosterone (A. A. Taylor and Davis 1971). However, the secretion of both steroids was markedly stimulated by the infusion of carp kidney extracts incubated with carp plasma. A stimulation of aldosterone secretion by frog renin had previously been observed in intact and hypophysectomized bullfrogs by Johnston et al. (1967a).

In vitro stimulation of aldosterone biosynthesis from endogenous precursors was first demonstrated in experiments with bovine adrenal tissue. Outer slices of bovine adrenal cortex ("zona glomerulosa") produced more aldosterone and corticosterone, inner slices ("zona fasciculata") more corticosterone and cortisol when angiotensin II (40 μg per gram of tissue) was added to the incubation medium (Kaplan and Bartter 1962). Similar effects were elicited by lower doses (0.1 and 1 μg/g) of the peptide (Kaplan 1965). Angiotensin II stimulated aldosterone output by incubated tissue slices of adrenocortical adenomas taken from patients with primary aldosteronism, but it had no steroidogenic effect on contralateral, adjacent, or normal human adrenocortical tissue (Melby et al. 1965). At concentrations of 2.4×10^{-8} to 2.4×10^{-5} mol/l, the peptide significantly enhanced aldosterone production by sheep adenocortical tissue (McDougall et al. 1976). Only at the concentration of 2.5×10^{-6} mol/l did angiotensin II stimulate the corticosterone output, whereas the cortisol production was not stimulated by any dose of the peptide. Aldosterone biosynthesis by rat adrenal tissue in vitro was not stimulated by angiotensin II, according to Gláz and Sugár (1962), Kaplan and Bartter (1962), and Spät et al. (1965). In order to compensate for the possible destruction of angiotensin II by adrenal tissue peptidases, the peptide was periodically replaced in the medium (Gröger-Korber 1966) or was added by constant infusion (Spät et al. 1969); but even with these precautions, angiotensin II did not stimulate aldosterone or corticosterone production by incubated rat adrenal tissue. However, a small but reproducible increase in aldosterone output was observed when quartered adrenals of sodium-deficient rats were incubated with large amounts of angiotensin II (3.1×10^{-4} mol/l; J. Müller 1965a). A similar dose of the peptide elicited significant steroidogenic responses in incubated capsular portions of rat adrenal glands without having any effect on the decapsulated portions of the glands (J. Müller 1970b; Fig. 8). It stimulated the production of aldosterone, corticosterone, and deoxycorticosterone in capsular adrenals of rats kept on a complete diet, but it enhanced only the production of aldosterone in tissue of sodium-deficient rats and only the production of deoxycorticosterone in tissue of potassium-deficient rats. Angiotensin II had no effect on corticosteroid production by superfused bisected adrenals of hypophysectomized rats (Baniukiewicz et al. 1968), but when added in a high concentration to superfused capsular adrenals of hypophysectomized-nephrectomized rats, it stimulated the production of aldosterone, 18-hydroxycorticosterone, and corticosterone, although only to a small extent (S. A. S. Tait et al. 1970). Angiotensin II failed to affect aldosterone production by duck subcapsular adrenal slices during stationary incubation (Klingbeil 1985b). When similar tissue slices were superfused, a significant and dose-dependent stimulation of aldosterone output was induced by angiotensin II concentrations ranging from 10^{-10} to 10^{-6} mol/l. Under the same experimental conditions, neither corticosterone output of subcapsular tissue nor aldosterone and corticosterone outputs of inner-zone tissue were influenced by the peptide. A synthetic

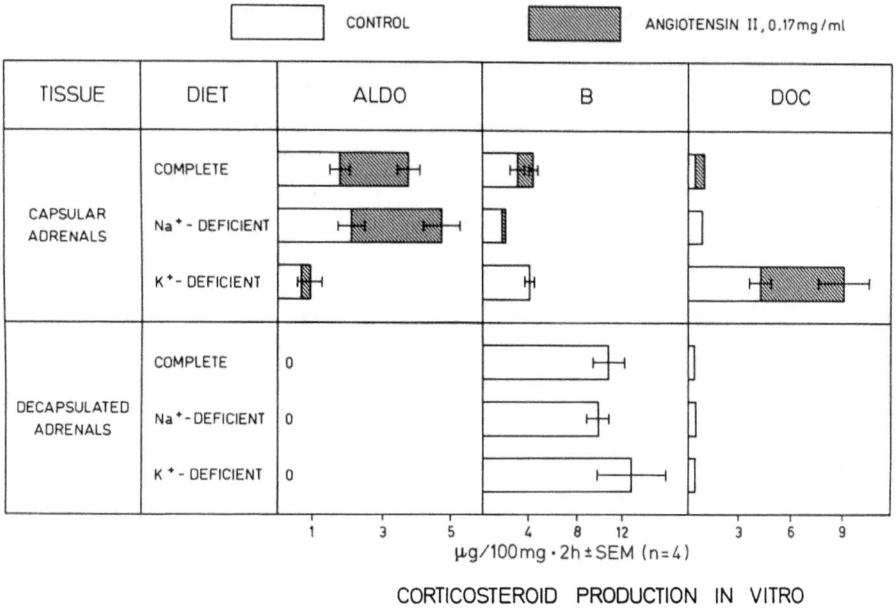

Fig. 8. Effect of angiotensin II (val^5-angiotensin II-asp^1β-amide) on in vitro production of aldosterone *(ALDO)*, corticosterone *(B)*, and deoxycorticosterone *(DOC)* by separate zones of adrenals of rats kept on different diets for 2 weeks. *White bars* represent baseline corticosteroid production, *hatched bars* increments due to addition of angiotensin II to the incubation medium. Mean values of two experiments ($n=4$) ± standard error of the mean. (Data from J. Müller 1970b)

angiotensin analogue, i.e., sar^1,-val^5-angiotensin II (10^{-7} mol/l), significantly stimulated aldosterone and corticosterone outputs by superfused frog adrenal tissue (Delarue et al. 1984).

Incubated dispersed adrenocortical cells have generally been found to be considerably more sensitive to the steroidogenic action of added angiotensin II than whole adrenocortical tissue. In a number of zona glomerulosa cell preparations, angiotensin II exerted a dose-dependent aldosterone-stimulating activity within physiological concentrations (Table 1). In some instances the dose response curve was biphasic, with a decreased response at high peptide concentrations (Fredlund et al. 1975a; Bravo et al. 1976a; G.H. Williams and Braley 1977), whereas in other studies a plateau in aldosterone output was observed. In rat adrenal zona glomerulosa cells, the maximum increase in aldosterone output due to angiotensin II was smaller than the maximum increases elicited by a high potassium concentration, serotonin, or ACTH (Haning et al. 1970; Bing and Schulster 1977; Braley and Williams 1977; Douglas et al. 1978a; Matsuoka et al. 1980a). In rabbit and guinea-pig adrenal cells, potassium ions and ACTH, respectively, also stimulated aldosterone production more extensively than angiotensin II (Chiu and Freer 1979; Pham-Huu-Trung et al. 1979). In normal human zona glomerulosa cells, these three stimulators had comparable maximum effects upon aldosterone output (G.H. Williams and Braley 1977; Belmega et al. 1983). The maximum increase in aldosterone output by normal human zona glomerulosa cells induced by ACTH was, however, substantially greater than the one induced by angiotensin II,

Table 1. Stimulation of aldosterone production by adrenocortical cells in response to added angiotensin II

Source of cells	Angiotensin II concentration (mol/l)		Ref.[a]
	Significant response	Maximal response	
Rat capsular adrenals	4.9×10^{-8}	4.9×10^{-5}	1
–	2×10^{-10}	2×10^{-8}	2
–	2.4×10^{-10}	2.4×10^{-8}	3
–	$3 \times 10^{\ 11}$	10^{-9}	4
Bovine outer adrenal cortex	10^{-9}	10^{-6}	5
Rabbit capsular adrenals	10^{-8}	5×10^{-7}	6
Dog capsular adrenals	10^{-11}	3×10^{-10}	7
Cat capsular adrenals	10^{-9}	10^{-8}	8
Guinea-pig total adrenals	2.4×10^{-12}	2.4×10^{-6}	9
Human (normal) capsular adrenals	–	2.4×10^{-10}	10
Human aldosteronoma	2.4×10^{-10}	2.4×10^{-9}	10

[a] 1 Haning et al. (1970)
2 Bing and Schulster (1977)
3 Braley and Williams (1977)
4 Douglas et al. (1978a)
5 Peytremann et al. (1974)
6 Chiu and Peach (1974)
7 Fredlund et al. (1975)
8 Bravo et al. (1976a)
9 Pham-Huu-Trung et al. (1979)
10 Williams and Braley (1977)

according to G. Brown et al. (1980). On the other hand, angiotensin II (10^{-7} mol/l) and ACTH (10^{-8} mol/l) stimulated to a similar extent the aldosterone production of cells from aldosterone-producing adenomas. At a medium potassium concentration of 5 mmol/l, ACTH and angiotensin II elicited similar maximum responses in aldosterone production by dog zona glomerulosa cells; under these conditions, the aldosterone production was higher than in a medium with a potassium concentration of 7.5 mmol/l without any additional stimulator (Fredlund et al. 1975, 1977). In cells originating from the outer 0.25 mm of the bovine adrenal cortex, angiotensin II had a marked stimulatory action upon aldosterone output, whereas ACTH was completely ineffective (unless zona fasciculata cells were added; Peytremann et al. 1974). Increases in rat zona glomerulosa cell aldosterone production induced by angiotensin II were accompanied by a parallel increase in corticosterone production (Haning et al. 1970; Braley and Williams 1977) and by a small but significant increase in 18-OH-DOC output (Moore et al. 1978). As shown by D. J. Campbell (1982a), the maximum stimulation of aldosterone output induced by angiotensin II and other agents in dispersed zona glomerulosa preparations is greatly dependent on the composition of the incubation medium, i.e., on its content of albumin and yet unidentified components.

Angiotensin II stimulated the production of aldosterone, 18-hydroxycorticosterone, and corticosterone by dispersed duck adrenocortical cells in a dose-dependent, biphasic manner in concentrations ranging from 10^{-11} to 10^{-7} mol/l (Vinson et al. 1979).

Studies regarding a possible steroidogenic effect of angiotensin II on zona fasciculata cells have yielded conflicting results. In concentrations between 10^{-9} and

10^{-7} mol/l, the peptide elicited a dose-dependent increase in corticosteroid[4] production by bovine zona fasciculata cells (Peytremann et al. 1973; Hepp et al. 1977). Very high concentrations of angiotensin II (1.9×10^{-4} to 1.5×10^{-3} mol/l) stimulated the corticosterone output of pure rat zona fasciculata cells as extensively as ACTH (S. A. S. Tait et al. 1974). However, according to Bing and Schulster (1977), only a relatively impure preparation of angiotensin II amide stimulated the corticosterone production by rat zona fasciculata-reticularis cells, whereas a pure preparation of angiotensin II free acid was completely inactive, even at very high concentrations. Doses of angiotensin II which stimulated the aldosterone output by capsular rat adrenal cells had no steroidogenic effect upon decapsulated adrenal cells (Haning et al. 1970; Braley and Williams 1977; Douglas et al. 1978a; Moore et al. 1978). In a high concentration (10^{-5} mol/l), angiotensin II consistently stimulated the cortisol output of cells originating from outer and from inner layers of human adrenal glands removed from patients with Cushing's syndrome or primary aldosteronism (McKenna et al. 1978a). In this context it may be mentioned that, according to Mason et al. (1979), angiotensin II infusions elicited dose-dependent increases in the plasma cortisol concentration in human beings receiving a low-dose, constant-rate infusion of ACTH. This observation sharply contrasts with the previous report by Rayyis and Horton (1971), who found that a pressor dose of angiotensin II reversed the cortisol-stimulating effect of an ACTH infusion. In untreated or dexamethasone-treated human beings, angiotensin II either had no effect upon the plasma cortisol concentration or even decreased it (Rayyis and Horton 1971; Mason et al. 1976, 1979; Scholer et al. 1973; McKenna et al. 1978a).

4.1.2 Angiotensin III

Angiotensin III[5] is the C-terminal heptapeptide derivative of angiotensin II and is a metabolite of both angiotensin II and (des-asp[1])-angiotensin I. It is a potent stimulator of aldosterone biosynthesis in vitro and of aldosterone secretion in vivo, despite its relatively weak blood pressure-raising activity. The original observation by Blair-West et al. (1971a) that the aldosterone-stimulating activities of angiotensin II and angiotensin III upon the autotransplanted sheep adrenal gland were almost equal has been confirmed by a large number of incubation studies with dispersed adrenal zona glomerulosa cells from various animal sources (for a comparative table see Coghlan et al. 1979a, p. 65 *P*). The finding by some investigators that the heptapeptide stimulated aldosterone production more rapidly (Chiu and Peach 1973) or more extensively (Goodfriend and Peach 1975) - although not confirmed by later studies - has fostered speculation about an exclusive or at least a very important physiological role of angiotensin III as the final mediator of the renin-angiotensin system in the control of aldosterone secretion (Goodfriend and Peach 1975; J. O. Davis and Freeman 1977). In view of the available evidence, such hypotheses appear to be rather controversial. On the one

---

[4] "Corticosteroids" were measured by fluorimetry or competitive protein binding and consisted mainly of cortisol and corticosterone.

[5] (Des-aspartic acid[1])-angiotensin II

hand, a substantial portion of endogenously produced angiotensin I may well exert its aldosterone-stimulating action in the final form of angiotensin III[6]. On the other hand, angiotensin II is an equally effective aldosterone stimulator on its own, and its conversion to angiotensin III in blood, tissues, or even within the target cell is apparently not requisite for its steroidogenic action.

When added to dispersed adrenal zona glomerulosa cells from rabbits (Chiu and Peach 1974), dogs (Fredlund et al. 1975a), and cats (Bravo et al. 1976a), the aldosterone-stimulating activities of angiotensin II and angiotensin III were identical, with superimposable dose-response curves. The two peptides were also equally active in promoting corticosteroid biosynthesis by bovine zona fasciculata cells (Hepp et al. 1977). In experiments with rat adrenal glomerulosa cells, equimolar concentrations of each peptide stimulated aldosterone production to the same extent, but only during the first 20 min of incubation (Aguilera et al. 1979). With longer incubation times, angiotensin III became progressively less active than angiotensin II, although the maximum response elicited by the two peptides remained the same. The two angiotensins had no cumulative stimulatory activity when each was added in a dose sufficient to yield a maximum response in aldosterone output on its own. Analysis of the peptides in the medium revealed extensive metabolism during 2 h of incubation, which might have been due partly to contamination with proteolytic enzymes of the albumin in the medium (Mendelsohn and Kachel 1980). A more rapid rate of degradation accounted for the lesser activity of the heptapeptide. According to Braley et al. (1983a), rat adrenal zona glomerulosa cells were less sensitive to angiotensin III (ED_{50} 4.6×10^{-9} mol/l) than to angiotensin II (ED_{50} 6.3×10^{-10} mol/l). Nevertheless, the two peptides were degraded to similar extents (angiotensin II - 36%, angiotensin III - 41.5%) during 1 h of incubation. Bell et al. (1984) found the relative activity of angiotensin III to be 30% that of angiotensin II in stimulating the corticosterone output of rat zona glomerulosa cells during 1 h of static incubation. Under these conditions, 46.5% of labeled angiotensin II were recovered as angiotensin II and 20.1% as unknown metabolites. After incubation with labeled angiotensin III, 34.8% angiotensin III and 30.7% unknown metabolites were recovered. When the peptides were added during superfusion to similar cells, angiotensin III had 82% of the steroidogenic activity of angiotensin II, and the degradation of either peptide was less than 5%. Goodfriend and Peach (1975) had also observed a more extensive degradation of angiotensin III during incubation with rabbit adrenal glomerulosa slices. In contrast, angiotensin degradation by dog adrenal cells was slower, and both peptides were metabolized at the same rate (Aguilera et al. 1979). During a 1-h incubation with rabbit capsular adrenals, 10% of the added angiotensin II was converted to angiotensin III (Goodfriend and Peach 1975). As shown by thin-layer chromatography of the peptides eluted from rat zona glomerulosa cells after incu-

[6] According to paper chromatography and a radioimmunoassay for angiotensin II with a 100% cross-reactivity for angiotensin III, 58% of the immunoreactive material in rat arterial and venous plasma (but only 12% in human plasma) was angiotensin III (Semple and Morton 1976). The ratio of endogenous angiotensins (II:III) was 8 and 4.3, respectively, in the plasma of sodium-replete and sodium-deficient dogs (Semple et al. 1978). In sheep, the arterial angiotensin III concentration was over 40% of the angiotensin II concentration; 63% of the arterial concentration of angiotensin III were derived from the pulmonary circulation (Coghlan et al. 1979a; Fei et al. 1981).

bation with radiolabeled angiotensin II, at least 90% of the bound radioactive material was unaltered octapeptide (Aguilera et al. 1979). In similar experiments with dog zona glomerulosa cells, angiotensin II accounted for most, angiotensin III for less than 15% of the bound radioactive material (Douglas et al. 1978b). In light of these findings, it seems unlikely that angiotensin II has to be converted to angiotensin III in order to exert its steroidogenic action.

Kono et al. (1975) found angiotensins II and III to be equipotent in raising plasma aldosterone when systemically infused into normal human beings. However, according to Carey et al. (1978), angiotensin III was consistently less effective than angiotensin II, and the relative aldosterone-stimulating activity of the heptapeptide depended on its infusion rate and on the sodium intake. During sodium restriction, a low dose (2 pmol/kg/min) of angiotensin III was nearly as effective as an equimolar dose of angiotensin II in raising plasma aldosterone. At higher infusion rates, the response to the heptapeptide remained constant, while the octapeptide elicited marked additional increments in aldosterone secretion. In dexamethasone-treated, nephrectomized, anesthetized dogs, the two angiotensins were equally effective in stimulating aldosterone and cortisol secretion (Bravo et al. 1976a), but angiotensin III raised the plasma aldosterone to a lesser extent than equal doses of angiotensin II (McCaa 1978). In dexamethasone-treated, intact, anesthetized dogs, angiotensin III was as effective as angiotensin II in stimulating aldosterone secretion (Freeman et al. 1976) and in raising plasma aldosterone (Beckerhoff et al. 1976). However, angiotensin III had a lesser effect on corticosterone and cortisol secretion (Freeman et al. 1976), and the plasma aldosterone dropped significantly faster in heptapeptide-treated dogs after cessation of the infusion (Beckerhoff et al. 1976). In intact, conscious, sodium-replete or sodium-deficient dogs, angiotensin III was equally or slightly more potent than angiotensin II in raising plasma aldosterone levels (Semple et al. 1978; Nicholls et al. 1978). In sodium-replete sheep, angiotensin II was four times as potent as angiotensin III in raising arterial blood pressure but only insignificantly more active in stimulating aldosterone secretion (Blair-West et al. 1980). In sodium-depleted sheep, the blood clearance rates of the two peptides were the same, and both substances were equally active in suppressing the plasma renin concentration (Fei et al. 1980, 1981). The two peptides had equal or very similar effects upon plasma aldosterone in conscious rabbits (Steele et al. 1976), urinary aldosterone excretion in sodium-deficient rats (Peach et al. 1976), and plasma aldosterone in conscious rats (W. B. Campbell et al. 1974). On the other hand, a high dose of angiotensin III (1 μg/min) stimulated aldosterone secretion less extensively than a high dose of angiotensin II in anesthetized rats treated with dexamethasone and morphine (Spielman et al. 1976).

When infused in high doses into normal untreated human subjects or into subjects treated with the converting-enzyme inhibitor captopril, the hexapeptide *des-asp¹-des-arg²-angiotensin II* had no effect on blood pressure, but raised plasma aldosterone and lowered plasma renin activity to a slight extent, which was statistically significant (Kono et al. 1982). In patients with Bartter's syndrome, it caused decreases in plasma renin activity and in plasma aldosterone (Kono et al. 1982, 1985b). In sodium-replete sheep, the hexapeptide did not stimulate aldosterone secretion when infused into the arterial supply of autotransplanted adrenal glands (Blair-West et al. 1971a).

4.1.3 Angiotensin I and Des-asp^1-angiotensin I

The aldosterone-stimulating activities of the decapeptide angiotensin I and of the nonapeptide des-asp^1-angiotensin I appear to depend largely but not entirely on the conversion of these substances to angiotensin II and angiotensin III respectively. According to a paper chromatographic separation, des-asp^1-angiotensin I accounted for 19% of the endogenous angiotensin I immunoactivity extracted from rat arterial blood (Garcia del Rio et al. 1981). In normal man, the systemic infusion of a high dose (300 ng/kg/min) of des-asp^1-angiotensin I led to a rise in plasma aldosterone, which was completely abolished by the oral intake of the converting-enzyme inhibitor SQ 14225 (Kono et al. 1980). In rats, treatment with other converting-enzyme inhibitors completely blocked a stimulation of urinary aldosterone excretion by des-asp^1-angiotensin I (Vaughan et al. 1977) but only partially attenuated rises in plasma aldosterone elicited by angiotensin I and des-asp^1-angiotensin I (W.B. Campbell et al. 1977). Apparently, the two peptides have a weak aldosterone-stimulating activity of their own, manifested in the presence of a converting-enzyme inhibitor during incubation of rat and dog capsular adrenal cells (Vaughan et al. 1977; Capponi and Catt 1979; Mendelsohn and Kachel 1980; Braley et al. 1981). Both peptides also stimulated the corticosteroid output of bovine zona fasciculata cells without being measurably converted to angiotensin II or III (Hepp et al. 1977). On the other hand, the addition of a converting-enzyme inhibitor completely blocked the increases in aldosterone output by rabbit zona glomerulosa cells in response to angiotensin I and des-asp^1-angiotensin I (Ackerly et al. 1977).

4.1.4 Synthetic Angiotensin Analogues

Of more than 200 angiotensin analogues with various agonistic and antagonistic properties or both (Khosla et al. 1974), relatively few have been used in studies on the control of aldosterone biosynthesis in vitro or in vivo.

Sarcosine^1-angiotensin II was ten times more active in promoting aldosterone biosynthesis by cat (Bravo et al. 1976) and dog zona glomerulosa cells (Fredlund et al. 1975b), but it was markedly less active than angiotensin II in its effects on aldosterone production by rabbit capsular adrenal cells (Peach and Chiu 1974). During static incubation of rat adrenal zona glomerulosa cells, sar^1-angiotensin II was twice as active as the natural asp^1-angiotensin II in stimulating corticosterone output (Bell et al. 1984). This was probably due mainly to a decreased degradation of the synthetic analogue, since the two peptides were nearly equipotent in their steroidogenic activity when tested in a system of superfused zona glomerulosa cells. During infusions into dexamethasone-treated nephrectomized, anesthetized dogs, sar^1-angiotensin II was twice as potent as the naturally occurring octapeptide in raising blood pressure but equally potent in its aldosterone-stimulating activity (Bravo et al. 1976b). Following cessation of the infusions, the plasma aldosterone concentration decreased more rapidly in dogs treated with asp^1-angiotensin II than in dogs treated with sar^1-angiotensin II.

Sarcosine^1, alanine^8-angiotensin II (Saralasin, P-113) has been widely used in studies on the role of the renin-angiotensin system in the control of aldosterone secretion. Experiments with canine zona glomerulosa cells showed it to be a potent inhibitor of the steroidogenic action of angiotensin II (although less active than sar^1,ile^8-angiotensin II on a molar basis), and it did not display any significant agonist activity (Saltman et al. 1976). In this preparation, the aldosterone-stimulating effects of angiotensin II and angiotensin III were identically and completely inhibited by the analogue. By contrast, in a preparation of rabbit capsular adrenal cells, the antagonist was 25 times more active in suppressing the aldosterone-stimulating action of angiotensin II than in suppressing the steroidogenic action of angiotensin III (Chiu and Peach 1974). In rat adrenal glomerulosa cells, sar^1,ala^8-angiotensin II completely suppressed the steroidogenic effect of angiotensin II; 50% inhibition was produced by an equimolar concentration of the antagonist (Bing and Schulster 1977). In another study with dispersed rat adrenal zona glomerulosa cells, sar^1,ala^8-angiotensin II inhibited the responses in aldosterone output to angiotensin II and to angiotensin III with equal efficacy (Braley et al. 1983a). In a system of superfused rat adrenal zona glomerulosa cells, sar^1,ala^8-angiotensin II inhibited the aldosterone response to angiotensin II but did not affect serotonin-, potassium-, or ACTH-stimulated aldosterone production (Boscaro et al. 1981). Subcutaneous administration of sar^1-ala^8-angiotensin II completely blocked the aldosterone-stimulating action of intravenously infused angiotensin II in the conscious rat and had a less extensive effect on the aldosterone-stimulating action of angiotensin III (W.B.Campbell et al. 1974). However, in sodium-replete, dexamethasone-treated, anesthetized dogs, the analogue (400 pmol/kg/min) completely reversed increases in plasma aldosterone induced by angiotensin III (20 pmol/kg/min) but only partially suppressed the aldosterone-stimulating activity of angiotensin II (Beckerhoff et al. 1976). At doses of up to 200 pmol/kg/min the analogue had no agonistic activity on aldosterone secretion in the dog (Beckerhoff et al. 1975). In sodium-replete sheep, sar^1,ala^8-angiotensin II equally suppressed the aldosterone-stimulating effects of angiotensin II and angiotensin III; infusions of the analogue alone had a minor agonistic action on aldosterone secretion (Blair-West et al. 1979). In sodium-replete rabbits, the analogue stimulated aldosterone secretion (Steele and Lowenstein 1974). When infused at a rate of 5 μg/kg/min, it blocked the pressor but not the steroidogenic action of simultaneously administered angiotensin II (0.2 μg/min/kg). In normal, sodium-replete human subjects sar^1,ala^8-angiotensin II had a weak, dose-independent agonistic effect upon plasma aldosterone concentrations (Hollenberg et al. 1976). However, during angiotensin II infusions it displayed a significant and dose-dependent antagonistic activity on blood pressure as well as on plasma aldosterone.

In dog adrenal glomerulosa cells, sarcosine^1, isoleucine^8-angiotensin II was the most potent competitive antagonist of angiotensin II-stimulated aldosterone output, but it was also a partial agonist at concentrations required to inhibit the steroidogenic effect of the naturally occurring octapeptide (Saltman et al. 1976). When given alone it also significantly stimulated aldosterone secretion in the dog in vivo (Beckerhoff et al. 1975). In rat capsular adrenal cells, sar^1,ile^8-angiotensin II had a weak agonistic effect when added alone and a potent antagonistic effect upon the steroidogenic action of angiotensin III (Sarstedt et al. 1975). Sar^1,ile^8-angiotensin II had a weak agonistic effect on aldosterone output by superfused rat zona glomerulosa cells (Boscaro et al. 1981). It blocked the increase in aldosterone output by this preparation not only when stimulated by angiotensin II but also when stimulated by ACTH. It did not, however, inhibit aldosterone production under stimulation by serotonin or a high potassium concentration. When added in equal concentrations, sar^1,-ile^8-angiotensin II partially blocked the aldosterone-stimulating action of angiotensin II on rabbit zona glomerulosa cells but did not suppress the stimulatory effect of angiotensin III (Goodfriend and Peach 1975). The heptapeptide analogue $isoleucine^7$-angiotensin III effectively blocked the steroidogenic actions of angiotensin II and of angiotensin III in this preparation. On the other hand, in dog zona glomerulosa cells, ile^7-angiotensin III (= des-asp^1,ile^8-angiotensin II) was much less potent than sar^1,-ile^8-angiotensin II as an inhibitor of the actions of both the heptapeptide and the octapeptide agonists (Saltman et al. 1976). Douglas et al. (1981) found a group of octapeptide antagonists to be better inhibitors of angiotensin II-induced aldosterone production by rat adrenal glomerulosa cells than the corresponding heptapeptide antagonists. Among the heptapeptide antagonists tested, des-asp^1,ala^8-angiotensin II and des-asp^1,ile^8-angiotensin II were the best inhibitors of angiotensin II-stimulated steroidogenesis, consistent with the previous observation by these investigators that they had the highest receptor-binding affinity among the heptapeptide antagonists (Douglas et al. 1980). Also, according to Braley et al. (1983a), ala^8-octapeptide analogues were better inhibitors of both angiotensin II and angiotensin III stimulation (of aldosterone production by rat adrenal glomerulosa cells) than the corresponding ala^8 heptapeptide analogues. The heptapeptide analogues inhibited angiotensin III stimulation preferentially. In man, ile^7-angiotensin III raised the plasma aldosterone concentration on its own and equally inhibited the steroidogenic actions of angiotensins II and III (Kono et al. 1977, 1978).

4.1.5 Adrenocortical Angiotensin Receptors

Specific binding of radiolabeled angiotensin II by adrenocortical tissue was first described by Lin and Goodfriend (1970). Among various rat tissues tested, capsular adrenals displayed the highest angiotensin II binding activity (Brecher et al. 1974). Autoradiography of rat adrenal glands labeled in vitro with $[^{125}I]$-angiotensin II, showed that binding sites were concentrated in a thin band in the outer cortex (over the cells of the zona glomerulosa) and in the adrenal medulla (Healy et al. 1985). Few sites were evident in the inner zones of the adrenal cortex. Bovine and rat adrenocortical angiotensin II receptors were found to be mainly localized in the plasma membrane fraction of tissue homogenates (Glossmann et al. 1974; Douglas et al. 1978a). Purified bovine adrenal cortical plasma-membrane vesicles contained not only a highly ACTH-sensitive adenylate-cyclase system but also high-affinity binding sites for angiotensin II (2 pmol/mg protein) with an apparent association constant of 2×10^7 (mol/l) at 12° C (Schlegel and Schwyzer 1977). Labeled angiotensin II is displaced from its receptor sites by angiotensin analogues in proportion to their known agonistic or antagonistic steroidogenic activities (Glossmann et al. 1974; Saltman et al. 1975; Capponi and Catt 1979; Douglas et al. 1980, 1985). Devynck et al. (1977) found two classes of angiotensin receptor sites in rat adrenal tissue. One of them preferably bound angiotensin II with an equlibrium dissociation constant of $3.3–5.2 \times 10^{-9}$ mol/l. The other one had a higher affinity for angiotensin III and an equilibrium dissociation constant of $1–3 \times 10^{-10}$ mol/l. In view of the equal aldosterone-stimulating activities of the two peptides, the biological significance of the specific angiotensin III receptors has yet to be elucidated. Douglas et al. (1985) found no high-affinity angiotensin III binding sites in rat adrenal glomerulosa preparations. Moreover, the order of potency of ten different analogues as competitive antagonists of angiotensin III receptor binding was similar to their order of potency as antagonists of angiotensin II receptor binding in the rat adrenal zona glomerulosa. This suggested that angiotensin II and angiotensin III were bound to the same receptor. Additional evidence for this view was the observation that the numbers of angiotensin II and angiotensin III binding sites, according to Scatchard analysis, were not significantly different. Moreover, analogues of angiotensin II and angiotensin III behaved as competitive antagonists of both angiotensin II and angiotensin III receptor binding, according to Lineweaver-Burk analysis. Finally, the majority of analogues showed similar potencies as antagonists of angiotensin III and angiotensin II.

Attempts to solubilize angiotensin II receptors from bovine adrenocortical plasma membranes with three different nonionic detergents, sodium deoxycholate, or the chaotropic ions ClO_4^- and Br^- failed because all these substances inhibited the binding of tritiated angiotensin II to the plasma membranes (Forget and Heisler 1979). Chang and Lotti (1981) succeeded in solubilizing angiotensin II receptors from bovine adrenal cortex membranes with a 1% digitonin solution. The solubilized receptors maintained the conformation necessary to recognize agonists and antagonists. Birabeau et al. (1984) used the zwitterionic detergent CHAPS to extract from the zona glomerulosa of rat and bovine adrenal glands angiotensin receptors which retained all the binding properties, affinity, and specificity of particulate receptors. In contrast to particulate receptors, solubilized receptors were affected neither by sodium and calcium ions nor by guanyl nucleotides.

Capponi and Catt (1980) permanently labeled angiotensin receptors from membranes of canine adrenal cortex and uterus with a photosensitive radioactive angiotensin analogue, [^{125}I]-2-nitro-5-azidobenzoyl angiotensin II. Analysis of the receptors solubilized with Triton X-100 by gel filtration chromatography and sucrose density gradient centrifugation showed two radioactive peaks with calculated molecular weights of 126000 and 64000, irrespective of the source of the receptors. Sodium dodecyl sulfate-polyacrylamide gel electrophoresis showed a single radioactive band with an apparent molecular weight of 65000 for adrenocortical receptors. This indicated that the receptor molecule consisted of two subunits with similar molecular weights. In studies with a different method of cross-linking, Paglin and Jamieson (1982) observed that the major angiotensin-binding polypeptide of rat adrenal gland membranes had a molecular weight of 116000 under reducing conditions. Using a photoaffinity labeling technique, Guillemette and Escher (1983) identified the bovine adrenocortical angiotensin receptor as a protein with a molecular weight of approximately 300000 by means of gel filtration under nonreducing and nondenaturating conditions. Under reducing and denaturating conditions, a labeled subunit with a molecular weight of 60000 was found by SDS-PAGE and gel filtration. These investigators also observed that photoaffinity labeling of the angiotensin receptor of bovine adrenocortical cells with sar^1,(4′-N_3)phe^8-angiotensin II resulted in a permanent stimulation of aldosterone production.

4.2 Monovalent Cations

4.2.1 Sodium

A direct influence of the ionic environment on aldosterone biosynthesis was first observed by Rosenfeld et al. (1956), who found a significant increase in the production of aldosterone-like material by perfused calf adrenals when the sodium concentration of the perfusing fluid was decreased from 149 to 125 mmol/l and the potassium concentration raised from 3.5 to 25 mmol/l. Denton et al. (1959) demonstrated that increases of potassium and decreases of sodium concentrations within physiological ranges in the arterial blood supply to autotransplanted sheep adrenals led to changes in the electrolyte composition of parotid saliva indicative of an increased secretion of aldosterone. Further studies by this group of investigators (Blair-West et al. 1962a) confirmed that local alterations of sodium and potassium concentrations in the adrenal arterial plasma significantly influenced aldosterone secretion - directly measured by a double-isotope dilution derivative assay. Although a simultaneous increase in potassium and decrease in sodium concentration was the most effective stimulus, independent changes in the concentration of either cation also influenced aldosterone production (Blair-West et al. 1963). A decrease in sodium concentration by 5-20 mmol/l significantly stimulated aldosterone secretion in sodium-replete sheep, without having a consistent effect on corticosterone or cortisol output. In hypophysectomized, nephrectomized dogs, infusion of isotonic glucose solution to the arterial supply of isolated adrenals, which lowered local plasma sodium concentration by 20 mmol/l, significantly increased aldosterone secretion (J. O. Davis et al. 1963).

Relatively small changes in the sodium concentration of the medium with which an isolated canine adrenal gland preparation was perfused in vitro caused significant alterations in aldosterone secretion in the absence of stimulators and affected aldosterone output even more extensively during stimulation by angiotensin II or a high potassium concentration (E. G. Schneider et al. 1984). The aldosterone secretion rate was 2.8 ng/min at a sodium concentration of 145 mmol/l; it rose to 8.8 ng/min and fell to 2.0 ng/min at sodium concentrations of 139 and 152 mmol/l respectively. Lowering the sodium concentration from 145 to 139 mmol/l induced a fivefold rise in potassium- or angiotensin II-stimulated aldosterone secretion, whereas an increase in the sodium concentration to 152 mmol/l decreased potassium- or angiotensin II-stimulated aldosterone outputs by a factor of 5. The suppressive effect of a high sodium concentration was greater at a lower than it was at a higher concentration of angiotensin II. According to a second study performed with the same experimental model, these effects of changes in the sodium (and chloride) concentrations on aldosterone secretion appeared to be due mainly to changes in osmolality (E. G. Schneider et al. 1985). Thus, increases in the osmolality of the perfusion medium by 8-10 mosmol/l by the addition of NaCl, sucrose, mannitol, or glucose inhibited angiotensin II- or potassium-stimulated secretion to a similar extent. On the other hand, iso-osmotic reduction in NaCl concentration by the addition of isotonic sucrose solution to the medium did not potentiate aldosterone stimulation by angiotensin II or potassium, whereas a hyposmotic reduction in NaCl concentration had a marked additional stimulatory effect. The observation that an increase in the osmolality by 9 mosmol/l by the addition of urea had no effect on angiotensin II-stimulated aldosterone production indicated that changes in intracellular volume or composition were crucial for the effect induced by changes in the extracellular sodium and chloride concentrations.

In sheep with established sodium deficiency, in which plasma sodium concentration was 9 mmol/l lower than in sodium-replete sheep, a local increase in the plasma sodium level by 8 mmol/l, obtained by the adrenal arterial infusion of a hypertonic NaCl solution, lowered aldosterone secretion by 25% for periods of 1-3 h (Blair-West et al. 1966). However, aldosterone secretion returned to its initial high rate after 3-6 h, even when the local sodium concentration was increased by 15 mmol/l. The infusion of a hypertonic $NaHCO_3$ solution for 4-6 h - at a concentration and rate sufficient to elevate the ambient plasma sodium and bicarbonate concentrations to sodium-replete levels - was follwed by a significant fall in the aldosterone secretion rate (Blair-West et al. 1971b). Under these conditions, a decrease in plasma potassium and increases in bicarbonate concentration and pH may have supported the supressive effect of sodium ions. Infusion of hypertonic mannitol into the adrenal artery had no effect on aldosterone secretion (Blair-West et al. 1966). At the onset of sodium deficiency the effects of high local sodium concentrations were more pronounced. Although the normalization of the small decrease in plasma sodium of 2-4 mmol/l was not sufficient to stop a rapid increase in the aldosterone secretion rate, an elevation of the adrenal arterial plasma sodium concentration by 6-15 mmol/l caused marked and sustained decreases in aldosterone production. In sodium-replete sheep, the aldosterone-stimulating effect of angiotensin II was significantly reduced when the peptide was infused in a hypertonic saline solution, leading to a mean increase of 10 mmol/l in the adrenal arterial plasma sodium level (Blair-West et al. 1965a).

Whereas relatively small changes in the potassium concentration of the incubation medium significantly influence aldosterone production by rat adrenocortical tissue or cells in vitro (see p. 38), even marked decreases in sodium concentration have either been ineffective or led only to small increments in aldosterone output. Thus, an isotonic decrease in the sodium concentration from 148 to 72 mmol/l resulted in only a 40% increase in aldosterone production by incubated quartered rat adrenals (J. Müller 1971; Fig. 9), whereas in the same preparation a 150% stimulation of the aldosterone output was observed when the potassium concentration was increased from 3.7 to 7.2 mmol/l, although the sodium/potassium ratio was in both instances reduced from 40 to 20. Aldosterone production by incubated adrenal glands of normal or sodium-restricted rats was not affected by either a high (165 mmol/l) or a low (125 mmol/l) ambient sodium concentration (Debreceni and Csete 1973). When outer slices of bovine adrenals were incubated in a medium with a sodium concentration of 112 mmol/l, aldosterone production increased by 37%, but corticosterone production decreased by 59% (Saruta et al. 1972). This finding suggested a direct effect of the sodium concentration upon the conversion of corticosterone to aldosterone. A high sodium concentration (160 mmol/l) had no effect on the steroidogenesis in this preparation. In capsular dog adrenal tissue, however, changes in aldosterone output in response to alterations in the sodium concentration were accompanied by homodirectional changes in corticosterone output (Lobo et al. 1978). In this tissue, increases and

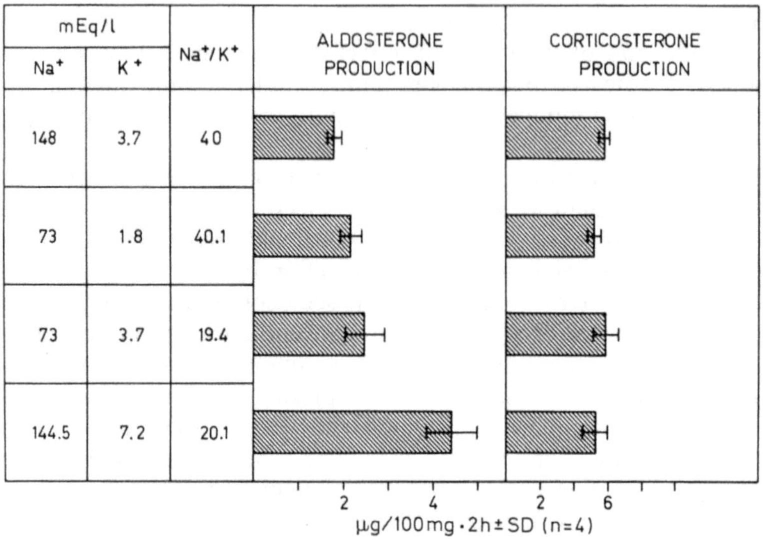

Fig. 9. Effect of sodium and potassium concentration in the incubation medium on aldosterone and corticosterone production by rat adrenal quarters in vitro. Mean values of two experiments ($n=4$) $\pm$ standard deviation. (From J. Müller 1971)

decreases in aldosterone production were induced by changes as small as 10 mmol/l sodium. According to these investigators, alterations in aldosterone output induced by increases or decreases in the extracellular sodium concentration ($\pm$20 mmol/l) were more extensive under basal conditions of incubation than under stimulation by angiotensin II or a high potassium concentration.

The reduction of the sodium concentration by 5-20 mmol/l - with or without osmotic changes - of the medium in which isolated rat adrenal glomerulosa cells were incubated did not influence aldosterone production under basal conditions or under stimulation by two different concentrations of angiotensin II (Enyedi and Spät 1981). This ineffectiveness of a reduced extracellular sodium concentration in stimulating aldosterone output in vitro contrasted with a close correlation between plasma sodium and aldosterone concentrations after sodium depletion of rats in vivo by peritoneal dialysis with 5% glucose (Balla et al. 1981). A much greater reduction of the extracellular sodium concentration, i.e., from 142 to 97, 52, or 7 mmol/l (by iso-osmotic dilution with sucrose or LiCl), led unexpectedly to decreases in aldosterone production by isolated rat adrenal glomerulosa cells under basal conditions (with the exception of a sodium concentration of 97 mmol/l, which was slightly stimulatory) and under stimulation by angiotensin II and potassium ions (Schiffrin et al. 1982). The stimulation of aldosterone production by ACTH, however, was reduced only by the lowest sodium concentration, 7 mmol/l.

Veratridine, a sodium-channel opener dissipating sodium gradients, completely blocked the increased aldosterone output of rat adrenal glomerulosa cells in response to potassium ions, angiotensin II, or ouabain and had a lesser but still significant suppressive effect on ACTH-stimulated aldosterone production (Schiffrin et al. 1982). Monensin, another natrium ionophore that translocates sodium across cellular membranes, had a potent inhibitory effect on aldosterone production by dispersed capsular rat adrenal cells both in the absence of stimulators and under stimulation by ACTH, angiotensin II, or dibutyryl cyclic AMP (Matsuoka et al. 1984). It did not affect the ACTH-induced cyclic AMP production by rat capsular adrenal cells and did not suppress aldosterone production by these cells during incubation in a sodium-free medium.

Alterations in the sodium concentration of the superfusion medium from 114 to 130 mmol/l did not affect aldosterone or corticosterone production by frog interrenal tissue (Lihrmann et al. 1985).

4.2.2 Potassium

Although it is generally accepted that the extracellular potassium concentration can directly influence aldosterone biosynthesis, the physiological importance of this factor in the control of aldosterone secretion is often underestimated. This may be due to the fact that some of the earlier studies demonstrated an increase in aldosterone output in response to unphysiologically high potassium concentrations (Rosenfeld et al. 1956; Giroud et al. 1956b). However, according to later studies, aldosterone secretion was markedly stimulated by minor increments in plasma potassium, which were not only well within physiological limits, but were in some cases as small as 0.1 mmol/l (Funder et al. 1969; Himathongkam et al. 1975). Since such changes are within the range of predictable analytical error of conventional flame spectrophotometrical methods, it is not surprising that, at least in one reported instance, an intravenous infusion of KCl stimulated aldosterone secretion but did not result in a measurable increase in the plasma potassium concentration (Birkhäuser et al. 1973). Thus, the ambient potassium concentration may be not only a very important determinant of aldosterone production but also an extremely elusive one.

As mentioned before, it was first shown in studies with sheep that small increases in the adrenal arterial plasma potassium concentration - with or without concomitant decreases in sodium concentration - could significantly stimulate aldosterone secretion (Blair-West et al. 1962a, 1963, 1970a). An elevation of the potassium concentration by 0.1-0.5 mmol/l was a potent stimulus of aldosterone production but had no effect on corticosterone or cortisol output (Funder et al. 1969). In hypophysectomized, nephrectomized dogs, the infusion of KCl or K_2SO_4 solution into the arterial supply of isolated adrenal glands repeatedly stimulated aldosterone secretion and occasionally also corticosterone secretion (J.O. Davis et al. 1963). Systemic infusion of potassium chloride into hypophysectomized dogs augmented the secretion rate of both aldosterone and corticosterone. The smallest increment in plasma potassium concentration eliciting a steroidogenic response was 1.3 mmol/l. In rats, even an extremely high serum potassium concentration of 8.4 mmol/l stimulated aldosterone biosynthesis, but it did not lead to any elevation in the serum corticosterone concentration (Komor and Müller 1979). Potassium-induced stimulation of aldosterone secretion by isolated rat adrenal glands perfused in situ was accompanied by an unchanged or even decreased corticosterone output (Hinson et al. 1985). In normal man, increases in serum potassium of 0.5-1.5 mmol/l led to significant rises in the plasma aldosterone concentration, but not in plasma corticosterone and cortisol concentrations (Dluhy et al. 1972). In normal human subjects on a low sodium intake, a potassium infusion of 0.17 mmol/min led to mean increments of 0.2 mmol/l in the plasma potassium and of 13 ng/dl in the plasma aldosterone concentrations (Himathongkam et al. 1975). In reverse, there were mean decreases of 0.3 mmol/l in plasma potassium and of 11 ng/dl in plasma aldosterone 90 min after the oral intake of glucose (Himathongkam et al. 1975). During the following 2 h the plasma aldosterone returned to its original level and above, despite a sustained low potassium concentration. This secondary rise in plasma aldosterone was associated with an increase in plasma renin activity. Similar effects of an oral glucose load were observed in sodium-replete, normal or borderline hypertensive human subjects by Beretta-Piccoli et al. (1980a). In anephric man, a fall in plasma potassium induced by insulin administration was accompanied by a more persistent decrease in plasma aldosterone (Cooke et al. 1973a). The information available on a possible steroidogenic effect of potassium ions in submammalian vertebrates is scarce (Vinson et al. 1979). In the American bullfrog, even a marked hyperkalemia did not stimulate aldosterone or corticosterone secretions (J.O.Davis et al. 1970). Similarly, increases in the extracellular potassium concentration from 5 to 15 mmol/l did not have any measurable stimulatory effect on aldosterone and corticosterone outputs by superfused frog interrenal tissue (Lihrmann et al. 1985).

In vitro studies with adrenocortical tissue or isolated zona glomerulosa cell preparations have confirmed a direct effect of ambient potassium concentration within physiological limits upon aldosterone biosynthesis. This was first demonstrated by Giroud et al. (1956b), who found that aldosterone production by incubated rat adrenals was 40% higher in a medium with a potassium concentration of 8.6 mmol/l than in a medium with a potassium concentration of 5.6 mmol/l. When quartered rat adrenals of sodium-deficient rats were incubated in buffers with different potassium concentrations, aldosterone production rose with increasing potassium levels to reach a maximum at between 7 and 9 mmol/l (J.Müller 1965a; Fig.10). At potassium concentrations above 10 mmol/l there was a significant decrease in aldosterone production. In this preparation, the corticosterone output was not significantly affected by the potassium concentration, but deoxycorticosterone output was stimulated by a high potassium concentration (J.Müller 1968). However, in capsular adrenals of sodium-replete rats, a high potassium concentration also elicited an increased corticosterone output (J.Müller 1970b; Fig.11). On the other hand, a high potassium concentration had no steroidogenic effect in the decapsulated portions of rat adrenals. Other studies with adrenocortical tissue or cell preparations from various sources have yielded similar findings, although a biphasic effect of potassium ions was not observed by all investigators. Biphasic dose-reponse curves similar to the one shown in Fig.10 were obtained in preparations of rat zona glomerulosa cells (Haning et al. 1970; Bing and Schulster 1977; Braley and Williams 1977), sheep adrenocortical tissue (McDougall et al. 1976), capsular dog adrenal tissue (Lobo et al. 1978), and cells from an aldosterone-producing human adrenocortical adenoma (G.H.Williams and Braley 1977). Aldosterone and corticosterone productions by bovine adrenal slices rose proportionally to the potassium concentration in the medium over a range between 0 and 14.5 mmol/l (Kaplan 1965). Dose-response curves which reached a plateau at potassium concentrations above 10 mmol/l were observed in incubation experi-

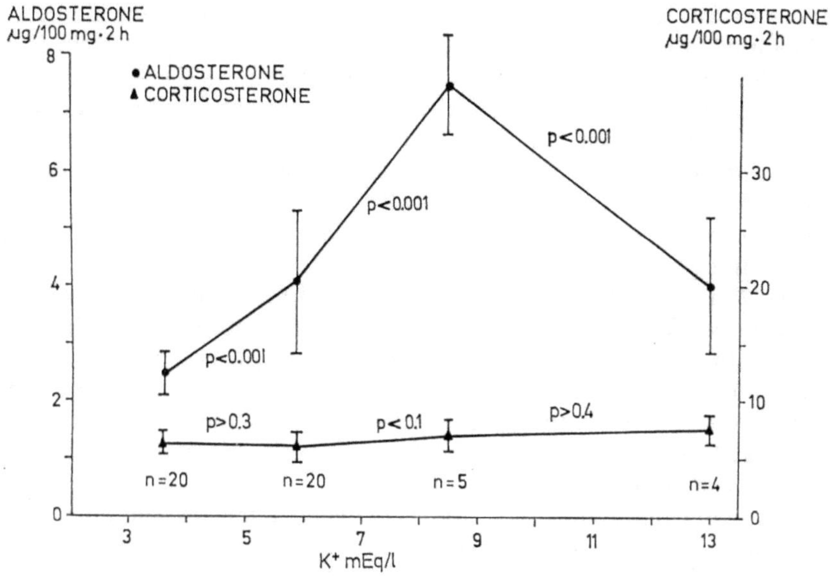

Fig. 10. Effect of potassium concentration in the incubation medium on in vitro production of aldosterone and corticosterone by rat adrenal quarters. Mean values ± standard deviation. *P* values of significance were calculated by *t*-tests. (From J. Müller 1965a)

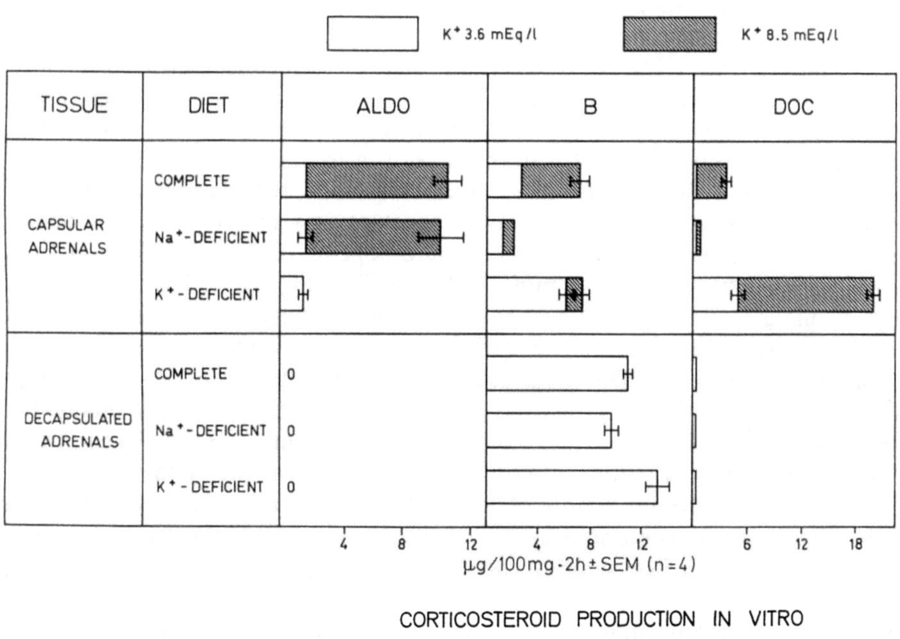

Fig. 11. Production of aldosterone *(ALDO)*, corticosterone *(B)*, and deoxycorticosterone *(DOC)* by separate zones of adrenals of rats kept on different diets for 2 weeks. *White bars* represent baseline corticosteroid production, *hatched bars* increments due to addition of KCl to the incubation medium. Mean values of two experiments ($n=4$) ± standard error of the mean. (Data from J. Müller 1970b)

ments with outer slices from dog adrenals (Burwell et al. 1969), dog zona glomerulosa cells (Fredlund et al. 1975; Fakundig et al. 1979), and guinea-pig (Pham-Hu-Trung et al. 1979) and duck adrenocortical cells (Vinson et al. 1979). According to three different groups of investigators, the smallest increment in potassium concentration which led to a significant increment in aldosterone output in vitro was 0.3 mmol/l (Boyd and Mulrow, 1972; Braley and Williams 1977; Pham-Huu-Trung et al. 1979). Studies with rat adrenal cells from separate zones have confirmed that the steroidogenic action of potassium ions is restricted to the zona glomerulosa (Haning et al. 1970; Moore et al. 1978). In bovine (Kaplan 1965; Saruta et al. 1972) and sheep (McDougall et al. 1976) mixed adrenocortical tissue, a high potassium concentration stimulated corticosterone production but had no effect on cortisol biosynthesis. In dog adrenocortical tissue slices, cortisol production was stimulated by potassium ions, according to Burwell et al. (1969) but not according to Lobo et al. (1978). In hypophysectomized dogs, the cortisol secretion rate was stimulated by an angiotensin II infusion but not by a potassium chloride infusion (A. A. Taylor et al. 1972a).

Besides influencing aldosterone biosynthesis on its own, the ambient potassium concentration also affects the steroidogenic response of the zona glomerulosa to other stimuli. In bovine adrenal tissue, increasing potassium concentrations in the medium potentiated the aldosterone-stimulating effect of a constant dose of angiotensin II (Kaplan 1965). In a potassium-free medium, the aldosterone-stimulating activity of serotonin upon rat adrenal tissue was diminshed but not abolished (J. Müller 1971). However, a high potassium concentration of 8.5 mmol/l did not further stimulate steroidogenesis in tissue maximally stimulated by serotonin. In experiments with rat zona glomerulosa cells (S. A. S. Tait et al. 1972), the addition of serotonin (10^{-4} mol/l) at different potassium levels further increased the outputs of aldosterone and corticosterone to reach a maximum at 5.9 mmol/l K^+. Addition of serotonin had no further stimulating effect at 8.4 and 13 mmol/l K^+. However, the combination of serotonin and a potassium concentration of 5.9 mmol/l led to a significantly higher output of both steroids than a potassium concentration of 8.4 mmol/l. In contrast, a constant dose of cyclic AMP or angiotensin II yielded approximately the same steroidogenic response at all potassium concentrations between 2 and 13 mmol/l (S. A. S. Tait et al. 1974; J. F. Tait and Tait 1976). Aguilera et al. (1980a) observed a complete absence of aldosterone stimulation by angiotensin II during incubations of rat adrenal glomerulosa cells in a potassium-free medium. The angiotensin II/aldosterone dose-response curves were translocated to the left when the potassium concentration of the incubation medium was increased from 4.5 to 5.0 mmol/l. Thus, small increases in the extracellular potassium concentration enhanced the sensitivity of zona glomerulosa cells to the steroidogenic action of angiotensin II. Neither angiotensin II nor ACTH stimulated aldosterone output by dog zona glomerulosa cells during incubation in a potassium-free medium (Fredlund et al. 1977). Increases in the extracellular potassium concentration from 2.5 to 5 and 7.5 mmol/l progressively potentiated the aldosterone-stimulating effects of both peptides over the whole range of the dose-response curves. Apparently, this interaction was not at the receptor level, since the concentration and binding affinity of angiotensin II receptor sites in the zona glomerulosa cells were not changed by increasing the potassium concentration from 0 to 7.5 mmol/l. In normal man, KCl infused at a rate of

10 mmol/h did not by itself raise plasma aldosterone, but it potentiated the stimulatory actions of angiotensin II or ACTH infusions (Gaillard et al. 1979).

There is at least some evidence indicating that the aldosterone-stimulating activity of potassium is not completely independent of the presence of other stimulators. As demonstrated in a study by D. B. Young et al. (1984), in which dogs on three different levels of potassium intake were treated or not with angiotensin II infusions, there was a multiplicative interaction between plasma potassium and angiotensin II levels in stimulating aldosterone secretion; i.e., higher angiotensin II concentrations increased the steepness of the plasma aldosterone/potassium regression line. In dexamethasone-treated conscious sheep with cervical adrenal autotransplants, constant local infusions of potassium raising adrenal venous potassium levels by 2 mmol/l elicited a phasic and poorly sustained response in aldosterone secretion in the absence of ACTH (Lun et al. 1979). The responsiveness of the adrenals to potassium was restored by doses of ACTH that were insufficient to stimulate aldosterone secretion alone. When added to rat adrenal glomerulosa cells incubated in Krebs-Ringer bicarbonate glucose albumin medium, small concentrations of angiotensin II, insufficient to stimulate steroidogenesis on their own, inhibited the stimulation of aldosterone production by a high potassium concentration (D. J. Campbell 1982a). When dogs were treated with an infusion of the converting-enzyme inhibitor captopril 1 h before an infusion of potassium chloride, there was no change in plasma aldosterone in response to rising plasma potassium concentrations (Pratt 1982). The response in plasma aldosterone to potassium administration was restored in captopril-treated dogs by angiotensin II infused at a rate which did not stimulate aldosterone secretion on its own. Superfused capsular adrenal glands of rats treated for 3 days with captopril no longer responded with an increased aldosterone output to small increments (0.5 or 1.0 mmol/l) in the potassium concentration of the medium (Parkinson et al. 1984). Greater increments in the ambient potassium concentration (2–6 mmol/l) stimulated the aldosterone production of adrenals from captopril-treated rats, but to a lesser extent than the aldosterone production of adrenals from untreated animals. Direct addition of captopril to the superfusion medium (10^{-6} mol/l) did not impair the aldosterone secretory response to a high potassium concentration in either capsular adrenals of control animals or capsular adrenals of captopril-treated rats. Dispersed zona glomerulosa cells of rats treated with captopril for a week produced 75% less aldosterone than zona glomerulosa cells of untreated rats during incubation in a medium with a potassium concentration of 4 mmol/l (Nakamura et al. 1985). Under in vitro stimulation by a high potassium concentration (12 mmol/l) the aldosterone output by zona glomerulosa cells of captopril-treated rats was moderately (−37%) but significantly lower than the aldosterone output by cells of control animals. The adrenal contents of angiotensin II and angiotensin III were markedly lower in captopril-treated rats, whereas adrenal angiotensin I and adrenal renin activity were elevated.

4.2.3 Ammonium, Rubidium, and Cesium

Addition of ammonium chloride or ammonium hydroxide to the incubation medium significantly stimulated aldosterone production in vitro by adrenal tissue of sodium-deficient rats and had only negligible effects on corticosterone produc-

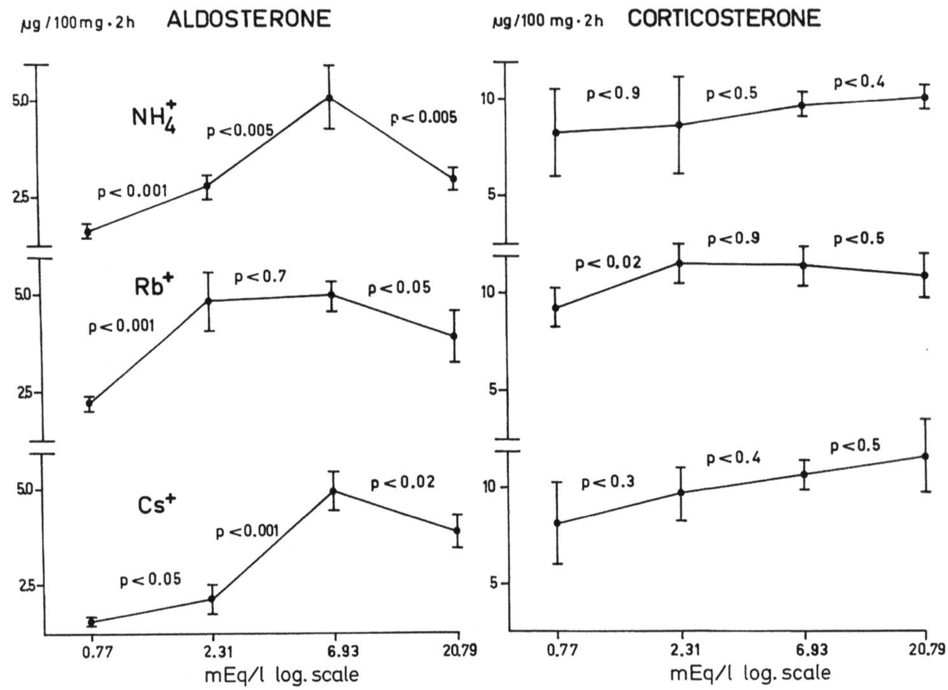

Fig. 12. Effect of increasing concentrations of NH_4Cl, RbCl, and CsCl in the incubation medium on in vitro production of aldosterone and corticosterone by rat adrenal quarters. Mean values of two experiments ($n = 4$) ± standard deviation. *P* values of significance were calculated by *t*-tests. (From J. Müller 1965c)

tion (J. Müller 1965b; Fig. 12). Rubidium chloride and cesium chloride had similar effects on corticosteroid biosynthesis (J. Müller 1965c). When increasing amounts of these electrolytes were added to the incubation medium, aldosterone production was highest at concentrations of approximately 7 mmol/l and was significantly lower at higher concentrations (Fig. 12). A similar dose-response curve had been observed with increasing potassium concentrations (Fig. 10). The effects of K^+ and NH_4^+ ions were additive only at low concentrations. Adrenal tissue maximally stimulated by potassium did not further respond to ammonium.

Aldosterone and corticosterone secretions by isolated dog adrenals in vivo rose in response to local RbCl and CsCl infusions (Bartter et al. 1964). However, rubidium ions had no aldosterone-stimulating effect on perfused adrenals of hypophysectomized dogs, according to Hilton et al. (1965). In concentrations above 1 mmol/l, ammonium chloride stimulated aldosterone secretion by autotransplanted sheep adrenals to the rate seen in sodium-deficient animals (Blair-West et al. 1968d). In normal human subjects on a low sodium intake, a 1-h systemic intravenous infusion of NH_4Cl (1.25 mmol/min) had no effect upon the plasma aldosterone and potassium concentrations, but it caused a significant drop in the plasma renin activity (Kisch et al. 1976). The oral intake of NH_4Cl (1.9 mmol per kg body weight) in sodium-depleted supine human subjects resulted in a significant rise in plasma aldosterone and a small but significant decrease in venous pH

(Perez et al. 1977a). It had no effect on the plasma renin activity and led to only a small increase in the plasma potassium concentration, which was statistically not significant. The plasma ammonium concentrations were not measured in these two studies.

Adrenals of rats that had received a diet supplemented with rubidium sulfate (15 mg Rb per day) for 2 weeks produced significantly more aldosterone than adrenals of control animals during incubation in vitro (Bach et al. 1967). Corticosterone output was unaltered.

4.2.4 Lithium

The substitution of lithium ions for sodium ions in the incubation medium resulted in a lowered output of aldosterone by quartered rat adrenals but did not affect the corticosterone production (J. Boyd et al. 1973). Addition of a small amount of LiCl (3.9 mmol/l) to the incubation medium did not influence aldosterone production by rat adrenal tissue in unpublished experiments carried out in this laboratory. At concentrations up to 4.3 mmol/l, lithium did not significantly affect ACTH-stimulated aldosterone production by incubated rat adrenal zona glomerulosa cells (O'Brien et al. 1979). When added to the medium with which isolated rat adrenal glomerulosa cells were superfused in a concentration of 10 mmol/l, lithium significantly inhibited angiotensin II-stimulated aldosterone production but did not affect ACTH-stimulated aldosterone output (Balla et al. 1984).

Treatment of manic-depressive patients with lithium led to only a transient (from day 2 to day 6) increase in urinary aldosterone excretion (Murphy et al. 1969). After 3–4 weeks of lithium administration, plasma aldosterone concentrations and plasma renin activity were unchanged when measured with the subjects in the supine position or after they had been standing (P. D. Miller et al. 1979). Urinary aldosterone excretion was also unaltered. In a longitudinal study before and during the first 3 months of lithium treatment, there was no consistent change in plasma aldosterone concentrations (E. B. Pedersen et al. 1977a, 1977b). However, in a cross-sectional study, carried out in patients who had been treated with lithium for 3 months to 20 years, E. B. Pedersen et al. (1977a) found a positive correlation between serum lithium and plasma aldosterone concentrations. High levels of plasma aldosterone might have been due to slight sodium depletion by nontoxic doses of lithium.

4.2.5 Hydrogen Ions

At all concentrations tested, angiotensin II elicited a significantly larger increment in aldosterone output in a medium with pH 6.5 than in a medium with pH 7.4, when added to incubated rabbit adrenal zona glomerulosa cells (Chiu and Freer 1979). On the other hand, lowering of the pH of the incubation medium from 7.4 to 7.1 or 6.8, as well as raising it to 7.7, resulted in marked decreases in aldosterone production (in the absence of stimulators) by rat adrenal zona glomerulosa cells (M. Gilchrist et al. 1983).

Alkalosis directly inhibited ACTH-stimulated aldosterone secretion by perfused canine adrenal glands (Radke et al. 1986). Within a pH range from 7.19 to 7.85, there was a positive correlation between H^+ concentration and the fractional secretion of aldosterone but a negative correlation between H^+ concentration and the fractional secretion of 18-hydroxycorticosterone and corticosterone in response to ACTH. These observations indicated an effect of H^+ concentration on late steps of aldosterone biosynthesis. On the other hand, neither acidosis nor alkalosis had a direct effect on basal or ACTH-stimulated cortisol secretion.

In patients with diabetic ketoacidosis, elevated plasma aldosterone levels were accompanied by hyperkalemia and increased plasma renin activity (Christlieb et al. 1975). Acute increases in plasma aldosterone in response to lactic acid infusion in dogs were probably due to elevated ACTH levels and were prevented by treatment with dexamethasone (Perez et al. 1980). Long-term feeding of sulfuric acid to dogs on a sodium- and chloride-deficient diet led to increases in aldosterone secretion and in plasma renin activity; these were most likely caused by urinary sodium loss (Kraut et al. 1982).

4.3 Divalent Cations

4.3.1 Calcium

ACTH stimulates the biosynthesis of corticosteroids in adrenal tissue in vitro only when the incubation medium contains calcium ions (Birmingham et al. 1953; Peron and Koritz 1958). Similarly, the steroidogenic response of the zona glomerulosa to all known stimulators critically depends on the extracellular calcium concentration. Thus, the in vitro response in aldosterone production by quartered rat adrenals to stimulation by potassium ions or angiotensin II was completely abolished in a calcium-free medium (J. Müller 1971; Table 2). Angiotensin II, potassium ions, or ACTH did not stimulate aldosterone production by rabbit adrenal zona glomerulosa cells when incubated in a calcium-free medium (Chiu and Freer 1979). The aldosterone-stimulating effects of angiotensin II and ACTH upon incubated rat capsular adrenal tissue became manifest only when calcium ions were added to the incubation medium (Shima et al. 1978, 1979). According to Fakunding et al. (1979), a physiological extracellular calcium concentration (1.2 mmol/l) was required for maximal increments in aldosterone output by rat and dog zona glomerulosa cells in response to angiotensin II, potassium ions, ACTH, serotonin, cholera toxin, or exogenous cyclic AMP. The increment in aldosterone output elicited by each of these stimulators was progressively reduced when the calcium concentration of the incubation medium was decreased from 1.2 to 0.5, 0.2, and 0 mmol/l. In addition, there was a gradual decrease in the sensitivity of the zona glomerulosa cells to ACTH (but not to the other stimulators), leading to increasing ACTH concentrations required for half-maximal aldosterone stimulation. However, according to Ganguly and Hampton (1985), a low concentration of 0.2 mmol/l calcium restored both basal and serotonin-stimulated aldosterone production of rat adrenal glomerulosa cells close to the level reached with a calcium concentration of 1.8 mmol/l in the incubation medium. Neither angiotensin II nor a high potassium concentration (12 mmol/l) stimulated the aldosterone output by bovine adrenal glomerulosa cells incubated in a medium without calcium or with

Table 2. Effect of calcium and magnesium ions on aldosterone and corticosterone production by rat adrenal quarters in vitro. Mean values of two flasks ± range. (For methodology see J. Müller 1965a)

Preincubation medium[a] (mEq/l			Incubation medium[a] (mEq/l)			Steroid production (μg/100 mg/2 h)	
EDTA[b]	Ca^{++}	Mg^{++}	Ca^{++}	Mg^{++}	Addition	Aldosterone	Corticosterone
+	0	0	2.6	2.4	-	0.62 ± 0.12	7.15 ± 0.73
+	0	0	2.6	2.4	KCl[c]	2.48 ± 0.25	8.79 ± 0.19
+	0	0	0	2.4	-	0.66 ± 0.03	6.51 ± 0.01
+	0	0	0	2.4	KCl	0.69 ± 0.07	6.92 ± 0.06
–	2.6	2.4	2.6	2.4	-	2.05 ± 0.32	6.27 ± 0.11
–	2.6	2.4	2.6	2.4	Angiotensin II[d]	4.46 ± 0.22	6.78 ± 0.40
+	0	0	0	2.4	-	1.13 ± 0.06	3.54 ± 0.13
+	0	0	0	2.4	Angiotensin II	1.20 ± 0.01	3.90 ± 0.16
+	0	0	2.6	2.4	-	1.05 ± 0.23	6.66 ± 0.23
+	0	0	2.6	2.4	KCl	2.27 ± 0.11	8.88 ± 0.04
+	0	0	2.6	0	-	0.92 ± 0.12	7.33 ± 0.21
+	0	0	2.6	0	KCl	2.29 ± 0.09	9.89 ± 0.42

[a] 3.6 mEq K^+/l
[b] Dihydrate of disodium ethylenediamine tetraacetate, 10^{-3} *M*. EDTA was not added to the medium used for the final incubation in these studies.
[c] 8.5 mEq K^+/l
[d] Val^5-angiotensin II-$asp^1\beta$-amide, 0.17 mg/ml

a calcium content of 0.25 mmol/l (Foster et al. 1981). The aldosterone response to both stimulators became maximal when the calcium concentration was increased to 0.5 mmol/l. At higher calcium concentrations (0.63 to 2.5 mmol/l) the response to potassium ions became smaller, whereas the response to angiotensin II remained constant. The binding of angiotensin II to zona glomerulosa cells was the same in the presence and absence of extracellular calcium.

It is still uncertain whether the extracellular calcium concentration can also influence aldosterone biosynthesis in the absence of other stimulators. The available experimental evidence is controversial. The absence of calcium from the incubation medium had no consistent effect upon baseline aldosterone and corticosterone production by quartered rat adrenals (J. Müller 1971). A gradual elevation of the ambient calcium concentration from 0 to 10 mmol/l did not result in alterations of the aldosterone output by rat adrenal tissue (Kotchen et al. 1977). In contrast, Shima et al. (1978, 1979) observed a progressive increase in aldosterone output by capsular rat adrenals in response to a gradual elevation of the extracellular calcium concentration from 0 to 0.1, 2.5, and 12.7 mmol/l, in the absence of other stimulators. At a concentration of 25.4 mmol/l, extracellular calcium became inhibitory. The addition of calcium (3 or 6 mmol/l) and of magnesium (1 mmol/l) to the medium in which capsular rat adrenals were incubated stimulated the aldosterone and 18-hydroxycorticosterone outputs but did not affect corticosterone, 18-OH-DOC, and deoxycorticosterone outputs of this tissue (control incubations were carried out in a medium without bivalent cations and containing EDTA; Vinson and Whitehouse 1982). The addition of increasing amounts of calcium to the medium with which frog interrenal tissue was superfused resulted in marked stimulation of aldosterone and corticosterone production, with calcium concentrations between 6 and 15 mmol/l yielding a maximum steroidogenic response (Lihrmann et al. 1985). Short-term infusion of calcium chloride into the arterial supply of autotransplanted sheep adrenals, leading to increases in the local calcium concentration by 1-5 mmol/l, did not influence the aldosterone secretion rate (Blair-West et al. 1968d). In normal human subjects kept on a low-sodium diet, intravenous infusions of calcium chloride (0.21 mmol/min) for 2 h had no effect upon the plasma aldosterone concentration, but it significantly decreased the plasma renin activity (Kisch et al. 1976). In four patients with chronic kidney disease and a high plasma potassium concentration (mean 6.0 mmol/l), a 2-h infusion of calcium gluconate elicited a marked and statistically significant rise in the plasma aldosterone concentration (Ølgaard et al. 1978). This was associated with an increase of the mean blood calcium concentration by 0.21 mmol/l, but not with significant changes in plasma potassium, cortisol, or renin activity. In anephric patients calcium infusions had no significant effect upon the plasma aldosterone concentration. An acute intravenous calcium load, raising serum calcium from 2.2 to 2.8 mmol/l, led to a marked increase in plasma aldosterone, ACTH, and cortisol concentrations without affecting plasma renin activity in normal human subjects (Isaac et al. 1984). In two subjects pretreated with dexamethasone, plasma cortisol and ACTH no longer increased in response to a calcium load, although the increase in plasma aldosterone was not abolished. A milder increase in plasma calcium from 2.25 to 2.5 mmol/l did not affect plasma aldosterone, either under basal conditions or under stimulation by angiotensin II (Bianchetti et al. 1982). In patients with hyperparathyroidism, plasma aldosterone concentration and plasma renin activity were normal and decreased normally in response to a saline infusion (Ganguly et al. 1982).

When calcium ions were substituted with strontium or barium ions in the incubation medium, aldosterone output by bovine adrenal zona glomerulosa cells increased in response to angiotensin II or to ACTH, but to a lesser extent than in the presence of calcium ions (Lobo and Marusic 1982). On the other hand, the zona glomerulosa cells did not respond to a high extracellular potassium concentration under these conditions. Other divalent cations such as manganese, nickel, and cobalt inhibited the stimulation of aldosterone production by angiotensin II, potassium ions, ACTH, or ouabain in the presence of calcium ions (Schiffrin et al. 1982). Addition of cobalt ions to the superfusion medium (4 mmol/l) inhibited aldosterone and corticosterone production by frog interrenal tissue (Lihrmann et al. 1985).

4.3.2 Substances Affecting Transport and Second-Messenger Functions of Calcium

The aldosterone-stimulating action of angiotensin II and ACTH on incubated rat capsular adrenal tissue could be blocked by *substances which inhibit calcium transport,* such as tetracaine, verapamil, and lanthanum (Shima et al. 1978, 1979). Similarly, the calcium antagonist D-600 (methoxyverapamil) blocked the steroidogenic effect of angiotensin II upon rabbit adrenal glomerulosa cells, but only when added in a high concentration (10^{-4} mol/l; Chiu and Freer 1979). Both verapamil and lanthanum inhibited the aldosterone-stimulating action of angiotensin II, ACTH, potassium ions, prostaglandin E_2, dibutyryl cyclic AMP, and ouabain on rat adrenal glomerulosa cells in a dose-dependent manner, but they did not influence basal aldosterone output (Fakunding and Catt 1980, Schiffrin et al. 1981; Balla et al. 1982a, 1982b). Methoxyverapamil had a similar inhibitory effect on angiotensin II- and potassium-stimulated aldosterone production by bovine adrenal glomerulosa cells (Foster et al. 1981). In another study with rat adrenal glomerulosa cells, verapamil markedly inhibited basal and serotonin-stimulated aldosterone production (Ganguly and Hampton 1985). Two other calcium-channel blockers, nifedipine and diltiazem, either had no effect on aldosterone output under these conditions or inhibited it to only a very small extent. Other dihydropyridine calcium antagonists, i.e., nimodipine, nitrendipine, and nisoldipine, blocked both potassium-stimulated and angiotensin II-stimulated aldosterone production by calf adrenal glomerulosa cells (K. Kojima et al. 1984). These drugs were more effective inhibitors of potassium effects than of angiotensin II effects, unless the cells were pretreated with dantrolene, a drug known to inhibit the release of calcium from the sarcoplasmic or endoplasmic reticulum. According to Finkel et al. (1984), nifedipine inhibited angiotensin II-stimulated and potassium-stimulated but not ACTH-stimulated steroidogenesis in dispersed rat adrenal glomerulosa cells; diltiazem was completely ineffective. These investigators found a single class of binding sites for tritiated nitrendipine in rat capsular adrenal membranes. Specific binding of this ligand was inhibited by unlabeled nitrendipine, nifedipine, and verapamil (but not by diltiazem). In the rat adrenal gland, nitrendipine binding sites were located in the capsular portion and in the medulla and were undetectable in the zona fasciculata. Nifedipine by itself lowered the cytosolic free calcium level in bovine zona glomerulosa cells (Capponi et al. 1984). This effect might be the indirect cause of the inhibitory action of the drug on angiotensin II-induced steroidogenesis. It had previously been shown by Fakunding and Catt (1980) that neither verapamil nor lanthanum inhibited the binding of radiolabeled angiotensin II to dog or rat adrenal glomerulosa cells.

Nisoldipine inhibited the stimulation of aldosterone secretion by potassium ions when infused concomitantly with KCl into the arterial blood supply of autotransplanted adrenal glands in sheep in vivo (E. I. M. Johnson et al. 1984). Verapamil and nisoldipine inhibited the stimulation of aldosterone secretion by angiotensin II. However, these two drugs, as well as methoxyverapamil and lanthanum, were completely ineffective in suppressing the elevated rate of aldosterone secretion by autotransplanted adrenals during established sodium depletion in sheep (E. I. M. Johnson et al. 1985). In contrast, long-term verapamil administration significantly lowered plasma aldosterone concentrations in rats kept on a low-NaCl or an NaCl-free diet (M. W. Roy et al. 1983). Plasma corticosterone was also markedly suppressed in NaCl-depleted verapamil-treated rats, whereas plasma renin activity and plasma potassium concentration were unchanged. In normal human subjects, an acute intravenous dose of verapamil, sufficient to blunt the pressor action of angiotensin II, did not blunt the increments in plasma aldosterone or cortisol in response to either angiotensin II or ACTH (Guthrie et al. 1983). However, after 5 days of oral administration of verapamil, the increments in plasma aldosterone induced by angiotensin II and in plasma cortisol induced by ACTH were significantly reduced, whereas the response in plasma aldosterone to ACTH remained unaltered. Long-term treatment of patients with essential hypertension with verapamil led to an increase in plasma aldosterone, both under basal conditions and under stimulation by ACTH (Zofková 1985). On the other hand, treatment of patients with essential hypertension with nifedipine for 6–8 weeks did not alter plasma aldosterone concentrations measured with patients in the supine or in the upright position, but significantly blunted the response in plasma aldosterone to angiotensin II infusions (Marone et al. 1985).

The *calcium ionophore* A-23187 markedly stimulated aldosterone production by bovine and rat adrenal zona glomerulosa cells on its own, but it also significantly inhibited the steroidogenic responses to angiotensin II and potassium ions (Foster et al. 1981; Fakunding and Catt 1982). As shown in experiments with superfused porcine and bovine zona glomerulosa cells, the stimulatory effect of this agent on aldosterone production was a transient one (I. Kojima et al. 1983, 1985a). The ionophore also markedly stimulated aldosterone and corticosterone production by superfused frog interrenal tissue (Lihrmann et al. 1985). BAY K 8644, another calcium ionophore, stimulated aldosterone production by calf adrenal glomerulosa cells at potassium concentrations between 3.5 and 8 mmol/l, but not at potassium concentrations of 10 and 12 mmol/l (K. Kojima et al. 1984). It had no additional effect on cells stimulated by angiotensin II unless they had been pretreated with dantrolene (see above). The effects of this agent on rat adrenal glomerulosa cells were partly different. At a potassium concentration of 4.5 mmol/l, BAY K 8644 stimulated aldosterone production, but only when added in high concentrations (10^{-7} and 10^{-6} mol/l; Hausdorff et al. 1986). At submaximally stimulating potassium concentrations (9 mmol/l) BAY K 8644 (3×10^{-8} mol/l) increased the sensitivity of zona glomerulosa cells to potassium without affecting the maximal response in aldosterone output. The same concentration of the drug augmented aldosterone output in the presence of angiotensin II at all concentrations studied. It increased the maximal response to the peptide by 70%, but did not change the sensitivity. BAY K 8644 did not affect ACTH-stimulated aldosterone production.

The *calmodulin inhibitor* trifluoperazine blocked the stimulation of aldosterone production in rat adrenal glomerulosa cells by angiotensin II, potassium, prostaglandin E_2, or ouabain, but it had less extensive inhibitory effects on ACTH- or dibutyryl-cyclic AMP-stimulated aldosterone production (Schiffrin et al. 1982; Balla et al. 1982a, 1982b). On its own, the drug had a weak aldosterone-stimulating activity (Balla et al. 1982a). Two other calmodulin inhibitors, penfluridol and pimozide, also stimulated aldosterone production in the absence of other stimulators while suppressing the stimulatory action of angiotensin II and dibutyryl-cyclic AMP (Balla and Spät 1982). According to J.X. Wilson et al. (1984), pimozide inhibited not only the increase in aldosterone output but – in the presence of cyanoketone – also the increase in pregnenolone output by zona glomerulosa cells in response to angiotensin II, ACTH, or A-23187 and did not affect the conversion of exogenous progesterone to aldosterone. However, the conversion of progesterone to aldosterone was markedly reduced by two other calmodulin inhibitors, W5 and W7.

4.3.3 Magnesium

When rat adrenal tissue was preincubated in a calcium- and magnesium-free medium containing EDTA, addition of calcium only to the medium used for final incubation restored the response in aldosterone production to stimulation by a high potassium concentration; the presence of magnesium ions was not required (J. Müller 1971; Table 2). In sheep, adrenal arterial infusions of magnesium chloride which elevated the local plasma magnesium concentration by up to 2.3 mmol/l had no effect on aldosterone secretion (Blair-West et al. 1968d). Local infusion of magnesium sulfate into the adrenal arterial blood supply of dogs induced a decrease in the adrenal venous plasma aldosterone concentration in two of six animals (Helber et al. 1972). In human subjects kept on a sodium-deficient diet, a systemic 2-h infusion of magnesium sulfate (0.21 mmol/min) led to a marked decrease in the plasma aldosterone concentration (Kisch et al. 1976). Perhaps this was due to a simultaneous decrease in the plasma potassium concentration. The plasma renin activity remained unaltered.

Long-term magnesium deficiency in rats has been found to elicit marked increases in aldosterone secretion (Ginn et al. 1967), in serum aldosterone concentration (Solounias and Schwartz 1975), in the zona glomerulosa width, and in the juxtaglomerular granulation index of the kidneys (Cantin and Veilleux 1968), as well as decreases in the peripheral plasma concentration and the adrenal content of corticosterone (Richer et al. 1968). The mechanism by which magnesium deficiency influences adrenocortical function and the renin-angiotensin system in the rat is unknown. It is possible that hypoproteinemia, resulting in edema and hypovolemia, plays an important role (Cantin 1970). In five of six dogs, dietary magnesium restriction led only to a transient increase in the secretion rate of aldosterone (Helber et al. 1972). It induced hypokalemia but did not affect the plasma renin activity.

4.4 ACTH, ACTH-related Peptides, and Other Pituitary Hormones

4.4.1 ACTH

Whereas the biosynthesis and secretion of glucocorticoids is almost exclusively controlled by ACTH[7], the production of aldosterone is relatively independent of ACTH, at least in mammalian organisms. Long-term increases in the secretion of cortisol or corticosterone due to excessive secretion of endogenous ACTH or administration of exogenous ACTH may be accompanied by unchanged or diminished aldosterone secretion (see p. 183). Aldosterone secretion and its adaptation to the body electrolyte status are not critically impaired even by the complete absence of ACTH (see p. 176). However, in short-term experiments, ACTH has consistently been found to stimulate aldosterone production as well as glucocorti-

[7] Synthetic β^{1-24}ACTH has been used in most of the more recent studies.

coid and adrenal androgen production. Stimulation by ACTH must always be considered when increases in aldosterone secretion are associated with simultaneous increases in glucocorticoid secretion. In particular, experimental stress (e.g., due to anesthesia, surgery, blood collection, restraining), leading to increases in ACTH secretion and thus to aldosterone stimulation, may severely jeopardize the validity of studies on the regulation of aldosterone secretion carried out in untrained animals with intact pituitary glands (Espiner et al. 1978; Fromm et al. 1983).

ACTH stimulated aldosterone production in vitro by adrenocortical tissue of man (Dyrenfurth et al. 1960), oxen (Kaplan and Bartter 1962), sheep (McDougall et al. 1976), rats (Giroud et al. 1956b; O.J. Lucis et al. 1961; Kaplan and Bartter 1962; J. Müller 1965a, (Fig. 13), ducks (Donaldson and Holmes 1965; Klingbeil 1985b), turtles (Macchi and Phillips 1966), and bullfrogs (Carstensen et al. 1961), by superfused frog interrenal tissue (Delarue et al. 1979a) and by perfused calf

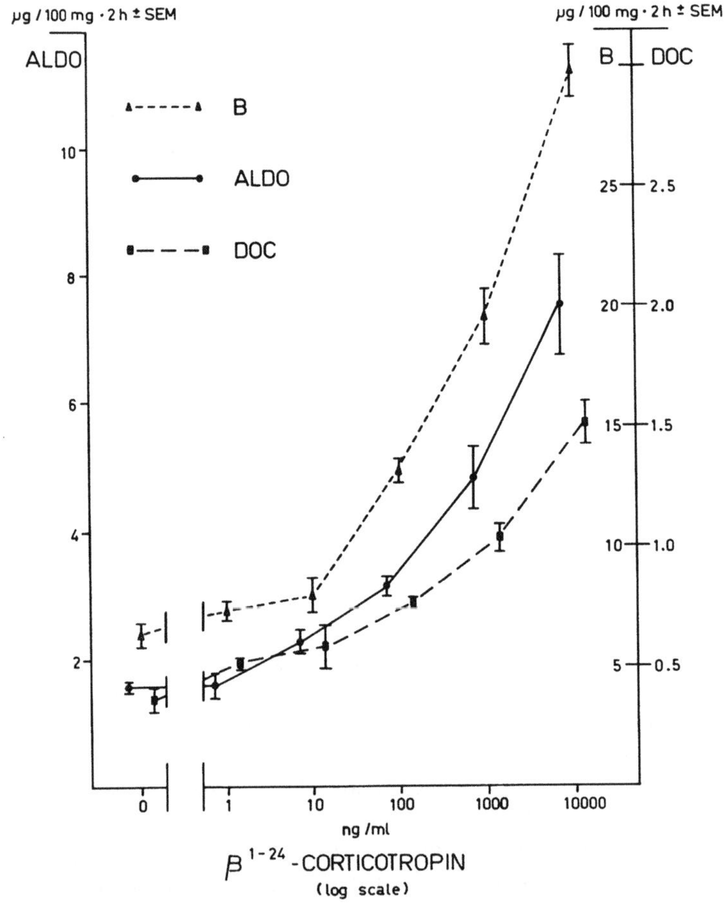

Fig. 13. Effect of increasing concentrations of ACTH in the incubation medium on in vitro production of corticosterone *(B)*, aldosterone *(ALDO)*, and deoxycorticosterone *(DOC)* by rat adrenal quarters. Mean values of three experiments ($n = 4$) ± standard error of the mean. (From J. Müller 1971)

(Rosenfeld et al. 1956) and dog adrenals (Greenway and Verney 1962), as well as by a perfused hyperplastic adrenal gland surgically removed from a patient with Cushing's disease (Schriefers et al. 1963). Since these experiments were carried out with mixed adrenocortical tissue, they are not entirely conclusive. Conversion to aldosterone of corticosterone produced by the zona fasciculata and entering the zona glomerulosa cells either by direct intracellular diffusion or from the incubation medium may have contributed considerably to the apparent responses in aldosterone output to ACTH. However, ACTH also significantly stimulated aldosterone and corticosterone outputs by rat adrenal zona glomerulosa cells highly purified by unit gravity sedimentation (J. F. Tait et al. 1972) or by column filtration (B. C. Williams et al. 1979). The sensitivity of rat adrenal zona glomerulosa and zona fasciculata cells to ACTH was the same with identical threshold concentrations of the peptide required for a significant steroidogenic response, i.e., 5×10^{-12} mol/l, according to Braley and Williams (1977). But ACTH elicited a much greater steroidogenic response in zona fasciculata than in zona glomerulosa cells. Thus, it stimulated the corticosterone output by a factor of 120 in pure zona fasciculata cells, but by a factor of only 2-3 in pure zona glomerulosa cells (J. F. Tait and Tait 1979). Accordingly, even a minor contamination by zona fasciculata cells might markedly augment the apparent response in aldosterone production to ACTH of a zona glomerulosa cell or tissue preparation. On the other hand, aldosterone secretion in vivo may more truly represent de novo synthesis of the hormone in the zona glomerulosa, since in the mammalian adrenal gland in situ the direction of the blood flow is mainly although not entirely centripetal (i.e., from the capsule to the medulla; Idelman 1978), and the conversion to aldosterone of corticosterone produced by the inner zones might be expected to be negligible.

ACTH enhanced aldosterone secretion or urinary excretion of aldosterone-18-glucuronide in vivo in man (Venning et al. 1956a; A. F. Muller et al. 1956; Crabbé et al. 1959; Tucci et al. 1967a), dogs (Farrell et al. 1955; Hilton et al. 1960; Mulrow and Ganong 1961), sheep (Blair-West et al. 1962a), rabbits (Braverman and Davis 1973), rats (Singer and Stack-Dunne 1955; Kinson and Singer 1968), opossums (Johnston et al. 1976b), and bullfrogs (Johnston et al. 1967a; Ulick and Feinholtz 1968). Exogenous ACTH raised plasma aldosterone concentrations in the echidna, a monotreme mammal (Sernia and McDonald 1981), and in a freshwater teleost (Reinking 1983). On the other hand, increases in glucocorticoid output induced by repeated exposure to ACTH of superfused adrenals of the Mongolian gerbil were not accompanied by increases in aldosterone output (Fenske 1985). In thoroughbred horses, bolus injections of ACTH that elicited significant rises in plasma cortisol and corticosterone concentrations brought about a continuous fall in plasma aldosterone (Guthrie et al. 1982). ACTH injections or infusions elicited marked rises in plasma aldosterone concentration in normal human subjects, according to - among others - Michelakis and Horton (1970), Oddie et al. (1972), Scholer et al. (1973) and Dluhy et al. (1974). In nephrectomized, hypophysectomized dogs, aldosterone secretion was stimulated only by a dose ten times higher than necessary for maximal glucocorticoid stimulation (Ganong et al. 1965a). Conversely, in stressed dogs, in which glucocorticoid production was already maximally stimulated by endogenous ACTH, a pharmacological dose of exogenous ACTH resulted in an increase of aldosterone production only

(A. F. Muller et al. 1964). In rabbits, the threshold dose of ACTH required to significantly raise the plasma glucocorticoid concentration was four times lower than the one required to raise plasma aldosterone (Grospietsch et al. 1980). In comparison, the human adrenal cortex appears to be more sensitive to the aldosterone-stimulating action of ACTH. Thus, in normal sodium-replete persons, a 1-h infusion of 1.25 μg of ACTH, which raised the plasma ACTH concentration within the physiological range (from 40–80 to 60–160 pg/ml) induced a 3.4-fold rise in plasma aldosterone and a 2.8-fold rise in plasma cortisol levels (Nicholls et al. 1975). In dexamethasone-treated, sodium replete man, a 30-min infusion of 12.5 mIU of ACTH raised the median plasma cortisol level from a value below 1.0 μg/dl to 3.1 μg/dl and the median plasma aldosterone concentration from 4.3 to 8.8 ng/dl (Kem et al. 1975).

The response in aldosterone output to ACTH can be experimentally modified by changes in sodium or potassium intake or by treatment with renin. ACTH elicited more extensive increases in the aldosterone secretion rate or more pronounced elevations in the plasma aldosterone concentration in persons on a low-sodium diet than in persons with an unrestricted or high sodium intake (A. F. Muller et al. 1959; Liddle et al. 1956; Venning et al. 1962; Tucci et al. 1967; G. H. Williams et al. 1970; Rayfield et al. 1973; Kem et al. 1975). A high potassium intake increased the aldosterone secretory response to ACTH in human subjects on a high sodium intake and potentiated the effect of a low sodium intake upon the aldosterone response to ACTH (G. H. Williams et al. 1970). In dogs kept on a sodium-deficient diet for 5 days prior to hypophysectomy and nephrectomy, ACTH stimulated aldosterone secretion at a lower dosage and to a greater extent than in sodium-replete animals (Ganong et al. 1965). Longer periods of sodium restriction further enhanced the aldosterone-stimulating effect of ACTH. On the other hand, sodium deficiency did not influence the response in 17-hydroxycorticoid output. The aldosterone-stimulating effect of ACTH was also enhanced in dogs that had been injected for 5 days with dog renin (Ganong et al. 1967 a). In dexamethasone-treated sheep, the infusion of ACTH at a low rate, which did not by itself stimulate aldosterone, potentiated and prolonged the aldosterone-stimulating effect of a simultaneous potassium chloride infusion (Lun et al. 1979). Injection of ACTH into hypophysectomized rats in vivo induced significant increases in aldosterone production by adrenal tissue in vitro only when the animals were kept on a sodium-deficient diet (Csánky et al. 1968). However, Kinson and Singer (1968) found that the effect of ACTH on aldosterone secretion in vivo in hypophysectomized, nephrectomized rats was the same whether the animals were kept on a normal or a sodium-deficient diet. ACTH stimulated aldosterone production in vitro more markedly in quartered adrenals of rats that had been kept on a sodium-deficient diet (J. Müller 1965 a) or that were uremic and hyperkalemic after bilateral nephrectomy (J. Müller and Huber 1969) than in the tissue of control animals. On the other hand, ACTH stimulated aldosterone production to a lesser degree in adrenal tissue of rats receiving sodium chloride solution as drinking fluid (J. Müller 1968) and did not stimulate aldosterone production in quartered or capsular adrenals of potassium-deficient rats (J. Müller and Huber 1969; J. Müller 1970 b; Fig. 14). Sodium and potassium intakes also influence the zona glomerulosa response in corticosterone and deoxycorticosterone outputs to ACTH added in vitro. Thus, ACTH stimulated aldosterone, corticosterone, and deoxycorticos-

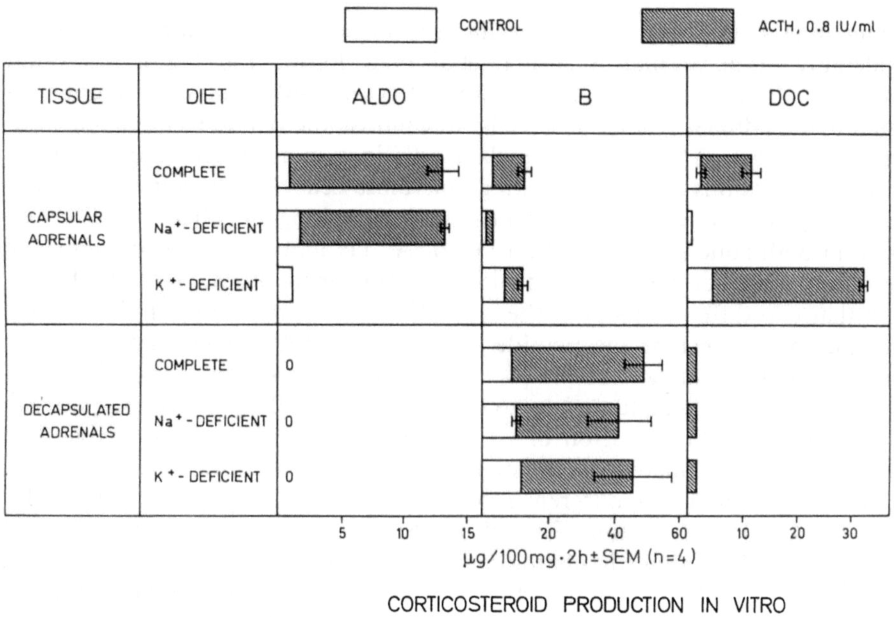

Fig. 14. Production of aldosterone *(ALDO)*, corticosterone *(B)*, and deoxycorticosterone *(DOC)* by separate zones of adrenals of rats kept on different diets for 2 weeks. *White bars* represent baseline corticosteroid production, *shaded bars* increments due to addition of ACTH to the incubation medium. Mean values of two experiments ($n=4$) ± standard error of the mean. (Data from J. Müller 1970b)

terone outputs to similar extents in capsular adrenals of rats kept on a complete diet (J. Müller 1970b; Fig. 14). In capsular adrenals of sodium-deficient rats, it stimulated only aldosterone production, whereas in capsular adrenals of potassium-deficient rats its main effect was on deoxycorticosterone output. The steroidogenic response to ACTH of decapsulated adrenals was independent of the sodium and potassium intakes.

4.4.2 ACTH-related Pituitary Peptides

α-MSH (alpha-melanocyte-stimulating hormone), a peptide with the 1-13 N-terminal amino acid sequence of ACTH, may play an important role in the control of cortisol biosynthesis by the fetal and newborn adrenal gland (Glickman et al. 1979; Llanos et al. 1979). α-MSH has also been identified as one of the active corticosteroid-stimulating components of a bovine pituitary extract and of a commercial antidiuretic hormone preparation (Vinson et al. 1980). It significantly stimulated the production of corticosterone, 18-hydroxycorticosterone, and aldosterone by capsular rat adrenal cells, with a linear log-dose/response curve at concentrations between 10^{-9} and 10^{-6} mol/l, but it had no steroidogenic effect upon zona fasciculata-reticularis cells. According to a later report by the same group of investigators (Vinson et al. 1983), α-MSH also stimulated corticosterone production in rat adrenal inner-zone cells but only at concentrations about 1000-fold higher than in zona glomerulosa cells. Other investigators found α-MSH to be equally potent in stimulating steroidogenesis in capsular and in decapsulated adrenal cells but considerably less active than ACTH (Li et al. 1982; Jornot et al. 1985). Szalay and Stark (1982) observed a biphasic stimulatory effect of α-MSH on aldosterone production by rat adrenal glomerulosa cells. In contrast, the peptide stimulated corticosterone production by zona fasciculata cells in a monophasic manner but at higher concentrations than in the zona glomerulosa cells. Besides stimulating steroidogenesis on its onw, α-MSH potentiated the steroidogenic effect of ACTH on cells from both zones, lowering the ED_{50} of ACTH and increasing the maximal steroid output elicited by it. Studies with short fragments of ACTH/α-MSH suggested that two different parts of the α-MSH sequence were responsible for the glomerulotropic effect of the peptide (Szalay et al. 1985). One important part seemed to be the sequence ACTH-(4-7), which induced steroid production equipotently in zona fasciculata and in zona glomerulosa cells. The second part was located within the 11-13 region of ACTH. ACTH-(11-13) was less active than ACTH-(4-7) but stimulated preferentially the zona glomerulosa cells, which suggested the involvement of different receptors. None of the fragments tested displayed the ACTH-potentiating effect of α-MSH. Experimental evidence of a possible in vivo aldosterone-stimulating activity of α-MSH is not conclusive. In a preparation of rat adrenal gland perfusion in situ, low doses of α-MSH selectively stimulated aldosterone output, whereas higher doses of the peptide were needed to stimulate corticosterone secretion (Hinson et al. 1985). Intravenous injection of α-MSH into sodium-deficient, dexamethasone- and phenobarbital-treated rats in vivo induced significant increases in the in vitro production of aldosterone and corticosterone by incubated adrenal glands (van der Wal and De Wied 1968). β-MSH and several ACTH fragments (1-16, 1-10, 7-10, 11-24) were ineffective. Treatment of sodium-deficient hypophysectomized rats combining α-MSH and growth hormone for 2 days resulted in a significant increase in adrenal weight and significantly enhanced in vitro production of aldosterone and corticosterone by incubated adrenal tissue, but it was not as effective as the combined treatment with ACTH and growth hormone (Page et al. 1974; see also p. 190). In hypophysectomized rats treated with dexamethasone and thyroxine and kept on a high sodium intake (0.9% NaCl as drinking fluid) the administration of α-MSH (8 μg/day) by subcutaneous osmotic minipumps for a week elicited a three fold increase in the mean plasma aldosterone concentration, which was, however, not statistically significant (Shenker et al. 1985a). In hypophysectomized rats kept on a sodium-deficient diet, α-MSH treatment significantly raised plasma aldosterone concentrations (+182%). α-MSH did not affect plasma aldosterone levels in intact sodium-replete or sodium-deficient rats and did not raise plasma corticosterone concentrations in either intact or hypophysectomized animals. Treatment of hypophysectomized sodium-replete or sodium-deficient rats with a dose of ACTH that significantly raised plasma corticosterone levels did not influence plasma aldosterone concentrations. In normal man, intravenous injection of α-MSH in dosage sufficient to stimulate growth hormone secretion did not induce any change in the plasma aldosterone concentration (Birkhäuser et al. 1975). However, the peptide elicited a marked rise in the plasma aldosterone concentration in a patient with panhypopituitarism and had more equivocal effects on plasma aldosterone in two other such patients. The plasma cortisol and corticosterone remained low, and the plasma renin activity did not change.

β-LPH (beta-lipotropic hormone) is derived from the same precursor molecule as ACTH and is itself a prohormone of several opioids as well as of β-MSH. According to Matsuoko et al. (1980b), β-LPH was a potent stimulator of aldosterone biosynthesis in dispersed rat capsular adrenal cells, with a linear log-dose/response curve at concentrations between 10^{-8} and 10^{-5} mol/l. It also stimulated the corticosterone output of decapsulated adrenal cells, but only at a concentration of 10^{-6} mol/l. Among β-LPH derivatives, met^3-enkephalin, α-endorphin, and β-endorphin had no steroidogenic activity when added to rat adrenal cells at concentrations between 10^{-8} and 10^{-4} mol/l, whereas β-MSH had the same aldosterone-stimulating activity as β-LPH (Matsuoka et al. 1981b). Gly^{10}-β-MSH, a synthetic analogue, inhibited the aldosterone-stimulating actions of β-MSH and β-LPH in a dose-dependent manner, but it had no effect on the aldosterone response elicited by ACTH or angiotensin II (Matsuoka et al. 1981b). The aldosterone-stimulating effects of β-LPH and β-MSH were also inhibited by an ACTH antagonist, ACTH-(7-38), and by an angiotensin antagonist, saralasin. These findings are somewhat controversial and need confirmation. Vinson et al. (1981) found β-MSH to be devoid of any corticosteroidogenic activity and β-LPH to stimulate corticosteroid output at equal concentrations in rat adrenal zona glomerulosa and fasciculata cells.

According to Li et al. (1982), β-LPH stimulated steroidogenesis by rat adrenal glomerulosa and fasciculata cells in a monophasic manner, whereas β-MSH had a biphasic stimulatory effect on aldosterone output by zona glomerulosa cells but a monophasic effect on corticosterone production by zona fasciculata cells. In another study with rat adrenocortical cells, β-LPH stimulated steroidogenesis at a lower concentration in zona fasciculata cells than in zona glomerulosa cells, whereas β-MSH was practically devoid of steroidogenic activity (Jornot et al. 1985). Yamakado et al. (1983) observed that sodium restriction of rats greatly increased the sensitivity of their zona glomerulosa cells to the stimulatory action of β-MSH added in vitro. Thus, the threshold concentration of the peptide needed to significantly stimulate aldosterone output was decreased by sodium restriction from 10^{-8} to 10^{-10} mol/l. According to these investigators, β-MSH did not affect corticosterone production by zona fasciculata cells, even at a concentration of 10^{-6} mol/l. On the other hand, Washburn et al. (1982) tested three different preparations of highly purified ovine and human β-LPH as well as the synthetic fragments β-LPH (37-58) and β-LPH (61-91) and concluded that the slight steroidogenic activity of purified β-LPH preparations could be accounted for partly by contamination with ACTH and partly by a weak intrinsic ACTH-like activity of β-LPH, most likely due to the amino acid sequence 47-53 shared with ACTH. β-Endorphin lowered the corticosterone output of rat adrenal fasciculata cells when added in low concentrations from 10^{-11} to 10^{-5} mol/l, but it did not influence steroid production by zona glomerulosa cells (Szalay and Stark 1981). The suppressive effect of the peptide on zona fasciculata cells was not affected by the addition of the opioid antagonist naloxone. At a concentration of 10^{-4} mol/l, β-endorphin stimulated steroidogenesis in cells of both zones. When added together with ACTH (1.6×10^{-7} mol/l), β-endorphin significantly blunted the steroidogenic responses to the former peptide by zona glomerulosa and zona fasciculata cells. When tested in a system of isolated cells from human aldosterone-producing adenomas, β-LPH stimulated aldosterone production on its own and potentiated the aldosterone-stimulating effect of ACTH (Pham-Huu-Trung et al. 1985). β-Endorphin did not affect steroidogenesis by these cells, either under basal conditions or under stimulation by ACTH. In contrast, met-enkephalin suppressed the production of aldosterone, corticosterone, and deoxycorticosterone by cells from a human aldosterone-producing adenoma in a dose-dependent manner in concentrations between 10^{-11} and 10^{-6} mol/l (Rácz et al. 1982). At a concentration of 10^{-5} mol/l, met-enkephalin also completely suppressed the steroidogenic response of these cells to ACTH (10^{-10} mol/l). In contrast, a stable analogue of met-enkephalin, FK 33824, markedly stimulated aldosterone and cortisol production by cells obtained from outer cortical slices of calf adrenals (Bevilacqua et al. 1982). The stimulatory effect of this substance was partially inhibited by naloxone but not by saralasin. Naloxone had a biphasic effect on ACTH-stimulated aldosterone production by superfused rat adrenals, with a high dose (10^{-6} mol/l) inhibiting and a low dose (10^{-10} mol/l) potentiating the action of the hormone (Lymangrover et al. 1983). Naloxone also potentiated the steroidogenic action of angiotensin II but did not influence potassium-stimulated aldosterone production. β-MSH did not influence aldosterone secretion when directly infused into the arterial supply of autotransplanted sheep adrenal glands (Blair-West et al. 1962a). β-Endorphin markedly stimulated aldosterone secretion by isolated adrenal glands of hypophysectomized, nephrectomized dogs perfused in situ and did not affect cortisol secretion (Güllner and Gill 1983). The stable met-enkephalin analogue FK 33824 was found to be inactive in this preparation. Bolus injections of increasing amounts of β-endorphin into normal human subjects on a normal sodium intake did not raise plasma aldosterone concentrations, according to Kem et al. (1985). Graded infusions of β-endorphin into sodium-restricted subjects elicited significant increases in plasma aldosterone concentration and in plasma renin activity, but also a significant fall in the plasma cortisol concentration (Rabinowe et al. 1985). Similar increases in plasma aldosterone and renin activity were induced by β-endorphin infusions in dexamethasone-treated subjects on a high sodium intake. The data suggested that the early rise in plasma aldosterone could have been secondary to an increase in renin release, but that renin could not have accounted for a sustained postinfusion elevation of plasma aldosterone. Marked increases in plasma aldosterone and renin activity accompanied by decreases in plasma cortisol and ACTH concentrations had been previously induced in normal man by dermorphin, a synthetic peptide with potent opiate-like activity (Degli Uberti et al. 1983b).

The common precursor protein of ACTH and β-LPH contains in its N-terminal region an MSH-like sequence, i.e. *γ-MSH,* which is partly homologous with the α- and β-MSH segments of the same prohormone. A synthetic peptide consisting of γ-MSH with a C-terminal extension of 15 additional residues, denoted, γ^3-MSH, stimulated the adrenocortical cholesterol ester hydrolase activity (but not the cholesterol side-chain cleavage activity) in hypophysectomized rats and synergistically enhanced the ACTH-induced rise in the serum levels of aldosterone and corticosterone (R.C. Pedersen et al. 1980a). A naturally occurring and circulating γ-MSH precursor (Pro-γ-$MSH_{1\text{-}77}$) potentiated the stimulatory effect of ACTH on aldosterone and corticosterone outputs by superfused rat adrenal zona glomerulosa and fasciculata cells as well as by human adrenocortical cells (Al-Dujaili et al. 1981, 1982a). These observations have not remained undisputed. According to

Moriwaki et al. (1985), neither a series of γ-MSH-related peptides, including synthetic γ_1-, γ_2-, γ_3-, and lys-γ_3-MSH, nor the N-terminal 16 K fragment of pro-opiomelanocortin (POMC) exhibited any ACTH-potentiating activity when tested in a stationary system of rat adrenal cell incubation. When added alone, γ_1-MSH did not stimulate corticosteroid output by rat adrenal zona glomerulosa or fasciculata cells (Vinson et al. 1981). The N-terminal 16 K POMC fragment was also found to be completely inactive when added without ACTH to rat adrenal glomerulosa or inner-zone cells (Jornot et al. 1985). On the other hand, N-terminal peptides of human or porcine POMC as well as γ_1-MSH, γ_2-MSH, and γ_3-MSH significantly stimulated aldosterone production by dispersed or cultured cells derived from human aldosterone-producing adenomas in the complete absence of ACTH or any other added stimulator (Seidah et al. 1981; Lis et al. 1981; Schiffrin et al. 1983; Pham-Huu-Trung et al. 1985). γ_3-MSH and the 16 K N-terminal fragment of POMC also potentiated the steroidogenic effect of ACTH on adenoma cells (Pham-Huu-Trung et al. 1985). However, not all human aldosterone-producing adenoma cells were found to be responsive to N-terminal POMC peptides. Thus, adenoma cells from a patient with a low plasma level of immunoreactive γ-MSH did not respond to lys-γ_3-MSH, whereas aldosterone production was stimulated by this peptide in adenoma cells from a patient with a high plasma level of immunoreactive γ-MSH (Griffing et al. 1985 a). Specific high-affinity binding sites for radioiodinated lys-γ_3-MSH were found in membrane particles of the rat adrenal cortex and other tissues (R.C. Pedersen and Brownie 1983 a). Lys-γ_3-MSH had 100 times greater affinity for these receptor sites than ACTH (1–24).

4.4.3 Prolactin

Ovine prolactin stimulated aldosterone production by quartered rat adrenals and had a lesser stimulatory effect on corticosterone output (Lichtenstein et al. 1976). Aldosterone production of isolated cells of two different aldosterone-producing adrenocortical adenomas was enhanced by ovine prolactin to a greater extent than by angiotensin II or angiotensin III or by a high potassium concentration, but to a lesser extent than by ACTH (Carroll et al. 1982). High doses (1–10 IU) of ovine prolactin stimulated corticosteroid production by superfused rat adrenal slices (Eldridge and Lymangrover 1984). Smaller doses (0.01–0.1 IU) of prolactin significantly potentiated the steroidogenic effect of ACTH in this preparation. In the absence as well as in the presence of ACTH, prolactin stimulated aldosterone production relatively more extensively than corticosterone, progesterone, and androstenedione outputs.

4.4.4 Aldosterone-Stimulating Factor

A protein fraction isolated from human urine, causing hypertension and abnormally high plasma levels of aldosterone when chronically administered to rats in vivo (Sen et al. 1977), stimulated aldosterone production in vitro by incubated cat (Bravo et al. 1980), rat (Sen et al. 1981 a), and rabbit (Saito et al. 1981) capsular adrenal cells. The active substance, called "aldosterone-stimulating factor" (ASF), was found to be glycoprotein with a molecular weight of 26000, according to equilibrium ultracentrifugation (Sen et al. 1981a). Treatment of this glycoprotein with neuroaminidase increased its aldosterone-stimulating activity. Proteolytic digestion and purification with high-performance liquid chromatography yielded a smaller aldosterone-stimulating molecule with a molecular weight of 4000 (Sen et al. 1981b). Immunofluorescence studies with a goat antibody against human ASF in normal dog tissues obtained from different organs at autopsy localized ASF immunoreactivity exclusively into cells of the anterior lobe of the pituitary gland (Sen et al. 1981b). The synthetic analogue sar^1,thr^8-angiotensin II markedly suppressed angiotensin II-stimulated aldosterone production by rabbit and cat zona glomerulosa cells, but did not affect ASF-stimulated aldosterone production (Saito et al. 1981; Bravo et al. 1980). Similarly, ile^9-ACTH inhibited aldosterone production under stimulation by ACTH but not under stimulation by ASF. According to these observations, ASF did not act at angiotensin or ACTH receptor sites but had its own separate receptors. Sodium depletion of rabbits increased the maximal aldosterone secretory responses of zona glomerulosa cells to angiotensin II and ACTH; responses to ASF remained unchanged (Hata et al. 1984). Large concentrations of ASF (10^{-7} mol/l) potentiated angiotensin II-induced aldosterone responses of adrenal capsular cells from sodium-deficient but not from sodium-replete rabbits. Similar concentrations of ASF inhibited ACTH-induced aldosterone production of adrenal capsular cells from both sodium-replete and sodium-deficient rabbits. Plasma ASF concentrations were found to be elevated in a group of patients with idiopathic hyperaldosteronism (Carey et al. 1984). In these patients as well as in normal human subjects, plasma ASF rose in response to 4 h of standing and did not fall in response to treatment with dexamethasone.

4.4.5 Vasopressin

When added to incubated rat adrenal zona glomerulosa cells, vasopressin did not affect aldosterone or corticosterone production (Vinson et al. 1980; Payet and Lehoux 1982b). However, vasopressin (10^{-11} mol/l) potentiated the aldosterone-stimulating effect of a low dose of ACTH (10^{-12} mol/l). In higher concentrations (10^{-10}–10^{-8} mol/l), it partially blocked the increases in aldosterone and corticosterone output in response to a high dose (10^{-7} mol/l) of ACTH (Payet and Lehoux 1982b). Superfused rat adrenal glomerulosa cells responded to arginine-vasopressin (10^{-8}–10^{-6} mol/l) with a transient increase in aldosterone output, according to Balla et al. (1985). These investigators also found high-affinity specific binding sites for tritiated arginine-vasopressin in a crude membrane fraction of rat adrenal capsules.

Vasopressin stimulated cortisol secretion directly, but it had no effect on aldosterone output by perfused adrenal glands of hypophysectomized dogs (Hilton et al. 1960). Adrenal arterial infusion of a very high dose of vasopressin stimulated aldosterone and corticosterone secretion by an autotransplanted sheep adrenal gland – but only to a relatively small extent – whereas oxytocin was ineffective (Blair-West et al. 1962a).

4.5 Amines

4.5.1 Serotonin

The steroidogenic effect of serotonin (5-hydroxytryptamine[8]) on adrenocortical tissue was discovered by Rosenkrantz and co-workers (Rosenkrantz 1959; Rosenkrantz and Laferte 1960; Connors and Rosenkrantz 1962). Small amounts of serotonin stimulated the in vitro production of a blue tetrazolium-reducing steroid, which was tentatively identified as aldosterone by its chromatographic properties and by its marked sodium-retaining biological activity. This corticosteroid-stimulating effect was observed during incubation of rabbit, guinea pig, rat, and cow adrenocortical tissue and of slices of a human adrenocortical tumor. Jouan and Samperez (Jouan 1963, 1967; Jouan and Samperez 1965) found that the steroidogenic effect of serotonin on rat adrenal tissue incubates was qualitatively similar to the one induced by pineal gland extracts and different from the one induced by ACTH. Serotonin and pineal gland extracts stimulated mainly aldosterone and 18-hydroxycorticosterone production and had only a small effect on corticosterone and 18-OH-DOC outputs; the main effect of ACTH was on the production of the latter two corticosteroid fractions. This specific aldosterone-stimulating effect of serotonin on rat adrenal incubates was confirmed by us (J. Müller and Ziegler 1968) when we isolated and characterized aldosterone-stimulating material from rat serum dialysate (J. Müller and Weick 1967) and found evidence that this activity was due to the presence of serotonin. Serotonin significantly stimulated aldosterone production by quartered adrenals of sodium-deficient rats at a concentration of 10^{-8} mol/l (J. Müller and Ziegler 1968; Fig. 15). The log-dose/response curve showed an approximately linear slope between serotonin concentrations of 10^{-8} and 10^{-6} mol/l and a plateau between concentrations of 10^{-6} and 10^{-4} mol/l. A similar but lesser effect was seen on deoxycorticosterone output, whereas the corticosterone production by mixed adrenocortical tissue was not stimulated by any dose of serotonin. The steroidogenic response of rat adrenal tissue to serotonin was dependent to a large extent on the sodium and potassium balance of the animals from which the adrenals were taken (J. Müller and Huber 1969; Fig. 16). Serotonin stimulated aldosterone production in adrenals of sodium-deficient and normal rats, but not in tissue of potassium-deficient rats, where it stimulated mainly the deoxycorticosterone output. In adrenals of rats that were uremic 40 h following bilateral nephrectomy, serotonin stimulated aldosterone production to a greater and deoxycorticosterone production to a lesser extent than it did in adrenals of sham-operated animals. The selective aldosterone-stimulating effect of serotonin upon mixed adrenocortical tissue was found to be due to a strict zone specificity of this stimulator. It stimulated steroidogenesis exclusively in the capsular portion of rat adrenals (J. Müller 1970b) and in capsular adrenal cells (Haning et al. 1970) and had no effect upon zona fasciculata tissue or cells. In preparations of rat adrenal zona glomerulosa cells, serotonin stimulated aldosterone production at the same concentration as in complete rat adrenocortical tissue,

[8] 5-Hydroxytryptamine creatinine sulfate was used in most experiments referred to in this chapter.

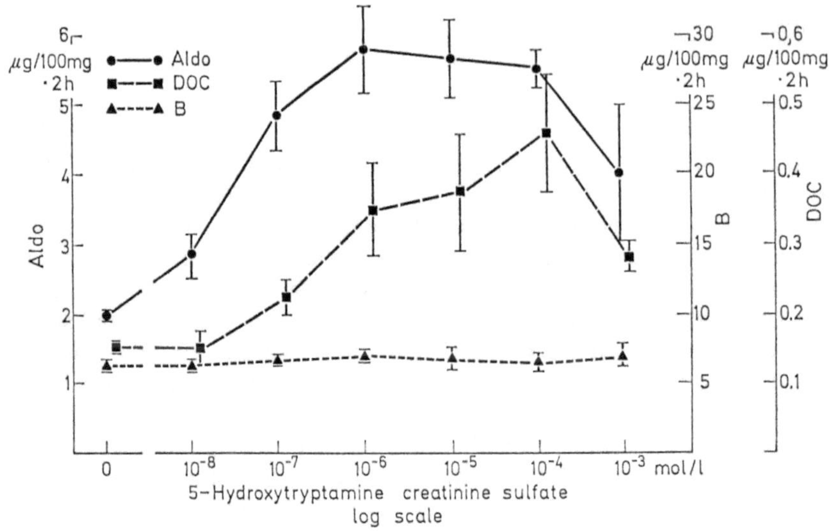

Fig. 15. Effect of increasing concentrations of serotonin (5-hydroxytryptamine creatinine sulfate) in the incubation medium on in vitro production of aldosterone *(ALDO)*, deoxycorticosterone *(DOC)*, and corticosterone *(B)* by rat adrenal quarters. Mean values of four experiments ($n=4$) ± standard error of the mean. (From J. Müller and Ziegler 1968)

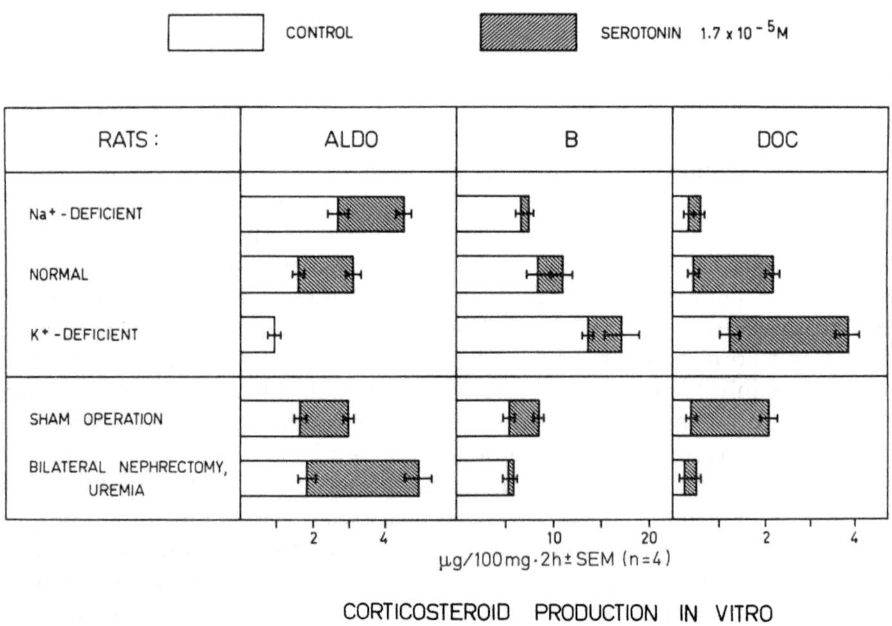

Fig. 16. Production of aldosterone *(ALDO)*, corticosterone *(B)*, and deoxycorticosterone *(DOC)* by quartered adrenals of rats kept on different diets for 2 weeks and rats uremic 40 h after bilateral nephrectomy. *White bars* represent basal corticosteroid production, *hatched bars* increments due to addition of serotonin (5-hydroxytryptamine creatinine sulfate) to the incubation medium. Mean values of two experiments ($n=4$) ± standard error of the mean. (Data from J. Müller and Huber 1969)

with log-dose/response curves similar to the one shown in Fig. 15 (Haning et al. 1970; Bing and Schulster 1977; Fujita et al. 1979; Shenker et al. 1985b). During stationary incubation of rat adrenal glomerulosa cells, serotonin stimulated aldosterone production significantly at a concentration of 10^{-9} mol/l (Matsuoka et al. 1985a). In a system of superfused rat zona glomerulosa cells, the lowest concentration of serotonin inducing a significant increase in aldosterone production was 10^{-12} mol/l (Al-Dujaili et al. 1982b). Whereas the aldosterone-stimulating activity of serotonin has been well established by in vitro studies with rat adrenal tissue and cells, there is as yet little evidence showing that serotonin has similar effects on adrenocortical tissue or cells from other animal species or on the human adrenal cortex. In unpublished experiments carried out in our laboratory, serotonin had no steroidogenic effect when added to incubated guinea pig, dog, human, or duck adrenocortical tissue. Serotonin did not stimulate aldosterone production in either autotransplanted sheep adrenals in vivo or sheep adrenal tissue in vitro (Blair-West et al. 1962a; McDougall et al. 1976). Serotonin significantly stimulated the outputs of aldosterone, 18-hydroxycorticosterone, and corticosterone of duck adrenocortical cells (Vinson et al. 1979), but it did not affect corticosteroid biosynthesis by incubated snake or bullfrog adrenocortical tissue (Macchi and Phillips 1966). Aldosterone and corticosterone outputs by superfused normal human adrenal cells were significantly enhanced by serotonin added in concentrations between 10^{-8} and 10^{-6} mol/l (Espiner et al. 1980). When added to the medium superfusing adrenal adenoma cells from a patient with primary aldosteronism, serotonin stimulated not only the production of aldosterone and corticosterone but also the biosynthesis of cortisol. During stationary incubation, serotonin (10^{-6} mol/l) stimulated aldosterone production by human normal zona glomerulosa or aldosterone-producing adenoma cells (Shenker et al. 1985b).

In normal human subjects, an infusion of serotonin (5 mg/h) induced a significant rise in plasma aldosterone; this was not accompanied by changes in plasma cortisol, plasma renin activity, or serum potassium (Mantero et al. 1982). Serotonin stimulated the secretion of cortisol and corticosterone by perfused adrenals of hypophysectomized dogs to the same extent as ACTH (Verdesca et al. 1961). A high aldosterone output was measured under serotonin stimulation; however, aldosterone production was not measured under control conditions. Farrell and McIsaac (1961) found serotonin 100 times less active than "adrenoglomerulotropin" (6-methoxytetrahydroharman[9]) in stimulating aldosterone secretion in the midcollicular brain-removed, hypophysectomized dog. Serotonin raised the plasma aldosterone but not the plasma corticosterone concentration when intraperitoneally administered to intact rats (Al-Dujaili et al. 1980). On the other hand, when serotonin was injected into hypophysectomized rats, it stimulated only the corticosterone output by incubated adrenal glands and did not affect the production of aldosterone (Gromova et al. 1967).

The interaction of radiolabeled serotonin with rat capsular adrenal tissue was characterized by a rapid uptake and extensive metabolic degradation to 5-hydroxyindoleacetic acid and another, unidentified metabolite (Trost and Müller

[9] This substance was found to have no aldosterone-stimulating activity in either the sheep (Blair-West et al. 1963) and the dog (Mulrow et al. 1963) in vivo or rat adrenal tissue in vitro (Jouan 1967; J. Müller and Ziegler 1968).

1976). A marked accumulation of radioactivity in the tissue was mainly due to delayed outflow of intracellularly formed metabolites, and there was no evidence of serotonin storage in the zona glomerulosa cells. A fast inactivation of serotonin by the zona glomerulosa may explain why relatively high concentrations of the amine are required for stimulating aldosterone biosynthesis. When considering a possible physiological role of serotonin in the control of aldosterone secretion it must also be kept in mind that the serotonin found in the serum is not derived from the plasma but is released from the blood platelets by clotting (Rand and Reid 1951). If serotonin is added to blood, it is readily taken up and stored by the platelets (Humphrey and Toh 1954). Thus, serotonin could probably not function as a generally circulating trophic hormone like ACTH or angiotensin II, but might possibly act as a neurotransmitter or an intracellular messenger.

4.5.2 Serotonin-Related Indole Derivatives

When testing a series of indole derivatives in equimolar concentrations, Rosenkrantz and Laferte (1960) found only bufotenine (5-hydroxy-*N,N*-dimethyltryptamine) to have a steroidogenic activity comparable to that of serotonin. According to Jouan and Samperez (1964), tryptamine, 5-hydroxytryptophan, *N*-methyltryptamine, 5-methoxytryptamine, melatonin (*N*-acetyl-5-methoxytryptamine), and psilocybin (4-phospho-*N,N*-dimethyltryptamine) had no steroidogenic activity; this is only partly consistent with our own findings (J. Müller and Ziegler 1968). We found bufotenine and 5-methoxytryptamine to stimulate rat adrenal aldosterone production by more than 100% above mean control values when added in a concentration of 10^{-6} mol/l, whereas L-5-hydroxytryptophan (i.e., a biological precursor of serotonin) had a similar activity in a 10^{-5} molar concentration. At this dose level, melatonin, 5-hydroxyindoleacetic acid, tryptamine, and L-tryptophan had no effect on aldosterone production (Table 3). In a more recent study, 5-methoxytryptamine, 5-methoxytryptophol, and melatonin stimulated the aldosterone output by superfused rat adrenal zona glomerulosa cells, with 5-methoxytryptamine being an even more potent stimulator than serotonin (Al-Dujaili et al. 1980). 5-Methoxytryptamine also raised the plasma aldosterone concentration in the rat in vivo. In a concentration of 10^{-6} mol/l, the serotonin precursor 5-hy-

Table 3. Aldosterone-stimulating activity of serotonin and chemically or pharmacologically related compounds (From J. Müller and Ziegler 1968)

Aldosterone-stimulating activity	No aldosterone-stimulating activity[b] at 10^{-5} mol/l	
Serotonin (10^{-7} *M*)[a]	Melatonin	Histamine
Bufotenine (10^{-6} *M*)	5-Hydroxy-3-indoleacetid acid	6-Methoxytetrahydro-harman
5-Methoxytryptamine (10^{-6} *M*)	L-tryptophan	Epinephrine
L-5-hydroxytryptophan (10^{-5} *M*)	Tryptamine	Norepinephrine
	Tyramine	L-thyroxine

[a] Values in parentheses are lowest concentration that stimulated aldosterone production of quartered rat adrenals by 100% above mean control values.

[b] Stimulated aldosterone production by less than 20%.

droxytryptophan did not affect aldosterone output by rat and human zona glomerulosa or adenoma cells (Shenker et al. 1985b). In normal dexamethasone-treated human subjects, the oral intake of 200 mg tryptophan led to an increase in plasma aldosterone, which became significant after 2 h and which was not accompanied by any alteration in plasma renin activity or in potassium, sodium, or ACTH concentrations (Shenker et al. 1985b). This was most likely due to a central nervous serotonergic stimulation, because the rise in plasma aldosterone elicited by 5-hydroxytryptophan was more extensive and of longer duration in subjects pretreated with the peripheral decarboxylase inhibitor carboxydopa. 5-Hydroxytryptophan also elicited increases in plasma aldosterone in four patients with idiopathic hyperaldosteronism and in two of four patients with aldosterone-producing adenomas. In bovine adrenal tissue, melatonin suppressed the conversions of progesterone to aldosterone and cortisol by a noncompetitive inhibition of microsomal 21- and 17α-hydroxylases (Mehdi and Sandor 1977). In rat adrenal tissue, melatonin suppressed the corticosterone output by stimulating 5α-reductase activity and thus the degradation of the hormone (Ogle and Kitay 1978). In vivo administration of melatonin to intact (but not to hypophysectomized) rats resulted in an increased corticosterone output and a diminished aldosterone production by incubated adrenal glands (Gromova et al. 1967).

4.5.3 Serotonin Antagonists

Methysergide (9,10-didehydro-*N*-[1-(hydroxymethyl)propyl]-1,6-dimethylergoline-8β-carboxamide) and BC-105[10], both in concentrations of 16.7 µg/ml, decreased the baseline aldosterone production of quartered rat adrenals by approximately 20% and completely inhibited the effect of serotonin (1.7 10^{-5} mol/l; Müller and Ziegler 1968; Fig. 17). They also blocked the stimulation of aldosterone production by rat serum but not by ACTH. Metergoline ([[(8β)-1,6-dimethylergolin-8-yl]-methyl] carbamic acid phenylmethyl ester) blocked the aldosterone response of superfused rat adrenal zona glomerulosa cells to serotonin and angiotensin II but not to potassium ions or ACTH (Boscaro et al. 1979). Both methysergide and metergoline suppressed the aldosterone output by rat adrenal zona glomerulosa cells under baseline conditions and completely blocked its stimulation by serotonin or human serum (Mendelsohn and Kachel 1981). Methysergide (but not metergoline) also partially blocked the response in aldosterone production to a high potassium concentration. Cyproheptadine (a blocker of serotonin type-1 and type-2 receptors) and ketanserin (a specific blocker of serotonin type-2 receptors) markedly suppressed aldosterone production by rat capsular adrenal cells under basal conditions of incubation and under stimulation by added serotonin (Matsuoka et al. 1985a). Cyproheptadine also completely blocked the stimulation of aldosterone production by angiotensin II but did not affect ACTH-stimulated aldosterone production. At a high concentration (10^{-5} mol/l), ketanserin partially but significantly suppressed angiotensin II- and ACTH-stimulated aldosterone production.

[10] Pizotyline, Sandoz Ltd., Basel; a serotonin and histamine antagonist structurally related to cyproheptadine

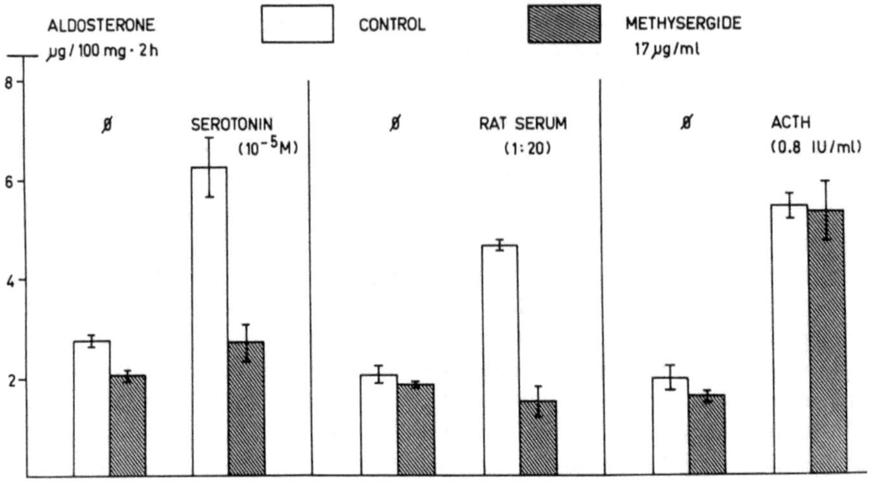

Fig. 17. Effect of methysergide bimaleinate on in vitro aldosterone production by rat adrenal quarters and its stimulation by serotonin, rat serum, and ACTH. Mean values of two flasks ± range. (Data from J. Müller and Ziegler 1968)

In supine normal human subjects, an intravenous bolus injection of ketanserin (10 mg) induced small increases in plasma aldosterone and renin activity; however, these were not significantly different from the increases induced by vehicle injection (Degli Uberti et al. 1982). According to Fagard et al. (1984), a rapid intravenous injection of ketanserin (10 mg) followed by a continuous infusion (2 mg/h) of the drug moderately lowered plasma aldosterone while not affecting plasma renin activity or plasma angiotensin II concentration in resting recumbent subjects. During exhausting physical activity on a bicycle ergometer, which markedly raised all three parameters, plasma aldosterone concentrations remained lower, whereas plasma renin activity and angiotensin II concentrations became significantly higher in ketanserin-treated than in untreated subjects. Neither ketanserin nor methysergide significantly blunted the increase in plasma aldosterone induced in normal human subjects by the oral intake of 5-hydroxytryptophan (200 mg; Shenker et al. 1985c). This indicated that the increase in aldosterone secretion was mediated by the central nervous system and was not due to a direct effect of the serotonin precursor on the zona glomerulosa.

4.5.4 Dopamine, Dopaminergic Agonists, and Dopamine Antagonists

The following paragraphs deal only with experiments in which direct effects of dopaminergic agonists and antagonists on zona glomerulosa cells or tissue were demonstrated. Experiments in which such agents were systemically administered to man or experimental animals in vivo will be reviewed in Sect. 5.7.

When added in a concentration of 10^{-5} mol/l to the incubation medium, *dopamine* (3-hydroxytryptamine) did not affect basal aldosterone production by capsular bovine adrenal cells, but it significantly diminished the response in aldosterone output to added angiotensin II (McKenna et al. 1979b). According to Sequeira and McKenna (1985), dopamine significantly inhibited aldosterone production by bovine zona glomerulosa cells at a concentration of 10^{-4} mol/l in the absence of stimulators and at a concentration of 10^{-6} mol/l during stimulation with angiotensin II (10^{-5} mol/l).

Short-term addition of dopamine or of dopamine 3-sulfate (10^{-4} mol/l) to cultured bovine adrenal zona glomerulosa cells suppressed angiotensin II-stimulated (10^{-8} mol/l) aldosterone output by 61% and 44% respectively (Rácz et al. 1984). Dopamine 4-sulfate was inactive. Apparently, dopamine 3-sulfate was active in the conjugated form, because less than 1% was hydrolyzed during the incubation period.

In dispersed rat adrenal zona glomerulosa cells, dopamine in concentrations of 10^{-10}–10^{-6} mol/l significantly suppressed angiotensin II- and serotonin-stimulated aldosterone outputs (Braley et al. 1983b). According to Aguilera and Catt (1984), such low concentrations of the amine were ineffective, whereas very high concentrations (10^{-5} and 10^{-4} mol/l) moderately inhibited basal and - more extensively - ACTH- and angiotensin II-stimulated aldosterone production by rat adrenal zona glomerulosa cells. However, in preparations of superfused rat adrenal zona glomerulosa cells, dopamine (10^{-6} and 10^{-5} mol/l) had no effect on aldosterone production on its own and did not alter the response in aldosterone output to serotonin, angiotensin II, ACTH, or a high potassium concentration (Boscaro et al. 1979; Edwards et al. 1980b). Finally, very high concentrations (10^{-4} and 10^{-3} mol/l) of dopamine significantly stimulated aldosterone production by quartered rat adrenals or capsular portions of rat hemiadrenals, without having a steroidogenic effect upon decapsulated adrenals (Table 4).

Dopamine (10^{-5}–10^{-3} mol/l) did not suppress aldosterone production by rabbit adrenal zona glomerulosa cells, either under basal conditions or under stimulation by angiotensin II or ACTH (Zanella and Bravo 1982).

In sodium-deficient sheep with a cervical adrenal autotransplant, an intra-adrenal arterial infusion of dopamine did not lower the aldosterone secretion rate (McDougall et al. 1981). In sodium-replete sheep, the response in adrenocortical aldosterone secretion to angiotensin II was not altered by a local infusion of dopamine (Lun et al. 1983).

Bromocriptine (2-bromoergocryptine) directly stimulated aldosterone output by superfused rat adrenal zona glomerulosa cells, whereas two other dopaminergic agonists, *lisuride* (3-(9,10-didehydro-6-methyl-ergolin-8α-yl)-1,1-diethylurea) and

Table 4. Effect of dopamine upon corticosteroid output by capsular portions of rat hemiadrenals (normal diet). Mean values of 4 experiments ± standard deviation ($n = 4$). (For methodology see J. Müller 1970b)

Dopamine-HCl concentration (mol/l)	Corticosteroid output in µg/100 mg per 2 h		
	Aldosterone	Corticosterone	Deoxycorticosterone
0	1.07±0.57	2.33±0.83	0.56±0.27
10^{-5}	1.70±0.36	2.91±0.06	0.59±0.10
10^{-4}	2.60±0.44[a]	3.20±0.50	0.73±0.18
10^{-3}	3.00±0.62[a]	3.30±0.19	1.16±0.08[a]

[a] Significantly different from control values ($p < 0.01$) according to t-tests

metergoline, were ineffective (Boscaro et al. 1979). However, all three agents blocked the aldosterone response to serotonin and angiotensin II, but not the response to ACTH and to potassium ions. Lisuride also completely blocked the increase in aldosterone output in response to 5-methoxytryptamine in superfused rat adrenal zona glomerulosa cells (Al-Dujaili et al. 1982b).

Metoclopramide (4-amino-5-chloro-*N*-[(2-diethylamino)ethyl]-2-methoxybenzamide), a drug with antidopaminergic activity, markedly stimulated aldosterone production by superfused rat adrenal zona glomerulosa cells in concentrations of between 10^{-8} and 10^{-5} mol/l (Edwards et al. 1980b). This stimulation was completely blocked by dopamine (10^{-5} mol/l). Under basal conditions of incubation, metoclopramide did not affect aldosterone production by dispersed rat adrenal zona glomerulosa cells, according to D.J.Campbell et al. (1981b), Braley et al. (1983), and Nagahama et al. (1983b). According to other reports, it even decreased the basal aldosterone and corticosterone outputs of rat adrenal glomerulosa cells, at least at high concentrations (Lauer et al. 1982; Aguilera and Catt 1984). However, it moderately stimulated aldosterone production in the presence of dopamine (10^{-10} mol/l; Lauer et al. 1982). On the other hand, metoclopramide significantly inhibited serotonin-, potassium-, angiotensin II-, and ACTH-stimulated aldosterone and corticosterone outputs by incubated rat adrenal zona glomerulosa cells (Lauer et al. 1982; Aguilera and Catt 1984). According to D.J.Campbell et al. (1981b) the sensitivity of rat adrenal zona glomerulosa cells to the aldosterone-stimulating action of increasing concentrations of serotonin was decreased by the addition of metoclopramide. Finally, metoclopramide stimulated aldosterone production of rat adrenal zona glomerulosa cells when added together with dopamine and angiotensin II or with dopamine and serotonin (Braley et al. 1983).

In sheep, metoclopramide stimulated aldosterone secretion when it was systemically infused but not when it was directly infused into the adrenal artery (McDougall et al. 1981; Lun et al. 1983). Metoclopramide did not stimulate aldosterone output by incubated sheep zona glomerulosa cells (T.A.Wilson et al. 1983b).

In a preparation of rabbit adrenal zona glomerulosa cells, metoclopramide did not affect aldosterone when it was added alone or when it was added together with angiotensin II or with ACTH (Zanella and Bravo 1982).

Aldosterone production by incubated minced adrenocortical adenomas taken from patients with primary aldosteronism was significantly stimulated by metoclopramide (R.D.Brown et al. 1981). These investigators also observed an increase in aldosterone output in response to metoclopramide by incubated nodular hyperplastic adrenocortical tissue surgically removed from a patient with idiopathic hyperaldosteronism. Metoclopramide also significantly stimulated aldosterone output by superfused cells from a human aldosterone-producing adrenocortical adenoma (Opocher et al. 1983). This response was partially suppressed by metergoline but not by bromocriptine.

Sulpiride, (5-(aminosulfonyl)-*N*-[(1-ethyl-2-pyrrolidinyl)methyl]-2-methoxybenzamide), another dopamine antagonist, stimulated the outputs of aldosterone and cyclic AMP by superfused bovine adrenal tissue but did not influence cortisol production by this preparation (Costa et al. 1980).

Specific *dopamine receptors* were found in the rat adrenal zona glomerulosa by M.G.Dunn and Bosmann (1981). The binding of [^{3}H]-spiroperidol by this tissue in the presence and absence of dopamine was saturable and of high affinity. According to its binding characteristics, the receptor belonged into the D-2 subclass of dopamine receptors. D-2 dopaminergic receptors were also identified in the bovine adrenal cortex; their density was higher in the zona fasciculata than in the zona glomerulosa (Missale et al. 1985).

Adrenocortical dopamine accounted for approximately 40% of the total adrenal gland content of this amine in rats (McCarty et al. 1984). The adrenocortical dopamine was not decreased by splanchnic nerve section or by treatment of rats with α-methyl-p-tyrosine, an inhibitor of catecholamine biosynthesis. Thus, adrenocortical dopamine appeared to be independent of the nervous supply and not to require de novo catecholamine biosynthesis.

4.5.5 Histamine and Histamine Antagonists

In hypophysectomized, nephrectomized dogs, the intravenous injection of histamine (0.1 mg) resulted in a marked increase in aldosterone secretion as well as in smaller but significant increases in cortisol and corticosterone secretion (Aikawa et al. 1979). No significant changes in the plasma potassium and sodium concentrations were observed. Histamine also directly stimulated cortisol and corticosterone production by incubated dog adrenal cells to a small but significant extent (T. Hirose et al. 1978). Marked increases in the secretion rates of aldosterone, corticosterone, and cortisol were induced by the infusion of histamine into the arterial supply of adrenal glands perfused in situ in hypophysectomized, nephrectomized dogs (Aikawa et al. 1981a).

When tested on superfused rat adrenal zona glomerulosa cells in vitro, the two H_2 receptor antagonists cimetidine and rametidine (10^{-6} and 10^{-7} mol/l) significantly stimulated basal aldosterone output and blocked the aldosterone response to angiotensin II but not to serotonin, ACTH, or potassium ions (Edwards et al. 1980a). The H_1 receptor antagonist chlorpheniramine (10^{-7} mol/l) enhanced the aldosterone response to angiotensin II and ACTH as well as the corticosterone response to ACTH. In normal human subjects, an intravenous injection of cimetidine did not affect the plasma aldosterone concentration on its own, but it significantly attenuated the aldosterone response to furosemide without altering the response in plasma renin activity (Edwards et al. 1980a). Treatment of normal human volunteers with ranitidine for 3 days resulted in significantly lowered plasma aldosterone concentrations as measured with subjects in the supine position or in the course of 2 h of ambulation (Sancho et al. 1984). Plasma renin activity and potassium concentration remained unchanged, whereas a small decrease in plasma cortisol was statistically significant only when measured in the supine position and after 30 min of ambulation.

4.5.6 Adrenergic Agonists and Antagonists

Epinephrine and norepinephrine significantly stimulated aldosterone production by bovine zona glomerulosa cell suspensions in concentrations of between 10^{-8} and 10^{-5} mol/l (McKenna et al. 1980). Aldosterone output by this adrenal cell preparation was also significantly enhanced by both phenylephrine, a specific α-adrenergic agonist, and isoproterenol, a specific β-adrenergic agonist. The α-

adrenergic receptor antagonist phentolamine significantly inhibited the aldosterone responses to phenylephrine and to isoproterenol, whereas the β-adrenergic receptor antagonist propranolol blocked only the aldosterone-stimulating action of isoproterenol. According to a later study by Sequeira and McKenna (1985), neither epinephrine nor norepinephrine (10^{-10}–10^{-6} mol/l) had any effect on aldosterone production by dispersed bovine adrenal glomerulosa cells. Aldosterone output by freshly dispersed bovine subcapsular adrenocortical cells was not responsive to ACTH, to angiotensin II, or to (−)isoproterenol (10^{-7} mol/l), according to De Léan et al. (1984c). But aldosterone production increased considerably in response to the same agents after the cells had been cultured at high density in monolayer for 3 days. Aldosterone production by cultured bovine adrenal glomerulosa cells was also stimulated by higher concentrations of other β-adrenergic agonists, such as (−)epinephrine, (−)norepinephrine, and (+)isoproterenol, but not by dopamine, serotonin, or metoclopramide. (−)Isoproterenol-stimulated aldosterone production was completely suppressed by the following β-adrenergic antagonists: (−)alprenolol, (+)alprenolol, atenolol, and IC 118551; the α-adrenergic blockers prazosin, yohimbine, or phentolamine were inhibitory only at very high concentrations. Isoproterenol (10^{-5} mol/l) significantly stimulated aldosterone and corticosterone production by incubated rat capsular adrenal tissue, but less extensively than ACTH (3.3×10^{-7} mol/l; Shima et al. 1984). The β-adrenergic agonist had no effect on corticosterone production by decapsulated adrenal tissue.

4.6 Prostaglandins and Inhibitors of Prostaglandin Biosynthesis

4.6.1 Prostaglandins

An ACTH-like activity of certain prostaglandins (PGE_1, PGE_2, $PGF_{2\alpha}$) was first demonstrated by Flack et al. (1969). When added to the medium with which bisected rat adrenals were superfused, these substances significantly stimulated corticosterone production. PGE_2 elicited a small but statistically significant increase in aldosterone production by incubated adrenal glands of dexamethasone-treated rats and had a more marked effect upon corticosterone production (Spät et al. 1971). However, prostaglandin H_2 stimulated aldosterone but not corticosterone output by superfused capsular adrenals of sodium-replete hypophysectomized rats (Spät and Józan 1975). In contrast, prostaglandin A_2 promoted only corticosterone output by the same tissue, but enhanced both aldosterone and corticosterone production by capsular adrenals of sodium-deficient hypophysectomized rats. At the same concentration, PGA_2 did not influence corticosterone production by the decapsulated portions of the glands. At a high concentration (10^{-4} mol/l) $PGF_{2\alpha}$ potentiated the responses to ACTH in aldosterone and corticosterone output, respectively, by capsular and decapsulated rat adrenals (Spät et al. 1977a). Prostaglandins E_2 and -$F_{2\alpha}$, prostacyclin (PGI_2), and arachidonic acid did not significantly influence aldosterone production by capsular or corticosterone production by decapsulated rat adrenal cells, when added in concentrations between 3×10^{-8} and 3×10^{-4} mol/l (Matsuoka et al. 1980a). According to

W. B. Campbell et al. (1980), PGE_2 stimulated aldosterone production by rat capsular adrenal cells, but the steroidogenic effects of PGE_2 and of angiotensin II were not additive. Addition of the prostaglandin endoperoxide PGH_2 or of arachidonic acid, in concentrations sufficient to markedly raise the outputs of PGE_2 and $PGF_{2\alpha}$, had no effect upon aldosterone production. R. T. Miller et al. (1980) and Enyedi et al. (1981) also found arachidonic acid to be inactive when added to rat adrenal glomerulosa cells at a concentration of 5×10^{-6} mol/l. However, at a higher concentration of 5×10^{-4} mol/l, it stimulated aldosterone production on its own and potentiated the aldosterone-stimulating effect of potassium ions, but inhibited angiotensin II-stimulated and ACTH-stimulated aldosterone outputs (Enyedi et al. 1981). These investigators observed a significant dose-dependent stimulation of aldosterone production by PGE_2 but not by $PGF_{2\alpha}$. According to W. B. Campbell and Gomez-Sanchez (1985), aldosterone output by rat zona glomerulosa cells was significantly stimulated by PGI_2, PGE_2, 6-keto-PGE_1, and 6-keto-PGH_2. PGI_2 and PGE_2 were active at the lowest concentrations, i.e., at 5×10^{-10} and at 5×10^{-9} mol/l respectively.

During stationary incubation of frog adrenal tissue, variable, mainly transient increases in aldosterone and corticosterone production were elicited by added PGA_1, PGA_2, PGB_1, PGB_2, PGE_1, $PGF_{1\alpha}$, and $PGF_{2\alpha}$ (Chavin et al. 1978). Several prostaglandins induced marked increments in aldosterone and corticosterone production in a system of superfused frog interrenal tissue (Delarue et al. 1981; Perroteau et al. 1984a). The most active stimulator of steroidogenesis was PGE_1, followed by PGI_2, PGE_2, PGB_2, PGB_1, and $PGF_{2\alpha}$.

When prostaglandin E_2 was directly infused to autotransplanted adrenal glands of sodium-deficient sheep, a marked drop in the aldosterone secretion rate resulted in a third of all experiments (Blair-West et al. 1971c). In the remainder, there were small and uncharacteristic changes in aldosterone secretion. Observed rises in cortisol and corticosterone secretion were not statistically significant.

In bovine adrenal tissue incubations, prostaglandins E_1 and E_2 enhanced the outputs of aldosterone, corticosterone, cortisol, and cyclic AMP, whereas prostaglandins $F_{1\alpha}$, $F_{2\alpha}$, and A were ineffective (Saruta and Kaplan 1972). According to Tsunoda et al. (1984), aldosterone production by bovine adrenal cells was stimulated by PGE_2 and 6-keto-PGE_1 but not by PGI_2.

The action of several prostaglandins upon corticosteroidogenesis by incubated human adrenal tissue varied greatly at different concentrations and was also dependent on the incubation time. High doses (100 μg/ml) of PGE_1 and PGE_2 significantly stimulated aldosterone output during 32-min incubations (Honn and Chavin 1976a). Low doses (1 and 10 μg/ml) of these substances suppressed aldosterone output during shorter incubations. Prostaglandins $F_{1\alpha}$ (1 and 10 μg/ml) and $F_{2\alpha}$ (10 and 100 μg/ml) significantly inhibited aldosterone production at all incubation times. Irrespective of incubation time and concentration, PGE_1 and PGE_2 stimulated cortisol output and tissue cyclic AMP content, whereas $PGF_{1\alpha}$ and $PGF_{2\alpha}$ both had an inhibitory action upon these parameters (Honn and Chavin 1976b). During 32-min incubations, high doses (10 and 100 μg/ml) of PGA_1 and PGA_2 significantly raised human adrenal cyclic AMP content and cortisol output, but low doses (1 μg/ml) depressed both parameters (Honn and Chavin 1977). PGA_1 stimulated aldosterone production at 1 μg/ml and inhibited it at higher concentrations. The low dose of PGA_2 suppressed aldosterone output; the higher doses were ineffective. PGB_1 and PGB_2 stimulated aldosterone production at 1 μg/ml and were ineffective at 10 and 100 μg/ml.

In a group of 13 normal human subjects, a 1-h infusion of prostaglandin A_1 induced a threefold rise in plasma aldosterone without significantly and consistently affecting plasma renin and cortisol (Fichman et al. 1972). However, according to a later study, prostaglandin A_1-induced increases in plasma aldosterone were closely correlated to parallel increases in plasma renin activity (Golub et al. 1976). In untreated and in indomethacin-treated patients with essential hypertension, graded infusions of prostaglandin E_1 led to parallel rises in plasma aldosterone and in plasma renin activity (Yoshimura et al. 1979). In propranolol-treated patients, plasma renin activity did not respond to PGE_1 infusions, whereas there was still a significant increase in plasma aldosterone. Infusions of prostacyclin (PGI_2) that lowered diastolic blood pressure and significantly raised plasma renin activity and angiotensin II concentrations did not affect plasma aldosterone concentrations (Miyamori et al. 1979).

6-Keto-$PGF_{1\alpha}$ was the prostaglandin most abundantly produced by rat adrenal glomerulosa cells during incubation under basal conditions and under stimulation by mellitin, whereas $PGF_{2\alpha}$ was the predominant prostaglandin of zona fasciculata cells (Swartz and Williams 1983). The production of endogenous 6-keto-$PGF_{1\alpha}$ by rat zona glomerulosa cells was not affected by increasing concentrations of ACTH, angiotensin II, or extracellular potassium. It had been previously shown by R.T. Miller et al. (1980) that the production of PGE_2 by rat zona glomerulosa cells was not increased in response to these three stimuli of aldosterone biosynthesis. W.B. Campbell and Gomez-Sanchez (1985) confirmed a lack of response in zona glomerulosa cell PGE_2 production to angiotensin II, but observed that the peptide stimulated the production of 6-keto-$PGF_{1\alpha}$ at 2×10^{-10} and at 2×10^{-9} mol/l. At higher concentrations of angiotensin II, there was a decrease in 6-keto-$PGF_{1\alpha}$ output toward baseline values, while aldosterone production continued to increase. In superfused frog interrenal tissue, the production of endogenous PGE_2 was stimulated to the same extent as aldosterone and corticosterone productions by sar^1,val^5-angiotensin II but was not affected by ACTH (Perroteau et al. 1984).

4.6.2 Inhibitors of Prostaglandin Biosynthesis

In the absence of any other stimulator, the addition of aspirin (salicyclic acid acetate; 100 μg/ml) to the incubation medium resulted in increased aldosterone and corticosterone outputs by incubated capsular portions of rat adrenals but in a diminished corticosterone production by decapsulated adrenals (Spät et al. 1977b); indomethacin (1 or 10 μg/ml) was ineffective. Both substances partially suppressed the response to ACTH in capsular adrenal corticosterone output without influencing the response in aldosterone production. Indomethacin potentiated the stimulatory action of a high dose of ACTH (5 μg/ml) upon corticosterone production in decapsulated adrenals. W.B. Campbell et al. (1979a, 1980) observed that indomethacin (1 μg/ml) inhibited the basal production of aldosterone as well as the aldosterone response to angiotensin II or III by rat capsular adrenal cells. Meclofenamate (1 μg/ml) also significantly attenuated the aldosterone response to angiotensin III, but it had no effect on the aldosterone response to angiotensin II or on the basal aldosterone output. However, according to Matsuoko et al. (1980a), indomethacin (3×10^{-6} mol/l = 1.1 μg/ml) had no effect on either basal aldosterone production or aldosterone responses to angiotensin II or ACTH by rat capsular adrenal cells. The inhibitor also did not influence the corticosterone output by cells from decapsulated rat adrenals under baseline conditions or under stimulation by ACTH. R.T. Miller et al. (1980) observed a significant suppression of aldosterone production by indomethacin under stimulation by angiotensin II, ACTH, or a high potassium concentration but not in the absence of stimulators. In contrast, meclofenamate inhibited aldosterone production in the absence as well as in the presence of the three stimulators. On the other hand, Enyedi et al. (1981) found indomethacin to have no effect on basal or on ACTH- or potassium-stimulated aldosterone production. The drug slightly but significantly suppressed aldosterone output under stimulation by angiotensin II. 5,8,11,14-Eicosatetraynoic acid, a competitive antagonist of arachidonic acid, did not affect angiotensin II-stimulated aldosterone production, but it moderately suppressed aldosterone production under stimulation by high concentrations of ACTH or potassium ions. U-51605, a prostacyclin synthetase inhibitor, did not suppress basal or angiotensin II-stimulated aldosterone output by rat adrenal glomerulosa cells (W.B. Campbell and Gomez-Sanchez 1985). Two different inhibitors of prostaglandin synthesis, indomethacin and meclofenamic acid, completely blocked the steroidogenic effect of sar^1,val^5-angiotensin II on superfused frog interrenal tissue, but did not inhibit the stimulation of corticosterone and aldosterone outputs by ACTH or dibutyryl cyclic AMP (Delarue et al. 1981; Perroteau et al. 1984). When added to incubated human adrenal tissue, indomethacin (10 μg/ml) depressed basal aldosterone production and the aldosterone response to ACTH (Honn and Chavin 1976a). The prostaglandin antagonist 7-oxa-13-prostynoic acid (50 μg/ml) had similar effects.

4.7 Aldosterone-Inhibiting Peptides

4.7.1 Atrial Natriuretic Peptides

Natriuretic peptides isolated from human or rat atria and their biologically active synthetic analogues are strong inhibitors of aldosterone production when added directly to adrenocortical zona glomerulosa cells or tissue during short-term incubation experiments. These substances are alternatively referred to as "atrial natriuretic factors" (ANF) and include auriculins A and B, atriopeptins I-III, and cardionatrin (for reviews see Cantin and Genest 1985; Palluk et al. 1985; Maack et al. 1985). They are structurally derived from the C-terminal part of a natural precursor peptide consisting of 151 (human) or 152 (rat) amino acids with an internal disulfide bridge essential for biological activity. In the rat, the circulating form of atrial natriuretic peptide contains 28 amino acids (Schwartz et al. 1985). However, smaller peptides consisting of 21-26 amino acids were found to have an equipotent natriuretic activity.

An atrial extract of rat heart markedly inhibited aldosterone production by rat adrenal capsular cells under basal conditions of incubation and partially inhibited the stimulation of aldosterone output by ACTH and angiotensin II (Atarashi et al. 1984). The extract had little effect on corticosterone output by decapsulated adrenal cells. Aldosterone production by dispersed rat adrenal glomerulosa cells was similarly inhibited by natural rat atrial natriuretic peptides (31 or 73 AA) as well as by a synthetic peptide (26 AA), but only when stimulated by angiotensin II (10^{-8} mol/l), ACTH (3×10^{-10} mol/l), or a high potassium concentration (15 mmol/l) and not in the absence of stimulators (Chartier et al. 1984a, 1984b). The inhibition was dose dependent, with an IC_{50} of between 10^{-9} and 10^{-8} mol/l. The synthetic peptide was no longer inhibitory after its disulfide bridge had been split by oxidation with performic acid (Chartier et al. 1984a). In other preparations of rat adrenal glomerulosa cells, synthetic atrial natriuretic peptides with 23, 24, and 28 amino acids extensively lowered aldosterone production in the absence of any stimulator, with a maximal effect at a concentration of 10^{-7} mol/l (Kudo and Baird 1984; Matsuoka et al. 1985b; Atarashi et al. 1985). Atriopeptin I was somewhat less potent than atriopeptins II and III (Atarashi et al. 1985). In contrast, rat atriopeptin II did not affect basal aldosterone and 18-hydroxycorticosterone outputs, though it slightly but significantly lowered corticosterone output when added at a concentration of 10^{-6} mol/l to rat capsular adrenal tissue (Fig. 18). In the same preparation, the peptide substantially suppressed the outputs of all three steroids under presumably maximal stimulation by serotonin, ACTH, angiotensin II, or a high potassium concentration. According to W. B. Campbell et al. (1985), atriopeptins I and III did not inhibit basal aldosterone production by zona glomerulosa cells of sodium-replete rats, but they moderately (−32%) suppressed aldosterone production by zona glomerulosa cells of sodium-restricted rats. In suspensions of bovine adrenal glomerulosa cells, auriculin A inhibited aldosterone production most extensively in the absence of stimulators at a potas-

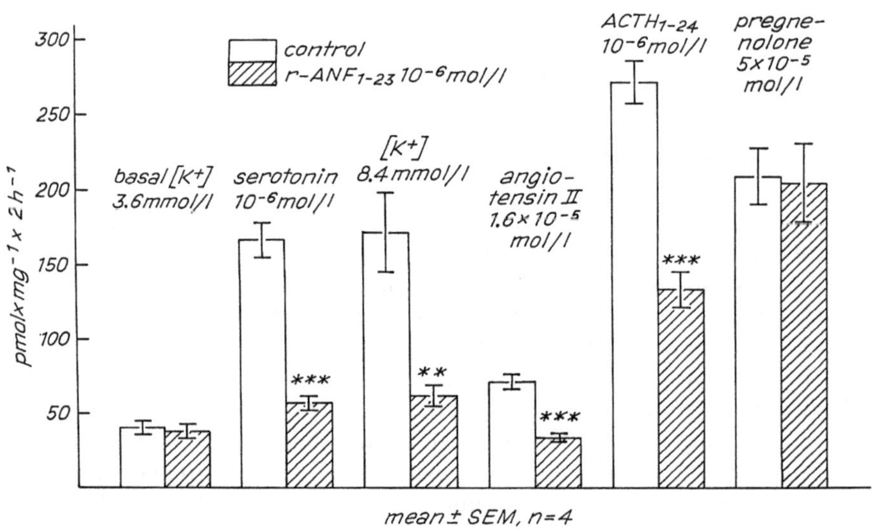

Fig. 18. Effect of an atrial natriuretic peptide (r-ANF_{1-23} = atriopeptin II) on aldosterone output by rat capsular adrenal tissue incubated in the absence or presence of various stimulators or the precursor pregnenolone. *Asterisks* refer to the statistical significance of ANF effects according to *t*-tests: ** $P<0.01$, *** $P<0.001$. (From J. Müller 1987)

sium concentration of 3.6 mmol/l, but it also suppressed the stimulatory effects of angiotensin II, dibutyryl cyclic AMP, and elevated potassium concentrations (Goodfriend et al. 1984). In cultured bovine adrenocortical glomerulosa cells, synthetic rat AFN (8-33) had only a minor (but statistically significant) suppressive effect on basal aldosterone output and inhibited aldosterone output more extensively during stimulation with angiotensin II, ACTH, prostaglandin E_1, or forskolin (De Léan et al. 1984a). The peptide also had a weak inhibitory effect on ACTH-, angiotensin II-, PGE_1-, or forskolin-stimulated cortisol production by cultured bovine zona fasciculata cells. In rat zona fasciculata cell (Kudo and Baird 1984; Atarashi et al. 1985) or tissue (Fig. 19) preparations, atrial natriuretic peptides did not significantly impair steroidogenesis in the absence or in the presence of ACTH. In dispersed cells from a human aldosterone-producing adrenocortical adenoma, synthetic α-human atrial natriuretic peptide (28 AA) inhibited both basal and potassium-stimulated aldosterone production but did not have a significant effect on aldosterone output under stimulation by ACTH or angiotensin II (Hirata et al. 1985).

In conscious rats on a normal sodium intake, a 30-min infusion of synthetic rat ANF (48-73) did not affect basal plasma aldosterone concentration or renin activity (Chartier et al. 1984b). However, both this peptide and atriopeptin II completely blocked a rise in plasma aldosterone in response to simultaneously infused angiotensin II (Chartier et al. 1984b; Atarashi et al. 1985). In rats with experimental 2-kidney, 1-clip Goldblatt renovascular hypertension, an auriculin A infu-

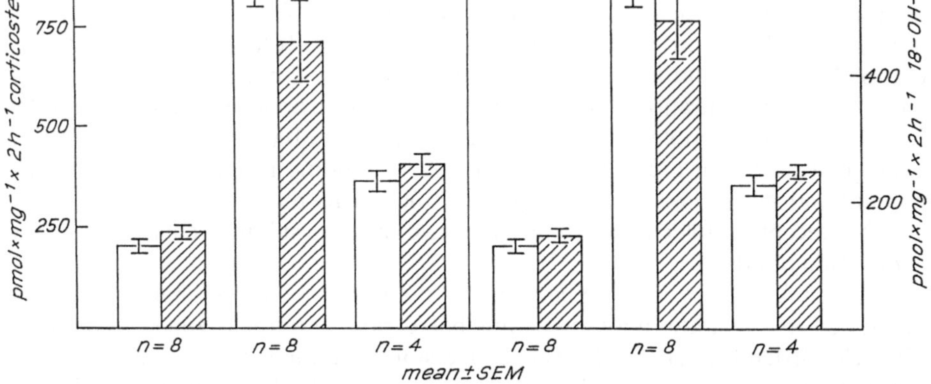

Fig. 19. Effect of an atrial natriuretic peptide ($r\text{-}ANF_{1\text{-}23}$ = atriopeptin II) on corticosterone and 18-hydroxy-11-deoxycorticosterone *(18-OH-DOC)* outputs by rat decapsulated adrenal tissue incubated in the absence or presence of ACTH or of the precursor pregnenolone. Methodology according to J. Müller (1970b) and to Meuli and Müller (1983). None of the effects of ANF were statistically significant

sion suppressed plasma aldosterone concentration (−31%) while raising plasma renin concentration (+39%; Volpe et al. 1985). In sodium-depleted rats with 1-kidney, 1-clip renovascular hypertension, ANF administration lowered both plasma aldosterone concentration and renin activity to a similar extent (−37%). In anesthetized normal dogs, a 1-h infusion of auriculin A significantly lowered plasma aldosterone concentration and renin activity as well as the renin secretion rate without affecting plasma cortisol concentration (Maack et al. 1984). According to Freeman et al. (1985), a suppressive effect of a 1-h synthetic rat atrial natriuretic peptide (28 AA) infusion on plasma aldosterone concentration and renin activity was observed in dogs with a constriction of the thoracic vena cava and ascites but not in normal conscious dogs. In patients with essential hypertension, a bolus injection of α-human natriuretic peptide induced a brief rise in plasma norepinephrine concentration and a more sustained drop in plasma epinephrine and aldosterone levels, while not affecting plasma renin activity or cortisol (Richards et al. 1985).

Radioiodinated ANF bound with high affinity to bovine zona glomerulosa membranes and was displaced by unlabeled ANF but not by angiotensin II or ACTH (De Léan et al. 1984a, b). On the other hand, ANF did not displace radioiodinated angiotensin II from its receptor sites on bovine adrenal glomerulosa cells (Goodfriend et al. 1984). De Léan et al. (1985a) tested various fragments of ANF as inhibitors of aldosterone production and as competitors of labeled ANF (arg^{101}-tyr^{126}) binding to zona glomerulosa cell membranes. Shortening or lengthening the N-terminal segment of ANF did not affect its biological activity and only marginally altered its receptor affinity. Removal of the C-terminal to cys^{121} or expansion up to arg^{128} resulted in a 1000-fold decrease in receptor affinity and activity. The potency of ANF (106-125), i.e., a linear fragment unable to form an intramolecular S-S bridge, was also 2-3 orders of magnitude lower than that of ANF (103-125) in its binding affinity to zona glomerulosa receptors and in its suppression of aldosterone production (Schiller et al. 1985). According to this observation, disulfide linkage is not an absolute requirement for biological activity of atrial natriuretic peptides, although it apparently stabilizes their bioactive conformation. An ANF receptor molecule was recently solubilized and partly purified from bovine adrenocortical tissue (S. Hirose et al. 1985). It had a dissociation constant of 1.8 nmol/l and a maximal binding capacity of 2 pmol per mg of protein. According to gel filtration chromatography and electrophoresis under nonreducing conditions, its molecular weight was between 130000 and 140000. During electrophoresis under reducing conditions, ANF binding activity migrated with a protein band with a molecular weight of 70000. This observation indicated that the ANF receptor molecule may consist of two or more subunits covalently linked by disulfide bonds.

Auriculin A did not alter the conversion to aldosterone of unlabeled progesterone added in substrate amounts to incubated bovine adrenal zona glomerulosa cells (Goodfriend et al. 1984). The conversion of added 25-hydroxycholesterol acetate to aldosterone was slightly lowered by the ANF, but only when this conversion had been stimulated by the simultaneous addition of angiotensin II. Auriculin A markedly suppressed the output of pregnenolone by bovine adrenal glomerulosa cells incubated in the presence of trilostane (see p. 80) with or without angiotensin II. Similarly, atriopeptin III inhibited the pregnenolone production by rat adrenal glomerulosa cells incubated with angiotensin II and cyanoketone (see p. 79; W. B. Campbell et al. 1985). Atriopeptin II did not inhibit the production of aldosterone, 18-hydroxycorticosterone, or corticosterone by rat capsular adrenal tissue incubated with substrate amounts of pregnenolone (Fig. 18). According to all this evidence, atrial natriuretic peptides do not block aldosterone biosynthesis from pregnenolone, but act mainly at an early biosynthetic step (see p. 90) preceding cholesterol side-chain cleavage. However, W. B. Campbell et al. (1985) reported that atriopeptin III completely suppressed the angiotensin II-stimulated conversion of tritiated corticosterone to aldosterone by rat adrenal cells incubated with cyanoketone; it did not affect this conversion in the absence of stimulators.

In spite of their inhibitory effects on zona glomerulosa cell steroidogenesis, atrial natriuretic peptides may operate as growth factors for the same cells. Thus, synthetic rat ANF (1-28) stimulated thymidine incorporation into DNA of bovine adrenal zona glomerulosa cells in a primary culture in a dose-dependent manner (Horiba et al. 1985).

4.7.2 Adrenal-Medullary Inhibitory Factor

A yet unknown factor, isolated from acid extracts of bovine adrenal medulla, markedly (−80%) suppressed PGE_1-stimulated aldosterone production by cultured bovine zona glomerulosa cells (De Léan et al. 1985 b). It did not significantly affect aldosterone production in the absence of added stimulators. Its log-dose/response curve closely paralleled the one obtained with synthetic atrial natriuretic factor (8-33). In a radioreceptor assay, the purified factor also displaced radioiodinated ANF from its binding sites. The factor was retained on cation exchange gels and could thus be separated from enkephalins, from catecholamines, and from 99% of the adrenal-medullary proteins. According to chromatography on Sephadex G-50, it had an approximate molecular weight of 5000.

4.7.3 Somatostatin

Somatostatin markedly inhibited angiotensin II-stimulated aldosterone production by dispersed rat adrenal zona glomerulosa cells, while not affecting aldosterone production, under stimulation by a high potassium concentration, ACTH, or 8-bromo-cyclic AMP (Aguilera et al. 1981 a). Whereas low concentrations of somatostatin had no effect on basal aldosterone output, higher concentrations (10^{-8} and 10^{-7} mol/l) moderately stimulated aldosterone production in the absence of other stimulators, while still markedly suppressing angiotensin II-stimulated aldosterone output. This agonistic effect was probably due to a weak interaction of the peptide with angiotensin receptors, because the stimulatory effect was inhibited by sar^1,ala^8-angiotensin II and because somatostatin at high concentrations displaced radioiodinated angiotensin II from its adrenal receptor sites. However, somatostatin exerted its inhibitory action on angiotensin II-stimulated aldosterone production at much lower concentrations, most likely through its own receptor. Specific high-affinity receptor sites for somatostatin were found in rat capsular adrenal membrane particles (Aguilera et al. 1981 a, 1982; Srikant and Patel 1985). The receptor-binding activities of several somatostatin analogues were proportional to their potencies as inhibitors of angiotensin-induced aldosterone production. Angiotensin II did not displace labeled somatostatin from its receptor sites. When tested in a system of superfused rat adrenal glomerulosa cells, somatostatin also blocked angiotensin II-stimulated aldosterone output and had no effect on serotonin-, potassium-, or ACTH-stimulated aldosterone outputs (Boscaro et al. 1982). In contrast, even a high somatostatin concentration (10^{-5} mol/l) did not impair the stimulation of aldosterone and corticosterone outputs induced by sar^1,val^5-angiotensin II (10^{-7} mol/l) in superfused frog interrenal tissue (Delarue et al. 1984).

Aguilera et al. (1981 a) measured a very high somatostatin-like immunoreactivity in the rat adrenal cortex: 1800 pg/mg in capsular and 146 ng/mg in decapsulated glands. However, this was not confirmed by Srikant and Patel (1985), who found only very low concentrations of somatostatin-like immunoreactivity in fresh rat adrenocortical tissue.

4.8 Specific Inhibitors of Steroidogenic Enzymes

In this section only a few inhibitors of steroidogenic enzymes that interfere with aldosterone biosynthesis will be discussed. For a review of other inhibitors of corticosteroid biosynthesis see Kahnt and Neher (1966b), Gaunt et al. (1965, 1968), and Gower (1974).

4.8.1 Metyrapone[11]

According to Kahnt and Neher (1962), metyrapone not only inhibits 11β-hydroxylation but is even more potent in suppressing the 18-hydroxylation step in aldosterone biosynthesis[12]. Aldosterone biosynthesis from endogenous precursors by a bovine adrenal homogenate was inhibited by a concentration of metyrapone 50% lower than the one needed to obtain a comparable suppression in the production of corticosterone, cortisol, and 18-hydroxycorticosterone. Stachenko and Giroud (1964) found that a low dose of metyrapone that was not sufficient to impair cortisol and corticosterone biosynthesis in bovine adrenal outer slices almost completely inhibited aldosterone production both from endogenous precursors and from added precursor steroids (progesterone, 11β-hydroxyprogesterone, deoxycorticosterone, corticosterone) and blocked the conversion of [^{14}C]-labeled corticosterone to aldosterone. The available evidence of an effect of metyrapone on the conversion of 18-hydroxycorticosterone to aldosterone is controversial. Sandor et al. (1972) observed that in duck adrenal mitrochondria metyrapone inhibited the conversion of corticosterone to aldosterone less extensively than the conversion of corticosterone to 18-hydroxycorticosterone. In another duck adrenal mitochondrial preparation, metyrapone did not affect the conversion of labeled 18-hydro-

[11] Metyrapone = Metopirone = Su-4885 = 2-methyl-1,2-bis-(3'pyridyl)-1-propanone

[12] Metyrapone also has a weaker inhibitory effect upon the conversion of cholesterol to pregnenolone (Cheng et al. 1974; Carballeira et al. 1976).

xycorticosterone to aldosterone when added in concentrations which suppressed the conversion of corticosterone to 18-hydroxycorticosterone by 70%–99% (Aupetit et al. 1979). On the other hand, metyrapone markedly inhibited the conversion of 18-hydroxycorticosterone to aldosterone by bovine zona glomerulosa mitochondria (I. Kojima et al. 1982, 1984a) and by a reconstituted cytochrome $P\text{-}450_{11\beta}$ system (Wada et al. 1985).

A suppression of aldosterone secretion in rats in vivo was observed after the intravenous injection of 0.1–0.3 mg/kg of metyrapone, whereas the doses needed to inhibit 18-OH-DOC and corticosterone secretion were 1 mg/kg and 3 mg/kg respectively (Kahnt and Neher 1962). The selective suppression of aldosterone secretion in rats by small amounts of metyrapone that had no effect upon corticosterone and 18-OH-DOC secretions was confirmed by De Nicola and Dahl (1971), although these investigators used considerably higher doses of the inhibitor. The aldosterone-suppressive activity of metyrapone has also been demonstrated in studies carried out in normal human subjects as well as in patients with primary or secondary aldosteronism and Cushing's syndrome (Coppage et al. 1958; Tucci et al. 1967b). Aldosterone secretion was markedly inhibited by metyrapone in all these subjects, even when their sodium intake was restricted. Sodium diuresis and potassium retention were observed when metyrapone was administered together with prednisone or dexamethasone in order to prevent an increase in deoxycorticosterone secretion. In a group of normal human subjects, a 5-h infusion of metyrapone elicited marked and sustained drops in the plasma concentrations of aldosterone and 18-hydroxycorticosterone (Schöneshöfer and Claus 1985).

According to Traikov et al. (1969), *reduced metyrapone* (2-methyl-1,2-bis3′-pyridyl)-1-propanol = Su-5236) is an even more selective inhibitor of aldosterone biosynthesis than metyrapone. When added at a low concentration to the medium in which rat adrenal glands were incubated, it strikingly inhibited the conversion of exogenous deoxycorticosterone to aldosterone and 18-hydroxycorticosterone, did not affect the conversion to corticosterone, and even increased the conversion to 18-OH-DOC. However, metyrapone and reduced metyrapone had very similar effects when injected into rats in vivo (De Nicola and Dahl 1971). Following the administration of small amounts of the two compounds (12 mg/kg) the secretion rate of aldosterone decreased, whereas the secretion rates of corticosterone and 18-OH-DOC remained unchanged. A higher dose of the inhibitors (50 mg/kg) suppressed the secretion of all three steroids.

4.8.2 Su-8000, Su-9055, and Su-10603[13]

In a bovine adrenal homogenate these three substances inhibited the production of aldosterone, 18-hydroxycorticosterone, and cortisol from endogenous precursors but enhanced the output of corticosterone (Kahnt and Neher 1962). Whereas Su-8000 and Su-10603 inhibited 17α-hydroxylation most actively, Su-9055 exerted a more extensive suppressive effect upon aldosterone biosynthesis. This substance also blocked the conversion of added pregnenolone and progesterone to aldoster-

[13] Su-8000 = 3-(6-chloro-3-methyl-2-indenyl)pyridine; Su-9055 = 3-(1,2,3,4-tetrahydro-1-oxo-2-naphthyl)-pyridine; Su-10603 = 7-chloro-3,4-dihydro-2(3-pyridyl)-1(2H)-naphthalenone

one and the incorporation of ^{14}C-labeled corticosterone into aldosterone. Addition of Su-8000 to rat adrenal glands resulted in a 50% decrease in aldosterone production, a small increase in corticosterone output, and a marked increase in deoxycorticosterone output (J. Müller 1971). In the presence of Su-8000, a high potassium concentration stimulated mainly deoxycorticosterone output rather than aldosterone output. When separate zones of adrenals from sodium-deficient rats were incubated with ACTH, the further addition of Su-8000 resulted in a markedly diminished aldosterone production (K. Baumann and Müller 1972b; Fig. 20). The output of deoxycorticosterone was increased in both portions, but more extensively in capsular than in decapsulated adrenals. The corticosterone output was unchanged in both portions of the adrenal cortex. Su-8000 and Su-10603 substantially inhibited the conversions of corticosterone to aldosterone and 18-hydroxycorticosterone and the conversion of 18-hydroxycorticosterone to aldosterone by bovine capsular adrenal mitochondria (I. Kojima et al. 1982, 1984a).

In normal human subjects on restricted sodium intake, in patients with cirrhosis of the liver with ascites, and in cases of primary aldosteronism, the administration of Su-9055 (4.8 g/day) resulted in a marked suppression of aldosterone secretion and a striking increase in corticosterone secretion - even when ACTH secretion was suppressed by dexamethasone - and in natriuresis (Bledsoe et al. 1964).

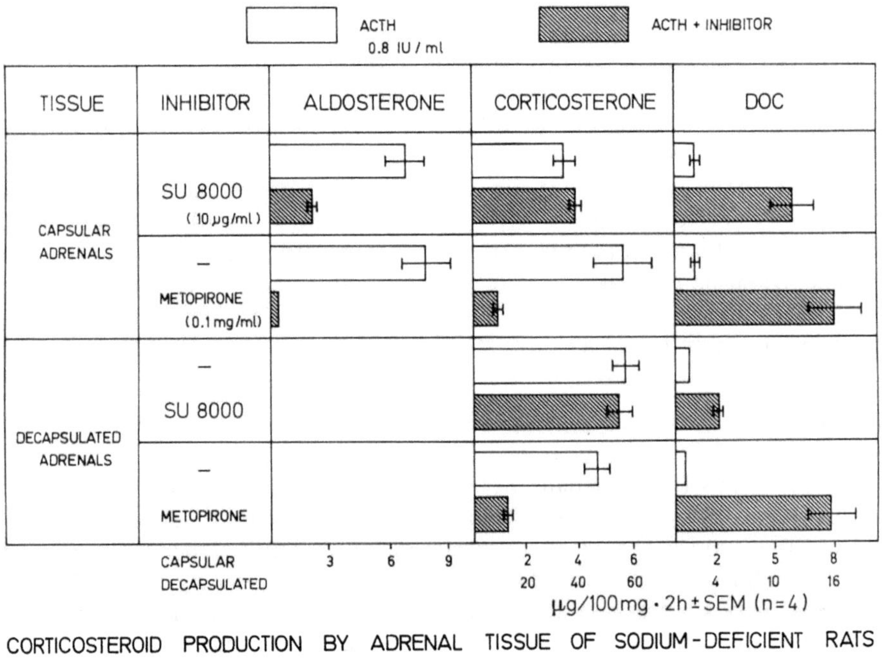

Fig. 20. Effects of Su-8000 and of metyrapone *(metopirone)* on the production of aldosterone, corticosterone, and deoxycorticosterone *(DOC)* by separate zones of rat adrenals incubated in vitro in the presence of ACTH. Rats had been kept on a sodium-deficient diet for 2 weeks. (Data from K. Baumann and Müller 1972b)

4.8.3 Aminoglutethimide

Aminoglutethimide equally inhibited the biosynthesis of cortisol, corticosterone, and aldosterone by bovine adrenal homogenates (Kahnt and Neher 1966b). This effect is due mainly to competitive inhibition of the conversion of cholesterol to pregnenolone in adrenocortical tissue, according to studies carried out in vivo and in vitro by Dexter et al. (1967). Aminoglutethimide directly interferes with adrenal enzymatic cholesterol side-chain fission, at least partially by inhibiting 20α-hydroxylation of cholesterol (Kahnt and Neher 1966b; M.P.Cohen 1968). However, aminoglutethimide also directly suppressed the conversion of corticosterone to 18-hydroxycorticosterone by sheep and human adrenal homogenates (Touitou and Legrand 1971; Touitou et al. 1975). According to indirect evidence obtained in a patient with Cushing's syndrome, aminoglutethimide may also inhibit 11β-hydroxylation (Faglia et al. 1971). In rat capsular adrenal cells, aminoglutethimide inhibited the conversion of exogenous corticosterone to aldosterone by 66%, but it had no effect upon the production of aldosterone from added deoxycorticosterone (Aguilera and Catt 1979). In dog adrenal zona glomerulosa cells, the inhibitor decreased aldosterone production from both corticosterone and deoxycorticosterone to an equal extent. This may point to the existence of an alternative pathway leading from deoxycorticosterone to aldosterone and bypassing corticosterone in the rat (see also p. 12).

In normal human subjects, the administration of aminoglutethimide resulted only in small decreases in the cortisol secretion rate and in plasma cortisol levels (Fishman et al. 1967). Consistently observed increases in plasma ACTH levels may have partially antagonized the cortisol-suppressing effects of aminoglutethimide; blocking of cortisol secretion was more clearly manifest in a patient with autonomous cortisol secretion by an adrenocortical adenoma. In all normal subjects, as well as in patients with Cushing's syndrome or primary aldosteronism, aminoglutethimide induced a striking drop in aldosterone secretion rates; this was accompanied by sodium diuresis and potassium retention.

4.8.4 Cyanosteroids

Cyanoketone[14] irreversibly inhibits the enzymatic conversion of C_{19}- and C_{21}-Δ^5-13β-hydroxysteroids to the corresponding Δ^4-3-ketosteroids in the adrenal cortex and in the gonads without having any effects on adrenal 21- and 11β-hydroxylase activities (Goldman et al. 1965; Goldman and Winter 1967; Neville and Engel 1968). At higher concentrations (above 10^{-4} mol/l), the substance also inhibited the cholesterol side-chain cleavage (Farese 1970). When added at a low concentration (10^{-6} mol/l) to rat or dog capsular adrenal cells, cyanoketone blocked the further conversion of endogenous pregnenolone and thus totally suppressed the stimulation of aldosterone production by angiotensin II, ACTH, or a high potassium concentration, but it did not alter the conversions of added progesterone, deoxycorticosterone, or corticosterone to aldosterone (Aguilera and Catt 1979).

---

[14] Cyanoketone = 2α-cyano-17β-hydroxy-4,4,17α-trimethyl-androst-5-en-3-one = cyanotrimethylandrostenolone = WIN 19,578

Trilostane,[15] a structurally related inhibitor of the conversion of pregnenolone to progesterone, decreased the aldosterone output by bovine adrenal zona glomerulosa cells under basal conditions as well as under stimulation by angiotensin II or potassium ions (McKenna et al. 1978b, c). It blocked the conversion to aldosterone of added pregnenolone but not of added progesterone or deoxycorticosterone (McKenna et al. 1978b). However, it significantly decreased the conversion of exogenous corticosterone to aldosterone by rat and dog capsular adrenal cells (Aguilera and Catt 1979). Treatment with trilostane resulted in marked decreases in the aldosterone secretion rate in five patients with primary aldosteronism (Hollifield et al. 1975).

4.8.5 Spirolactones

Spirolactones[16] are not only potent mineralocorticoid antagonists, competitively displacing aldosterone from its cytoplasmatic receptor sites in target cells; they can also directly interfere with the biosynthesis of aldosterone and other corticosteroids in the adrenal cortex. This side effect could - at least theoretically - support the main action of these drugs.

An inhibitory action of a spirolactone upon aldosterone biosynthesis was first observed by Erbler (1972), who added spironolactone at concentrations of between 6×10^{-6} and 10^{-4} mol/l to quartered adrenal glands of sodium-deficient rats. Spironolactone significantly suppressed aldosterone production under baseline conditions as well as under stimulation by dibutyryl cyclic AMP, ACTH, angiotensin II, or increasing ambient potassium concentrations in a dose-dependent manner, whereas the corticosterone output (as measured by a fluorimetric assay) remained unchanged. However, spironolactone also had a weak but significant suppressive effect upon the corticosterone production when the steroid was measured by a more specific method (Erbler 1973a). At the same concentrations as spironolactone, canrenone inhibited the production of corticosterone, 18-OH-DOC, 18-hydroxycorticosterone, and aldosterone by quartered rat adrenals (Erbler 1973a). During its interaction with the adrenal tissue, this spirolactone was converted to two different metabolites. According to a mass spectrometrical analysis, the metabolites had hydroxyl groups in two different positions, possibly at C-11 and C-18. The production of the metabolites was reduced when corticosteroid production was stimulated by dibutyryl cyclic AMP. These findings suggested that the inhibition of steroid hormone production was due to a competition between the spirolactone and intermediate products of corticosteroid biosynthesis for mitochondrial hydroxylases. The selective inhibition of aldosterone biosynthesis by an 11β-hydroxylated spirolactone, 11β-OH-SC-5233, further supported this

[15] Trilostane = 2α-cyano-17β-hydroxy-4α,5-epoxy-5α-androstan-3-one = WIN 24540

[16] Spironolactone = Aldactone = 3'-(3-oxo-7α-acetylthio-17β-hydroxy-4-androsten-17α-yl)propionic acid-γ-lactone = SC-9420.
Canrenone = Aldadiene = 3'-(3-oxo-17β-hydroxy-4,6-androstadien-17α-yl)propionic acid-γ-lactone = SC-9376. SC-5233 = 3'-(3-oxo-17β-hydroxy-4-androsten-17α-yl)propionic acid-γ-lactone.
Canrenoate-K = Soldactone = 3'-(3-oxo)-17β-hydroxy-4,6-androstadien-17α-yl)propionic acid potassium salt = SC-14266.
Prorenoate-K = 3'-(3-oxo-17β-hydroxy-6,7β-methylene-4-androsten-17α-yl)propionic acid potassium salt

hypothesis (Erbler 1973b). The parent compound SC-5233 inhibited the production of both aldosterone and corticosterone from endogenous precursors and decreased the conversions of ^{14}C-labeled progesterone to all aldosterone, 18-hydroxycorticosterone, and corticosterone. In contrast, at concentrations up to 10^{-4} mol/l, 11β-OH-SC-5233 suppressed only the production of aldosterone and inhibited only the conversions to aldosterone and 18-hydroxycorticosterone. Spironolactone markedly inhibited the aldosterone production from endogenous precursors, as well as from added progesterone, deoxycorticosterone, and corticosterone, by rat adrenal zona glomerulosa cells (Aguilera and Catt 1979). Spironolactone substantially inhibited the conversions of corticosterone to 18-hydroxycorticosterone and aldosterone as well as the conversion of 18-hydroxycorticosterone to aldosterone by sonicated bovine zona glomerulosa mitochondria (I. Kojima et al. 1982, 1984a). Several spirolactones - among them spironolactone, canrenoate K, and prorenoate K - blocked aldosterone output by superfused frog adrenal tissue; however, not all spirolactones with antimineralocorticoid activity inhibited aldosterone production in this preparation (Delarue et al. 1979b; Netchitailo et al. 1982, 1983, 1985; Perroteau et al. 1984b). In mitochondrial preparations from separate zones of bovine adrenocortical tissue, as well as from adrenal adenomas or hyperplastic adrenocortical tissue removed from patients with primary aldosteronism, spironolactone, canrenone, and canrenoate-K bound to cytochrome P-450 (producing type-I difference spectra) and underwent hydroxylation (Cheng et al. 1976). All three agents inhibited 11β-hydroxylation in bovine mitochondria, and - less efficiently - in human mitochondria. In bovine and in human mitochondria they inhibited the 18-hydroxylation of corticosterone at low concentrations that had no effect upon 11β-hydroxylation. The order of potency of inhibition was found to correlate with the order of affinity of these agents for cytochrome P-450. However, the inhibition of corticosteroid biosynthesis by spirolactones is not limited to mitochondrial hydroxylations. Canrenone and spironolactone also directly inhibited 21-hydroxylation of 17α-hydroxyprogesterone by a microsomal preparation from guinea pig adrenals (Greiner et al. 1978).

4.9 Inhibitors of RNA or Protein Synthesis

4.9.1 Actinomycin D

When added to the medium in which quartered rat adrenals were incubated, actinomycin D (10 µg/ml) inhibited both the incorporation of ^{14}C-labeled leucine into adrenal protein and the production of aldosterone by 38% (Burrow et al. 1966). At an equal concentration, the inhibitor decreased the basal production of aldosterone by rat adrenal tissue by an average of 37% and the basal output of corticosterone by 10% (J. Müller 1971; Table 5). However, actinomycin D did not impair the stimulation of aldosterone production by a high potassium concentration or the stimulation of the production of both steroids by ACTH. It had previously been shown by Ferguson and Morita (1964) and by Halkerston et al. (1965) that actinomycin D did not block the steroidogenic response to ACTH by rat adrenals in vitro, whereas it completely suppressed the stimulatory effect of ACTH on corti-

Table 5. Effect of actinomycin D (10 μg/ml) on aldosterone and corticosterone production by rat adrenal quarters in vitro. Actinomycin D was added to the preincubation medium and the incubation medium. Mean values of two flasks ± range. (From J. Müller 1971)

Addition to incubation medium[a]	Steroid production (μg/100 mg/2 h)			
	Aldosterone		Corticosterone	
	Control	Actinomycin D	Control	Actinomycin D
–	1.61 ± 0.05	0.96 ± 0.04	7.25 ± 0.39	6.26 ± 0.19
KCl[b]	3.84 ± 0.13	2.94 ± 0.22	10.86 ± 0.23	9.43 ± 0.20
–	1.72 ± 0.19	1.13 ± 0.14	8.89 ± 0.45	8.07 ± 0.39
ACTH[c]	3.74 ± 0.35	3.91 ± 0.60	57.68 ± 0.78	53.95 ± 1.86

[a] 3.6 mEq K^+/l
[b] 8.5 mEq K^+/l
[c] 5 IU/flask

costeroid production by cow adrenal slices (Farese 1966). When added to duck adrenal mitochondria, actinomycin D did not influence the conversion of labeled corticosterone to aldosterone and 18-hydroxycorticosterone (Sandor et al. 1972). In a system of superfused rat mixed-zone adrenocortical cells, actinomycin D (4×10^{-6} mol/l) did not affect ACTH-induced corticosterone or aldosterone outputs, but it completely inhibited the potentiating effect of the human γ-MSH precursor *N*-POMC (1-76) on ACTH-stimulated steroidogenesis (Al-Dujaili et al. 1982a). A similar inhibitory effect was produced by two other inhibitors of DNA transcription, mithramycin and cordycepin. Increasing doses of *N*-POMC (1-76) elicited a progressive increase in total [^{3}H]-uridine incorporation during static incubation of adrenal cells; this effect was inhibited by actinomycin D.

4.9.2 Cycloheximide

The addition of cycloheximide (10^{-3} mol/l) to rat adrenal incubates resulted in a 84% inhibition of amino acid incorporation into adrenal protein and a 63% decrease in aldosterone production (Burrow et al. 1966). We found that cycloheximide induced a 40%-50% decrease in the basal outputs of aldosterone, corticosterone, and deoxycorticosterone by quartered rat adrenals and completely inhibited the steroidogenic response to serotonin and potassium ions (J. Müller 1971; Table 6). When pregnenolone was added in substrate amounts to the incubation medium, similar increments in the output of all three corticosteroids were observed in the absence and in the presence of cycloheximide. Cycloheximide significantly suppressed the basal outputs of aldosterone and corticosterone by rat capsular adrenal tissue and blocked the respective steroidogenic responses to ACTH and to a high potassium concentration (G. H. Williams et al. 1972a). In the same preparation it significantly decreased the conversion of labeled corticosterone (tracer amounts) to aldosterone under baseline conditions of incubation as well as during stimulation with ACTH or potassium ions. However, it did not significantly affect the conversion to aldosterone of substrate amounts of corticosterone. Addition of cycloheximide did not influence the conversion of tritiated corticosterone to 18-hydroxycorticosterone and to aldosterone by duck adrenal

Table 6. Effect of cycloheximide (10^{-3} *M*) on corticosteroid production by rat adrenal quarters in vitro. Cycloheximide was added only to the medium used for the final incubation. Mean values of two flaske ± range. (From J. Müller 1971)

Addition to incubation medium[a]	Steroid production (µg/100 mg/2 h)					
	Aldosterone		Corticosterone		Deoxycorticosterone	
	Control	Cycloheximide	Control	Cycloheximide	Control	Cycloheximide
–	2.19 ± 0.09	1.29 ± 0.07	8.87 ± 0.03	4.65 ± 0.38	0.37 ± 0.03	0.25 ± 0.01
Serotonin[b]	5.12 ± 0.18	1.39 ± 0.06	8.73 ± 0.31	5.07 ± 0.43	0.48 ± 0.03	0.26 ± 0.01
–	2.76 ± 0.02	1.48 ± 0.07	7.17 ± 0.50	4.02 ± 0.10		
KCl[c]	5.36 ± 0.37	1.62 ± 0.04	7.18 ± 0.06	3.72 ± 0.52		
–	4.14 ± 0.22	2.28 ± 0.30	5.02 ± 0.29	3.12 ± 0.17	0.39 ± 0.20	0.18 ± 0.02
Pregnenolone[d]	8.28 ± 1.31	5.78 ± 0.19	12.78 ± 2.50	10.13 ± 1.83	3.47 ± 0.13	3.31 ± 0.78

[a] 3.6 mEq K^+/l
[b] 5-Hydroxytryptamine creatinine sulfate, 1.7×10^{-5} *M*
[c] 8.5 mEq K^+/l
[d] 300 µg/flask

(Sandor et al. 1972) and rat zona glomerulosa mitochondria (Kramer et al. 1980). These results indicated that cycloheximide does not directly interfere with late steps of aldosterone biosynthesis. Most likely, the effects of the inhibitor on the conversion of trace amounts of corticosterone to aldosterone observed by G. H. Williams et al. (1972a) were secondary to a decrease in endogenous corticosterone output (see p. 92). According to Elliott and Goodfriend (1984), a low concentration (7×10^{-6} mol/l) of cycloheximide completely suppressed the stimulation of aldosterone production in bovine adrenal glomerulosa cells by angiotensin II (10^{-7} mol/l), but it did not interfere with the receptor binding of angiotensin II and did not affect the increase in labeled phosphate incorporation into phosphatidylinositol or the decrease of the influx of calcium ions induced by angiotensin II. Mitochondria from angiotensin II-treated glomerulosa cells produced more pregnenolone than mitochondria from untreated cells. When added to the cells together with angiotensin II, cycloheximide blocked this effect of the peptide. On the other hand, cycloheximide had no effect on pregnenolone production when added directly to mitochondria.

Administration of cycloheximide to rats in vivo blocked the increase in adrenal corticosterone production in response to ACTH at the level of the conversion of cholesterol to pregnenolone (W. W. Davis and Garren 1968). When cycloheximide was given to angiotensin II-treated hypophysectomized rats 25 min before they received a final injection of angiotensin and 30 min before they were killed, it completely inhibited the stimulatory action of the peptide on the conversion of corticosterone to 18-hydroxycorticosterone and aldosterone but had no effect on the promotion of zona glomerulosa cholesterol side-chain cleavage (Kramer et al. 1980).

4.9.3 Puromycin and Chloramphenicol

Puromycin (7.3×10^{-4} mol/l) significantly decreased the in vitro production of aldosterone, corticosterone, and deoxycorticosterone, but not of 18-OH-DOC, by rat adrenal tissue (Kittinger 1964). It also completely inhibited the stimulation of corticosterone, 18-OH-DOC, and deoxycorticosterone outputs by ACTH. Aldosterone production was not stimulated by ACTH in these experiments, even in the absence of puromycin. In bovine adrenal outer slices, puromycin (10^{-3} mol/l) significantly suppressed aldosterone and corticosterone outputs under basal conditions of incubation and completely blocked the respective steroidogenic effects of angiotensin II and a high potassium concentration (Saruta et al. 1972). According to Burrow et al. (1966), puromycin (10^{-3} mol/l) inhibited the basal aldosterone production from endogenous precursors by 88% and the corticosterone output by 26%. It also blocked the conversion of added progesterone to aldosterone and 18-hydroxycorticosterone, whereas it slightly enhanced the conversion of progesterone to corticosterone. These findings indicated that puromycin interfered particularly with the 18-hydroxylation step of aldosterone biosynthesis. However, the addition of large amounts of corticosterone (200 μg/ml) completely counteracted the inhibition of aldosterone production by puromycin. Moreover, puromycin did not impair a marked increase in aldosterone output in response to added NADP and glucose-6-phosphate. Puromycin (5×10^{-4} mol/l) and chloramphenicol

(10^{-4} mol/l) inhibited the conversions of corticosterone to aldosterone and of deoxycorticosterone to corticosterone by a rat adrenal mitochondrial preparation incubated with a Krebs-cycle intermediate, according to Marusic and Mulrow (1972). In the presence of NADPH, chloramphenicol was no longer inhibitory, whereas puromycin still inhibited the conversion of corticosterone to aldosterone. The authors concluded that the inhibition of the conversion of corticosterone to aldosterone by puromycin was due to a direct effect on the enzymes involved and not to the inhibition of mitochondrial protein synthesis. Chloramphenicol did not affect the conversion of corticosterone to 18-hydroxycorticosterone and aldosterone by duck adrenal mitochondria incubated in the presence of an NADPH-generating system (Sandor et al. 1972).

4.10 Steroid Hormones

4.10.1 Corticosteroids and Gestagens

The conversion of corticosterone to aldosterone and 18-hydroxycorticosterone by sheep adrenal mitochondria was inhibited by small amounts of *18-hydroxycorticosterone* but not by aldosterone or by 18-OH-DOC (Raman et al. 1966). The inhibitory action of 18-hydroxycorticosterone and the lack of effect of aldosterone were also observed in a duck adrenal mitochondrial preparation (Sandor et al. 1972). Sturcz et al. (1966) found that aldosterone production from endogenous precursors by rat adrenal tissue was not influenced by changes in the volume of the incubation medium, and they concluded that aldosterone biosynthesis was independent of the ambient extracellular concentration of aldosterone or other secretory products. The conversion of tritiated corticosterone (50 nmol/ml) by rat capsular adrenal tissue was not affected by the addition of aldosterone or 18-hydroxycorticosterone (9 nmol/ml) to the incubation medium (K. Baumann and Müller 1972a).

A large number of natural and synthetic C_{21}-steroids were found to decrease by 50% or more the conversion of trititated cholesterol to aldosterone by bovine adrenal homogenates when added at high concentrations (150 μg/ml; Kahnt and Neher 1966a). These steroids had varying effects on the biosynthesis of cortisol and corticosterone. *Cortisol* and *cortisone* were the most active corticosteroids in suppressing aldosterone biosynthesis; they were active at a concentration of 30 μg/ml. Cortisol (200 μg/ml) inhibited the production of aldosterone in quartered rat adrenals by 45%, according to Burrow et al. (1966). Cortisol (100 μg/ml) also blocked the conversion of labeled corticosterone to aldosterone by a rat adrenal mitochondrial preparation during incubation with succinate or with Ca^{2+} and NADPH (Burrow 1969). Smaller concentrations (2.5 μg/ml) of cortisol or 11-deoxycortisol did not influence aldosterone production by quartered rat adrenals (Weinstein and Kliman 1972). Dexamethasone (42 μg/ml) did not alter aldosterone and corticosterone outputs by incubated quartered rat adrenals, either under basal conditions or under stimulation by serotonin (J. Müller 1970a). The addition of *18-OH-DOC, corticosterone,* or cortisol (55 nmol/ml) to monolayers of rat adrenal zona glomerulosa cells, cultured in a medium with a potassium concentration of 18 mmol/l, resulted within days in markedly decreased conversions of tritiated deoxycorticosterone to aldosterone and 18-hydroxycorticosterone with correspondingly increased conversions to corticosterone and 18-OH-DOC, i.e., in a change from a "glomerulosa-like" pattern to a "fasciculata-like" pattern of deoxycorticosterone metabolism (Hornsby and O'Hare 1977; see also p. 189). The addition of aldosterone to the culture medium was ineffective.

When added to quartered rat adrenals, *17α-hydroxyprogesterone* or *17α-hydroxypregnenolone* (25 μg/ml) significantly suppressed the production of aldosterone but had no effect upon corticosterone and deoxycorticosterone output (Weinstein and Kliman 1972). However, during incubation of separate zones of rat adrenal glands, 17α-hydroxyprogesterone inhibited not only aldosterone biosynthesis but also the capsular and decapsulated adrenal outputs of corticosterone and deoxycorticosterone. The addition of 17α-hydroxyprogesterone or *medroxyprogesterone acetate* (50 μg/ml) to the medium in which bovine adrenal slices were incubated with a mixture of ^{3}H-labeled pregnenolone and ^{14}C-labeled progesterone resulted in a striking shift of the biosynthetic pathway leading from pregnenolone to corticosterone and aldosterone (Patwardhan and Lanthier 1971). The $^{3}H/^{14}C$ ratios of aldosterone and corticosterone formed in the presence of these two compounds were markedly greater than the final $^{3}H/^{14}C$ ratio of progesterone.

Of 70 different naturally occurring steroids tested by Campanile and Goodfriend (1981), 30 inhibited the binding of labeled angiotensin II and III to bovine adrenal glomerulosa cells, but the concentrations required for inhibition were relatively high. The most potent inhibitors were 3α,5β-tetrahydroaldosterone and -tetrahydrodeoxycorticosterone. According to kinetic analysis, inhibition was mostly competitive. Tetrahydrodeoxycorticosterone inhibited angiotensin II-stimulated and ACTH-stimulated aldosterone production by zona glomerulosa cells at concentrations that inhibited angiotensin binding.

4.10.2 Androgens and Estrogens

Small amounts of natural androgens and estrogens did not influence the conversion of corticosterone to aldosterone and 18-hydroxycorticosterone by sheep adrenal mitochondria (Raman et al. 1966). The biosynthesis of aldosterone from tritiated cholesterol by bovine adrenal homogenates was blocked by several natural and synthetic estrogens and androgens (Kahnt and Neher 1966a). The most active inhibitors among these compounds were *19-nortestosterone* and *19-hydroxytestosterone,* which decreased aldosterone production at concentrations of 2 μg/ml and 6 μg/ml respectively. A low concentration of *17α-methyltestosterone* (17 nmol/ml) selectively inhibited the conversion of labeled progesterone and deoxycorticosterone to aldosterone and 18-hydroxycorticosterone by rat adrenal quarters (Rembiesa et al. 1967). At a higher concentration (67 nmol/ml) this steroid also inhibited the incorporation of ^{14}C-progesterone into corticosterone and 18-OH-DOC. However, it did not significantly decrease the conversion of corticosterone to aldosterone at either dose level. *Testosterone* (347 nmol/ml) inhibited the conversion of labeled corticosterone to aldosterone and 18-hydroxycorticosterone by rat adrenal mitochondria in a competitive manner (Burrow 1969). When dog adrenals were perfused in vivo with *dehydroepiandrosterone* (30 μg/ml) the conversions of radioactive deoxycorticosterone and 11-deoxycortisol to corticosterone and cortisol respectively were markedly decreased, whereas the conversion of deoxycorticosterone to aldosterone was not affected (Fragachan et al. 1969). However, slices of adrenals which had been perfused with dehydroepiandrosterone in vivo produced less aldosterone in vitro than slices of control adrenals. A competitive inhibition of the 11β-hydroxylation and 18-hydroxylation of deoxycorticosterone by natural and synthetic androgens (Sharma et al. 1962; Rembiesa et al. 1968; Colby et al. 1970) and a decrease in adrenocortical mitochondrial cytochrome P-450, both resulting in overproduction of deoxycorticosterone, appear to be the basis of experimental androgen-induced hypertension in rats (Hyde and Daigneault 1968; Brownie et al. 1968, 1970; Colby et al. 1970). No information is available on aldosterone biosynthesis in such animals. Treatment with testosterone of castrated male rats resulted in increased plasma renin activity; however, this was associated with unaltered plasma aldosterone concentrations (Katz and Roper 1977).

Bovine adrenal zona glomerulosa cells that had been preincubated with *testosterone hemisuccinate* or with *dehydroepiandrosterone sulfate* bound labeled angiotensin II and angiotensin III with a markedly increased affinity and produced moderately (but significantly) increased amounts of aldosterone in the absence or presence of angiotensins II and III (Caroll and Goodfriend 1984). The effect was seen only if the steroids were applied to the cells and then washed away.

4.11 Ouabain

Ouabain, a selective inhibitor of (Na-K)-ATPase, has been found to directly influence aldosterone biosynthesis in adrenal zona glomerulosa tissue and cell preparations. The reported effects of the glycoside are mainly inhibitory. However, under certain conditions, stimulatory effects have also been observed. These effects varied among different animal species and depended on the extracellular potassium concentration and on the presence or absence of aldosterone-stimulating agents.

When added to the medium in which dog adrenocortical slices were incubated, ouabain (5.5×10^{-5} mol/l) significantly inhibited aldosterone production (Cushman 1969). It also prevented the stimulation of aldosterone output induced by raising the potassium concentration in the incubation medium from 5.8 to 9.8 mmol/l. Under the experimental conditions used, ouabain also caused decreases in the tissue uptake of ^{42}K, in the tissue potassium content, and in the potassium exchange between tissue and medium. Wellen and Benraad (1969a) found that at low concentrations (10^{-7} mol/l) ouabain markedly inhibited the in vitro production of corticosterone by outer slices of calf adrenals but not by a mitochondrial preparation of this tissue. At the same time, it blocked active cation transport across cell membranes, resulting in a loss of intracellular potassium and a gain in intracellular sodium. At a high potassium concentration (10 mmol/l), ouabain decreased corticosterone production in outer, but not in inner slices of calf adrenals (Benraad and Wellen 1971). In rat adrenal tissue, ouabain was found to be less active; at a concentration of 10^{-4} mol/l, it decreased corticosterone production by only 13% (Wellen and Benraad 1969b). However, some marked effects

Table 7. Effect of ouabain (5×10^{-4} *M*) on corticosteroid synthesis by rat adrenal quarters in vitro. Ouabain was added only to the medium used for final incubations. Mean values of two flasks ± range. (From J. Müller 1971)

Addition to incubation medium[a]	Steroid production (μg/100 mg/2 h)			
	Aldosterone		Corticosterone	
	Control	Ouabain	Control	Ouabain
–	1.24±0.12	1.47±0.07	8.55±1.15	8.27±0.25
Serotonin[b]	3.33±0.44	1.48±0.01	10.06±0.40	8.80±0.07
–	1.13±0.01	1.91±0.02	5.92±0.12	7.92±0.37
KCl[c]	4.04±0.49	1.95±0.12	6.19±0.78	8.64±0.46
–	1.46±0.04	2.06±0.29	6.54±0.16	7.33±0.21
ACTH[d]	6.69±0.86	3.30±0.34	69.52±0.22	54.65±1.15

[a] 3.6 mEq K^+/l
[b] 5-Hydroxytryptamine creatinine sulfate, 1.7×10^{-5} *M*
[c] 8.1 mEq K^+/l
[d] 5 IU/flask

of a very high dose of ouabain (5×10^{-4} mol/l) on aldosterone biosynthesis by quartered rat adrenals were observed in our laboratory (J. Müller 1971; Table 7). Thus, under basal conditions of incubation (K^+ 3.6 mmol/l), ouabain moderately stimulated aldosterone production by an average of 40%. However, it completely blocked a further increase in aldosterone production in response to a high potassium concentration (8.1 mmol/l) in the medium or to added serotonin, and it partially suppressed the aldosterone-stimulating action of ACTH. On the other hand, ouabain had only marginal effects on corticosterone production both under baseline conditions and under stimulation by ACTH. According to Szalay (1971), ouabain had a very minor but statistically significant aldosterone-stimulating effect when added in a concentration of 10^{-8} mol/l to quartered rat adrenals incubated in media with potassium concentrations of 4 and 8 mmol/l. When added in a higher dose (10^{-4} mol/l), ouabain stimulated aldosterone output at a potassium concentration of 4 mmol/l but prevented an increase in response to higher potassium concentrations (6 and 8 mmol/l). In a dose of 10^{-3} mol/l, ouabain was ineffective at a potassium concentration of 4 mmol/l but again blocked the aldosterone-stimulating effects of higher potassium concentrations. J. Boyd et al. (1973) found ouabain to exert only inhibitory effects on rat adrenal tissue. At a concentration of 5×10^{-4} mol/l it significantly decreased the aldosterone output by rat adrenal quarters or capsules, under basal conditions (K^+ 1 and 3 mmol/l) as well as during stimulation by potassium ions, ACTH, or cyclic AMP. It also blocked an increase in capsular adrenal cyclic AMP content in response to an extracellular potassium concentration of 6 mmol/l. In rat adrenal zona glomerulosa cells, ouabain elicited a biphasic response in corticosterone output, with low doses (10^{-7}–10^{-6} mol/l) being slightly stimulatory (+10%) and high doses (10^{-4} and 10^{-3} mol/l) being markedly inhibitory (Mendelsohn and Mackie 1975). In contrast, ouabain consistently decreased the intracellular potassium content of these cells in a dose-dependent manner between 10^{-6} and 10^{-3} mol/l. At the dose of 5×10^{-4} mmol/l, ouabain inhibited not only the stimulation of the corticosterone output by serotonin or a high potassium concentration in zona glomerulosa cells, but also the stimulation by ACTH in zona fasciculata cells. Partially different effects of ouabain on rat adrenal zona glomerulosa cells were observed by Braley and Williams (1978). In the absence of other stimulators (K^+ 3.7 mmol/l), ouabain stimulated aldosterone and corticosterone outputs by about 60% when added in a dose of 10^{-5} mol/l; at higher and lower concentrations it was ineffective. When added together with a constant dose of angiotensin II or ACTH (eliciting near-maximal responses on their own), ouabain influenced aldosterone output in a biphasic manner, with an additive stimulation at 10^{-6} and 10^{-5} mol/l and a marked inhibition at 10^{-4} and 10^{-3} mol/l. At the higher doses, ouabain also decreased the corticosterone output. At a potassium concentration of 5.9 mmol/l, ouabain had inhibitory effects on aldosterone and corticosterone outputs at 10^{-4} and 10^{-3} mol/l but was ineffective at 10^{-6} and 10^{-5} mol/l. At very high potassium concentrations of 15 and 20 mmol/l (which in this preparation yielded a maximal steroidogenic effect), ouabain was ineffective at 10^{-4} but inhibitory at 10^{-3} mol/l. At 10^{-4} mol/l, ouabain suppressed the steroidogenic response to ACTH in zona glomerulosa but not in zona fasciculata cells. When added to dog outer adrenal slices, a very high dose of ouabain (2.5×10^{-3} mol/l) stimulated aldosterone production at a potassium concentration of 3.0 mmol/l but was ineffective and inhibitory at potassium concentrations of 4.7 and 7.5 mmol/l respectively (Lobo et al. 1978). It also blocked the stimulatory effect of angiotensin II on aldosterone - but not on cortisol - biosynthesis. At a concentration of 10^{-5} mol/l it significantly reduced the aldosterone-stimulating effect of angiotensin II (Foster et al. 1979) in the same tissue preparation. The aldosterone-stimulating action of exogenous dibutyryl cyclic AMP (10^{-2} mol/l) was also completely abolished by simultaneously added ouabain (10^{-4} mol/l).

Blaine et al. (1974a) infused ouabain into the arterial supply of autotransplanted adrenal glands of moderately sodium-deficient sheep. During the infusion of a high dose of the glycoside (10^{-5} mol/l) there was a significant rise in adrenal venous plasma potassium concentration (from 4.2 to 5.1 mmol/l); this was not associated with any change in aldosterone, corticosterone, or cortisol secretion. However, 30 min after the infusion, when the plasma potassium concentration had returned to normal, there was a significant drop in aldosterone secretion (−80%) together with a marked increase in cortisol and corticosterone secretion. These changes persisted during the second hour after the end of the ouabain infusion. Thirty minutes after the infusion had ended, the adrenal began taking up potassium, but it did not recover its normal potassium content within an hour (Blaine et al. 1974b).

4.12 Miscellaneous Biological Substances with Direct Stimulatory or Inhibitory Effects on Aldosterone Biosynthesis

Porcine and chicken vasoactive intestinal peptide (VIP) stimulated aldosterone production and - less extensively - corticosterone production by superfused frog interrenal tissue in a dose-dependent manner, but it was several orders of magnitude less active than human ACTH (Leboulenger et al. 1983). In concentrations ranging from 3×10^{-10} to 10^{-7} mmol/l, porcine vasoactive peptide failed to influence aldosterone output by rat adrenal glomerulosa cells both in the absence of stimulators and under submaximal stimulation by angiotensin II (Enyedi et al. 1983).

Sera obtained from five patients with manic-depressive psychosis significantly inhibited ACTH-, serotonin-, or angiotensin II-stimulated aldosterone production when added together with these agents to the medium in which rat adrenal zona glomerulosa cells were incubated (O'Brien et al. 1979).

According to Mendelsohn and Kachel (1981), normal human serum markedly stimulated aldosterone production by rat adrenal glomerulosa cells under basal conditions of incubation, as well as under maximal stimulation by serotonin, angiotensin II, or a high potassium concentration. The serum had no effect on basal or ACTH-stimulated corticosterone output by zona fasciculata cells. The active serum component was ultrafiltrable and dialyzable and was not inactivated by chymotrypsin digestion. The aldosterone-stimulating activity appeared to be partially due to a serotonin-like substance, since it was inhibited by methysergide and metergoline (see p. 63).

A fourfold increase in the concentration of rat adrenal zona glomerulosa cells substantially enhanced aldosterone production per incubated cell in the absence of stimulators and even more extensively in the presence of maximally steroidogenic concentrations of potassium ions, angiotensin II, ACTH, serotonin, and cyclic AMP (D.J. Campbell 1982b). The cell concentration had a lesser stimulatory effect on corticosterone output. Addition of substrate amounts of pregnenolone, progesterone, deoxycorticosterone, or corticosterone stimulated aldosterone output independently of the cell concentration. The cell concentration-associated increase in aldosterone production was greatest during the first 30 min of incubation. These observations indicated that the cell concentration effect was due to a novel aldosterone-stimulating factor released by the cell and acting on an early step of aldosterone biosynthesis, preceding the formation of pregnenolone.

4.13 Site of Action of Stimulators or Inhibitors in the Biosynthetic Pathway

4.13.1 "Early Steps" and "Late Steps"

It is generally accepted that ACTH exerts its short-term steroidogenic action on zona fasciculata cells mainly by facilitating cholesterol side-chain cleavage and thus by stimulating the formation of pregnenolone (Stone and Hechter 1954; Karaboyas and Koritz 1965; Simpson 1979). The biochemical events involved in the conversion of cholesterol to pregnenolone will, in the following, be called "early steps". The term will not be restricted to the three intramitochondrial reactions catalyzed by cytochrome P-450_{SCC}, i.e., cholesterol 22R-hydroxylation, 20α-hydroxylation, and side-chain fission, but will include preceding rate-limiting reactions, such as adrenal cholesterol uptake from plasma lipoproteins (Gwynne et al. 1976), hydrolysis of cholesterol esters (Vahouny et al. 1978), and intracellular movement of cholesterol (Crivello and Jefcoate 1980; Conneely et al. 1984), all of which appear to be under ACTH control. The term "late steps" will refer to the reactions involved in the conversion of deoxycorticosterone to aldosterone. According to the available experimental evidence, most of the short-term aldosterone-stimulating action of angiotensin II, potassium ions, ACTH, and serotonin appears to be due to an activation of early biosynthetic steps in zona glomerulosa cells. Whereas such an effect on early steps is undisputed, the evidence supporting an additional effect on late steps is more equivocal. Stimulation of late steps observed during short-term incubation experiments was generally less extensive than simultaneous effects on early steps and depended on the medium composition (G. H. Williams et al. 1972a). In certain instances, the stimulation of late steps seemed to be, at least in part, secondary to the stimulation of early steps. No agent has as yet been found to stimulate only late steps. However, dopamine has been reported to exert an inhibitory action only on late steps of aldosterone biosynthesis (McKenna et al. 1979b).

4.13.2 Endogenous Steroid Output

The addition of stimulators such as angiotensin II, potassium ions, serotonin, or ACTH to incubated zona glomerulosa tissue or cells enhances not only the production of aldosterone but also the output of corticosterone, deoxycorticosterone, and 18-OH-DOC, i.e., intermediate products or by-products of aldosterone biosynthesis (Kaplan and Bartter 1962; Kaplan 1965; J. Müller 1970b; Haning et al. 1970; McDougall et al. 1976; G. H. Williams and Brailey 1977; Moore et al. 1978; Braley and Williams 1979). These findings are consistent with an effect on early steps of aldosterone biosynthesis. However, when zona glomerulosa cells from sodium-replete rats were incubated with various concentrations of stimulatory agents, the relative responses in aldosterone output were greater than the relative responses in corticosterone output (Haning et al. 1970; J. F. Tait et al. 1980a). Generally, the relative increase in aldosterone output was equal to the square of the relative increase in corticosterone output, which indicated a stimulation of the conversion of corticosterone to aldosterone by the intermediate product. Stimulation of zona glomerulosa cells with serotonin, angiotensins II or III, and ACTH resulted in ratios of (relative increase in aldosterone output):(relative increase in corticosterone output)^2 equal or similar to 1.0 (J. F. Tait et al. 1980a). In contrast, high potassium concentrations (5.3, 5.9, and 13 mmol/l) elicited values of this ratio significantly above 1.0, indicating a direct effect of potassium ions on late steps of aldosterone biosynthesis. However, in capsular adrenals of potassium-deficient rats, all these agents had a major effect on deoxycorticosterone output and did not significantly enhance aldosterone production (J. Müller 1970b). In normal human adrenal glomerulosa cells, angiotensin II stimulated the outputs of aldosterone, corticosterone, 18-hydroxycorticosterone, and 18-OH-DOC to the same relative extent (Braley and Williams 1979).

Atriopeptin II not only inhibited aldosterone and 18-hydroxycorticosterone production by rat capsular adrenal glands incubated in the presence of various stimulators (potassium ions, serotonin, angiotensin II, ACTH), but also suppressed to a similar extent the output of corticosterone. In cultured bovine zona glomerulosa cells stimulated by prostaglandin E_1, ANF suppressed the outputs of aldosterone, deoxycorticosterone, and progesterone to similar extents (Rácz et al. 1985). Under the same experimental conditions, the production of progesterone and corticosterone by cultured zona fasciculata cells was suppressed more markedly by ANF than was the production of cortisol.

A low sodium concentration (112 mmol/l) in the incubation medium led to a 37% increase in aldosterone production but to a 59% decrease in corticosterone output by bovine adrenal outer slices (Saruta et al. 1972). This observation, which suggested a selective effect on late steps, has not yet been confirmed (see also p. 35).

Brecher et al. (1973) did not observe any change in the contents of free and esterified cholesterol of rat zona glomerulosa and fasciculata cells during incubation with ACTH. However, during prolonged incubation (8 h) of rat zona glomerulosa cells there was a significant progressive decrease in the esterified cholesterol content, whereas the free cholesterol content remained constant (D. J. Campbell 1982c). The depletion of esterified cholesterol was accelerated during incubation in a medium with a high potassium concentration (8.6 mmol/l).

4.13.3 Conversion of Labeled Precursors

When quartered adrenals of sodium-deficient rats were incubated with trace amounts of tritiated corticosterone, deoxycorticosterone, progesterone, or pregnenolone, the addition of potassium, ammonium, angiotensin II, or ACTH to the medium did not influence the incorporation of tritium into aldosterone, though it stimulated the biosynthesis of unlabeled aldosterone from endogenous precursors (J. Müller 1966; Table 8). However, all these agents stimulated the conversion of tritiated cholesterol to aldosterone. Serotonin also stimulated the incorporation of tritiated cholesterol but not of tritiated pregnenolone into aldosterone by rat adrenal tissue (J. Müller and Ziegler 1968; Table 9). The conclusion drawn from these experiments, that during short-term incubation, aldosterone stimulators acted only on early and not on late biosynthetic steps, was later challenged by Haning et al. (1970) and G. H. Williams et al. (1972a). These investigators found that serotonin, angiotensin II, extracellular potassium, and ACTH stimulated not only the production of endogenous corticosterone and aldosterone but also the conversion of trace amounts of tritiated corticosterone to aldosterone (but not to 18-hydroxycorticosterone) by rat adrenal zona glomerulosa cells or tissue, provided that the medium contained albumin (which by itself inhibited the conversion). The effects of the stimulators on the conversion of tritiated corticosterone to aldosterone were proportional to the induced increments in corticosterone output and could be partly reproduced by the addition of small amounts of unlabeled corticosterone to the incubation medium. Since the stimulation of the conversion of tritiated corticosterone to aldosterone became manifest immediately after the addition of unlabeled corticosterone and was not inhibited by cycloheximide, it was probably not due to substrate enzyme induction. Dexamethasone did not reproduce the effect of corticosterone on the conversion.

The percentage conversion of tritiated corticosterone to aldosterone in vivo by autotransplanted sheep adrenal glands remained constant or even decreased significantly during stimulation of endogenous aldosterone secretion by angiotensin II (Baniukiewicz et al. 1968). On the other hand, there was a moderate increase in the percentage conversion when aldosterone secretion was stimulated by an elevation in local plasma potassium concentration (Blair-West et al. 1973a).

Table 8. Incubation of rat adrenal quarters with trace amounts of tritiated precursors. Mean values of two flasks ± range. (From J. Müller 1966)

Substrate	Addition to incubation medium	Conversion to Aldosterone-$-^{3}H$ (cpm/100 mg × 10^{-3})	Conversion to Corticosterone-$-^{3}H$ (cpm/100 mg × 10^{-3})	Production of corticosterone[a] (µg/100 mg)
Corticosterone-1,2-^{3}H (945 600 cpm per 100 mg tissue)	-	16.2 ±0.9		9.1±0.6
	NH_4Cl[b]	16.7 ±3.0		9.0±1.2
	NADP + Glc-6-P[c]	20.6 ±1.4		63.7±3.7
	ACTH[d]	13.6 ±3.7		92.3±6.5
	-	21.9 ±0.1		13.3±0.4
	KCl[e]	25.7 ±1.4		16.1±0.7
	Angiotensin II[f]	17.4 ±0.7		13.1±0.4
Pregnenolone-7α-^{3}H (1 505 000 cpm per 100 mg tissue)	-	69.6 ±3.2	393 ± 3	10.8±0.7
	NH_4Cl	66.6 ±3.5	370 ±22	16.6±0.4
	NADP + Glc-6-P	74.0 ±9.8	637 ±60	71.4±3.2
	ACTH	60.8 ±3.7	390 ± 8	93.6±0.4
	-	71.2 ±3.0	318 ± 6	14.7±0.6
	KCl	73.2 ±9.4	334 ±13	18.2±0.9
	Angiotensin II	74.1 ±4.1	302 ±17	15.4±0.9
Cholesterol-7α-^{3}H (21.4 × 10^{6} cpm per 100 mg tissue, added in Tween 80)	-	1.34±0.11	3.40± 0.08	7.7±0.3
	NH_4Cl	2.40±0.26	4.37± 0.35	9.0±0.5
	NADP + Glc-6-P	27.64±1.74	197.53± 5.96	48.5±0.6
	ACTH	3.94±0.30	15.25± 1.80	64.3±5.2
	-	0.36±0.03	2.54± 0.89	11.2±1.3
	KCl	3.26±0.80	4.82± 0.92	14.7±1.3
	Angiotensin II	0.80±0.02	2.44± 0.16	11.6±0.1

[a] Fluorimetric assay
[b] 7.7 mmol/l
[c] NADP-Na + glucose-6-phosphate-Na_2; 10.5 µmol/flask
[d] 5 IU/flask
[e] 8.1 mEq K^+/l
[f] Val[5]-angiotensin II-asp[1] β-amide; 2 mg/flask

Table 9. Incorporation of tritiated precursor steroids into aldosterone and corticosterone by quartered rat adrenals. Mean values of two flasks ± range. (From J. Müller and Ziegler 1968)

Substrate	Addition to incubation medium	Conversion to Aldosterone-^{3}H (cpm/100 mg × 10^{-3})	Conversion to Corticosterone-^{3}H (cpm/100 mg × 10 ^{3})
Cholesterol-7$\alpha^{3}H$ (74.5 × 10^{6} cpm per 100 mg tissue)	-	2.23± 0.03	12.9± 0.5
	Serotonin (10^{-5} *M*)	5.14± 0.35	16.8± 0.8
Pregnenolone-7$\alpha^{3}H$ (1.31 × 10^{6} cpm per 100 mg tissue)		87.5 ± 4.2	369 ±11
	Serotonin (10^{-5} *M*)	104.5 ±13.5	349 ±12

A high potassium concentration in the incubation medium stimulated the conversion of tritiated progesterone to aldosterone and of tritiated deoxycorticosterone to corticosterone but did not significantly affect the conversion of tritiated corticosterone to aldosterone by outer slices of dog adrenal cortex in vitro (Burwell et al. 1969).

Angiotensin II increased the incorporation of ^{14}C-labeled acetate into cortisol, corticosterone, and aldosterone by bovine adrenal slices but had no effect upon the conversion of labeled progesterone or cholesterol to these corticosteroids (Lommer and Wolf 1966).

ACTH, potassium, or angiotensin II did not influence the percent conversion of tritiated corticosterone to aldosterone and 18-hydroxycorticosterone by human capsular adrenal cells (G.H. Williams and Braley 1977). The addition of unlabeled corticosterone to the incubation medium resulted in significant decreases in these conversions.

4.13.4 Experiments with Unlabeled Precursors

When bovine outer adrenocortical slices were incubated in the presence of unlabeled progesterone or corticosterone (20 μg/g tissue), angiotensin II did not further stimulate aldosterone production, whereas the peptide had a significant stimulatory effect in the presence of unlabeled cholesterol (Kaplan and Bartter 1962). ACTH enhanced aldosterone production in the presence of unlabeled cholesterol but decreased it in the presence of deoxycorticosterone or corticosterone.

The effect of added rat plasma lipoproteins on steroidogenesis by dispersed rat adrenal zona glomerulosa cells was investigated by D.J. Campbell (1982d). Neither very low density lipoprotein (VLDL) nor high-density lipoprotein (HDL) induced any increase in basal (3.8 mmol/l K^+) or potassium-stimulated (8.6 mmol/l K^+) aldosterone production. Low-density lipoprotein (LDL) did not affect basal aldosterone production, but it had a biphasic effect on potassium-stimulated aldosterone production, causing a significant enhancement at a concentration of 50–60 μg/ml HDL-cholesterol. According to Nagy et al. (1984), both human LDL and HDL had dose-dependent enhancing effects on ACTH-stimulated aldosterone production by rat adrenal glomerulosa cells. The addition of 25-hydroxycholesterol or 22S-hydroxycholesterol (2.5 or 5×10^{-5} mol/l) to rat adrenal glomerulosa cells resulted in marked increases in aldosterone production under basal conditions of incubation. A high potassium concentration (8.4 mmol/l) significantly enhanced the aldosterone-stimulating effects of the two sterols. However, asp^1,ile^5-angiotensin II or asn^1,val^5-angiotensin II (10^{-8} mmol/l) did not further stimulate aldosterone production in the presence of 25-hydroxycholesterol or 22S-hydroxycholesterol. The addition of deoxycorticosterone (6.1×10^{-6} mol/l) stimulated aldosterone output to a considerably greater extent than the addition of sterols, potassium ions, or angiotensin II. In the presence of deoxycorticosterone, aldosterone production was not further enhanced by a high potassium concentration, angiotensin II, 25-hydroxycholesterol, or 22S-hydroxycholesterol. In rat adrenal capsular tissue, atriopeptin II did not affect aldosterone, 18-hydroxycorticosterone, or corticosterone outputs under stimulation by added substrate amounts of pregnenolone (Fig. 18). In the presence of substrate amounts of added

25-hydroxycholesterol, angiotensin II had an additional stimulatory effect on aldosterone production by bovine adrenal glomerulosa cells (Goodfriend et al. 1984). This effect of angiotensin II was partially inhibited by the atrial natriuretic peptide auriculin A.

A high potassium concentration in the incubation medium stimulated the conversion to aldosterone of unlabeled added progesterone and the conversion to corticosterone of unlabeled deoxycorticosterone by outer slices of dog adrenal cortex, but it did not further stimulate aldosterone output in the presence of a "saturating" concentration of corticosterone (Burwell et al. 1969).

4.13.5 Experiments with Inhibitors of Steroidogenic Enzymes

In the presence of Su 8000 (see p. 78), a high potassium concentration stimulated mainly the deoxycorticosterone output by quartered adrenals of sodium-deficient rats (J. Müller 1971, p. 41). When the conversion of pregnenolone to progesterone was blocked by trilostane or cyanoketone, ACTH, angiotensin II, a high potassium concentration, β-MSH, or prostaglandin E_2 stimulated the production of pregnenolone from endogenous cholesterol by rat, dog, and bovine adrenal glomerulosa cells severalfold (McKenna et al. 1978b, 1978c; Aguilera and Catt 1979; Yamakado et al. 1985; W. B. Campbell et al. 1986). These observations demonstrated a major effect of the stimulators on early biosynthetic steps. Histamine also stimulated cortisol production in canine adrenocortical cells by an action on an early biosynthetic step (T. Hirose et al. 1982). When 17α-hydroxylation was blocked by Su 10603, the amine stimulated mainly the output of corticosterone. Consistent with an action of angiotensin II and potassium on early steps are the results of an in vivo study by R. D. Brown et al. (1972). According to these investigators, angiotensin II infusion or the oral intake of potassium chloride induced significant elevations in the plasma deoxycorticosterone concentrations in normal human subjects treated with dexamethasone and metyrapone but did not affect the plasma 11-deoxycortisol level.

Angiotensin II stimulated the production of aldosterone from added deoxycorticosterone to a small (30%) but statistically significant extent by bovine glomerulosa cells in which the formation of endogenous pregnenolone was blocked by aminoglutethimide (McKenna et al. 1978b). However, the peptide did not stimulate the conversion of added pregnenolone, progesterone, or corticosterone to aldosterone under these conditions. When extracellular potassium concentration was increased from 0 to 6 mmol/l, there was a significant stimulation of the conversion of exogenous corticosterone to aldosterone by bovine or canine adrenal glomerulosa cells incubated with aminoglutethimide (McKenna et al. 1978c). McKenna et al. (1979a) also demonstrated a direct effect of ACTH on late steps of cortisol synthesis. In bovine zona fasciculata cells incubated with aminoglutethimide, ACTH stimulated the formation of cortisol from added progesterone, 17α-hydroxyprogesterone, or 11-deoxycortisol.

In the presence of cyanoketone, the conversion of added corticosterone (but not of added progesterone or deoxycorticosterone) to aldosterone by rat adrenal glomerulosa cells was stimulated by angiotensin II, ACTH, β-MSH, and a high potassium concentration (Aguilera and Catt 1979; Yamakado et al. 1985;

W. B. Campbell et al. 1986). In contrast, the conversion of tritiated corticosterone to aldosterone by rat adrenal zona glomerulosa cells incubated with cyanoketone was not significantly stimulated by a dose of prostaglandin E_2 that significantly enhanced the production of pregnenolone under these conditions (W. B. Campbell et al. 1986). Angiotensin II and potassium ions (but not ACTH) had a similar effect on the conversion of corticosterone to aldosterone by dog adrenal glomerulosa cells incubated with cyanoketone (Aguilera and Catt 1979). The fractional increases induced in the conversion by the stimulators were relatively small (23%–65%) although statistically significant. Similar increments in the conversion of added corticosterone to aldosterone were also elicited by angiotensin II, ACTH, and urinary aldosterone-stimulating factor in rabbit capsular adrenal cells incubated with cyanoketone (Saito et al. 1981; Saito and Saruta 1982). In rat zona glomerulosa cells incubated with cyanoketone, atriopeptin III completely blocked the angiotensin II-induced increase in endogenous pregnenolone production as well as the angiotensin II-induced increase in the conversion of tritiated corticosterone to aldosterone (W. B. Campbell et al. 1985). In the presence of trilostane, auriculin A partially inhibited angiotensin II-stimulated pregnenolone production, but it had no effect on the conversion of substrate amounts of progesterone to aldosterone by bovine zona glomerulosa cells (Goodfriend et al. 1984).

4.14 Primary Effects and Intracellular Mediation

4.14.1 General Considerations

According to a generally accepted hypothesis, which is supported by a large body of experimental evidence, ACTH initiates its steroidogenic action at the level of the plasma membrane of the adrenocortical cells. Its specific reversible binding to a membrane-bound hormone receptor results in the activation of membrane-bound adenylate cyclase. This enzyme catalyses the formation of cyclic AMP, which in turn acts as a second messenger mediating the stimulation of corticosteroid biosynthesis within the cell (Haynes and Berthet 1957; Grahame-Smith et al. 1967; Garren et al. 1971; Sala et al. 1979; J. F. Tait et al. 1980b). The bulk of available evidence clearly indicates that angiotensins, potassium ions, and serotonin also act primarily at the level of the plasma membrane when stimulating steroidogenesis in the zona glomerulosa cell. Specific angiotensin binding sites have been found in zona glomerulosa cell plasma membranes, and their hormone receptor function has been unequivocally established. Added serotonin is rapidly taken up by incubated rat capsular adrenals (Trost and Müller 1976). A high extracellular potassium concentration leads to an increased intracellular potassium content of zona glomerulosa cells (Mendelsohn and Mackie 1975). Thus, an additional action of these agents at an intracellular site cannot be excluded.

A rise in cytosolic free calcium concentration is associated with the action of a number of regulatory peptides and amines that are considered to act on plasma membrane receptors but not to activate adenylate cyclase (Rasmussen and Barrett 1984). Binding of these hormones to their receptor leads to the activation of phospholipase C, a membrane-bound phosphodiesterase that catalyzes the breakdown of phosphatidylinositol-4,5-bisphosphate (PIP_2), leading to the formation of myoinositol-1,4,5-trisphosphate (IP_3) and 1,2-diacylglycerol (DAG). Both of these products can function as second messengers (Nishizuka et al. 1984; Berridge 1984; Exton 1985). IP_3 enters the cytosol and releases calcium ions from endoplasmic reticulum and perhaps other stores. DAG activates protein kinase C in the cell membrane. This enzyme phosphorylates substrates in and near the membrane or may be released into the cytosol. The activation of protein kinase C by DAG requires the presence of calcium ions and (membrane) phospholipids. In several cell systems, the physiological response to an extracellular messenger is considered to be mediated by the synergistic effects of calcium mobilization and protein kinase C activation. Thus, according to I. Kojima et al. (1983, 1984b), the sustained aldosterone-stimulating effect of angiotensin II on superfused porcine adrenal glomerulosa cells was mimicked by the combined actions of the calcium ionophore A-23187 and of the phorbol ester TPA, which activates protein kinase C. Diacylglycerol can be converted to phosphatidic acid (PA) by diacylglycerolkinase and be recycled into phosphatidylinositol. Alternatively, DAG can - outside of the cell membrane - be hydrolyzed by a diacylglycerol lipase to form monoglyceride and arachidonic acid, which in turn can be a precursor of leukotrienes, prostaglandins, and thromboxanes.

There are multiple possible interactions between the cyclic AMP and the calcium messenger systems (for a review see Rasmussen and Barrett 1984). On the one hand, calcium (via calmodulin) can activate phosphodiesterase and can activate or inhibit adenylate cyclase. On the other hand, cyclic AMP can increase calcium influx through the plasma membrane via voltage-gated channels, but at the same time it can stimulate calcium efflux directly through activation of the calcium pump or indirectly through activation of the sodium-potassium pump. Moreover, cyclic AMP increases calcium uptake into the endoplasmic reticulum and inhibits phospholipase C.

According to Farese (1983), steroidogenic agents may stimulate not only phosphatidylinositol breakdown in the plasma membrane of target cells but also de novo synthesis of phosphatidic acid from fatty acyl coenzyme A and glycerol-3′-phosphate in inner-cell membranes (endoplasmic reticulum). This effect requires calcium ions and is inhibited by cycloheximide.

4.14.2 Cyclic AMP

Cyclic AMP (cyclic adenosine 3′,5′-monophosphate) has the same steroidogenic and adrenal weight-maintaining activities as ACTH in vitro and in vivo (Haynes et al. 1959; Hilton et al. 1961; Imura et al. 1965; Ney 1969). It also resembles ACTH in its direct stimulatory effect upon aldosterone biosynthesis by incubated adrenal zona glomerulosa cells or tissue. Cyclic AMP and ACTH stimulated aldosterone production by bovine adrenal slices to similar extents (Kaplan 1965). There was no cumulative effect when these two stimulators were added together. On the other hand, cyclic AMP potentiated the aldosterone-stimulating effect of angiotensin II. Like ACTH, cyclic AMP stimulated steroidogenesis in the capsular as well as in the decapsulated portions of rat adrenal glands (Müller 1970b; Table 10). When graded concentrations of dibutyryl cyclic AMP were added to dog adrenal

Table 10. Effect of dibutyryl cyclic AMP on corticosterone production by adrenal tissue of sodium-deficient rats. Mean values of two experiments ($n=4$) ± standard deviation in μg/100 mg/2 h. *P* values of significance were calculated by *t*-tests and refer to differences between adjacent lines. (From J. Müller 1970b)

Tissue	Addition to incubation medium	Steroid production	
		Aldosterone	Corticosterone
Capsular adrenals	–	1.71 ± 0.42 ($P<0.001$)	1.84 ± 0.44 ($P<0.001$)
	Cyclic AMP[a]	11.31 ± 2.44	5.42 ± 0.35
Decapsulated adrenals	–		11.79 ± 2.67 ($P<0.001$)
	Cyclic AMP[a]		50.72 ± 11.17

[a] Dibutyryl adenosine-3′,5′-monophosphate, 10^{-3} *M*

zona glomerulosa cells, a significant increase in aldosterone output was observed at 10^{-3} mol/l and maximal response at 10^{-2} mol/l (Douglas et al. 1978d). In rat adrenal zona glomerulosa cells, a significant increase in corticosterone output was elicited by cyclic AMP (10^{-2} mol/l) at all potassium concentrations between 2 and 13 mmol/l (Tait and Tait 1976). In the presence of 8-bromo cyclic AMP (10^{-2} mol/l), potassium ions (15 mmol/l) and angiotensin II had small but significant additional stimulatory effects on the aldosterone production by dog zona glomerulosa cells, whereas ACTH was ineffective (Fujita et al. 1979).

The production of cyclic AMP was stimulated by added ACTH in a dose-dependent manner, not only in zona fasciculata cells but also in pure rat and dog zona glomerulosa cell preparations (Albano et al. 1974; Fujita et al. 1979). In incubated calf adrenal zona glomerulosa cells, ACTH raised the intracellular cyclic AMP content in a dose-dependent manner (I. Kojima et al. 1985a). However, the ED_{50} for intracellular cyclic AMP was ten times as high as that for aldosterone production. In contrast, forskolin (an activator of adenylate cyclase) stimulated aldosterone production with a dose dependence that correlated well with the dose-dependence curve of forskolin-induced cyclic AMP accumulation. In a superfusion system, the effect of ACTH (10^{-9} mol/l) on aldosterone output was much greater than that of forskolin (10^{-6} mol/l), although these two stimulators induced identical increases in intracellular cyclic AMP. Cyclic AMP production by rat adrenal glomerulosa cells was also stimulated by added serotonin (Albano et al. 1974; Fujita et al. 1979). However, the response in cyclic AMP output to serotonin was smaller than the response in cyclic AMP output to ACTH at concentrations of the two stimulators which elicited comparable responses in steroid output.

In incubated rat capsular adrenal membranes, adenylate cyclase activity was stimulated by ACTH and by isoproterenol; the effects of the two agents were not additive (Shima et al. 1984). The stimulatory or inhibitory effects of prostaglandins on the cyclic AMP content of incubated human adrenocortical tissue have been mentioned previously in this review (Honn and Chavin 1976b; p. 69). In statically incubated frog adrenal tissue, PGE_2 raised the cyclic AMP content more extensively than it stimulated aldosterone and corticosterone production (Chavin et al. 1978). Other prostaglandins had only a minor effect on intracellular cyclic AMP, which correlated poorly with the effects on corticosteroid production.

The previously available experimental evidence about a possible stimulatory effect of angiotensin II or a high extracellular potassium concentration on zona glomerulosa cyclic AMP production was partly conflicting. More recently, however, there seems to be a general agreement that angiotensin II inhibits and high extracellular potassium concentrations activate adenylate cyclase. Whereas angiotensin II stimulated cyclic AMP production while enhancing cortisol output by incubated bovine zona fasciculata cells (Peytremann et al. 1973), the aldosterone output by zona glomerulosa cells was stimulated by angiotensin II without any detectable concomitant change in cyclic AMP production (Peytremann et al. 1974). Angiotensin II did not influence cyclic AMP production by bovine adrenal outer slices (Saruta et al. 1972). A high dose of the peptide (10^{-5} mol/l) had no effect on cyclic AMP formation by capsular rat adrenals, but it stimulated phosphodiesterase activity in this tissue (Shima et al. 1978). High concentrations of a commercial preparation of angiotensin II stimulated the output of cyclic AMP by rat adrenal glomerulosa cells, according to Albano et al. (1974). This might have been due to contamination by an unknown peptide with ACTH-like activity, since in a later study performed in the same laboratory pure angiotensin II or III did not stimulate cyclic AMP production by rat zona glomerulosa cells; rather, they inhibited the output of the nucleotide to a small (but significant) extent (Bell et al. 1981). No effects of angiotensin II or III on the cyclic AMP production by dog or rat adrenal glomerulosa cells were observed by Douglas et al. (1978d) or by Fujita et al. (1979). The cyclic AMP production of cat capsular adrenal cells was also not affected by angiotensin II (Bravo et al. 1980). In contrast, Bing and Schulster (1978) found angiotensin II to elicit parallel increments in the outputs of aldosterone and cyclic AMP by rat adrenal glomerulosa cells. In purified membranes from the zona glomerulosa of bovine adrenocortical tissue, angiotensin II inhibited adenylate cyclase activity in a dose-dependent manner (Marie and Jard 1983). The effect required the presence of GTP. Angiotensin II also inhibited adenylate cyclase activity under stimulation by ACTH. A marked inhibitory effect of angiotensin II on adenylate cyclase activity was also observed in rat adrenal glomerulosa cells (Woodcock and Johnston 1984). At higher concentrations, angiotensin I and III displayed a similar activity. The antagonistic compound sar^1,ile^8-angiotensin II reversed the effect of angiotensin II.

According to Albano et al. (1974), a rise in extracellular potassium concentration yielded a biphasic response in cyclic AMP production by rat adrenal glomerulosa cells. The output of the nucleotide decreased when K^+ was increased from 2 to 3.6 mmol/l, remained constant at 5.9 mmol K^+/l, but increased significantly at higher potassium levels (8.4 and 13 mmol/l). Bell et al. (1980) also observed a consistent increase in the cyclic AMP output of pure rat adrenal glomerulosa cells when the extracellular potassium concentration was raised from 3.6 to 8.4 mmol/l. A significant increase in the rat capsular adrenal content of cyclic AMP was induced by raising the potassium concentration in the incubation medium from 3 to 6 mmol/l (J. Boyd et al. 1973). The cyclic AMP production of bovine adrenal slices was enhanced by an increase in the extracellular potassium from 5.2 to 9.2 mmol/l (Saruta et al. 1972). On the other hand, the stimulation of aldosterone production of dog or rat adrenal glomerulosa cells induced by raising the potassium concentration in the incubation medium from 5 to 15 mmol/l was not accompanied by any significant change in cyclic AMP output (Douglas et al.

1978; Fujita et al. 1979). A high potassium concentration in the superfusion medium (8.0 mmol/l) moderately raised the cyclic AMP content of bovine adrenal glomerulosa cells, whereas the addition of the ionophore A-23187 to a medium with a potassium concentration of 3.5 mmol/l had the opposite effect on intracellular cyclic AMP (while at least transiently stimulating aldosterone production; I. Kojima et al. 1985b). The sustained stimulatory action of a high potassium concentration on aldosterone production in this preparation could be mimicked by the combined addition of forskolin and A-23187.

In a number of rat tissue preparations (aorta washed particles, renal artery homogenate, heart sarcolemma, cultured mesenteric artery smooth muscle cells, anterior and posterior pituitary homogenate), a synthetic atrial natriuretic peptide (arg^{101}-tyr^{126}) inhibited adenylate cyclase activity under basal conditions or under stimulation by appropriate hormones (Anand-Srivastava et al. 1984, 1985). The inhibitory effects of angiotensin II and ANF on the enzyme were additive. Synthetic α-human atrial natriuretic peptide slightly but significantly inhibited cyclic AMP production by rat capsular adrenal cells (unstimulated), but at a higher concentration than the one needed to suppress aldosterone production (Matsuoka et al. 1985b).

4.14.3 Cyclic GMP

Angiotensin II, angiotensin III, a high potassium concentration (15 mmol/l), and ACTH did not significantly influence the output of cyclic GMP (guanosine 3′,5′-monophosphate) by dog adrenal glomerulosa cells (Douglas et al. 1978d). Angiotensin II did not induce an increase in cyclic GMP output in rat adrenal glomerulosa cells either (Bell et al. 1981). Evidence suggesting that cyclic GMP rather than cyclic AMP might be the physiological mediator of ACTH-induced steroidogenesis in zona fasciculata cells has been presented by Perchellet et al. (1978). However, this hypothesis was not supported by the results of a study by Hayashi et al. (1979).

A synthetic atrial natriuretic peptide (ANF[1]; Merck) significantly activated particulate guanylate cyclase obtained from a number of rat tissue homogenates (kidney, aorta, lung, liver, intestines, testes) but had no effect on soluble guanylate cyclase from these tissues (Waldmann et al. 1984). Particulate guanylate cyclase from brain or heart homogenate was not affected by the peptide. ANF also stimulated the accumulation of cyclic GMP in incubated kidney and aorta tissue. A dose-dependent activation of Triton X-100-dispersed particulate guanylate cyclase from bovine adrenal cortex by a synthetic rat atrial natriuretic peptide (arg^{101}-tyr^{126}) has been observed by Tremblay et al. (1985). Another atrial natriuretic peptide (α-human ANP) markedly stimulated cyclic GMP accumulation in rat capsular adrenal cells at lower concentrations (3×10^{-9} mol/l) than those needed to inhibit aldosterone production (3×10^{-8} mol/l; Matsuoka et al. 1985b). However, an inhibition of aldosterone biosynthesis by exogenous cyclic GMP has not yet been demonstrated (Fig. 21).

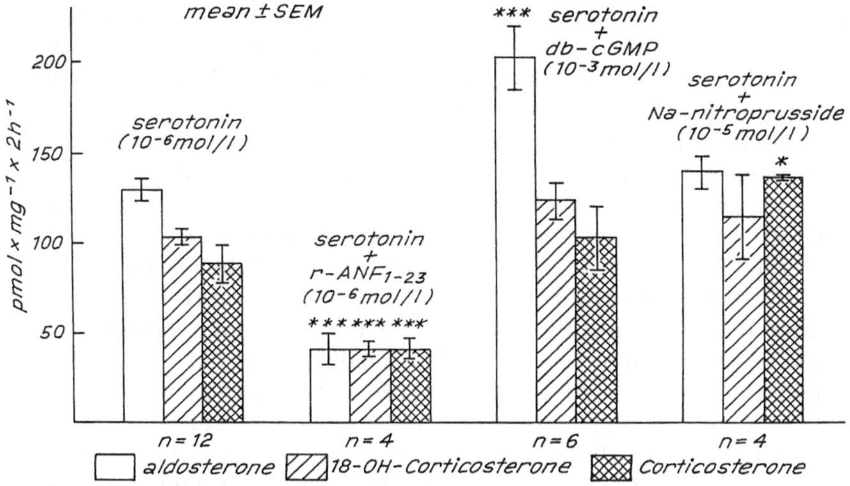

Fig. 21. Effects of an atrial natriuretic peptide (r-$ANF_{1\text{-}23}$ = atriopeptin II), dibutyryl cyclic GMP, and sodium nitroprusside on the output of corticosteroids by rat capsular adrenal tissue incubated in the presence of serotonin. *Asterisks* refer to the statistical significance of these effects according to *t*-test: * $P < 0.05$, *** $P < 0.001$. (From J. Müller 1987)

4.14.4 Calcium

Calcium ions are known intracellular mediators of hormone effects and play an important role in coupling excitation and response in a variety of cells (Rasmussen and Goodman 1977). In some hormone target cells, calcium fluxes are induced by cyclic AMP. In other instances, calcium ions act independently of cyclic AMP and have the function of second messengers. Thus, calcium ions are assumed to directly mediate the responses to angiotensin II in smooth muscle and liver cells (Angles d'Auriac et al. 1972; Freer 1975; Garrison et al. 1979). It has been clearly established that the steroidogenic response of the zona glomerulosa cell to any stimulator critically depends on the presence of extracellular calcium ions (see p. 44). Moreover, there is evidence indicating that the aldosterone-stimulating action of some of these agents is at least partially mediated by calcium ions.

Experiments with the radioactive isotope ^{45}Ca have generally revealed an influence of angiotensin II - and in some instances also of potassium ions and ACTH - on calcium flux across the plasma membrane of zona glomerulosa cells. Some apparently conflicting observations about the direction of stimulator-induced alterations in calcium flux can be explained by differences in the experimental protocol, particularly by differences in the calcium concentration of the extracellular medium used. Mackie et al. (1978) observed a delayed and small (but significant) decrease in the uptake of $^{45}Ca^{2+}$ by rat adrenal glomerulosa cells when they were preincubated and incubated in a medium with a potassium concentration of 8.4 mmol/l. The efflux rate of radioactive ions from zona glomerulosa cells preloaded with $^{45}Ca^{2+}$ was significantly diminished in a high-potassium incubation medium. No such effects of the extracellular potassium concentration on the calcium uptake and efflux were seen in corresponding experiments with zona fasciculata cells. A rapid and dose-dependent stimulation of radioactive calcium efflux from superfused rat adrenal glomerulosa cells was elicited by angiotensin II and by angiotensin III but not by a high potassium concentration (8.4 mmol/l), ACTH, serotonin, or cyclic AMP (B.C. Williams et al. 1981). Neither angiotensin II nor ACTH had any effect on the calcium efflux rate from superfused zona fasciculata cells. In a preparation of bovine adrenal glomerulosa cells incubated in a buffer containing only trace amounts of labeled calcium, small doses of angiotensin II not only stimulated aldosterone production, but also had marked and significant effects on $^{45}Ca^{2+}$ flux which became apparent within 2 min (Elliott and Goodfriend 1981; Elliott et al. 1982). The peptide stimulated the release and suppressed the influx of radioactive calcium ions. The angiotensin II-induced alteration in calcium flux was markedly diminished by the calmodulin inhibitor trifluoperazine and by the calcium channel blocker nifedipine but not by the protein synthesis inhibitor cycloheximide (Elliott et al. 1982). A synthetic atrial natriuretic peptide did not affect the angiotensin II-induced decrease in radioactive calcium uptake (Goodfriend et al. 1984). However, in another preparation of bovine adrenal glomerulosa cells, both angiotensin II and a high extracellular potassium concentration (12 mmol/l) significantly stimulated the uptake of radioactive calcium under steady-state conditions during incubation in a calcium-containing medium (Foster et al. 1981). Methoxyverapamil inhibited the angiotensin-induced increase in calcium uptake without affecting the basal rate of calcium uptake. According to the investigations of I. Kojima et al. (1985a, 1985b, 1985c, 1985e), both calcium efflux and calcium influx were accelerated during stimulation of the zona glomerulosa by angiotensin II, potassium ions, or ACTH. In superfused porcine adrenal glomerulosa cells prelabeled with ^{45}Ca, angiotensin II induced a 3.5-fold increase in the fractional efflux ratio of radiocalcium, whereas a high potassium concentration (12 mmol/l) caused only a 1.5-fold increase in this parameter in spite of having a similar steroidogenic effect (I. Kojima et al. 1985c). When the medium contained no calcium or when the calcium ions were replaced by strontium ions, the effects of a high potassium concentration on both calcium efflux and aldosterone production were completely abolished. Under these conditions, angiotensin II still caused a normal increase in the fractional efflux ratio of radiocalcium and elicited 25%–50% of the normal increase in aldosterone output. However, addition of dantrolene (a drug that inhibits the release of calcium ions from endoplasmic reticulum) to a calcium-containing medium diminished the angiotensin II-induced increase in the fractional efflux ratio of radiocalcium and delayed the onset and reduced the maximal rate of aldosterone secretion in response to the peptide. Dantrolene did not affect potassium-stimulated aldosterone production or calcium efflux. In a medium without calcium or in a medium with strontium ions in place of calcium ions, dantrolene completely inhibited the responses in aldosterone production and in calcium efflux to angiotensin II. According to these results, the potassium-activated aldosterone secretion is almost totally dependent on the uptake of calcium from the extracellular pool. However, during angiotensin II stimulation, approximately 50% of the calcium needed for stimulus-response coupling appears to come from a dantrolene-sensitive intracellular pool and 50% from the extracellular pool. According to a hypothesis by the same investigators (I. Kojima et al. 1985d), the initial transient rise in the zona glomerulosa cytosolic free calcium concentration is due to the mobilization of calcium by inositol trisphosphate. This initial rise in cytosolic free calcium triggers the calmodulin-dependent increase in aldosterone production on the one hand and activates on the other hand - also via calmodulin - the cell membrane calcium pump, thus accelerating calcium efflux. Moreover, the transient rise in cytosolic free calcium, together with the sustained increase in the diacylglycerol content of the cell membrane, leads to the conversion of the calcium-activated phospholipid-dependent C-kinase from its calcium-insensitive form to its calcium-dependent, membrane-bound form. During the sustained phase of the cellular response to angiotensin II there is a high rate of calcium cycling across the cell membrane.

Calcium influx increased rapidly during stimulation of calf adrenal zona glomerulosa cells by a high potassium concentration (8 mmol/l; +240%) and - less extensively - also under stimulation by angiotensin II (+130%) or ACTH (+60%; I. Kojima et al. 1985e). The calcium influx induced by the three stimulators critically depended on the extracellular calcium concentration, with apparent K_m values between 0.32 and 0.46 mmol/l. The increase in calcium influx elicited by angiotensin II or ACTH was also dependent on the extracellular potassium concentration. At a potassium concentration of 2 mmol/l, neither agent had any effect on this parameter. Conversely, the two stimulators caused a greater increase of calcium influx at a potassium concentration of 6 mmol/l than at a potassium concentration of 4 mmol/l. When calcium influx was maximally stimulated by a potassium concentration of 10 mmol/l, the two peptides had no additional effect. According to these observations and to the results of experiments with the calcium channel blocker nitrendipine and the calcium channel agonist BAY K 8644, all three agents, i.e., angiotensin II, potassium ions, and ACTH, stimulated calcium influx by a common calcium channel but acted by different mechanisms to change its function. The dependence on potassium concentration would indicate that the stimulator-induced calcium influx occurs via voltage-dependent channels. It has been shown by Foster et al. (1982) and by Capponi et al. (1984) that increases in the extracellular potassium concentration as small as 1 mmol/l and within the physiological range lead to a depolarization of bovine zona glomerulosa cells. This effect was dependent on the extracellular calcium concentration and was not observed in zona fasciculata cells. In rat adrenal glomerulosa cells, voltage-dependent calcium channels appear to be involved in angiotensin II- and potassium-stimulated aldosterone production, according to studies with the antagonists nifedipine and nitrendipine (Aguilera and Catt 1986). Although angiotensin II and nitrendipine bound to separate membrane receptors, there was a close correlation between the concentrations of angiotensin II receptors and nitrendipine binding sites in membranes of dissected zones of the adrenal gland in three different animal species, which suggested a functional relationship between angiotensin II receptors and voltage-dependent calcium channels.

Whereas the total calcium content of bovine adrenal glomerulosa cells was decreased by angiotensin II (Elliott et al. 1985), the cytosolic free calcium concentration was moderately but significantly raised in a dose-dependent manner by angiotensin II as well as by elevated extracellular potassium in bovine and rat adrenal glomerulosa cells loaded with the fluorescent calcium indicator quin 2 (Capponi et al. 1984; Braley et al. 1984). The potassium-induced, but not the angiotensin II-induced rise in cytosolic calcium was blocked by nifedipine (Capponi et al. 1984).

Calmodulin-like activity and calcium-dependent phosphodiesterase were found in rat adrenal glomerulosa and fasciculata cells (Koletsky et al. 1983). According to the activation of calmodulin-deficient phosphodiesterase by boiled cell sonicates and to a radioimmunoassay, zona fasciculata cells contained significantly more calmodulin than zona glomerulosa cells.

4.14.5 Phospholipid Metabolism

Potassium ions and angiotensin II caused dose-dependent increases in the rat capsular adrenal tissue content of diphosphoinositide (PI-4-P), triphosphoinositide (PI-4,5-P_2), phosphatidylinositol (PI), and phosphatidic acid (PA), according to Farese et al. (1980). Angiotensin II and potassium ions also enhanced the incorporation of ^{32}P-labeled phosphate into PA and PI by rat adrenal glomerulosa cells without affecting the phosphorylation of other phospholipid fractions (Farese et al. 1981). These investigators also elicited a similar stimulation of in vitro capsular adrenal phosphate incorporation into PA and PI by in vivo treatment of rats with the diuretic furosemide 2 h before killing them. In cells preincubated with [^{32}P]-phosphate, angiotensin II significantly stimulated PI hydrolysis. According to these observations, the stimulatory effect of angiotensin II on the incorporation of phosphate into phospholipids could have been due to the stimulation of two separate processes, i.e., to PI hydrolysis (with a subsequent increase in PA synthesis) and to de novo synthesis of PA. The effect of angiotensin II on de novo synthesis of PA (and thus also of PI) was also exerted by potassium ions, serotonin, ACTH, and cyclic AMP (Farese et al. 1983). On the other hand, a rapid decrease of prelabeled PI-4,5-P_2 (manifest within 5 s) was elicited by angiotensin II but not by other aldosterone-stimulating agents (Farese et al. 1984).

According to Whitley et al. (1984), angiotensin II, sar^1-angiotensin II, and angiotensin III stimulated the incorporation of ^{32}P-labeled phosphate into PI and PA but not into phosphatidylcholine (PC) in zona glomerulosa cells from rat adrenals. In this preparation, serotonin and ACTH displayed no such effects on phospholipid metabolism. An extracellular potassium concentration of 8.4 mmol/l slightly stimulated phosphate incorporation into PI but not into PA. In rat zona fasciculata-reticularis cells, ACTH weakly stimulated the incorporation of labeled phosphage into all three phospholipid fractions, whereas angiotensins II and III (which did not stimulate steroidogenesis in these cells) stimulated the incorporation of [^{32}P]-phosphate into PI and less extensively into PA, but not into PC.

Hunyady et al. (1982) had previously observed that angiotensin II (but not a high potassium concentration, ACTH, prostaglandin E_2, or dibutyryl cyclic AMP) stimulated the incorporation of labeled phosphate into PI in rat adrenal glomerulosa cells. This effect was detected within 2.5 min and was still maintained after 60 min. Cycloheximide abolished the effect of angiotensin II on aldosterone production but not on PI synthesis. In zona glomerulosa cells prelabeled with [^{32}P]-phosphate, the radioactivity in PI (relative to the radioactivity in other phospholipid fractions) decreased in response to angiotensin II within 5 min. Neither the stimulation of PI breakdown nor the stimulation of phosphate incorporation into PA and PI required the presence of extracellular calcium ions (Hunyady et al. 1983). Lithium, a cation which inhibits the activity of the enzyme *myo*-inositol-

1-phosphatase, completely blocked an increased incorporation of radioactive phosphate into PI in response to angiotensin II and led to a simultaneous accumulation of radioactivity in PA (Balla et al. 1984). On the other hand, lithium did not interfere with the angiotensin II-induced breakdown of PI. In rat adrenal zona glomerulosa cells preloaded either with ^{32}P-labeled phosphate or with ^{3}H-labeled inositol, angiotensin II induced a 40% reduction in the radioactivity of PI-4,5-P_2 within 15 s (Enyedi et al. 1985). Only a slight increase in the labeling of PI-4,5-P_2 was observed during the following 30 min. Changes in the radioactivity of PI-4-P showed similar kinetics. Angiotensin II also decreased the pool size of labeled PI-4,5-P_2 and PI-4-P. In cells prelabeled with tritiated inositol, the production of inositol trisphosphate and inositol bisphosphate showed a manifold increase during the first seconds of stimulation with angiotensin II and remained enhanced for several minutes. This observation indicated that the rate of resynthesis of polyphosphoinositides increased shortly after the activation of phospholipase C. Hunyady et al. (1985) observed that angiotensin II stimulated the incorporation of both [^{3}H]-arachidonate and [^{32}P]-phosphate into PI. These effects were abolished by lithium ions but not by the phospholipase A_2 inhibitor mepacrine, which suggested that the increased tritium labeling of PI occurred via CDP-diacylglycerol and not via an enhanced deacylation-reacylation cycle. Angiotensin II decreased the incorporation of [^{3}H]-arachidonate into phosphatidylcholine and phosphatidylethanolamine, thus inducing a shift in phospholipid synthesis favoring the production of phosphoinositides.

In bovine adrenal glomerulosa cells, angiotensin II elicited marked dose-dependent increases in the incorporation of [^{32}P]-phosphate into PI (Elliott et al. 1982). Such an effect was not induced in this preparation by an aldosterone-stimulating dose of dibutyryl cyclic AMP (Elliott et al. 1983). Neither angiotensin II nor dibutyryl cyclic AMP significantly altered the mass of PI in bovine zona glomerulose cells. Whereas neither EGTA nor TMB-8 (a substance considered to be an antagonist of intracellular calcium) prevented a marked increase in phosphate labeling of PI in response to angiotensin II on their own - while completely suppressing the steroidogenic action of the peptide, the combined addition of both substances to the incubation medium totally blocked the effect on phospholipid metabolism. The angiotensin-induced increase in phosphate incorporation into PI was not affected by a dose of an atrial natriuretic peptide (auriculin A) which substantially inhibited the aldosterone-stimulating effect of the peptide (Goodfriend et al. 1984). In calf adrenal glomerulosa cells prelabeled with [^{32}P]-phosphate, angiotensin II caused a rapid (manifest within 10 s) drop in the levels of labeled PI-4,5-P_2 and PI-4-P, whereas the level of labeled PI remained constant (I. Kojima et al. 1984b). In similar cells, prelabeled with tritiated inositol and incubated in the presence of lithium ions, angiotensin II elicited a rapid and persistent labeling of inositol disphosphate and inositol trisphosphate. Finally, in cells prelabeled with [^{3}H]-arachidonic acid, the peptide rapidly induced a sustained elevation in the formation of tritiated diacylglycerol. The effects of a high extracellular potassium concentration (8 mmol/l) on these parameters of phospholipid metabolism in calf adrenal glomerulosa cells were rather equivocal (I. Kojima et al. 1985b). Thus, in cells prelabeled with [^{3}H]-inositol, a high potassium concentration led only to a small decrease in their content of PI-4,5-P_2, which was not accompanied by a measurable increase in inositol trisphosphate concentration. Moreover, in cells prelabeled with [^{3}H]-arachidonic acid, a high potassium concentration did not induce an increase in [^{3}H]-diacylglycerol formation. In bovine adrenal zona fasciculata cells, three different agents which acutely stimulated steroidogenesis, i.e., ACTH, angiotensin II, and acetylcholine, also stimulated the incorporation of radioactive phosphate into PA and PI (Hadjian et al. 1984). However, whereas angiotensin-induced or acetylcholine-induced alterations in phospholipid metabolism preceded any measurable increase in cortisol output by 1 or 2 min, ACTH-induced increases in the phosphate incorporation into PA and PI were only observed 5 min after the onset of an increased cortisol production. Moreover, ACTH stimulated phospholipid metabolism to a lesser relative extent than cortisol production.

4.14.6 Intracellular Potassium and (Na,K)-ATPase

The hypothesis that intracellular potassium in the zona glomerulosa might be a common direct regulator of aldosterone biosynthesis was first proposed by Baumber and co-workers (1971), who had observed significant increases in the adrenocortical (total) content of potassium in four experimental groups of dogs treated with different stimuli of aldosterone secretion, i.e., angiotensin II infusion, sodium depletion, ACTH injections, and potassium chloride infusion. These changes in adrenocortical potassium content were not accompanied by any significant changes in the electrolyte composition of the psoas muscle. Later, J. Boyd et al. (1973) presented the following supportive evidence suggesting an important role for intracellular potassium in the regulation of aldosterone biosynthesis: (a) Capsular rat adrenals had a higher potassium content than decapsulated adrenals; (b) (Na,K)-ATPase activity was higher in capsular than in decapsulated adrenals; (c) ouabain blocked the steroidogenic actions of ACTH or added cyclic AMP in capsular but not in decapsulated adrenals; (d) stimulation of aldosterone secretion by sodium restriction in rats was blocked by simultaneous potassium depletion. The possibility that intracellular potassium could be an important mediator of aldosterone secretion was also suggested by the results of a study by Birkhäuser et al. (1973), who observed that during infusion of angiotensin II to normal human subjects the administration of minor amounts of potassium chloride stimulated aldosterone secretion without altering the plasma potassium concentration. However, Mendelsohn and Mackie (1975) and Mackie et al. (1977) - using two different techniques of separating cells from extracellular fluid (Mendelsohn and Warren 1975; Mackie et al. 1977) - found that neither serotonin nor angiotensin II raised the intracellular potassium content of rat adrenal glomerulosa cells while stimulating corticosterone production. Angiotensin II even induced a small decrease in intracellular potassium. Previously, Szalay (1969) had observed an inhibitory effect of angiotensin II on potassium accumulation by dog adrenocortical tissue. Angiotensin II had no effect on the potassium content of outer dog adrenal slices (Lobo et al. 1978) and did not significantly affect active or passive uptake of $^{86}Rb^+$ by this tissue (Foster et al. 1979). A high extracellular potassium concentration (8.4 mmol/l) led to a small but significant increase in the intracellular potassium content of rat adrenal glomerulosa and fasciculata cells (Mendelsohn and Mackie 1975; Mackie et al. 1977). According to determinations of adrenocortical sodium and potassium by X-ray electron microprobe, neither sodium restriction nor sodium loading of rats resulted in any significant change in the zona glomerulosa potassium content (Szalay et al. 1975; Decorzant et al. 1977). Whereas Szalay et al. (1975) had found a consistently higher potassium content in zona glomerulosa cells than in zona fasciculata cells, this was not confirmed by Decorzant et al. (1977). The sodium content was found to be higher in the zona glomerulosa than in the zona fasciculata by both groups of investigators. Wiederholt et al. (1984) measured cell membrane potential and intracellular potassium activity with microelectrodes filled with ion-sensitive liquid ion exchanger in the zona glomerulosa of superfused hemiadrenals of rats kept on different diets, and found that these parameters were not influenced by a high or low sodium intake or by a high potassium intake.

A direct stimulatory effect of angiotensin II upon the (Na,K)-ATPase activity of a microsomal preparation from outer layers of bovine adrenal cortex was observed by Gutman and co-workers (1972). However, other investigators found no direct influence of angiotensin II upon total or sodium- and potassium-dependent ATPase activity in whole homogenates or plasma membrane fractions from dog or rat adrenal glomerulosa (Douglas et al. 1978d; Foster et al. 1979; Meuli and Müller 1982). Serotonin and ACTH did not affect (Na,K)-ATPase or rat capsular adrenal microsomes, either (Meuli and Müller 1982). In the same preparation, potassium ions influenced the enzyme in a monophasic manner, yielding half maximal stimulation at 0.5 and maximal stimulation at 5 mmol/l. Ouabain also exerted a monophasic inhibitory action on rat capsular adrenal microsomal (Na,K)-ATPase.

5 Alterations in Aldosterone Biosynthesis and Secretion in Long-Term Experiments and Diseases

5.1 Alterations in Sodium Intake or in Sodium Balance

5.1.1 "Sodium Deficiency"

"Sodium deficiency" is generally considered the most potent stimulus of aldosterone production. However, as used in aldosterone physiology, the term "sodium deficiency" is often quite ambiguous and may refer in turns to either decreased sodium intake, to a decreased total body sodium content, or to a sodium depletion of circulating plasma.

Restriction of sodium intake, loss of body sodium due to pregnancy, lactation, diuresis, sweating, diarrhea, or hemorrhage, and spontaneous, therapeutic, or experimental removal or sequestration of sodium (e.g., by diuretics, hemodialysis, peritoneal dialysis, collection of saliva, edema) lead to marked increases in aldosterone secretion and in the plasma aldosterone concentration. In reverse, increased sodium intake or parenteral administration of sodium result in diminished aldosterone production and in lower plasma aldosterone levels. Since sodium retention is the primary biological effect of aldosterone, the vital importance of the adaptation of aldosterone secretion to sodium intake and the body sodium status is obvious. It allows survival of man and animals in sodium-deficient environments and in many diseases.

The close correlation between sodium intake and aldosterone secretion was first discovered in man (Luetscher and Axelrad 1954) and has since been observed in a great number of different animal species, including toads (Crabbé 1963), bullfrogs (Ulick and Feinholtz 1968), lizards (Bradshaw and Genot 1976), fowl (Skadhauge et al. 1983)[1], duck (Klingbeil 1985a), rats (Eisenstein and Hartroft 1957), rabbits (Blair-West et al. 1968a), guinea pigs (Whipp et al. 1976), dogs (Rosnagle and Farrell 1956), sheep (Blair-West et al. 1963), opossum (Johnston et al. 1967b), and wombats and kangaroos (Coghlan and Scoggins 1967b). Generally, an increased aldosterone secretion maintains sodium balance and prevents the occurrence of a sizable sodium deficit during periods of low sodium intake. Nevertheless, it is widely assumed that the adaptation of aldosterone secretion to dietary sodium restriction is mediated by mechanisms which are also operative during spontaneous or contrived sodium losses with resulting sodium deficits. This assumption may be justified, but it certainly has its limitations. It must be kept in mind that

[1] Previously, Taylor et al. (1970) had reported that sodium depletion did not affect aldosterone secretion in the chicken.

severe sodium depletion leads to a number of physiological strains not occurring during low sodium intake, such as decreased extracellular fluid volume, decreased total and circulating plasma volume, decreased arterial blood pressure, decreased arterial blood flow, potassium retention, and alterations of electrolyte concentrations in the extra- and intracellular fluids. The mode of experimental sodium removal is probably also important. Thus, treatment with a diuretic agent (in contrast to sodium depletion by peritoneal dialysis) leads to an increased distal tubular sodium flow, which could by itself be a more active stimulus to renin secretion than the resulting sodium deficit. Sequestration of parotid saliva, a model of experimental sodium depletion extensively used in sheep, permits the study of aldosterone secretion and biosynthesis during states of extremely high sodium deficit. However, a comparable sodium deficit may have a greater impact on a nonruminant animal such as the dog than on a sheep, which has available a considerable amount of sodium stored in its ruminal fluid.

5.1.2 Morphological and Histochemical Changes

In man (Peschel and Race 1954), as well as in the rat (Deane et al. 1948) and other mammalian animal species such as the opossum (Johnston et al. 1967b), the mouse (Shelton and Jones 1971), the rabbit, the wombat, and the kangaroo (Blair-West et al. 1968a, Scoggins et al. 1970), sodium deficiency results in zona glomerulosa hypertrophy, hyperplasia, or both. An increased zona glomerulosa width was found in rats that had been kept on a sodium-deficient diet for only 2 days (Hartroft and Eisenstein 1957). The zona glomerulosa thickness was found to be increased by approximately 100% within 1-2 weeks of sodium restriction (Deane et al. 1948; Hartroft and Eisenstein 1957; Marx and Deane 1963; J. Müller and Huber 1969). Zona glomerulosa hypertrophy increases successively for several weeks. However, the zona glomerulosa of rat adrenals was found to be wider after 3 weeks of sodium deprivation than after 48 weeks (R. B. Cohen 1965). Reported effects of sodium deficiency on the inner zones and adrenal weight are conflicting. Whereas Deane et al. (1948), J. Müller and Huber (1969), and Kenyon et al. (1978a) found the zona fasciculata width to be normal, Hartroft and Eisenstein (1957), Marx and Deane (1963), and Blair-West et al. (1968a) found it to be decreased to the same extent as the zona glomerulosa was increased. In severe sodium deficiency of young animals, decreases in the width of the inner zones and in absolute adrenal weight may reflect overall growth retardation (Eisenstein and Hartroft 1957).

Sodium restriction of mice led to an early depletion of lipid droplets in the zona glomerulosa (Shelton and Jones 1971). Lipid droplets were restored, however, after 3 weeks of sodium depletion. Other ultrastructural features were a transient increase in rough endoplasmic reticulum and a progressive increase in smooth endoplasmic reticulum and Golgi apparatus. According to Kobayashi (1974a, 1974b), 1 or 2 weeks of sodium restriction led to a progressive increase in the area occupied by mouse adrenal zona glomerulosa mitochondria. The spherical and enlarged mitochondria were characterized by long, tubular cristae. Following sodium loading of mice, there was an atrophy of cytoplasmic organelles, an increase in size, number, and osmophilia of lipid droplets, and the appearance of

β-glycogen granules (Shelton and Jones 1971). Eventually, there was also an increase in the size and number of lysosomes.

In rats, also, sodium deficiency results in time-dependent alterations in the lipid content of zona glomerulosa cells. Elema et al. (1968a) depleted rats of sodium by peritoneal dialysis with 5% glucose solution. After 3 h there was a marked lipid depletion in all zones of the adrenal cortex; after 27 h lipid depletion continued in the zona glomerulosa, whereas repletion occurred in the two inner zones. Depending on the severity or duration of sodium deficiency, lipid droplets in the zona glomerulosa became smaller or disappeared completely (Deane et al. 1948). On the other hand, R.B. Cohen (1965) found the zona glomerulosa cells to be filled with large lipid droplets in rats which had been kept on a sodium-deficient diet for 48 weeks. It appears to be very difficult to correlate lipid content with steroidogenic activity, since both decreased and increased secretory activity of the zona glomerulosa may be associated with lipid depletion (Deane et al. 1948). In zona glomerulosa cells of sodium-depleted rats, Giacomelli et al. (1965) observed numerous, greatly elongated mitochondria with parallel arrays of tubules and amorphous intramitochondrial deposits. According to Fisher and Horvat (1971), some zona glomerulosa mitochondria of sodium-deficient rats exhibited laminated cristal elements. An increase in smooth endoplasmic reticulum was seen by both groups of investigators. A quantitative ultrastructural study on the effect of a high sodium intake on the adrenal cortex was carried out in two strains of rats, that were either resistant or sensitive to the hypertensive effects of sodium chloride, by Nickerson (1977). Sodium loading caused an increased nuclear volume in zona glomerulosa cells in both strains. Only in the resistant animals, it led to significant decreases in the volumes of mitochondria and smooth endoplasmic reticulum, as well as in the surface areas of smooth endoplasmic reticulum, mitochondrial membranes, and cristae.

Histochemical changes which accompany increases in zona glomerulosa width during sodium deficiency include increased activity of Δ^5-3β-hydroxysteroid dehydrogenase, glucose-6-phosphate dehydrogenase, isocitrate dehydrogenase, and NADPH tetrazolium reductase (Marx and Deane 1963; Kuhn and Kissane 1964; R.B. Cohen 1965; Elema et al. 1968a). These changes in enzyme activity could be observed 27 (but not 3) h after sodium depletion by peritoneal dialysis (Elema et al. 1968a) and were prevented by a preceding bilateral nephrectomy (Elema et al. 1968b). According to measurements with an electron microprobe, neither sodium restriction nor sodium loading of rats led to significant changes in zona glomerulosa potassium content (Szalay et al. 1975, Decorzant et al. 1977). An increased sodium content of the zona glomerulosa due to sodium loading was observed by Szalay et al. (1975) but not by the other group of investigators.

5.1.3 Altered Sensitivity of the Zona Glomerulosa to Aldosterone-Stimulating Substances

According to a great number of investigators, the response of the adrenal cortex in vivo or of adrenocortical tissue or cells in vitro to stimulators such as angiotensins II and III, ACTH, potassium ions, or serotonin is critically dependent on sodium intake or sodium balance (A.F. Muller et al. 1956b, Thorn et al. 1957; Venning et al. 1962; Blair-West et al. 1962a, 1970c; J. Müller 1965a, 1968, 1970b; Cannon et al. 1966; Csanky et al. 1968; Kinson and Singer 1968; J. Müller and Huber 1969; Oelkers et al. 1974a; Hollenberg et al. 1974; Kem et al. 1975; Cowley and McCaa 1976; G. H. Williams and Braley 1977; Aguilera et al. 1978; Nicholls et al. 1978; Kenyon et al. 1978a; R. E. McCaa 1978; W. B. Campbell et al. 1979b; R. D. Gordon et al. 1980; Nicholls et al. 1982). Generally, sodium deficiency enhances and sodium loading diminishes the increment in adrenocortical aldosterone output in response to an identical dose of one of these stimulators. With an increasing degree of sodium deficiency the dose-response curves become steeper and end in a higher plateau. Ganong and co-workers (Ganong et al. 1965a, 1966; Ganong and Boryczka 1967) demonstrated that a comparable increment in aldosterone secretion was elicited by a lower dose of angiotensin or ACTH in dogs kept on a sodium-deficient diet than in sodium-replete dogs. The increased sensitivity became manifest after 5 days of sodium restriction; at that time, hypertrophy of the zona glomerulosa was not yet detectable (Ganong et al. 1967a). Increased responsiveness in aldosterone secretion was not associated with altered responsiveness in 17-hydroxycorticosteroid output to ACTH or angiotensin.

There are a few exceptions. Thus, sodium deficiency in rats enhanced the response of aldosterone secretion to angiotensin II but not the response to ACTH (Kinson and Singer 1968). Moreover, the sensitivity of rat adrenal zona glomerulosa cells to the steroidogenic action of ACTH was not altered by sodium restriction, according to Vinson et al. (1983). In contrast, sodium depletion markedly increased the sensitivity of zona glomerulosa cells to α-MSH. In sheep, the adrenocortical responsiveness to angiotensin II was not only decreased by sodium loading; it was also virtually abolished by moderate sodium depletion (Blair-West et al. 1970c; Scoggins et al. 1981). However, in animals which were first sodium-depleted and then nephrectomized, very small doses of angiotensin II stimulated aldosterone secretion to its maximal rate (Coghlan et al. 1979a). This indicated that aldosterone secretion in intact sodium-depleted sheep may no longer have been responsive to exogenous angiotensin II because it had already been stimulated by endogenous angiotensin to the highest level compatible with the sodium status of the animals.

Although the sodium status has opposite effects on the angiotensin sensitivity of the zona glomerulosa and of smooth muscular tissue (vascular and uterine), it seems to influence the angiotensin receptors of both tissues in a similar manner. According to the displacement of angiotensin II dose-response curves of rabbit aorta and rat zona glomerulosa cells by a small dose of the antagonist sar^1, ala^8-angiotensin II, a high sodium intake increased the receptor affinity in both tissues (G. H. Williams et al. 1976b). Treatment of rats with a low-sodium diet for 6 weeks resulted in a decreased receptor affinity to angiotensin II in zona glomerulosa cells, as shown by ^{125}I-angiotensin II binding assays (Douglas and Catt 1976). This was accompanied by a significant increase in the number of specific angiotensin binding sites, which became already apparent after 1 week of sodium restriction. Devynck et al. (1979) observed that a low-sodium diet in rats raised the number of angiotensin II receptors in adrenocortical as well as in myometrial tissue, whereas a high-sodium diet induced an increase in the number of uterine receptors without affecting adrenocortical receptors.

Sodium intake also has a marked influence on the stimulation of aldosterone secretion or biosynthesis by long-term administration of angiotensin. A 48-h infusion of angiotensin II induced only a transient increase in aldosterone secretion in hypophysectomized dogs that were kept on a high-sodium diet (200 mmol/day), in spite of a sustained increase in blood pressure (W. W. Davis et al. 1969). A similar dose of angiotensin led to a sustained elevation of aldosterone, corticosterone, and cortisol secretion in animals on a normal sodium intake (60 mmol/day). These findings were partly confirmed by R. E. McCaa and co-workers (McCaa et al. 1975, Cowley and McCaa 1976; McCaa 1978), who, however, found angiotensins II and III to have a biphasic effect on the aldosterone secretion of dogs, irrespective of sodium intake. In dogs on a high-sodium diet (120 mmol/day), a low (5 ng/kg/min) or a high (23 ng/kg/min) dose of angiotensin II elicited only a temporary rise in plasma aldosterone. Plasma aldosterone returned to normal within a few hours and remained normal during 2 weeks of continued angiotensin infusion. In dogs on a sodium intake of 40 mmol/d, angiotensin also elicited a sustained moderate elevation of plasma aldosterone that depended on the dose of the peptide. The chronic steady-state response to long-term infusion of angiotensin II or III was markedly higher in dogs kept on a sodium-deficient diet (5 mmol/d).

A 6-day infusion of angiotensin II (3 ng/kg/min) into two normal human subjects kept on a constant sodium intake of 135 mmol/d elicited sustained increases in the plasma aldosterone concentration and in the aldosterone excretion rate (Oelkers et al. 1978a). Toward the end of the infusion there was a tendency for aldosterone secretion to decrease; this could have been due to a positive sodium balance or a drop in plasma potassium. When gross sodium retention was prevented by an appropriate dietary restriction, plasma and urinary aldosterone increased during 3 days of angiotensin II infusion into sodium-replete human subjects (Oelkers et al. 1978b). When the subjects were first sodium depleted (furosemide administration and low-sodium diet), a 3-day angiotensin II infusion led to marked and sustained additional increases in both parameters of aldosterone secretion.

A high sodium intake prevented an increase in adrenal aldosterone and corticosterone content in response to 3 weeks of angiotensin II administration to rats (Dufau et al. 1969). Subcutaneous infusion of a high dose of angiotensin II for a week into potassium-deficient rats stimulated late steps in aldosterone biosynthesis to an extent that depended critically on the concurrent sodium intake (J. Müller et al. 1984; Fig. 22).

5.1.4 Sites of Action in the Biosynthetic Pathway

Although experimental evidence shows that "early" as well as "late" steps in the aldosterone biosynthetic pathway (see p. 90) are influenced by the body sodium status, observed long-term alterations in late biosynthetic steps are of particular interest, because they are pronounced and seem to indicate a specific qualitative change in zona glomerulosa function which could at least partially account for the alterations in the adrenocortical responsiveness to aldosterone stimulators described in the preceding section. It has been previously pointed out that the short-term actions of stimulators are mainly directed at early steps (p. 91).

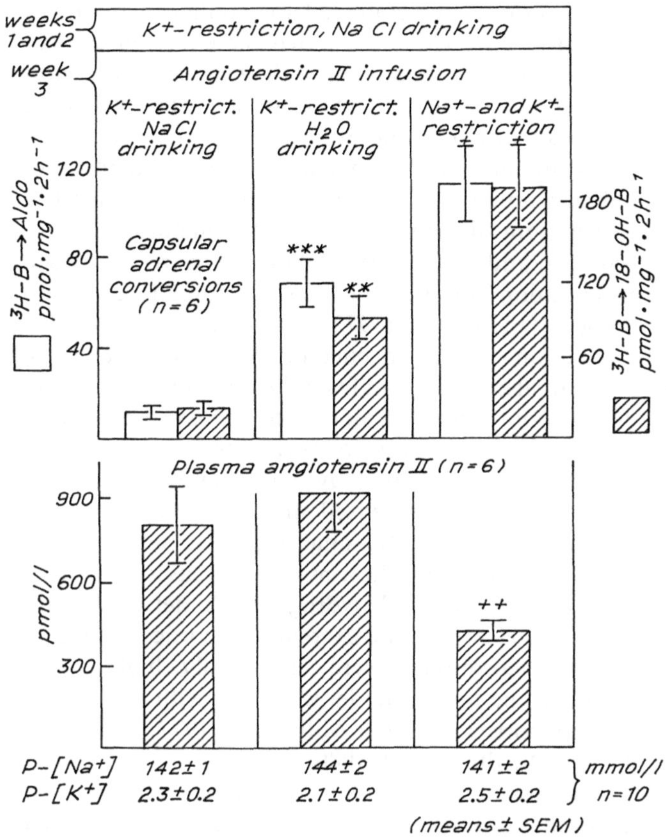

Fig. 22. Effect of sodium intake during angiotensin II infusion in potassium-deficient rats on the capsular adrenal conversions of tritiated corticosterone to aldosterone *(ALDO)* and 18-hydroxy-corticosterone *(18-OH-B)* and on the concentrations of angiotensin II, sodium, and potassium in the plasma. All animals were kept on a potassium-deficient diet and isotonic saline as drinking fluid for 2 weeks. During the 3rd week they were treated by an s.c. infusion of angiotensin II (750 $ng \cdot kg^{-1} \cdot min^{-1}$). Mean values of n observations ± standard error of the mean. Statistical significance (t-tests) of differences from values in *left panel:* *** $P<0.001$, ** $P<0.01$; from values in *middle panel:* $^{++}P<0.01$, $^{+}P<0.05$. (From J. Müller et al. 1984)

In intact dogs, sodium depletion resulted in a selective increase of aldosterone secretion without any alteration in glucocorticoid output; in hypophysectomized dogs, sodium deficiency induced also a significant increase in corticosterone and cortisol production (Binnion et al. 1965; W.W. Davis et al. 1968). This indicated that sodium deficiency stimulated steroid production in the zona fasciculata as well as in the zona glomerulosa and acted at an early stage of steroid biosynthesis, common to the production of aldosterone and the two glucocorticoids, i.e., at a biosynthetic step preceding the formation of progesterone. Studies carried out in man, sheep, and rats have yielded conflicting evidence on the influence of sodium balance on steroidogenesis in the zona fasciculata, but have largely confirmed that sodium deficiency stimulates an early step in the aldosterone biosynthetic pathway. In human subjects whose endogenous ACTH secretion was suppressed by a constant dose of dexamethasone, sodium depletion led to parallel increases in aldosterone and corticosterone production without a noticeable effect on cortisol secretion rate (Bledsoe et al. 1966). During treatment with dexamethasone and metyrapone, sodium depletion elicited consistent increases in deoxycorticosterone production. Urinary tetrahydrodeoxycorticosterone excretion was not affected by sodium restriction in normal human subjects, but increased in two patients with pituitary insufficiency (Schambelan and Biglieri 1972). Plasma deoxycorticosterone rose after 5 days on a low-sodium diet and 2 days of dexamethasone treatment in normal human subjects (Tan and Mulrow 1975). Sodium restriction raised the plasma corticosterone concentration in normal persons according to Tuck and Mayes (1980), but not according to Oddie et al. (1972) or to Mason et al. (1977). Indirect evidence for a site of action of sodium balance in the early stages of aldosterone biosynthesis in man can also be derived from the data obtained in patients with sodium loss due to an inborn error in aldosterone biosynthesis (for a review see Veldhuis and Melby 1981). In the cases of corticosterone methyl oxidase type-1 (18-hydroxylase) deficiency described by Visser and Cost (1964), Degenhart et al. (1966), and Jean et al. (1969), increased amounts of corticosterone were secreted during severe sodium deficiency. In some instances, high amounts of urinary deoxycorticosterone were also found. In a number of cases with corticosterone methyl oxidase type-2 ("18-hydroxydehydrogenase") deficiency, the plasma level, secretion rate, and urinary metabolite excretion of corticosterone (in some instances also of deoxycorticosterone) were abnormally high during restriction of sodium intake and became normal under treatment with a mineralocorticoid (Ulick et al. 1964; Rappaport et al. 1968; Milla et al. 1977; Reiter and Teree 1979; Katznelson et al. 1979; Veldhuis et al. 1980).

Continuous superfusion and perfusion studies with rat adrenal tissue in vitro and autotransplanted sheep adrenal glands in vivo respectively indicated that aldosterone production was affected by changes in sodium balance at two different sites of the biosynthetic pathway, one before and one after the formation of corticosterone (Baniukiewicz et al. 1968; Blair-West et al. 1970b). Capsular portions of adrenal glands ("zona glomerulosa") taken from hypophysectomized rats that had been drinking sodium chloride solution for 2 days produced less aldosterone, 18-hydroxycorticosterone and corticosterone than corresponding tissue from hypophysectomized animals that had been drinking water, during the first 30 min of in vitro superfusion. During the later periods of superfusion, tissue from sodium-loaded rats produced less aldosterone and 18-hydroxycorticosterone but more corticosterone and 18-OH-DOC than tissue from control rats. The conversion of trace amounts of tritiated corticosterone to aldosterone by sheep adrenal autotransplants rose proportionally with the increasing production of aldosterone from endogenous precursors during the onset of mild sodium depletion until the aldosterone secretion rate had reached a value of 10–12 μg/h. In sheep more severely depleted of sodium, the aldosterone secretion rose to higher values, but the conversion of tritiated corticosterone to aldosterone decreased again to baseline levels. This latter finding could have been due either to an increase in zona glomerulosa corticosterone production during severe sodium depletion or to the switching over to another biosynthetic pathway not involving corticosterone as an intermediate product (Blair-West et al. 1971d). The type of sodium depletion may also be important for the biosynthetic site of aldosterone stimulation. According to the pattern of aldosterone and corticosterone output by superfused capsular rat adrenals, dietary sodium restriction increased the formation of aldosterone from corticosterone, whereas diuretic treatment enhanced the formation of corticosterone only (Spät et al. 1979).

Effects of sodium balance on late steps in aldosterone biosynthesis were first described by Vecsei et al. (1966), who found an increased conversion of radiolabeled progesterone to aldosterone by the adrenal tissue of sodium-restricted rats. Marusic and Mulrow (1967b) observed that adrenal mitochondria of rats kept on a sodium-deficient diet had a greater capacity for converting corticosterone to aldosterone and 18-hydroxycorticosterone than mitochondria of sodium-replete rats. The increased activity of the final steps in aldosterone biosynthesis became manifest after 1 day of sodium restriction. Adrenals of rats that had been drinking sodium chloride solution produced less aldosterone and more corticosterone when incubated in a medium with a low (3.6 mmol/l) or a high (8.5 mmol/l) potassium concentration, converted less added unlabeled progesterone, 11β-hydroxyprogesterone, deoxycorticosterone, and corticosterone to aldosterone, and converted less tritiated pregnenolone, progesterone, deoxycorticosterone, and corticosterone to aldosterone than adrenals of control animals (J. Müller 1968). Adrenals of sodium-restricted rats produced more aldosterone (under basal conditions of incubation and when stimulated by serotonin, potassium ions, or ACTH) and less corticosterone (when stimulated by serotonin or ACTH) and converted more added progesterone, deoxycorticosterone, and corticosterone to aldosterone than adrenals of sodium-replete rats (J. Müller and Huber 1969; Figs. 16, 23). Simi-

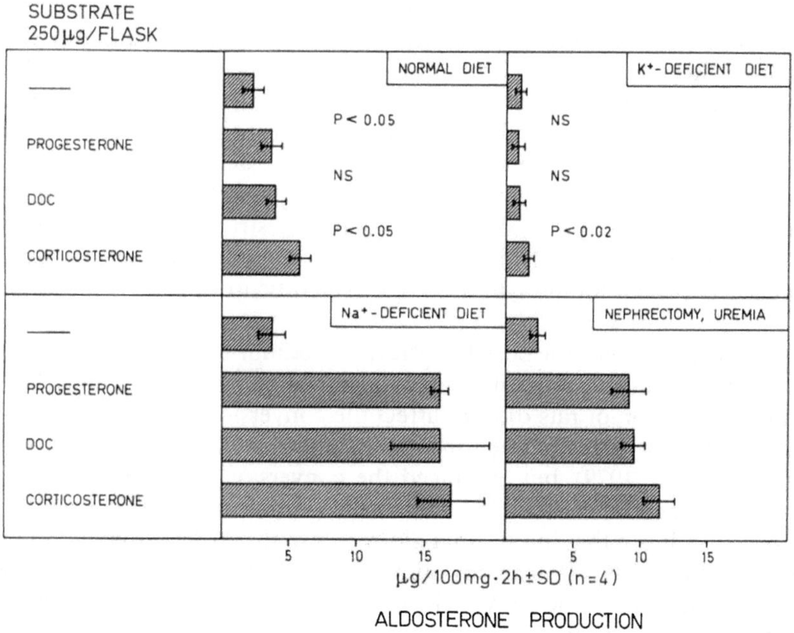

Fig. 23. Aldosterone production by adrenal tissue of various groups of rats incubated with substrate amounts of unlabeled precursor steroids. Mean values of two experiments ± standard deviation. *P* values of significance were calculated by *t*-tests and refer to differences between adjacent bars. *NS*, not significant. (Data from J. Müller and Huber 1969)

lar observations were made in experiments with dog adrenocortical tissue or human zona glomerulosa cells. Adrenal cortical slices of sodium-depleted dogs produced significantly more aldosterone but less corticosterone than adrenal tissue of sodium-replete dogs; cortisol production was equal in both tissues (W.W. Davis et al. 1968). Adrenal tissue of sodium-deficient dogs also converted more added unlabeled or tritiated progesterone and corticosterone to aldosterone. Four days of sodium depletion led to a threefold increase in the conversion of corticosterone to aldosterone by dog adrenocortical mitochondria (Marusic et al. 1973a). Adrenal glomerulosa cells from human subjects kept on a sodium-deficient diet for 3 days produced significantly more aldosterone and less corticosterone from endogenous precursors and converted more tritiated corticosterone to aldosterone (but not to 18-hydroxycorticosterone) than zona glomerulosa cells from sodium-replete subjects (G.H. Williams and Braley 1977).

Provided that the conversion of corticosterone to aldosterone is really a two-step reaction and that 18-hydroxycorticosterone in some way represents the immediate precursor of aldosterone (see p. 8), most of the available information indicates that sodium intake equally affects corticosterone methyl oxidations 1 and 2. Baniukiewicz et al. (1968) found that decreases in 18-hydroxycorticosterone production by superfused rat capsular adrenals due to drinking of saline solution paralleled decreases in aldosterone production. Sodium restriction of rats induced parallel increases in the adrenal mitochondrial conversions of corticosterone to aldosterone and 18-hydroxycorticosterone (Marusic and Mulrow 1976b; R.E. Kramer et al. 1979) and stimulated the capsular adrenal conversion of added tritiated 18-hydroxycorticosterone to aldosterone (Fattah et al. 1977; J. Müller 1980a). In man, also, the fractional increases in the secretion rates and plasma concentrations of aldosterone and 18-hydroxycorticosterone due to sodium deficiency are equal (Ulick et al. 1964a; Touitou et al. 1970; Tuck and Mayes 1980). However, sodium depletion raised the plasma level of 18-hydroxycorticosterone more extensively than the plasma level of aldosterone in normal man, according to Mason et al. (1977). Also, the conversion of tritiated progesterone to aldosterone by rat adrenal tissue was more markedly decreased by sodium loading than the conversion to 18-hydroxycorticosterone; sodium restriction induced only an increased incorporation of tritiated progesterone into aldosterone and did not influence the incorporation into 18-hydroxycorticosterone (Vecsei et al. 1966).

The available evidence about a possible effect of sodium balance on the 11β-hydroxylation step in the aldosterone biosynthetic pathway is somewhat ambiguous. Sodium restriction of rats did not affect the conversion of deoxycorticosterone to corticosterone by zona glomerulosa mitochondria (Marusic and Mulrow 1967; R.E. Kramer et al. 1979), but enhanced the conversion of 18-OH-DOC to 18-hydroxycorticosterone by capsular adrenal tissue (Fattah et al. 1977; J. Müller 1980b). This might indicate that in the zona glomerulosa the 11β-hydroxylation of deoxycorticosterone and the 11β-hydroxylation of 18-OH-DOC are catalyzed by two different enzymes, only one of which is influenced by the sodium status. According to the output of endogenous deoxycorticosterone by the zona glomerulosa in vitro, 11β-hydroxylation appears to be a rate-limiting step during sodium repletion but not during sodium deficiency of rats (J. Müller 1968, 1970; J. Müller and Huber 1969). Adrenals of rats that had been drinking saline produced

100%–200% more deoxycorticosterone than adrenals of control animals when stimulated in vitro by potassium ions or ACTH. In reverse, the adrenal tissue of rats that had been kept on a sodium-deficient diet produced significantly decreased amounts of deoxycorticosterone when incubated with or without aldosterone stimulators (Figs. 8, 11, 14, 16). Moreover, in adrenals of sodium-deficient rats, aldosterone was stimulated to the same extent by the addition of substrate amounts of progesterone, deoxycorticosterone, and corticosterone, whereas in adrenals of rats receiving a complete diet aldosterone production was highest when corticosterone was added (Fig. 23).

Both the early (i.e., cholesterol side-chain cleavage) and the late (i.e., conversion of corticosterone to aldosterone and 18-hydroxycorticosterone) steps of aldosterone biosynthesis that are stimulated by sodium deficiency are catalyzed by mitochondrial cytochrome P-450 mixed-function oxidases. Sodium restriction of rats did not lead to a measurable change in the total mitochondrial content of cytochrome P-450 in the zona glomerulosa, but it increased the magnitude of heat-generated and corticosterone-induced type-I and pregnenolone-induced type-II absorbance changes (R. E. Kramer et al. 1979). This indicated that the increased enzyme activities were mediated at least in part by enhanced binding of cholesterol and corticosterone to mitochondrial cytochrome $P\text{-}450_{SCC}$ and cytochrome $P\text{-}450_{CMO}$ respectively.

5.1.5 Role of the Renin-Angiotensin System

Alterations in aldosterone biosynthesis and secretion in response to changes in sodium intake or sodium balance are widely considered to be mainly or even exclusively mediated by the renin-angiotensin system (Gross et al. 1965; Ganong et al. 1966; J. O. Davis 1967, 1975; G. H. Williams and Dluhy 1972; Aguilera and Catt 1978). This hypothesis is based on the following evidence:

1. Sodium restriction and sodium depletion lead to parallel increases in aldosterone production and in the activity of the renin-angiotensin system. In particular, changes in plasma aldosterone concentration during alterations of sodium balance are accompanied by simultaneous and commensurate changes in plasma renin activity and plasma angiotensin II concentration.
2. Short-term and long-term effects of sodium deficiency on aldosterone biosynthesis can be imitated by exogenous renin or angiotensin.
3. Stimulation of aldosterone production by sodium depletion is absent in bilaterally nephrectomized man and animals.
4. Pharmacological or immunological blockade of the renin-angiotensin system impedes or reverses the stimulatory effects of sodium deficiency on aldosterone production.

A vitally important role of the renin-angiotensin system in aldosterone homeostasis during alterations in sodium balance has been firmly established by the bulk of such evidence. However, the hypothesis that "increased activity of the renin-angiotensin system is the primary mechanism leading to hyperaldosteronism during sodium depletion," as stated by Binnion et al. (1965), is debatable in several respects. In particular, experimental evidence obtained in sheep seems to indicate that the role of the renin-angiotensin system is a permissive rather than a regulatory one (Coghlan et al. 1979a).

In the following paragraphs, only the parallelism of changes in aldosterone production and in the activity of the renin-angiotensin system will be discussed. Experiments with administration of exogenous renin and angiotensin, bilateral nephrectomy, and blockade of the renin-angiotensin system will be treated in separate chapters of this review.

A low sodium intake or sodium depletion leads to increased peripheral plasma levels of renin in man (J.J. Brown et al. 1964a; G.H. Williams et al. 1972b), the dog (Binnion et al. 1965), sheep (Blair-West et al. 1967) and the rat (Gross et al. 1965; Douglas et al. 1978c). Sodium deficiency also induced an increased renin content of the kidneys in rats and rabbits (Gross et al. 1965; Blair-West et al. 1968a). Increases in the plasma angiotensin II concentration were observed during sodium restriction in man by J.J. Brown et al. (1972) and in rats by Aguilera et al. (1980a). However, sodium restriction raised plasma aldosterone concentration and plasma renin acitivity but not plasma angiotensin II concentration in normal human subjects, according to Best et al. (1971) and to Mendelsohn et al. (1972). In another study, marked increases in the plasma aldosterone concentration were accompanied by relatively minor increases in the plasma angiotensin II concentration; much higher plasma angiotensin II concentrations were required to elicit similar increases in plasma aldosterone when angiotensin II was infused into sodium-replete human subjects (G.W. Boyd et al. 1972). Similarly, small increments in the plasma renin activity contrasted with marked increments in the plasma aldosterone concentration during sodium restriction of normal human subjects (Michelakis and Horton 1970). When rats were kept on a sodium-deficient diet for 6 weeks, a 20-fold increase in plasma aldosterone was associated with a less than twofold increase in plasma renin activity (Douglas et al. 1978c). On the other hand, 6 weeks of a high-sodium diet lowered the plasma aldosterone to 54% and the plasma renin activity to 22% of the respective control values.

In normal man, Michelakis and Horton (1970) observed parallel diurnal variations in plasma renin acitivity and in plasma aldosterone under normal sodium intake. But, during sodium restriction, a high plasma level of aldosterone at noon was accompanied by a low level of plasma renin activity. This was not confirmed by Armbruster et al. (1975), who found a closer correlation between plasma aldosterone peaks and plasma renin activity peaks in sodium-restricted than in sodium-replete supine subjects.

In sheep, progressive depletion of sodium up to 900 mmol over 69 h resulted in a progressive and proportional increase in the aldosterone secretion rate, peripheral blood concentrations of aldosterone and angiotensin II, and the plasma renin level (Blair-West et al. 1971e). However, during the rapid correction of the sodium deficit by voluntary drinking of hypertonic $NaHCO_3$ solution, aldosterone secretion fell faster into the normal range than blood angiotensin II or plasma renin levels.

In rats, alterations in sodium intake or sodium balance generally led to homodirectional changes in the activity of the final steps of aldosterone biosynthesis and of the renin-angiotensin system, which were, however, often not proportional. Thus, a 300% increase in the conversion of corticosterone to aldosterone induced by 2 days of dietary sodium restriction in potassium-deficient rats was accompanied by a 50% increase in the plasma renin activity, whereas 2 days of treatment with a natriuretic drug (furosemide) resulted in equal increases (+ 200%) of both

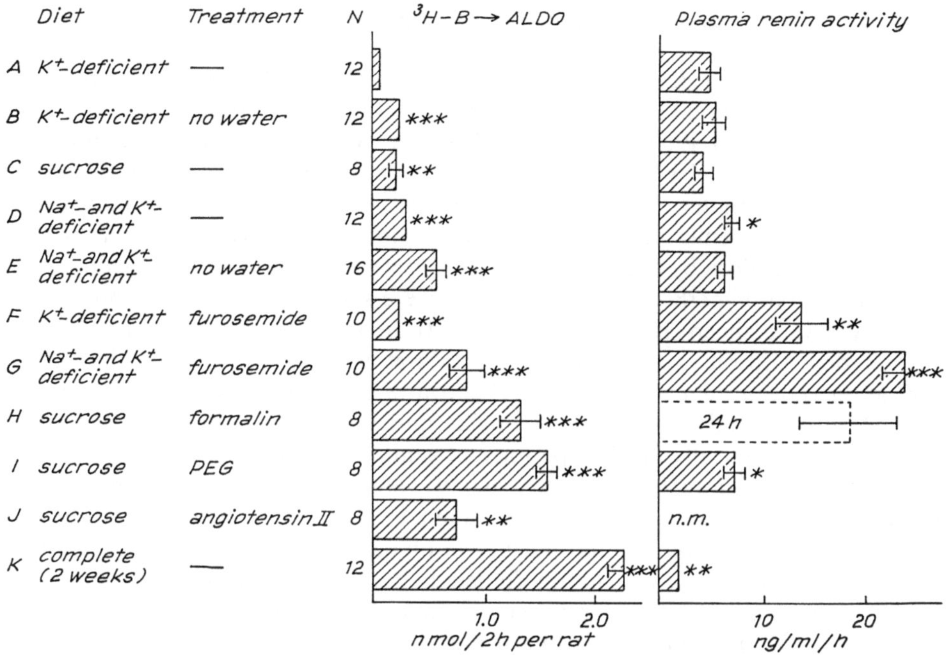

Fig. 24. Effect of various treatments for 2 days in potassium-deficient rats on the conversion of tritiated corticosterone (*^{3}H-B*, 300 nmol/flask) to aldosterone *(ALDO)* by incubated quartered adrenals and on the plasma renin activity. Mean values ± SEM. Capital letters refer to experimental groups of animals. In group *H* the plasma renin activity was measured 24 h after the injection of formalin. Animals from group *K* had been kept on a complete diet for 2 weeks. Statistical significance (*t*-tests) of differences from group *A*: $*P<0.05$, $**P<0.01$, $***P<0.001$. *PEG*, polyethylene glycol; *n.m.*, not measured. (From Haldy and Müller 1981b)

parameters (Haldy and Müller 1981b; Fig. 24). Addition of sodium chloride to the drinking fluid of rats that had been kept on a sodium- and potassium-deficient diet elicited a precipitous drop in plasma renin activity within a day, but the capsular adrenal conversion of corticosterone to aldosterone remained normal for the first 24 h (J. Müller et al. 1984; Fig. 50). A gradual suppression of aldosterone biosynthesis became manifest after 2 days and reached statistical significance after 3 days of sodium repletion.

The suppressive effect of sodium chloride intake or administration on the renin-angiotensin system seems to be at least partly due to chloride ions (Kotchen et al. 1976; Stephens et al. 1978; Abboud et al. 1979; Kotchen et al. 1980). Short-term infusion of NaCl into sodium-restricted normal human subjects elicited a significant drop in plasma renin activity, which was not accompanied by a corresponding change in plasma aldosterone (Julian et al. 1982). In contrast, an $NaHCO_3$ infusion had no significant effect on the plasma renin acitivity while markedly lowering the plasma aldosterone concentration.

5.1.6 Role of Plasma Potassium Concentration

Sodium restriction or sequestration induced significant elevations in the plasma potassium concentration in several experimental studies carried out in rats or dogs (J. Müller and Huber 1969; Aguilera et al. 1980a; Stricker et al. 1979; Nicholls et al. 1978, 1982; Lohmeier et al. 1980; DeForrest et al. 1980). In view of the known short-term and long-term aldosterone-stimulating effects of potassium ions, even moderate hyperkalemia could markedly contribute to the overall stimulation of the zona glomerulosa by sodium deficiency. However, sodium deficiency does not regularly lead to hyperkalemia. Thus, even severe sodium depletion did not cause hyperkalemia in sheep (Blair-West et al. 1974). In a number of studies performed in man, dogs and rats, sodium restriction did not alter the plasma potassium concentration (Lowenstein and Steele 1977; Swartz et al. 1980; R. D. Gordon et al. 1980; Spielman and Davis 1974; Douglas and Catt 1976). Hyperkalemia may occur only periodically during periods of sodium restriction. After 2 days of sodium restriction of rats, J. E. Boyd et al. (1971) found an increased plasma potassium concentration in blood samples taken between 9:30 p.m. and midnight (when the animals were feeding) but not in blood samples taken in the morning. Rapid sodium repletion of sodium-depleted sheep by the drinking of $NaHCO_3$ solution was in most instances associated with a drop in plasma potassium (Blair-West et al. 1971e). Sodium repletion of rats that had been kept on a sodium- and potassium-deficient diet led to a significant drop in the plasma potassium concentration (J. Müller et al. 1984). On the other hand, the suppression of plasma aldosterone in sodium-restricted human subjects by sodium chloride infusion was not accompanied by any significant alteration in plasma potassium (G. H. Williams et al. 1972c).

The stimulatory effect of sodium deficiency and the inhibitory effect of potassium deficiency on aldosterone biosynthesis appear to neutralize each other during combined deficiency of both cations. In normal human subjects, urinary aldosterone excretion rose at a slower rate and to a lesser extent during combined sodium and potassium restriction than during simple sodium restriction (B. B. Johnson et al. 1957). In rats, simultaneous sodium and potassium restriction did not affect the aldosterone secretion rate (Catalona et al. 1972) or the capsular adrenal conversions of deoxycorticosterone and corticosterone to 18-hydroxycorticosterone and aldosterone (J. Müller and Baumann 1973). It resulted in moderately decreased plasma potassium and sodium concentrations and in a very high plasma renin activity (J. Müller et al. 1984; Fig. 25). When simple potassium restriction was followed by a period of combined sodium and potassium restriction, the aldosterone secretion rate remained very low (Catalona et al. 1972), whereas the activity of late steps in aldosterone biosynthesis significantly increased, although not to the level seen in normally fed animals (J. Müller and Baumann 1973; Fig. 26).

5.1.7 Role of Plasma Volume

According to traditional models of sodium homeostasis, alterations in the body sodium status of the mammalian organism influence aldosterone secretion through associated changes in circulating blood volume, which are registered by

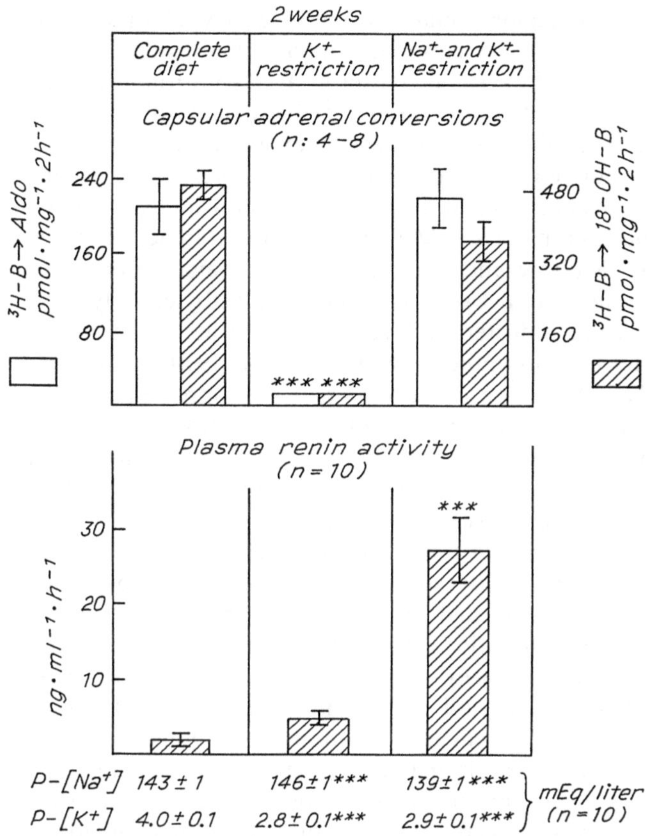

Fig. 25. Effects of 2 weeks of various diets on capsular adrenal conversions of tritiated corticosterone (*^{3}H-B*, 300 nmol/flask) to aldosterone *(ALDO)* and 18-hydroxycorticosterone *(18-OH-B)* and on plasma renin activity, sodium, and potassium in rats. Mean values ± SEM. Statistical significance (*t*-tests) of differences from rats on complete diet: *** $P < 0.001$. (Data partially from J. Müller et al. 1984)

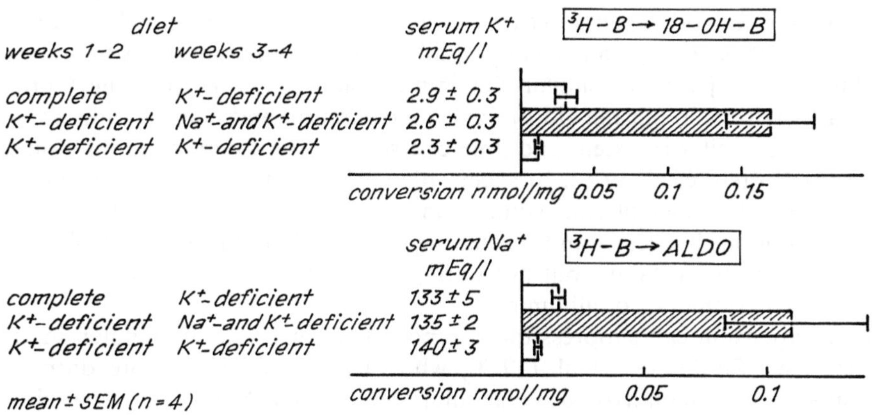

Fig. 26. Effect of various diets on serum electrolytes and on the conversion of tritiated corticosterone (*^{3}H-B*, 300 nmol/flask) to 18-hydroxycorticosterone *(18-OH-B)* and aldosterone *(ALDO)* by incubated rat capsular adrenals. Mean values of two experiments. (Data from J. Müller and Baumann 1973)

"volume receptors", e.g., by the juxtaglomerular cells in the kidneys (Binnion et al. 1965). Even before the discovery of the close functional relationship between aldosterone and the renin-angiotensin system, Bartter et al. (1956) found urinary aldosterone excretion to be dependent mainly on extracellular fluid volume. Later studies have confirmed that, in the absence of changes in external sodium balance, aldosterone production is stimulated by hypovolemia and suppressed by general or central blood volume expansion. Thus, aldosterone secretion increases markedly in response to hemorrhage in man (A.F. Muller 1963) and in intact as well as in hypophysectomized dogs (Mulrow and Ganong 1961). In rats, hypovolemia due to extracellular fluid sequestration by subcutaneous polyethylene glycol or formalin leads to elevated plasma aldosterone levels and to an increased activity of the final steps in aldosterone biosynthesis (Stricker et al. 1979; Haldy and Müller 1981 a, b). Conversely, aldosterone secretion in man is suppressed by blood transfusion (A.F. Muller 1963), by dextran infusion (Tuck et al. 1974), or by central hypervolemia due to immersion in water up to the neck (M. Epstein and Saruta 1971; Crane and Harris 1974). Although severe sodium depletion leads to both hypovolemia and hyperaldosteronism, several studies carried out in man or sheep did not confirm the general importance of plasma volume changes in the adaptation of aldosterone secretion to the body sodium status. Bull and co-workers (1970) induced sodium depletion in normal human subjects by dietary sodium restriction and by treatment with a diuretic. This resulted in a reduction in blood volume, which was paralleled by increases in plasma renin activity and in the aldosterone secretion rate. During a period of sodium repletion, blood volume was maintained below normal by repeated phlebotomy and plasmapheresis. Nevertheless, aldosterone secretion and plasma renin acitivity were suppressed to control or near-control levels. According to Dalakos and Streeten (1975), a low sodium intake did not result in any decrease in the plasma volume of normal human subjects while eliciting a fourfold increase in the aldosterone secretion rate. In the sodium-deficient subjects, the expansion of plasma volume by albumin infusion did not suppress aldosterone secretion. In contrast, acute blood volume expansion by dextran infusion into normal human subjects on a low-sodium diet led to significant decreases in both plasma aldosterone concentration and renin activity (Tuck et al. 1974). These effects became manifest after 2 h, whereas the suppression of these two parameters by a saline infusion became significant after only 10 min. In subjects on a similar diet, immersion in water up to the neck produced marked drops in plasma aldosterone concentration and plasma renin activity, which were similar in extent and time course to those observed during saline infusion (M. Epstein et al. 1979). In sheep, Blair-West et al. (1976b) observed an inverse relation between plasma volume and the aldosterone secretion rate in established sodium deficiency and during the subsequent intravenous infusion of 6% dextran in isotonic saline, but not during the intravenous infusion of saline (isotonic or hypertonic) or of albumin.

Volume expansion can suppress the adrenocortical response to angiotensin II, as reported by Birkhäuser et al. (1973), who observed an immediate drop in plasma aldosterone when hyperosmolar solutions of NaCl, $NaHCO_3$, or mannitol were infused into normal human subjects during constant infusion of angiotensin II. Hyperoncotic solutions of dextran in isotonic NaCl also caused an immediate drop in plasma aldosterone, whereas dextran in isotonic glucose solution had a delayed suppressive effect. Dextran infusion did not measurably alter the plasma potassium concentration, but the addition of small amounts of KCl to dextran solutions completely blocked their aldosterone-suppressing action. Volume depletion by treatment with a diuretic (ethacrynic acid) during constant infusion of angiotensin II elicited an additional rise in plasma aldosterone concentration (Gaillard et al. 1976). This effect appeared to have been due partly to a stimulation of ACTH secretion (it was accompanied by an increase in plasma cortisol and was prevented by dexamethasone infusion) and partly to a decrease in the metabolic clearance rate of aldosterone.

Stimulation of aldosterone production by hypovolemia is critically dependent on the kidneys. In most instances, hemorrhage did not stimulate aldosterone secretion in nephrectomized dogs (Ganong and Mulrow 1962). The stimulation of the last steps in aldosterone biosynthesis in potassium-deficient rats by subcutaneous polyethylene glycol or formalin was prevented by bilateral nephrectomy (Haldy and Müller 1981a; Fig. 27). A polyethylene glycol-induced increase in plasma aldosterone of nephrectomized potassium-replete rats observed by Stricker et al. (1979) may have been mediated by ACTH and hyperkalemia. A rise in plasma aldosterone provoked by hemorrhage in nephrectomized and decapitated dogs was probably due to a marked increase in the plasma potassium concentration (R. E. McCaa et al. 1973a).

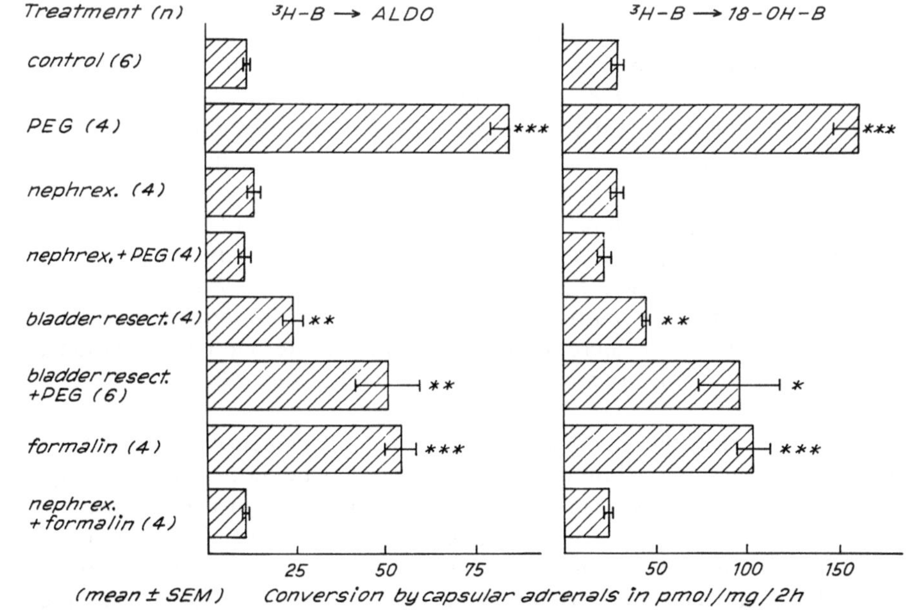

Fig. 27. Effects of sodium sequestration by subcutaneous polyethylene glycol *(PEG)* or formalin in intact, bilaterally nephrectomized *(nephrex.)* and uremic (bladder resection, kidneys intact) potassium-deficient rats on the capsular adrenal conversions of tritiated corticosterone (*^{3}H-B*, 300 nmol/flask) to aldosterone *(ALDO)* and 18-hydroxycorticosterone *(18-OH-B)*. Duration of treatment, 1 day. Statistical significance (*t*-tests) of differences from control animals: $*P<0.05$, $**P<0.01$, $***P<0.001$. (From Haldy and Müller 1981a)

5.1.8 Role of Plasma Sodium Concentration

During the rapid correction of sodium deficits in sheep by either voluntary drinking of hypertonic $NaHCO_3$ solution or intravenous infusion of hypertonic NaCl solutions, the observed drop in the aldosterone secretion rate correlated better with increasing plasma sodium concentrations than with increases in plasma volume or decreases in plasma renin or angiotensin II concentrations (Blair-West et al. 1967b, 1971e). According to experiments in which hypertonic $NaHCO_3$ was infused into the adrenal artery of sodium-deficient sheep, the direct adrenal action of the ionic changes in the plasma was relatively small and did not fully account for the aldosterone-suppressive effect of sodium repletion (Blair-West et al. 1971b). Altogether, these studies pointed to a major effect of plasma sodium concentration on an as yet unknown extraadrenal aldosterone control system.

Evidence for the existence of an unknown aldosterone-stimulating principle acting in response to hyponatremia in rats has been presented by Spät and co-workers, who induced an acute sodium depletion - not accompanied by water depletion - by injecting isotonic glucose solution into the peritoneal cavity (Spät et al. 1979b; Balla et al. 1981). Induced increases in the serum aldosterone concentration correlated significantly with decreasing plasma sodium concentrations but not with increasing levels of plasma renin activity. The rise in serum aldosterone was not affected by treatment of the animals with the converting-enzyme inhibitor teprotide (SQ 20881) or with the prostaglandin-synthesis inhibitor indomethacin. The ratio of serum aldosterone to plasma renin activity did not change when plasma sodium fell from 145 to 132 mmol/l, but it rose steeply at serum sodium concentrations below 132 mmol/l. Apparently, this was not due to a direct action of low ambient sodium concentrations on the adrenal cortex, because isosmotic or hyposmotic reduction of the sodium concentration in the incubation medium from 145 to 125 mmol/l did not affect aldosterone production of rat zona glomerulosa cells, either under basal conditions or under stimulation by angiotensin II (Enyedi and Spät 1981). On the other hand, nephrectomy or treatment with the converting-enzyme inhibitor captopril completely blocked the stimulation of aldosterone biosynthesis in potassium-depleted rats by subcutaneous polyethylene glycol, despite a persistent decrease in plasma sodium (J. Müller et al. 1982). Water deprivation stimulated the conversion of corticosterone to aldosterone in potassium-deficient rats and led to a significant increase in the plasma aldosterone concentration in normal rats, irrespective of hypernatremia (Haldy and Müller 1981b; Lucke et al. 1980).

Experiments carried out in dogs have also yielded evidence pointing to an important role of plasma sodium concentration - independent of volume changes and the renin-angiotensin system - in the physiological control of aldosterone secretion. Thus, chronic hypotonic expansion of body fluids induced by the administration of vasopressin and unrestricted water intake resulted in a drop in plasma sodium concentration to an average of 115 mmol/l and in marked increases in the plasma concentration and urinary excretion of aldosterone (Cohen et al. 1976). In another study, in which water intake was limited and sodium intake was restricted to 10 mmol/d, a 4-day infusion of vasopressin lowered the plasma sodium from 140 to 137 mmol/l but did not affect plasma aldosterone concentration or renin activity (Young and McCaa 1981b). A sustained hypernatremia, induced in previously sodium-depleted dogs by means of a high sodium intake, a fixed water intake, and a continuous infusion of vasopressin, led to a sustained suppression of plasma aldosterone concentration, but to only a transient fall in plasma renin activity (D. B. Young and McCaa 1981 a). Infusion of hypotonic or hypertonic NaCl solutions into sodium-depleted dogs for 22 h brought about a progressive fall in plasma aldosterone concentration but not in plasma renin activity (Childers and Schneider 1981). Irrespective of equal sodium loads, plasma aldosterone dropped to significantly lower levels during hypertonic than during hypotonic NaCl infusion. Neither water deprivation nor resumption of water intake induced any change in plasma aldosterone in dogs on a normal sodium intake, although plasma renin acitivity rose significantly during dehydration and returned to control levels upon rehydration (Zucker et al. 1982). Most likely, the effects of volume changes on aldosterone production had been offset by opposite changes in the plasma sodium concentration.

In man, plasma volume changes seem to be more important than plasma sodium concentration in the adaptation of aldosterone secretion to the body sodium status, according to several investigators. Treatment of human subjects on a normal or low sodium intake with vasopressin caused decreases in urinary aldosterone secretion in a period of hyponatremia, whereas dehydration by water restriction stimulated aldosterone secretion while raising the plasma sodium concentration (Bartter et al. 1956; A. F. Muller et al. 1956 a; Cox et al. 1961). However, evidence for a control mechanism that stimulates aldosterone secretion in response to hyponatremia - in the absence of volume changes and independent of the renin-angiotensin system - has been found in anephric human subjects by R. E. McCaa et al. (1973 b). In three nephrectomized patients, reduction of plasma sodium by hemodialysis from 138 to 125 mmol/l without a change in body fluid volume or plasma potassium concentration led to a threefold increase in the plasma aldosterone level.

5.1.9 Involvement of the Central Nervous System

The following experimental evidence obtained in sheep and dogs points to the involvement of the central nervous system in yet unknown control mechanisms of aldosterone secretion during sodium deficiency.

When sodium deficiency of sheep was compensated by voluntary drinking of $NaHCO_3$ solution, the aldosterone secretion rate decreased in a biphasic manner (Denton et al. 1977). It fell rapidly within 15-36 min following satiation of sodium appetite. After 1 h, aldosterone secretion returned toward its pre-drinking rate. In the course of the next 2-5 h it dropped again into the sodium-replete range as a result of sodium absorption from the gut. When the same amount of $NaHCO_3$ was given through a rumen tube, aldosterone secretion decreased slowly in a monophasic manner. The primary inhibition of aldosterone secretion by satiation of sodium appetite was not due to changes in plasma sodium, potassium, angiotensin II, or ACTH concentrations. Most likely, the secretion of an unknown inhibitor had been elicited by a gustatory-alimentary reflex mechanism. Perfusion of the brain ventricles by indwelling cannulae with an artificial cerebrospinal fluid with a high sodium content (170 mmol/l) in conscious sodium-deficient sheep induced a progressive drop in the blood aldosterone concentration, which became statistically significant after 5 h (Abraham et al. 1976). The drop in aldosterone secretion could not have been caused by an increase in plasma sodium or decreases in plasma potassium, renin, or angiotensin II concentrations. It was also not due to ACTH suppression, because the decrease in blood aldosterone was accompanied by increases of cortisol and corticosterone levels.

In anesthetized sodium-deficient dogs, the intracerebro-ventricular infusion of angiotensin II lowered the plasma aldosterone concentration, but it also decreased the plasma renin activity and potassium concentration (Brooks and Malvin 1981). Additional evidence for an involvement of endogenous brain angiotensin II in the regulation of aldosterone secretion was found by the same investigators in experiments in which the converting-enzyme inhibitor teprotide (SQ 20881) or the angiotensin antagonist sar^1, ala^8-angiotensin II was intraventricularly infused. Both agents elicited a significant rise in plasma aldosterone, which occurred independent of concomitant changes in plasma renin acitivity or in plasma sodium, potassium, or cortisol concentrations.

5.2 Alterations in Potassium Intake or in Potassium Balance

5.2.1 General Considerations

Since aldosterone is not only a sodium-retaining but also a kaliuretic hormone, the physiological importance of an adaptation of aldosterone secretion to the potassium status of the organism is obvious. The influence of potassium balance on aldosterone biosynthesis and secretion has not been studied as extensively as the influence of sodium balance. However, all the available evidence indicates that potassium intake and total body potassium may be as important as sodium intake and total body sodium in the overall control of aldosterone production. In general, homodirectional changes in sodium balance and in potassium balance have opposite effects on aldosterone biosynthesis. Thus, increased ingestion or parenteral administration of potassium stimulate, while potassium depletion or dietary potassium deprivation suppress aldosterone secretion in man and in experimental animals (Singer and Stack-Dunne 1955; Falbriard et al. 1955; Laragh and Stoerk 1957; B.B.Johnson et al. 1957; Gann et al. 1962, 1964; Eilers and Peterson 1964; Cannon et al. 1966).

According to A.F.Muller et al. (1958a), potassium balance could influence aldosterone production indirectly by its effect on sodium balance and associated changes in the intravascular volume. It is known that potassium loading results in sodium loss (Bunge 1873), potassium depletion in sodium retention (Black and Milne, 1952; Womersley and Darragh 1955). However, long-term potassium loading generally stimulates aldosterone secretion while suppressing the plasma renin acitivity (see p. 133). Thus, sodium losses due to long-term increases in potassium intake could stimulate aldosterone secretion only by mechanisms not involving the renin-angiotensin system.

Alterations in the total body potassium content are usually accompanied by homodirectional changes in the plasma potassium concentration. Because it has been amply documented that extracellular potassium ions exert direct short-term effects on zona glomerulosa cells, either by stimulating aldosterone biosynthesis on their own or by influencing the steroidogenic response to other stimulators (see pp. 37-41), it is at least very tempting to assume that long-term effects of potassium balance on aldosterone biosynthesis are also directly mediated by the extracellular potassium concentration. However, with the notable exception of the cell culture study by Hornsby and O'Hare (1977), there is no direct evidence supporting this assumption. Cannon et al. (1966) have proposed that long-term changes in aldosterone production depend rather on the intracellular potassium content of the adrenal cortex than on the extracellular potassium concentration. Experimental evidence obtained in dogs by Gann et al. (1962) indicated that potassium ions could act at some intracranial site to stimulate aldosterone production by an as yet unknown extra-adrenal mechanism.

5.2.2 Morphological Changes

The zona glomerulosa width of rat adrenals was found to be decreased by approximately 30% after 2, 5, or 10 weeks of dietary potassium restriction (Deane et al. 1948; J. Müller and Huber 1969). In reverse, parenteral administration of potassium resulted in a marked increase of the zona glomerulosa width within 12 h (Deane et al. 1948). Addition of rubidium sulfate or cesium chloride to a sodium- and potassium-deficient diet elicited hypertrophy of the zona glomerulosa of rat adrenals, according to Bach et al. (1960).

An acute potassium load in hypophysectomized rats caused within 36 h some distinct ultrastructural changes in the mitochondria of the zona glomerulosa cell, i.e., mitochondrial vacuolation and open-ended mitochondria (Lever 1956). According to Domoto et al. (1973), both a low-sodium and a high-potassium diet led within 8 days to the same distinct ultrastructural changes in the zona glomerulosa of the rat adrenal, consisting of a loss of the mitochondrial matrix and a change from platelike to tubulovesicular cristae. Lustyik et al. (1977) observed two types of mitochondria in the zona glomerulosa cells of control rats and of potassium-treated rats, i.e., mitochondria with tubular cristae and mitochondria with tubulosaccular cristae. Potassium loading for 4 days resulted in an increased number of mitochondria with tubular cristae.

Potassium loading for 2 days did not result in any significant alteration in the zona glomerulosa and zona fasciculata sodium or potassium contents in rats (Decorzant et al. 1977).

5.2.3 Duration of Stimulatory Effects of Potassium

A high potassium intake or long-term parenteral administration of potassium leads to sustained elevations in aldosterone secretion rate or in plasma aldosterone levels. In normal human subjects, an increase of potassium intake from 40 to 200 mmol/day for 8-10 days during a constantly high sodium intake (200 mmol/day) resulted in a minor elevation in the plasma potassium concentration from 4.0 to 4.2 mmol/l; this was, however, accompanied by a substantial increase in plasma aldosterone from 4.1 to 23 ng/dl (Dluhy et al. 1972). In dogs, a prolonged intravenous infusion of KCl (250 mmol/d) for 16 days led to a sustained elevation of the plasma aldosterone concentration from 6.8 to 21.8 ng/dl, associated with an elevation of the plasma potassium concentration from 4.1 to 4.9 mmol/l (R. E. McCaa et al. 1975). In rats, oral potassium loading stimulated the aldosterone secretion rate from 6.0 to 15.6 ng/min within 2 days and to 17.3 ng/min within 8 days (J. E. Boyd et al. 1971). The respective plasma potassium concentrations were 3.9 (control), 4.5 mmol/l (2 days), and 4.1 mmol/l (8 days). In response to an intravenous KCl infusion into potassium-deficient rats, the serum potassium concentration rose from 2.9 to 8.4 mmol/l within 12 h and returned to 5.3 mmol/l within 48 h (Komor and Müller 1979). The serum aldosterone rose from 6 to 142 ng/dl within 12 h and remained stable during the next 36 h of potassium infusion.

5.2.4 Alterations in Zona Glomerulosa Sensitivity

In man, long-term alterations in potassium intake significantly modify the response in plasma aldosterone or in the aldosterone secretion rate to the short-term action of various stimulators. A high potassium intake stimulated not only the basal secretion rate of aldosterone but also the response of this parameter to a continuous 48-h infusion of ACTH (G. H. Williams et al. 1970). The rise in plasma aldosterone elicited by a KCl infusion was greater in subjects on a 200 mmol/day potassium diet than in subjects on a 40 mmol/day potassium diet (Dluhy et al. 1972). The plasma aldosterone/plasma potassium curve was steeper in subjects with the high than in subjects with the low potassium intake. In contrast, a low sodium intake did not increase the steepness of this curve. Similarly, during angiotensin II infusions, the plasma aldosterone/plasma angiotensin II curve was steeper in subjects with a potassium intake of 200 mmol/day than in subjects with potassium intakes of 100 and 40 mmol/day (Hollenberg et al. 1975).

In rats, 3 days of dietary potassium deprivation markedly attenuated the in vivo response in plasma aldosterone to increasing doses of exogenous angiotensin II or angiotensin III, whereas 3 days of a high potassium intake raised the plasma aldosterone concentration and enhanced the increase of this parameter in response to the two peptides (W. B. Campbell and Schmitz 1978).

Treatment of rats with a low-potassium diet led to a slight decrease in the zona glomerulosa cell angiotensin II receptor affinity within 6 weeks and to a decreased number of binding sites within 1 week (Douglas and Catt 1976). A high-potassium diet caused a significant decrease in angiotensin II receptor affinity, which was manifest after 1 week but no longer after 6 weeks, and a marked increase in the zona glomerulosa angiotensin II binding sites, which was apparent after 1 week and 6 weeks. Shorter periods of increased potassium intake, i.e., 3–7 days, lowered the association constant and raised the concentration of angiotensin II receptors in zona glomerulosa cells (Douglas 1980a). The dose-response curve of aldosterone production by incubated zona glomerulosa cells elicited by increasing concentrations of angiotensin II was shifted to the right; the maximal aldosterone output was increased by a factor of 1.9. By contrast, the ACTH: aldosterone dose-response curve was shifted to the left by the high potassium intake. The effects of a high potassium intake induced on uterine angiotensin receptors were directly opposite to the ones induced on the zona glomerulosa angiotensin receptors (Douglas 1979). Thus, potassium loading led to an increase in the equilibrium association constant of myometrial angiotensin receptors but to a significant decrease in their number. Simultaneous treatment with the angiotensin converting-enzyme inhibitor captopril prevented most of the effects of a high potassium intake on smooth muscle and adrenal angiotensin receptors, but it did not block the increase in the number of adrenal receptors (Douglas 1980b). When a high potassium diet was combined with an angiotensin II infusion, the two treatments yielded cumulative stimulatory effects on the number of adrenal angiotensin receptors.

5.2.5 Site of Action in the Biosynthetic Pathway

Whereas potassium ions exert their short-term steroidogenic action mainly on early steps and have only minor effects on late steps of aldosterone biosynthesis (see p. 90), long-term alterations in potassium balance or intake strikingly affect late steps of the biosynthetic pathway. Quartered adrenals of rats kept on a potassium-deficient diet for 2 weeks produced significantly less aldosterone but more corticosterone and deoxycorticosterone than adrenals of rats kept on a complete diet (J. Müller and Huber 1969). According to incubations of separated zones of the adrenal cortex, the increased output of corticosterone and deoxycorticosterone derived entirely from the zona glomerulosa (J. Müller 1970b). Capsular adrenals of potassium-depleted rats also produced more 18-OH-DOC but less 18-hydroxycorticosterone than capsular adrenals of potassium-repleted rats (Meuli and Müller 1983b). In quartered adrenals or capsular hemiadrenals of potassium-deficient rats, aldosterone production no longer responded to stimulation by serotonin, potassium ions, angiotensin II, or ACTH, whereas the response in deoxycorticosterone production to these stimulators was greatly increased (J. Müller and Huber 1969; J. Müller 1970b, Figs. 8, 11, 14, 16). However, potassium deficiency did not affect either the basal or the ACTH-induced deoxycorticosterone output by the decapsulated portion of the adrenal gland. Altogether, these observations indicated that potassium deficiency affected corticosteroid biosynthesis only in the zona glomerulosa, left aldosterone biosynthesis intact up to the formation of deoxycorticosterone, and suppressed the activity of one or more of the enzymes involved in the conversion of this intermediate product to aldosterone.

When potassium chloride was added to the drinking fluid of rats for 2 weeks, the adrenal tissue paradoxically produced less aldosterone under basal conditions of incubation than quartered adrenals of control animals, but equal amounts under in vitro stimulation by potassium ions, ACTH, or added progesterone, 11β-hydroxyprogesterone, deoxycorticosterone, or corticosterone (J. Müller 1968). However, in these experiments the rats were kept on a diet with a relatively low sodium (50 mmol/kg) and high potassium (187 mmol/kg) content, and aldosterone biosynthesis may already have been maximally stimulated under control conditions. According to Regöly-Mérey and Solyom (1975) and to Debreceni and Csete (1975), quartered adrenals of potassium-loaded rats produced increased amounts of aldosterone when incubated in a medium with a potassium concentration of 5.9 mmol/l. Under basal conditions of incubation, aldosterone production by capsular adrenals or zona glomerulosa cells of potassium-loaded rats was normal or slightly decreased (Regöly-Mérey and Solyom 1975; Kenyon et al. 1978a; Douglas 1980a). However, an increased capacity for aldosterone production became apparent under in vitro stimulation by potassium ions, ACTH, or added corticosterone. On the other hand, zona glomerulosa cells from normal human subjects with a high potassium intake produced increased amounts of aldosterone when incubated with or without stimulators (G.H. Williams and Braley 1977).

Oral potassium loading of rats for 2 days resulted in an increased adrenal mitochondrial conversion of corticosterone to aldosterone, even when potassium-induced sodium losses were compensated (J.E. Boyd et al. 1971). The increased conversion became apparent within 4 h, reached a maximum within 48 h, and was maintained for 8 days. Incubated capsular adrenals or capsular adrenal mitochondria of potassium-deficient rats converted 20–30 times less tritiated corticosterone to aldosterone and 18-hydroxycorticosterone than capsular adrenals or capsular adrenal mitochondria of potassium-replete rats (K. Baumann and Müller 1972a; Fig. 28). This difference was demonstrable over a wide range of substrate concentrations. Conversion to both corticosteroid fractions increased within 4 h and became normal within 24 h of resumed potassium intake. A statistically significant increase in the conversion of corticosterone to aldosterone was observed after 12 h of oral (K. Baumann and Müller 1972a) and after 6 h of intravenous potassium loading (Komor and Müller 1979). Kinetic data, as well as the time course of increasing conversion during potassium repletion (Fig. 49), suggested that the potassium-induced stimulation of the final steps of aldosterone biosynthesis was most likely due to de novo enzyme synthesis. Incubated capsular adrenal tissue or mitochondria of potassium-replete rats converted 80% more 11-deoxycortisol to

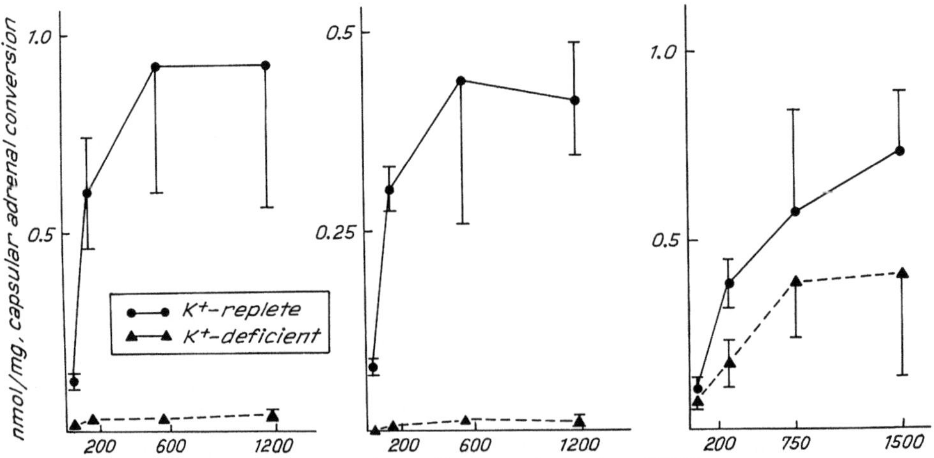

Fig. 28. Effect of 2 days of potassium repletion of potassium-deficient rats on the capsular adrenal conversions of substrate amounts of tritiated corticosterone *(^{3}H-B)* to 18-hydroxycorticosterone *(18-OH-B)* and aldosterone *(ALDO)* and of tritiated 11-deoxycortisol *(^{3}H-S)* to cortisol *(F)*. Mean values of three experiments ($n=3$) ± standard deviation. (Data from K. Baumann and Müller 1972a, 1972b)

cortisol than capsular adrenal tissue or mitochondria of potassium-deficient animals (K. Baumann and Müller 1972b; Fig. 28). The conversion of tritiated deoxycorticosterone to 11β-hydroxylated steroids was 130% higher in capsular adrenals of potassium-replete than in capsular adrenals of potassium-deficient rats. However, this difference was due entirely to an increased conversion of the precursor to aldosterone and 18-hydroxycorticosterone, whereas the capsular adrenal conversions to corticosterone and to 18-OH-DOC remained unchanged. Potassium restriction of rats led to marked and statistically significant decreases in the capsular adrenal conversions of 18-hydroxycorticosterone (18-OH-B) and 18-OH-DOC to aldosterone and of 18-OH-DOC to 18-OH-B (J. Müller 1980a, 1980b; Figs. 29, 30).

The metabolism of tritiated deoxycorticosterone in primary monolayer cultures of rat adrenal glomerulosa cells was markedly influenced by the potassium concentration of the medium (Hornsby and O'Hare 1977). When the potassium concentration of the culture medium was 8, 12, or 40 mmol/l, the metabolism followed a "glomerulosa-like" pattern, with the major part of the precursor being converted to aldosterone and 18-OH-B. In media with a high potassium concen-

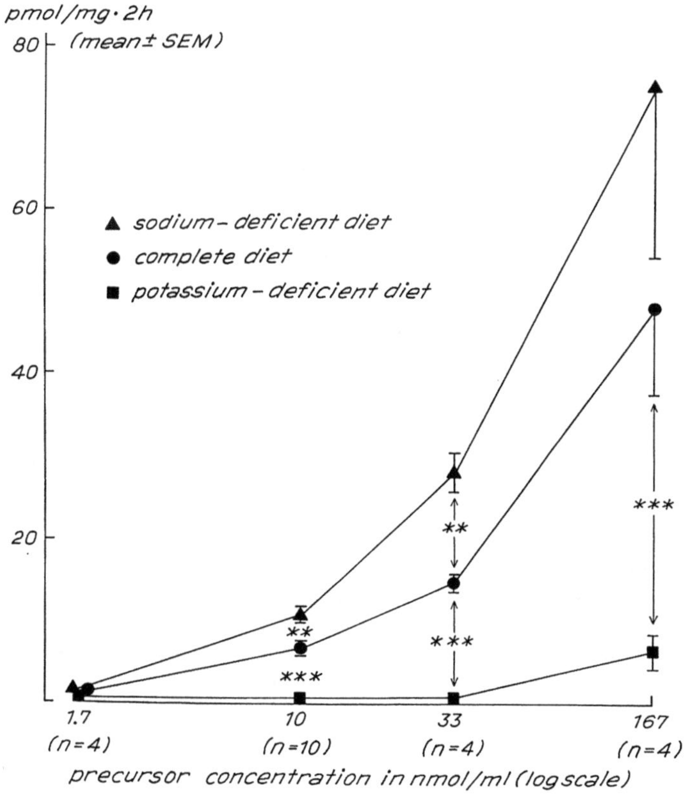

Fig. 29. Effects of sodium-deficient and potassium-deficient diets 2 weeks) on the conversion of tritiated 18-hydroxycorticosterone (^{3}*H-18-OH-B)* to aldosterone by rat capsular adrenals. Statistical significance (*t*-tests) of the difference between values on adjacent curves: ** $P<0.01$, *** $P<0.001$. (From J. Müller 1980a)

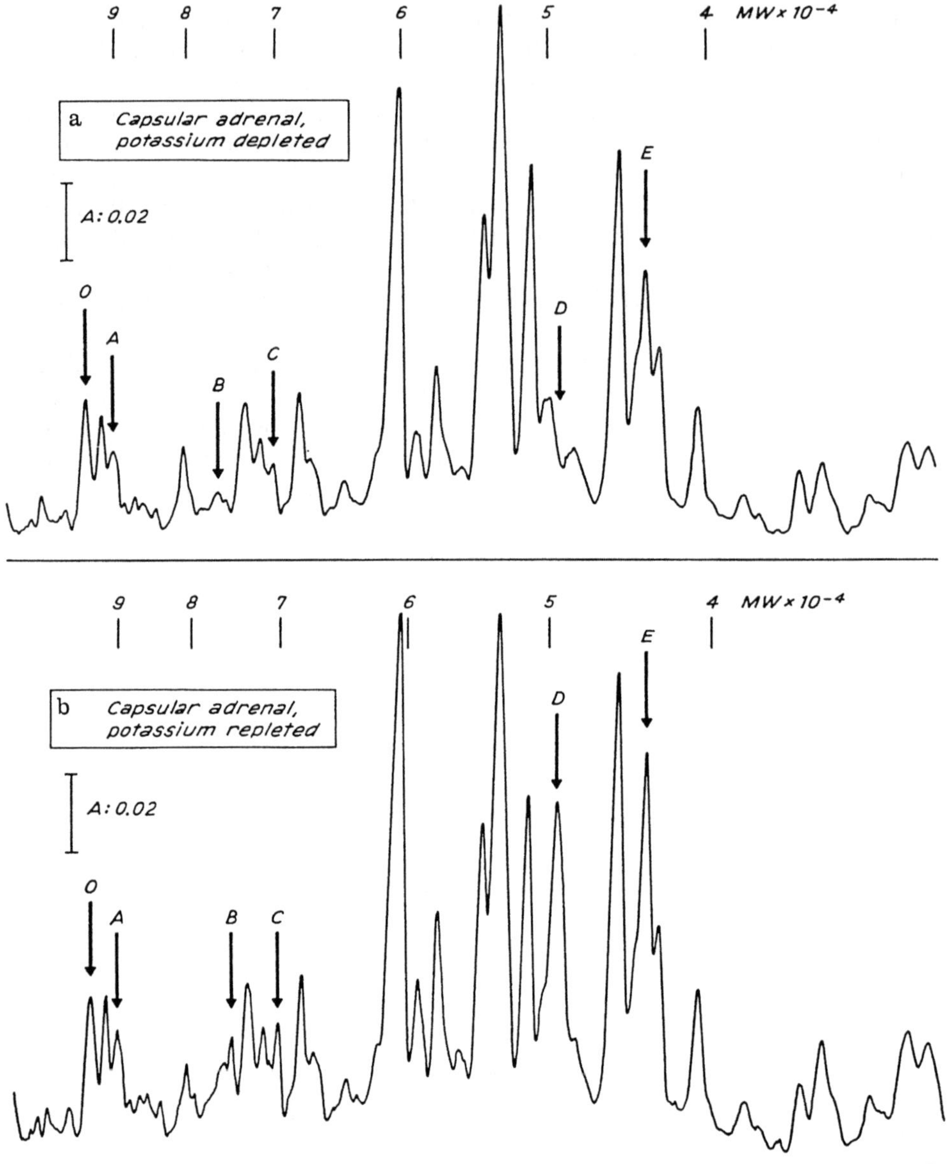

Fig. 31. SDS-PAGE of capsular adrenal mitochondrial proteins shown by densitometric scanning of the Coomassie blue-stained gel. *A)*, rats kept for 2 weeks on a low-potassium diet; *B)*, rats given KCl in their drinking fluid for 2 days after 2 weeks on a low-potassium diet. *Letters A to E* designate protein bands that were reproducibly increased on potassium repletion. (From Meuli and Müller 1983)

mitochondria of potassium-replete but not of potassium-depleted rats. In these mitochondria, a type-I-like absorbance change could be induced by 18-OH-DOC but not by 18-OH-B. Apparently, potassium repletion of potassium-depleted rats specifically increased the binding of corticosterone and perhaps also of 18-OH-DOC to a mitochondrial P-450 enzyme complex. This increased binding could have been due either to an increase in enzyme protein concentration or to a greater availability of the substrate to the enzyme.

Provided that the conversions of corticosterone to 18-OH-B and to aldosterone and the conversion of 18-OH-DOC to 18-OH-B are catalyzed by the same mitochondrial mixed-function oxidase ("steroid 18-methyl oxidase"), an increased activity of this enzyme could account for most of the changes in zona glomerulosa corticosteroid biosynthesis induced by potassium repletion. On the other hand, potassium intake apparently has only a minor influence on the ("11β-hydroxylase") activity of mitochondrial cytochrome P-$450_{11\beta}$ and does not affect cytochrome P-450_{SCC}.

5.2.6 Potassium-Induced Mitochondrial Protein

The stimulation of late steps in aldosterone biosynthesis by potassium repletion of potassium-depleted rats was accompanied by the appearance of a protein in the mitochondria of the zona glomerulosa (Meuli and Müller 1983b). According to SDS gel electrophoresis, five different capsular adrenal mitochondrial protein fractions significantly increased upon potassium repletion (Fig. 31). Among them, "protein D" showed the greatest increase. Its apparent molecular weight of 49000 was very similar to the molecular weights of two known adrenocortical mitochondrial mixed-function oxidases, cytochromes P-450_{SCC} and P-$450_{11\beta}$. Protein D was not found, even in traces, in mitochondria of the decapsulated portion of rat adrenals. In capsular adrenal mitochondria, its concentration rose in parallel with the activity of the final steps of aldosterone biosynthesis during potassium repletion. All these observations indirectly indicated that protein D might represent the steroid 18-methyl oxidase (see p. 19), which catalyzes the conversion of corticosterone to aldosterone and 18-hydroxycorticosterone. However, this assumption needs confirmation. The role of the other potassium-dependent capsular adrenal mitochondrial proteins A, B, C, and E is as yet completely unknown.

5.2.7 Potassium Intake and the Renin-Angiotensin System

In certain situations, potassium intake stimulates aldosterone secretion while suppressing the plasma renin activity, as was first observed by Veyrat et al. (1967a, 1967b) when they gave potassium supplements to sodium- and potassium-deficient human subjects. Such a dissociation between plasma renin activity and aldosterone secretion rate was also seen by Driessen et al. (1969) and by Dluhy et al. (1970) when they were studying the effects of oral potassium loading during dietary sodium restriction in normal subjects and in patients with essential or renovascular hypertension. According to Brunner et al. (1970), a high potassium intake suppressed the plasma renin activity in 18 of 28 studies performed in normal human subjects or hypertensive patients; in some instances, this suppression

Kotchen et al. (1976) treated rats first with a sodium- and chloride-deficient, normal-potassium diet for 1 week and then added various salts to their drinking fluid while keeping them on the same diet for another 8–9 days. Addition of KCl but not of $KHCO_3$ to the drinking fluid suppressed the plasma renin activity and the renal renin concentration. This observation indicated an important role for the chloride ion in the suppression of the renin-angiotensin system by KCl intake. However, Kirchner and Mueller (1982) found that 90-min intravenous infusions of potassium chloride, potassium acetate, potassium bicarbonate, and potassium nitrate into rats kept on a sodium- and chloride-deficient diet induced similar decreases in the plasma renin activity. Addition of $KHCO_3$ to the drinking fluid of rats kept on a sodium-, potassium-, and chloride-deficient diet substantially lowered the excessively high plasma renin activity but was less effective than the addition of KCl (J. Müller 1986). Stephens et al. (1978) had previously observed that renal arterial infusions of potassium lactate or potassium sulfate also suppressed the renin secretion in dogs with a constriction of the thoracic caval vein.

The suppressive effect of potassium intake on renin release does not seem to be due to a direct action of potassium ions on the juxtaglomerular epitheloid cells. Thus, the renin release by rat kidney slices was not affected by a low (2.4 mmol/l) or a high (9.5 mmol/l) potassium concentration in the incubation medium (Aoi et al. 1974). However, renin release by incubated rat renal cortical slices was abolished in a potassium-free medium or in a medium with a potassium concentration of 1 mmol/l (Churchill and Churchill 1979). A near maximal renin release occurred in a medium with a potassium concentration of 2 mmol/l, and further increases in the potassium concentration had only a slight additional stimulatory effect.

5.3 Exogenous Angiotensin or Renin

If the renin-angiotensin system plays a primary role in the regulation of aldosterone secretion under physiological conditions as well as in spontaneous or experimental secondary hyperaldosteronism, as has been postulated by Ganong et al. (1966) and J.O. Davis (1967), several criteria should be fulfilled, according to Ames et al. (1965):

1. Angiotensin should stimulate aldosterone secretion in nonpressor doses or mild pressor doses, since secondary hyperaldosteronism is generally not accompanied by arterial hypertension.
2. Angiotensin should stimulate aldosterone secretion selectively, since hyperaldosteronism is generally not accompanied by increased glucocorticoid secretion.
3. Aldosterone secretion should remain elevated as long as the blood angiotensin level is elevated.

occurred in spite of marked sodium diuresis. In the other ten studies the plasma renin activity remained unchanged. Dietary potassium restriction led to an increase in plasma renin activity in all of six studies. Short-term graded KCl infusions into normal subjects on a low-sodium diet led to an increase in plasma aldosterone, which preceded a measurable increase in the plasma potassium concentration, but they did not affect the plasma renin activity (Himathongkam et al. 1975). On the other hand, oral glucose loading of these subjects induced a rise in plasma renin activity, which was probably due to a drop in the plasma potassium concentration and was accompanied by a decrease in plasma aldosterone. In a later study by Dluhy et al. (1977), an intravenous KCl infusion raised the plasma aldosterone concentration and left the plasma renin activity unchanged when given alone to subjects on a low-sodium diet. When KCl was infused into subjects already receiving an NaCl infusion, it led to a further decrease in the plasma renin activity while moderately raising the plasma aldosterone level.

Bauer and co-workers (1977) induced a marked increase in plasma renin activity in normal human volunteers by giving them a virtually electrolyte-free diet for 4 days. The increase in plasma renin activity was not prevented when 50 mmol/day of KCl were added to the diet. The plasma renin activity rose to even higher levels when greater amounts of KCl (150 or 300 mmol/day) were consumed together with the sodium-free diet (Bauer and Gauntner 1979). This was probably due to a greater sodium loss in the urine. In patients with Addison's disease, kept on constant cortisone (37.5 mg/day) and 9α-fluorocortisol (0.1 mg/day) medication, oral potassium loading for a few days led to sodium losses and to increases in plasma renin activity (P.D. Miller et al. 1975). When the sodium losses were compensated by sodium supplements, the plasma renin activity remained unchanged. These observations indicate that the long-term suppression of renin secretion by potassium intake depends on increased aldosterone secretion and on a minimum intake of sodium.

Short-term infusion of KCl into the renal artery of normal or acutely sodium-depleted dogs induced a marked suppression of renin release into the renal vein (Vander 1970; E.G. Schneider et al. 1972). The renin release by the contralateral kidney was also suppressed, although to a lesser extent (Flamenbaum et al. 1975).

Sealey et al. (1970) studied the influence of diets with various sodium and potassium contents on the plasma renin activity in rats. The highest levels of plasma renin activity were observed in animals that had been kept on a low-sodium, low-potassium diet for 7 weeks. In sodium-deficient rats, the plasma renin activity was inversely related to the plasma potassium concentration. An increase in plasma renin activity in response to sodium restriction was virtually abolished by a high potassium intake. According to Flamenbaum et al. (1970), a high potassium intake led within 2 weeks not only to a low plasma renin activity in rats, but also to a significant decrease in the renal renin content. Marked increases in plasma renin activity and in the plasma angiotensin II concentration were induced in rats by treating them for 6 weeks with a potassium-deficient diet (Douglas and Catt 1976; Douglas et al. 1978c). In reverse, a high-potassium diet led to a significant suppression of these two parameters of the renin-angiotensin system. On the other hand, acute oral potassium loading of rats led to an increase in the plasma renin concentration, unless sodium was given at the same time (Corvol et al. 1977).

Kotchen et al. (1976) treated rats first with a sodium- and chloride-deficient, normal-potassium diet for 1 week and then added various salts to their drinking fluid while keeping them on the same diet for another 8-9 days. Addition of KCl but not of $KHCO_3$ to the drinking fluid suppressed the plasma renin activity and the renal renin concentration. This observation indicated an important role for the chloride ion in the suppression of the renin-angiotensin system by KCl intake. However, Kirchner and Mueller (1982) found that 90-min intravenous infusions of potassium chloride, potassium acetate, potassium bicarbonate, and potassium nitrate into rats kept on a sodium- and chloride-deficient diet induced similar decreases in the plasma renin activity. Addition of $KHCO_3$ to the drinking fluid of rats kept on a sodium-, potassium-, and chloride-deficient diet substantially lowered the excessively high plasma renin activity but was less effective than the addition of KCl (J. Müller 1986). Stephens et al. (1978) had previously observed that renal arterial infusions of potassium lactate or potassium sulfate also suppressed the renin secretion in dogs with a constriction of the thoracic caval vein.

The suppressive effect of potassium intake on renin release does not seem to be due to a direct action of potassium ions on the juxtaglomerular epitheloid cells. Thus, the renin release by rat kidney slices was not affected by a low (2.4 mmol/l) or a high (9.5 mmol/l) potassium concentration in the incubation medium (Aoi et al. 1974). However, renin release by incubated rat renal cortical slices was abolished in a potassium-free medium or in a medium with a potassium concentration of 1 mmol/l (Churchill and Churchill 1979). A near maximal renin release occurred in a medium with a potassium concentration of 2 mmol/l, and further increases in the potassium concentration had only a slight additional stimulatory effect.

5.3 Exogenous Angiotensin or Renin

If the renin-angiotensin system plays a primary role in the regulation of aldosterone secretion under physiological conditions as well as in spontaneous or experimental secondary hyperaldosteronism, as has been postulated by Ganong et al. (1966) and J.O. Davis (1967), several criteria should be fulfilled, according to Ames et al. (1965):

1. Angiotensin should stimulate aldosterone secretion in nonpressor doses or mild pressor doses, since secondary hyperaldosteronism is generally not accompanied by arterial hypertension.
2. Angiotensin should stimulate aldosterone secretion selectively, since hyperaldosteronism is generally not accompanied by increased glucocorticoid secretion.
3. Aldosterone secretion should remain elevated as long as the blood angiotensin level is elevated.

5.3.1 Aldosterone Stimulation by Pressor and Nonpressor Doses

In two normal human subjects, 8-h infusions of angiotensin II, at a dose which had no effect on blood pressure, elicited a marked increase in urinary aldosterone excretion (Genest 1961). In other instances, pressor doses of angiotensin II were needed to stimulate aldosterone secretion in normal human subjects or patients with essential hypertension during short-term or long-term infusion studies, although the increases in blood pressure were often very small (Laragh et al. 1960; Biron et al. 1961; 1962; Ames et al. 1965). In a normal human subject, a graded response in aldosterone secretion was induced by pressor doses of human renin (Bryan et al. 1964). According to Carey et al. (1978), the relative pressor effects of angiotensins II and III were greater in subjects with a high sodium intake than in subjects with a low sodium intake. Also, angiotensin II had a more marked pressor effect than angiotensin III. However, even in sodium-restricted human subjects, angiotensin III stimulated aldosterone secretion only when given at a dose which also raised blood pressure. Prolonged infusions of low doses of angiotensin II in sodium-replete and sodium-depleted human subjects led not only to sustained increases in plasma and urinary aldosterone levels but also to a sustained increase in blood pressure (Oelkers et al. 1978a, 1978b). A subpressor dose of angiotensin II, infused for 1 h into normal human subjects, stimulated the aldosterone secretion rate to the same extent as a pressor dose (Messerli et al. 1977). A greater effect of the pressor dose of the peptide on the plasma aldosterone concentration was due mainly to a more extensive suppression of the metabolic clearance rate of the steroid (most likely induced by a decreased splanchnic blood flow). In normal subjects kept on a low-sodium diet, graded infusions of angiotensin III which raised the plasma aldosterone concentration had a minor effect on systolic but not on diastolic blood pressure and led to a small but statistically significant decrease in the metabolic clearance rate of aldosterone (Zager et al. 1980).

In intact dogs, prolonged infusions of angiotensin II at doses which had no effect on blood pressure led to sustained increases in urinary aldosterone excretion and to changes in electrolyte balance that were very similar to those observed during prolonged treatment with mineralocorticoids, i.e., transient sodium retention, polyuria, and fecal potassium loss (Urquhart et al. 1963). These observations were not confirmed by Cowley and McCaa (1976), who infused angiotensin II for periods of 2 weeks into dogs with various sodium intakes. In dogs with a normal (40 mmol/day) or a high (120 mmol/day) sodium intake, a low dose of angiotensin II did not lead to a sustained increase in plasma aldosterone but to a sustained elevation of blood pressure. In sodium-depleted dogs (low-sodium diet and diuretics) even a high dose of angiotensin II did not raise blood pressure. However, in these animals, plasma aldosterone was already very high under control conditions and did not rise further in response to the peptide. In dogs kept on a low-sodium diet, long-term infusions of angiotensin II or III which led to a sustained increase in plasma aldosterone did not raise blood pressure (R.E. McCaa 1978). In dogs with a normal sodium intake, a long-term angiotensin II infusion (15 ng/kg/min) which raised plasma aldosterone twofold also raised the mean arterial blood pressure by 39 mmHg, whereas an infusion of angiotensin III (30 n/kg/min) which had a similar effect on plasma aldosterone did not affect the blood pressure.

In sodium-replete sheep, systemically infused angiotensin II or renin stimulated aldosterone secretion in pressor doses only (Blair-West et al. 1962a, 1967a). However, according to a later study performed in similar animals (Blair-West et al. 1981), the lowest doses of angiotensin II or III which stimulated aldosterone secretion had no effect on blood pressure. Angiotensins II and III were equipotent in stimulating aldosterone secretion, but angiotensin II was four times more potent in raising blood pressure than angiotensin III.

In rats, angiotensin and renin either had no effect on aldosterone production (Eilers and Peterson 1964) or stimulated aldosterone secretion in pressor doses (Cade and Perenich 1965; Marieb and Mulrow 1965; Bojesen 1966; Dufau and Kliman 1968a, 1968b; Kinson and Singer 1968; Masson and Travis 1968; Coleman et al. 1974). Long-term treatment with angiotensin II or renin, eliciting increased in vitro production of aldosterone and elevated secretion of aldosterone respectively, also resulted in chronically elevated arterial blood pressure (Marx et al. 1963; Masson and Travis, 1968).

5.3.2 Selectivity of Aldosterone Stimulation

The specificity of the aldosterone-stimulating action of angiotensin in short-term incubation experiments has been discussed in Sect. 4.1. In general, angiotensin II and angiotensin III act exclusively or preferentially on zona glomerulosa cells and stimulate mainly an early biosynthetic step preceding the formation of pregnenolone. Accordingly, in a majority of experiments performed in normal man or in intact experimental animals, systemically infused angiotensins selectively raised the secretion rates and plasma concentrations of aldosterone and its by-product 18-hydroxycorticosterone, whereas the secretion rates and plasma concentrations of corticosteroids that are exclusively (cortisol) or predominantly (corticosterone, 18-OH-DOC, deoxycorticosterone) produced by the inner zones of the adrenal cortex were unaffected or remained at least within physiological limits. Occasionally observed increases in plasma cortisol or urinary 17-hydroxycorticoid excretion might have been due to (a) a direct stimulatory action of the peptide on the zona fasciculata, (b) a conversion to cortisol in the zona fasciculata of precursor steroids produced by the zona glomerulosa, or (c) an angiotensin-elicited secretion of ACTH by the pituitary gland. Reported observations of angiotensin-induced drops in plasma cortisol (accompanied by increased or decreased levels in plasma ACTH) remain as yet unexplained.

In normal human subjects, prolonged infusions of angiotensin II stimulated cortisol secretion only occasionally and only at very high doses (Ames et al. 1965). In patients with cirrhosis of the liver and ascites, very high doses of angiotensin II did not stimulate aldosterone secretion but did stimulate the secretion rate of cortisol. This effect might have been due to ACTH secretion since it was blocked by dexamethasone. Prolonged infusions of a low dose of angiotensin II into sodium-replete or sodium-depleted normal human subjects did not induce any significant changes in plasma cortisol or plasma ACTH concentrations, according to Oelkers et al. (1978a, 1978b). In a later study, however, the same investigators (Schöneshöfer and Oelkers 1980) observed some minor but statistically significant effects of angiotensin II on plasma levels of glucocorticoids in the course of 3-day infusion experiments. In sodium-replete subjects there was a decrease in 9 a.m. plasma cortisol on days 1 and 2, with a return to normal on day 3. In sodium-depleted subjects, there were small rises in plasma cortisol and corticosterone on days 2 and 3. Rayyis and Horton (1971) induced a significant increase in plasma ACTH by the short-term infusion of angiotensin II into normal human subjects, which was accompanied by a (normal diurnal) drop in plasma cortisol. When these investigators infused angiotensin II into hypopituitary patients who had already been treated by an infusion of ACTH, they observed a decrease in the plasma cortisol concentration. No alteration in plasma cortisol was observed when angiotensin II was infused into otherwise untreated hypopituitary subjects. On the other hand, according to Semple et al. (1979), a 1-h infusion of angiotensin II elicited consistent and dose-dependent decreases in the plasma ACTH concentration in six normal human subjects, which were in two instances accompanied by a decrease in plasma cortisol. In untreated and dexamethasone-treated normal human subjects, angiotensin II infusions selectively raised the plasma concentrations of aldosterone and 18-hydroxycorticosterone (Mason et al. 1976, 1979). However, during a constant infusion of ACTH into dexamethasone-treated human subjects, graded infusions of angiotensin II also led to dose-dependent increases in the plasma levels of cortisol, corticosterone, 18-OH-DOC, and deoxycorticosterone (Mason et al. 1979).

In two conscious dogs with arteriovenous fistulas, prolonged angiotensin II infusions for 3 days led to progressive increases in aldosterone secretion, which were accompanied by small increases in corticosterone secretion; however, corticosterone secretion rates were highest on the first day of angiotensin administration and returned to normal on the third day of the experiment (Urquhart et al. 1964). During angiotensin II infusions for 3 h into intact anesthetized dogs, increases in aldosterone secretion were accompanied by comparable fractional increases in corticosterone and deoxycorticosterone secretion (J.O. Davis et al. 1968). In two nephrectomized, hypophysectomized dogs, a 4.5-h infusion of angiotensin II stimulated aldosterone secretion significantly, had an equivocal effect on corticosterone secretion, and did not affect 17-hydroxycorticoid output (Mulrow et al. 1962). In intact conscious dogs, long-term infusions of angiotensin II or III did not induce any significant alteration in the plasma cortisol concentration, irrespective of the sodium intake (Cowley and McCaa 1976; R.E. McCaa 1978). Nicholls et al. (1978) observed that graded short-term angiotensin II infusions into conscious sodium-replete or sodium-depleted dogs did not induce any significant change in the plasma cortisol concentration. However, according to Ramsay et al. (1978), angiotensin II, when infused at a rate of 10 or 20 ng/kg/min into conscious trained dogs, significantly raised the plasma ACTH and 11-hydroxycorticosteroid concentrations. These effects became manifest not before 30 min after the start of the infusion. When very low doses (500 pg/min) of angiotensin II were directly infused into the anterior pituitary gland of conscious dogs, they elicited a transient peak in cortisol secretion within 10–20 min, which was comparable to the peak in cortisol secretion induced by the systemic infusion of the peptide at a rate of 1 µg/min (Maran and Yates 1977). The angiotensin II-induced increase in cortisol secretion was abolished by dexamethasone pretreatment.

When directly infused into the adrenal arterial blood supply of sodium-replete sheep, angiotensins II and III had no consistent effect on cortisol secretion at rates which stimulated aldosterone secretion (Blair-West et al. 1963, 1971a). Systemic infusions of the two peptides had a biphasic effect, with lower doses decreasing and higher doses stimulating cortisol secretion (Blair-West et al. 1980). However, cortisol secretion rates did not exceed the normal range in these experiments. Angiotensin II did not inhibit cortisol production by the autotransplanted sheep adrenal during stimulation with low or high doses of ACTH (Espiner et al. 1978b).

Chronic administration of angiotensin II to hypophysectomized rats for 36 days had no effect on aldosterone secretion, but at the highest dosage (75 µg/day), a small increment in corticosterone secretion was observed (Cade and Perenich 1965). Using a higher dose of angiotensin II (250 µg/day) and intact rats, Dufau and Kliman (1968a) induced a significant increase in corticosterone but not in aldosterone secretion within 1 week of treatment. After 2 weeks of treatment the secretion rates of both steroids were significantly elevated. At that time, the total adrenal aldosterone content was increased nine fold, the corticosterone content 2.5-fold. In rats kept on a normal sodium and potassium intake, a 4-day intravenous infusion of angiotensin II (100 ng/kg/min) which induced a transient rise of serum aldosterone on day 1 did not affect either the serum corticosterone level or the adrenal weight (Komor and Müller 1979). In potassium-depleted rats, the subcutaneous infusion of angiotensin II (750 ng/kg/min) for 1 or 2 days had no effect on these parameters either (Haldy and Müller 1981b, J. Müller et al. 1982). Significant increases in plasma corticosterone and in the adrenal weight were observed in potassium-depleted rats after a 1-day sc infusion of a very high dose of angiotensin II (3000 ng/kg/min; J. Müller et al. 1982).

Infusion of a homologous renin preparation into dexamethasone-treated or hypophysectomized opossums led to significant increases in the secretion rates of aldosterone, corticosterone and cortisol (Johnston et al. 1967b). Renin preparations or kidney extracts stimulated corticosterone secretion in dexamethasone-treated turtles (Nothstine et al. 1971) and corticosterone as well as aldosterone secretion in hypophysectomized bullfrogs (Johnston et al. 1967a; Taylor and Davis 1971).

5.3.3 Persistence of Aldosterone Stimulation

Studies on the duration of effects of long-term administration of angiotensin II on aldosterone biosynthesis and secretion in man and experimental animals have yielded many apparently conflicting findings. Some of these might have been due to differences in the sodium status of the objects studied. The marked influence of sodium balance and sodium intake on the sensitivity of the zona glomerulosa to stimulators has already been discussed in Sect. 5.1.3. Long-term stimulation of aldosterone secretion would also be expected to cause changes in sodium and potassium balance which would eventually lower the sensitivity of the zona glomerulosa to the peptide. In addition, alterations in water balance, possibly due to antidiuretic effects of angiotensin II, stimulation of thirst, and the release of antidiuretic hormone might result in plasma volume changes which could by some yet unknown mechanism affect the responsiveness of the zona glomerulosa. Other controversies might be due to species differences.

In normal human subjects with a normal or high sodium intake, increased secretion rates of aldosterone were sustained during prolonged angiotensin II infusions for periods of up to 11 days (Ames et al. 1965). The sensitivity to the pressor effects of angiotensin increased with prolonged treatment, but there was no concomitant increase in the sensitivity to the aldosterone-stimulating effect of the peptide. In contrast, in a later study by Oelkers et al. (1975), prolonged infusion of a low dose of angiotensin II (2 ng/kg/min) into normal human subjects for 3 days led to an increased sensitivity in the response of plasma aldosterone but not in the response of blood pressure to short-term graded infusions of angiotensin II. However, these subjects were kept on a low sodium intake (15-20 mmol/day) during the experimental period. When a low dose of angiotensin II (3 ng/kg/min) was infused into two normal human subjects with a daily sodium intake of 135 mmol for 132 h, there was a sustained elevation of the plasma aldosterone level; whereas urinary aldosterone excretion was also high but tended to decrease toward the end of the experiment (Oelkers et al. 1978a).

In normal dogs and in dogs with arteriovenous fistulas, prolonged angiotensin II infusions stimulated aldosterone excretion and secretion, respectively, for periods of several days (Urquhart et al. 1963, 1964). According to later experiments by R.E. McCaa and co-workers (McCaa et al. 1975; Cowley and McCaa 1976; McCaa 1978), the aldosterone response to long-term infusions of angiotensin II and angiotensin III in dogs was biphasic and depended on sodium intake as well as on the dose of the peptide. In dogs with a high sodium intake (120 mmol/day), even a high dose of angiotensin II had only a transient effect on plasma aldosterone. In dogs with a normal sodium intake (40 mmol/day), a low dose of angiotensin II (5 ng/kg/min) had a transient effect on plasma aldosterone with a decrease to normal after 2 days. In the second week of infusion, there was a gradual secondary increase to 60% above control values. At higher doses of angiotensin (15 or 23 ng/kg/min ANG II; 30 ng/kg/min ANG III) an elevated plasma aldosterone concentration was maintained over the whole period of infusion.

In sodium-replete sheep, continuous adrenal arterial or systemic infusion of angiotensin II stimulated aldosterone secretion only for a few hours, despite a sustained effect of the peptide on blood pressure (Blair-West et al. 1962a). Progressively increasing doses were necessary in order to keep aldosterone secretion at a constant rate for 5 h.

In rats, long-term treatment with high doses of renin or angiotensin stimulated aldosterone secretion to the same extent as short-term treatment (Masson and Travis 1968; Dufau and Kliman 1968a). According to Bojesen (1966), plasma aldosterone was still elevated after 24 h of continuous infusion of angiotensin II or renin into conscious rats, but to a somewhat lesser extent than after 2 h. In a study carried out in our laboratory (Komor and Müller 1979), continuous intravenous infusion of a relatively high dose of angiotensin II (100 ng/kg/min) into conscious rats kept on a fixed sodium and potassium intake elicited a significant increase in serum aldosterone and in the adrenal conversion of corticosterone into aldosterone after 24 h. With continued infusion of the peptide, serum aldosterone returned to baseline values after 48 h, whereas the conversion returned to normal after 96 h and blood pressure remained elevated. Angiotensin did not lead to sodium retention or hypokalemia in these experiments. In potassium-deficient rats kept on sucrose and water, a subcutaneous angiotensin II infusion (750 ng/kg/min) caused a small but statistically significant increase in plasma aldosterone within 48 h (Haldy and Müller 1981b). When angiotensin II was continuously infused into sodium-replete rats from intraperitoneally implanted osmotic minipumps, plasma aldosterone was kept at the same high level after 1.5, 4, and 6 days (Aguilera et al. 1980a).

5.3.4 Effects on Angiotensin II Receptors and on Steroidogenic Enzymes

The maximum response in aldosterone output by incubated zona glomerulosa cells to angiotensin II added in vitro was increased by a factor of 2.8 when the rats from which the adrenals were taken had been treated in vivo by an intraperitoneal infusion of angiotensin II for a period of 36 h to 6 days (Hauger et al. 1978). The angiotensin II concentration required to achieve a half-maximal aldosterone response was not changed. The increased responsiveness of the zona glomerulosa cells was accompanied by a significant increase in the concentration of their angiotensin II receptors, whereas the association constant of the angiotensin II receptors remained unaltered. A similar effect of long-term angiotensin II infusion on the zona glomerulosa angiotensin II receptors was observed in hypophysectomized rats by Aguilera et al. (1980b). According to Douglas and Brown (1982), a 7-day low-dose angiotensin II infusion into intact rats induced a significant increase in the dissociation constant of zona glomerulosa angiotensin II receptors, together with an increase in the receptor concentration. In the same animals, the dissociation constant of uterine smooth-muscle angiotensin II receptors was decreased by the angiotensin II treatment, whereas the receptor concentration remained constant.

Short-term infusions of angiotensin II or renin into the lumboadrenal artery of sodium-replete dogs for periods of 100–180 min did not influence the adrenal mitochondrial conversion of corticosterone to aldosterone (Aguilera and Marusic 1971). On the other hand, prolonged treatment of intact or hypophysectomized dogs with repeated subcutaneous injections of renin for 4 days resulted in a significant increase of this parameter of late steps of aldosterone biosynthesis. In sodium-restricted hypophysectomized and nephrectomized dogs, a short-term (3 h) adrenal arterial infusion of angiotensin II produced a statistically significant stimulation (+36%) of the incorporation of tritiated corticosterone into aldosterone by whole-adrenal mitochondria (Aguilera et al. 1975).

During prolonged intravenous infusion of angiotensin II into sodium- and potassium-replete rats, the conversion of corticosterone to aldosterone by quartered adrenal glands rose significantly to 64% and 46% above control values after 1 and 2 days of treatment respectively (Komor and Müller 1979). Putting potassium-restricted rats on a diet of sucrose and water for 2 days resulted in a three fold increase in the adrenal conversion of corticosterone to aldosterone; a simultaneous angiotensin II infusion (750 ng/kg/min, sc) yielded an 11-fold increase in this parameter (Haldy and Müller 1981b; Fig. 24). In intact and hypophysectomized rats kept on a normal sodium and potassium intake, a continuous angiotensin II infusion for 6 days brought about a more than 100% increase in the conversion of corticosterone to aldosterone by zona glomerulosa mitochondria (Aguilera et al. 1980b). The basal rate of this conversion was higher in hypophysectomized than in intact animals. On the other hand, the zona glomerulosa mitochondrial cholesterol content, pregnenolone production, and 11β-hydroxylase activity, as well as the zona glomerulosa microsomal cytochrome P-450 content and 21-hydroxylase activity were found to be markedly suppressed in hypophysectomized rats. All these parameters of early steps in aldosterone biosynthesis were significantly stimulated by angiotensin II treatment in hypophysectomized but not in intact rats. These observations were only partially confirmed by R.E.Kramer et al. (1980), who also treated hypophysectomized rats with angiotensin II, although by repeated subcutaneous injections over 4 days. Microsomal and mitochondrial cytochrome P-450 levels remained unchanged. Mitochondrial 11β- and 18-hydroxylation of deoxycorticosterone and microsomal 21-hydroxylation of progesterone were also not affected. In contrast, angiotensin II increased the rate of cholesterol side-chain cleavage in zona glomerulosa mitochondria. This effect correlated with an increase in the heat-generated type-I and the pregnenolone-induced type-II absorbance changes. Angiotensin II also stimulated the rate of conversion of corticosterone to 18-hydroxycorticosterone and aldosterone. Further, it enhanced the corticosterone-induced type-I absorbance change in zona glomerulosa mitochondria. The effects of angiotensin II on the final steps of aldosterone biosynthesis became apparent within 3 h and reached a plateau within 48 h. Treatment of the rats with cycloheximide inhibited the effects of angiotensin II on late steps but not on early steps of aldosterone biosynthesis. Apparently, angiotensin II induced the de novo synthesis of a peptide that promoted the association of corticosterone with a cytochrome P-450 ("18-methyl oxidase") and increased its conversion to 18-hydroxycorticosterone and aldosterone.

5.3.5 Effects on Sodium Balance

The interpretation of studies on the effect of exogenous renin or angiotensin on aldosterone secretion is sometimes difficult, because angiotensins can have a direct natriuretic or antidiuretic effect on the kidney (for reviews see Gross et al. 1965; Peart 1965; J.J.Brown et al. 1966a; Navar and Langford 1974; Thurau 1974). Thus, changes in aldosterone secretion could be due to angiotensin-induced alterations in sodium balance as well as to the direct action of the peptide on the adrenal cortex. However, in the majority of studies cited above, aldosterone secretion was modified by exogenous angiotensin or renin, independent of alterations in sodium balance.

Short-term infusions of angiotensin II stimulated aldosterone excretion to the same extent in normal human subjects, where it induced sodium retention, and in patients with essential hypertension, where it induced sodium diuresis (Biron et al. 1962). In normal human subjects, prolonged angiotensin II infusions stimulated aldosterone secretion equally well during the initial 4 days of sodium retention and during the following period of re-established sodium balance (Ames et al. 1965). Marked sodium diuresis induced by high doses of angiotensin II in patients with cirrhosis of the liver and ascites did not result in a further increase in aldosterone secretion (Ames et al. 1965). Subjects receiving long-term infusions of angiotensin II in the studies by Oelkers et al. (1975, 1978b) were generally in sodium balance.

In dogs as well, long-term angiotensin II treatment stimulated aldosterone secretion during periods of sodium retention, sodium balance, and sodium loss (Urquhart et al. 1963, 1964). Treatment of dogs with homologous renin for 5 days resulted in an increased responsiveness in aldosterone secretion to stimulation by ACTH or angiotensin II but was not accompanied by sodium diuresis (Ganong et al. 1967a). Treatment of dogs with renin for 4 days stimulated the conversion of corticosterone to aldosterone but did not affect sodium balance (Aguilera and Marusic 1971).

In sodium-replete sheep, replacement of the sodium loss induced by an angiotensin II infusion did not prevent a transient increase in aldosterone (Blair-West et al. 1962a).

In the normal rat, the sodium-diuretic effect of angiotensin II wears off after 1 h according to Gross et al. (1965). Infusion of high doses (up to 750 ng/kg/min) of angiotensin II into normally fed or potassium-depleted rats, which significantly stimulated the final steps of aldosterone biosynthesis, did not induce a significant increase in sodium excretion (Komor and Müller 1979; Haldy and Müller 1981b; J. Müller et al. 1982). At an even higher infusion rate (3000 ng/kg/min), angiotensin II caused a fall in plasma sodium, a rise in the plasma protein concentration, and an increase in urinary sodium excretion (J. Müller et al. 1982). A 7-day intravenous angiotensin II infusion (30 ng/kg/min) into normal rats, which induced significant increases in the concentration and dissociation constant of zona glomerulosa angiotensin II receptors (but not in plasma aldosterone), did not affect urinary sodium excretion (Douglas and Brown 1982).

5.3.6 Renal Arterial Infusion of Angiotensin II

The continuous infusion of a low dose of angiotensin II into the sole renal artery of unilaterally nephrectomized sheep with a unilateral parotid fistula prevented the usual rise in plasma renin concentration in response to salivary sodium loss without having any effect on systemic blood angiotensin II concentration, blood pressure, plasma potassium concentration, or renal sodium excretion (Blair-West et al. 1971f). During mild sodium deficiency caused by withdrawal of $NaHCO_3$ solution for 24 h, blood aldosterone concentration increased seven- to eightfold, whereas the peripheral blood angiotensin II showed only a minor (but statistically significant) increase and remained within the normal sodium-replete range (Blair-West et al. 1973b). The plasma cortisol and potassium concentrations remained unchanged. The mean plasma sodium decreased by 5 mmol/l. These findings suggested the existence of an unidentified factor in aldosterone control that either acted directly on the zona glomerulosa or altered the sensitivity of the adrenal cortex to angiotensin II.

5.3.7 Immunization Against Renin or Angiotensin

Prolonged treatment of dogs with pig renin induces the formation of circulating antibodies capable of neutralizing the pressor activity of both pig and dog renin (Helmer 1958; Wakerlin 1958). Ganong et al. (1963) incubated pig and dog renin with plasma of dogs immunized with pig renin prior to injection into nephrectomized, hypophysectomized dogs and observed that this treatment not only inhibited the pressor activity of the enzyme but also its aldosterone- and 17-hydroxycorticosteroid-stimulating activities. Passive transfer of antirenin-globulin resulted in an acute decrease of the aldosterone secretion rate of dogs on a sodium-deficient diet but not of dogs on a normal sodium intake (T.C. Lee et al. 1965). However, in dogs actively immunized with pig renin, the aldosterone secretion rate was normal on a high-sodium as well as on a low-sodium intake (Ganong et al. 1965b). The immunized dogs conserved sodium normally and responded with an increased aldosterone secretion but only with a slight arterial blood pressure increase when they were hypophysectomized and subjected to a constriction of the aorta above the renal arteries. The zona glomerulosa of the immunized dogs was significantly narrower than normal. The kidney contained substantially increased amounts of renin, the magnitude of the renal renin content correlating with the plasma antirenin titer. Active immunization against renin in seven dogs did not result in any consistent alteration in the urinary aldosterone excretion rate, according to Gomoll and Schmid (1965).

In rabbits actively immunized against angiotensin II, blood pressure and plasma aldosterone responses to exogenous angiotensin II were markedly attenuated (Lowenstein et al. 1972). In these animals, sodium depletion (by 2 weeks of dietary sodium restriction and 3-5 days of furosemide treatment) resulted nevertheless in a normal rise in plasma aldosterone. Beckerhoff et al. (1975b) also found a normal increase in plasma aldosterone in response to sodium depletion of rabbits immunized against angiotensin II. However, in their immunized animals only the pressor response to exogenous angiotensin II was blocked, whereas the response in plasma aldosterone was undiminished.

5.3.8 Morphological and Histochemical Alterations of the Adrenal Cortex

Prolonged infusion of angiotensin II into intact rats for 7-12 days did not alter zona glomerulosa width, according to Marieb and Mulrow (1965). Chronic subcutaneous injections of high doses of angiotensin II (4 mg/kg/day) into rats resulted in a moderate increase of zona glomerulosa width within 2 weeks and in a marked increase within 4 weeks (Marx et al. 1963). In the hypertrophic zona glomerulosa but not in the inner zones of the adrenal cortex there was an enhanced activity of Δ^5-3β-hydroxysteroid dehydrogenase, glucose-6-phosphate dehydrogenase, and succinic acid dehydrogenase (Marx et al. 1963; Lamberg et al. 1964). A significant increase in zona glomerulosa width was induced in rats by treatment with pig renin for 12 days (Masson and Travis 1968). Treatment of normal female rats with renin for 2 days significantly stimulated the mitotic activity in the zona glomerulosa of the adrenal cortex, whereas treatment with angiotensin II was ineffective (Payet and Isler 1976). The ultrastructural effects of angiotensin II treatment of rats on the zona glomerulosa and the zona fasciculata have been evaluated using morphometric techniques by Mazzocchi et al. (1980) and by Nussdorfer et al. (1981) respectively. Short-term administration of angiotensin II induced a lipid depletion in the zona glomerulosa cells, whereas long-term treatment with daily intraperitoneal injections of the peptide resulted eventually in a significant increase in the volume of lipid droplets. This was accompanied by significant increases of the following morphometric parameters of zona glomerulosa cells: volume of cells, volume of nuclei, volume of mitochondrial compartment, surface of mitochondrial cristae, and surface of smooth endoplasmic reticulum. Chronic administration of angiotensin II to rats treated with dexamethasone and maintenance doses of ACTH also resulted in significant, although somewhat less extensive increases in all these morphometric parameters of the zona fasciculata cells. In both zones, angiotensin II treatment also led to hypertrophy of the Golgi apparatus. In addition, chronic angiotensin II administration resulted in an increase in the number of dense bodies in the zona glomerulosa cells (Nussdorfer and Mazzocchi 1982). Since the intracellular activity of lysosome-marker enzymes was unaltered, the nature of these organelles is as yet unknown. Perhaps they were secretory organelles, because the increase in their volume density was proportional to the increase in the intracellular concentration of aldosterone.

5.4 Pharmacological Blockade of the Renin-Angiotensin System

5.4.1 Angiotensin Antagonists

Experiments with synthetic angiotensin antagonists have often produced limited or ambiguous information on the role of endogenous antiotensins II and III in the physiological regulation of aldosterone secretion. This has been due mainly to the considerable intrinsic agonist activity these compounds display in vivo, particularly at their adrenocortical target sites. Thus, in several instances "angiotensin antagonists" stimulated aldosterone secretion while lowering blood pressure and raising plasma renin activity.

In normal sodium-replete human subjects, saralasin (sar^1,ala^8-angiotensin II) induced a significant increase in the plasma aldosterone concentration (Hollenberg et al. 1976). This indicated that in the sodium-replete state the zona glomerulosa was not under the tonic influence of angiotensin II. The same dose of saralasin significantly reduced the plasma aldosterone response to simultaneously infused angiotensin II. According to Noth et al. (1977a), saralasin raised the plasma aldosterone concentration in supine subjects on an unrestricted sodium intake, but had no significant effects in sodium-depleted or erect subjects. An increase in plasma aldosterone was also elicited by saralasin in normal subjects on a high or an unrestricted sodium intake as well as in sodium-restricted patients with low-renin essential hypertension or primary aldosteronism (G.H. Anderson et al. 1980). However, these investigators observed a significant fall in plasma aldosterone induced by saralasin in sodium-restricted normal subjects and in patients with cirrhosis of the liver and ascites. G.H. Williams et al. (1978) had previously observed dose-dependent decreases in plasma aldosterone by saralasin infusions into sodium-restricted normal human subjects. Saralasin induced a drop in plasma aldosterone in approximately half of 41 recumbent sodium-depleted hypertensive patients (essential or renovascular hypertension) but raised plasma aldosterone in the other half (R.D. Brown et al. 1978). In the second group, baseline plasma aldosterone and renin activity were significantly lower than in the first group.

These observations indicated that in some hypertensive human beings angiotensin II may not be the dominant regulator of aldosterone secretion during sodium depletion. Fagard et al. (1981), who also studied the effects of saralasin in hypertensive patients, observed a significant increase in plasma aldosterone in subjects on a normal sodium intake as well as in sodium-restricted subjects treated with the converting-enzyme inhibitor captopril. In sodium-restricted subjects not on captopril, saralasin lowered the plasma aldosterone concentration, although not to a statistically significant extent. These findings suggested that the rise of angiotensin II in response to sodium restriction, rather than sodium restriction itself, permits the antagonistic action of saralasin on the adrenocortical receptors.

In anesthetized sodium-depleted dogs and in anesthetized dogs with a constricted thoracic inferior vena cava, saralasin infusion induced a striking decrease in the aldosterone secretion rate (J. A. Johnson and Davis 1973). In intact, sodium-replete dogs, saralasin completely blocked the aldosterone-stimulating action of concurrently infused angiotensin II. Infusion of saralasin also blocked an increase in plasma aldosterone in response to acute sodium depletion induced in conscious normal dogs by ethacrynic acid (Stephens et al. 1977). In chronically sodium-depleted dogs (diuretic and low-sodium diet) saralasin infusions suppressed the plasma aldosterone level for several hours (Stephens et al. 1977). However, in dogs in whom sodium deficiency had been induced by dietary restriction alone, the infusion or saralasin had only a weak and transient aldosterone-suppressing effect, which was no longer manifest after 1 day (R. E. McCaa 1977). The infusion of another inhibitory analogue, sar^1, ile^8-angiotensin II, into chronically sodium-restricted dogs led to a significant and sustained additional elevation in plasma aldosterone, which contrasted with a marked blood pressure-lowering effect of the peptide (R. E. McCaa 1977).

In sodium-deficient sheep, saralasin infusions of up to 1000 µg/h for 1-2 h had no consistent effect on aldosterone secretion (Blair-West et al. 1979). In sodium-replete sheep, however, saralasin effectively blocked aldosterone responses to high circulating levels of angiotensins II and III produced by systemic intravenous or direct adrenal arterial infusion.

In sodium-restricted rabbits, the infusion of saralasin at a rate of 1 µg/kg/min induced a significant additional rise in plasma aldosterone (Steele and Lowenstein 1974). At a higher infusion rate of the peptide, i.e., 5 µg/kg/min, the plasma aldosterone remained at the high sodium-deficient control level. In rabbits in whom sodium depletion had been brought about by the combination of a diuretic with dietary restriction, a very high dose of saralasin (50 µg/kg/min) suppressed plasma aldosterone to the sodium-replete range (Rowe et al. 1980).

In anesthetized sodium-replete or sodium-restricted rats, saralasin infusions (10 µg/kg/min) did not affect the aldosterone secretion rate (Spielman and Davis 1974). However, in sodium-replete animals, it effectively blocked the aldosterone-stimulating effect of simultaneously infused angiotensin II. In sodium-restricted but not in sodium-replete rats, a competitive antagonist of angiotensin III, ile^7 angiotensin III, caused a marked decrease in urinary aldosterone excretion but did not affect plasma renin activity (Sarstedt et al. 1975). In contrast, sar^1, ile^8 angiotensin II had no effect on aldosterone excretion but raised the plasma renin activity.

5.4.2 Converting-Enzyme Inhibitors

Converting-enzyme inhibitors, i. e., substances which block the production of angiotensin II from angiotensin I, have been very useful in studies on the role of the renin-angiotensin system in the control of aldosterone secretion. Since the angiotensin-converting enzyme is identical with kininase II (Erdös 1979), pharmacological effects of these drugs may also be due to impeded inactivation of bradykinin. Although there is as yet no evidence pointing to a direct involvement of bradykinin in the regulation of aldosterone biosynthesis, bradykinin might indirectly influence aldosterone secretion through its sodium-diuretic and blood pressure-lowering activities.

In normal human subjects on a low-sodium diet, a 3-min infusion of a threshold dose of teprotide (SQ 20881) caused transient decreases in the plasma concentrations of angiotensin II and aldosterone, manifest from 5 to 10 min and from 10 to 40 min respectively (Mersey et al. 1977). Bolus injections of teprotide induced dose-dependent decreases in plasma aldosterone, which were marked and statistically significant in sodium-restricted but not in sodium-replete human subjects (G. H. Williams et al. 1978 a). Treatment of sodium-restricted human subjects with the orally active converting-enzyme inhibitor captopril (SQ 14225) for 26 h resulted in a suppression of both plasma angiotensin II and aldosterone concentrations to sodium-replete levels (Swartz et al. 1980). Aldosterone secretion was also suppressed by captopril treatment of healthy normotensive persons on a fixed sodium intake of 120 mmol per day (MacGregor et al. 1981). Plasma aldosterone fell to a third of its control value within 2 h and remained at the same low level for 5 days of continued captopril administration. The urinary aldosterone excretion rate decreased moderately (−27%) during the first 4 days of captopril treatment but returned to normal on the 5th day. The combined treatment of normal human subjects with captopril and furosemide for 1 day did not result in any change in plasma aldosterone concentration, in spite of sodium depletion (Kigoshi et al. 1980). Nevertheless, the response in plasma aldosterone to ACTH stimulation was markedly higher after the combined treatment than under control conditions. This indicated that angiotensin II did not play an important role in enhancing the aldosterone-stimulating activity of ACTH during acute sodium depletion. By contrast, Oelkers et al. (1982) observed that treatment with captopril alone significantly blunted the elevation of plasma aldosterone elicited by a constant low-dose ACTH infusion for 34 h in normal human subjects. They concluded that angiotensin II partially mediated the aldosterone-stimulating effect of ACTH or that ACTH required normal plasma angiotensin II levels for a full effect on aldosterone secretion. In patients with congestive heart failure, kept on a constant digoxin and furosemide medication, treatment with captopril induced a fall in plasma aldosterone which paralleled a fall in plasma angiotensin II, in spite of a positive cumulative potassium balance and a significant rise in plasma potassium (Nicholls et al. 1981). This indicated that angiotensin II was more potent than potassium in regulating aldosterone secretion in these patients. In a group of normal human subjects on a low sodium intake, the treatment with captopril for 70 h lowered the plasma concentrations of angiotensin II and aldosterone into the sodium-replete range, but did not decrease the slope of the aldosterone/angiotensin II dose-response curve obtained during angiotensin II infusion (Dawson-Hughes et al.

1981). However, captopril enhanced the blood pressure response to angiotensin II in sodium-restricted subjects. According to these observations, the plasma angiotensin II concentration regulates the vascular but not the adrenal responsiveness to the peptide during sodium restriction in normal men.

Studies on prolonged effects of captopril in man have been carried out mainly in hypertensive patients. In a group of patients with essential or renovascular hypertension on a low sodium intake, the acute administration of captopril resulted within 75 min in a significant drop in plasma aldosterone (−79%), which was more extensive than the simultaneous drop in plasma angiotensin II (−53%), according to Lijnen et al. (1980). Prolonged suppression of aldosterone secretion for a few days by captopril treatment of hypertensive patients has been observed by Brunner et al. (1979) and by Atlas et al. (1979). This was accompanied by a negative sodium balance and positive potassium balance. Generally, these effects were greater in patients with high-renin hypertension than in patients with low-renin hypertension. In the course of long-term administration of captopril, aldosterone secretion tended to increase again, despite sustained decreases in blood pressure and plasma angiotensin II concentration. During chronic captopril treatment (up to 18 months), the plasma aldosterone concentration did not further decrease along with progressive increases in the plasma renin activity, but rather slightly increased again (Johns et al. 1980). A biphasic response in plasma aldosterone during long-term treatment of seven hypertensive patients with captopril was observed by Staessen et al. (1981). The mean plasma aldosterone concentration decreased from 74 to 21 pg/ml after 1 month, but increased to 165 pg/ml within a year of continued converting-enzyme inhibition. On the other hand, Havelka et al. (1982) found a significant decrease in plasma aldosterone after chronic treatment (up to 18 months) of patients with severe essential or renovascular hypertension. However, control values were obtained after a period of combined antihypertensive therapy with other drugs, and captopril treatment was in most instances combined with additional antihypertensive medication.

The long-term infusion of teprotide into conscious dogs prevented an increase in plasma aldosterone in response to constriction of the thoracic inferior vena cava (Watkins et al. 1976). In conscious sodium-depleted dogs (dietary sodium restriction and treatment with a diuretic), an infusion of teprotide resulted within 15 min in a striking decrease (−60%) in plasma aldosterone (Stephens et al. 1977). Long-term infusion of teprotide for 7 days into conscious sodium-restricted dogs lowered the plasma aldosterone to less than half the control level (R. E. McCaa 1977). This was accompanied by small but statistically significant decreases in the plasma sodium (from 143.5 to 138.5 mmol/l) and potassium (from 4.4 to 3.8 mmol/l) concentrations. Long-term oral administration of captopril had similar effects on plasma aldosterone and sodium concentrations in sodium-restricted dogs but did not affect the plasma potassium concentration (R. E. McCaa et al. 1978). During 2 weeks of converting-enzyme inhibition, plasma aldosterone was kept at a constant low level, which was, however, still above the sodium-replete range. In another study, a constant infusion of captopril into sodium-restricted dogs for 8 days significantly suppressed plasma aldosterone, but also to values above the sodium-deplete level (Hall et al. 1979). Short-term infusion of captopril into sodium-loaded dogs caused a significant decrease in plasma aldosterone within 30 min (Tree and Morton 1980). Short-term infusion of captopril into dexamethasone-treated dogs on an unrestricted sodium intake completely prevented an increase in plasma aldosterone in response to a 15-min infusion of potassium chloride, which raised the plasma potassium concentration by 1.4 mmol/l (Pratt 1982). Moreover, captopril treatment blunted the response in plasma aldosterone to an injection of ACTH. The aldosterone response to potassium was not impaired when angiotensin II was simultaneously infused with captopril. According to these observations, the short-term stimulation of aldosterone secretion in vivo by potassium ions and, to a lesser extent, by ACTH critically depended on the concurrent level of circulating angiotensin II.

In chronically sodium-depleted rabbits (diuretic and sodium-deficient diet), an infusion of teprotide for 1 h led to a significant decrease (−26%) in plasma aldosterone (Rowe et al. 1980). In sodium-replete dexamethasone-treated rabbits, bolus injections of captopril resulted in small but statistically significant drops in plasma aldosterone which correlated well with simultaneous decreases in plasma angiotensin II (Eguchi et al. 1980). Captopril injections completely blocked the aldosterone-stimulating effect of angiotensin I, but did not impair the rise in plasma aldosterone in response to infusions of angiotensin II, potassium chloride or ACTH. In the same study, the infusion of bradykinin in dexamethasone-treated rabbits elicited a small (but statistically significant) rise in plasma aldosterone, which was accompanied by proportional increases in the plasma renin activity and in the plasma angiotensin II concentration.

In anesthetized sodium-depleted rats, treated with dexamethasone alone or in combination with morphine, the intravenous administration of teprotide had no effect on the aldosterone secretion rate (Spielman and Davis 1974). However, teprotide significantly lowered the secretion rates of both aldosterone and corticosterone in hypophysectomized sodium-depleted rats. In dexamethasone-treated conscious rats, a marked rise in plasma aldosterone in response to the injection of furosemide was completely blocked by the simultaneous administration of teprotide (Spät et al. 1979c). In contrast, the converting-enzyme inhibitor did not impede a similarly marked increase in plasma aldosterone elicited by peritoneal dialysis with isotonic glucose solution in dexamethasone-treated rats (Spät et al. 1979b). In rats that had been kept on a sodium-deficient diet for a week, a 1-day infusion of teprotide from subcutaneous osmotic minipumps significantly lowered the plasma aldosterone concentration (−61%) and the capsular adrenal conversion of tritiated corticosterone to aldosterone (−29%), according to Tarjan et al. (1980). Similar decreases in both parameters were induced by the converting-enzyme inhibitor in sodium-replete rats. The long-term treatment with teprotide (with subcutaneous injections, twice a day) of sodium-replete rats failed to induce any change in the ultrastructure of the adrenal cortex or in the serum concentration of aldosterone within 3 or 10 weeks (Weaver et al. 1981). Aguilera and Catt (1978) treated rats with continuous subcutaneous captopril infusion during 4 days of dietary sodium restriction. Converting-enzyme inhibition completely abolished the increases in (a) plasma aldosterone concentration, (b) pregnenolone production by zona glomerulosa mitochondria incubated with cyanoketone (see p. 79), (c) cholesterol content of zona glomerulosa mitochondria, (d) conversion of corticosterone to aldosterone by zona glomerulosa mitochondria, and (e) the number of angiotensin II binding sites of a zona glomerulosa membrane preparation. Treatment of previously sodium-restricted rats with captopril (addition of the drug to the drinking fluid) for a week suppressed the plasma aldosterone to the sodium-replete level (R. E. McCaa et al. 1979). However, the plasma aldosterone concentration was not altered by captopril treatment in rats on a normal sodium intake. Oral administration of captopril to rats during 12 days of sodium restriction effectively blocked any increase in plasma aldosterone or in the width of the zona glomerulosa (McKeever et al. 1983). Infusion of captopril for 1 week into rats with established sodium and potassium deficiency (2 weeks of dietary restriction) did not affect the capsular adrenal conversion of corticosterone to aldosterone and 18-hydroxycorticosterone (J. Müller et al. 1984; Fig. 32). On the other hand, captopril blocked the stimulation of these conversions by 1 week of sodium restriction of potassium-depleted rats (Fig. 33). This indicated that high levels of angiotensin II are necessary for the induction but not for the maintenance of a high activity of late steps in aldosterone biosynthesis during combined

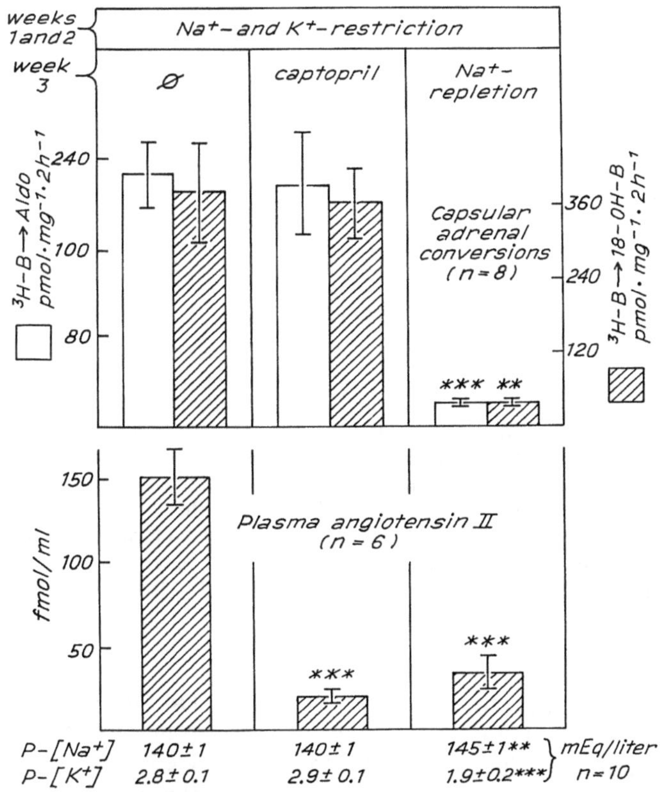

Fig. 32. Effect of treatment with captopril or of sodium repletion of sodium- and potassium-deficient rats on the capsular adrenal conversions of tritiated corticosterone (*^{3}H-B;* 300 nmol/flask) to aldosterone *(ALDO)* and 18-hydroxycorticosterone *(18-OH-B)* and on the plasma concentrations of angiotensin II, sodium, and potassium. All animals were kept on a sodium- and potassium-deficient diet and demineralized water for 2 weeks. During the 3rd week, groups 1 and 2 were kept on the same dietary conditions, whereas group 3 was given isotonic saline as drinking fluid. In addition, group 2 received a captopril infusion during the 3rd week. Mean values of n observations ± SEM. Statistical significance (t-tests) of differences from control animals *(left panel):* *** $P<0.001$, ** $P<0.01$. (From J. Müller et al. 1984)

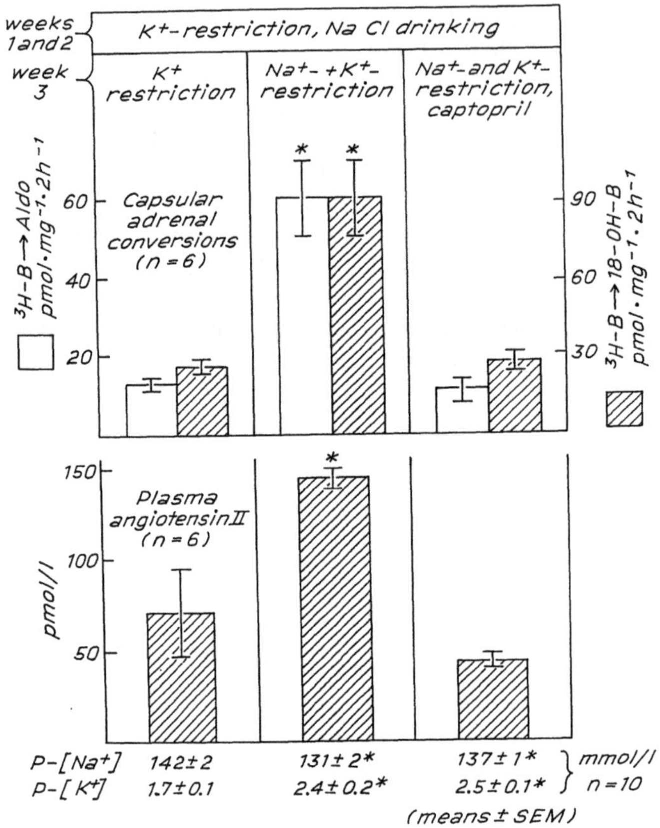

Fig. 33. Effect of sodium restriction with or without captopril treatment of potassium-deficient rats on the capsular adrenal conversions of tritiated corticosterone (*^{3}H-B;* 300 nmol/flask) to aldosterone *(ALDO)* and 18-hydroxycorticosterone *(18-OH-B)* and on the plasma concentrations of angiotensin II, sodium, and potassium. All animals were kept on a potassium-deficient diet and isotonic saline as drinking fluid for 2 weeks. During the 3rd week, all were given demineralized water as drinking fluid. Group 1 continued on the potassium-deficient diet, whereas groups 2 and 3 received a sodium- and potassium-deficient diet. In addition, group 3 received a captopril infusion during the 3rd week. Mean values of *n* observations ± SEM. Statistical significance (*t*-tests) of differences from control animals *(left panel):* *** $P<0.001$, ** $P<0.01$. (From J. Müller et al. 1984)

sodium and potassium deficiency. Treatment of potassium-depleted rats with captopril effectively blocked the stimulation in late steps of aldosterone biosynthesis by 1-day sodium sequestration induced by subcutaneous polyethylene glycol (J. Müller et al. 1982; Fig. 34). The stimulation of aldosterone biosynthesis by polyethylene glycol was fully restored by a simultaneous infusion of angiotensin II. However, combined treatment with captopril and angiotensin II stimulated aldosterone biosynthesis to a significantly greater extent in rats with edema than in rats without edema. These findings implied that angiotensin II was an essential, but not the only stimulator of the zona glomerulosa operative during hypovolemia.

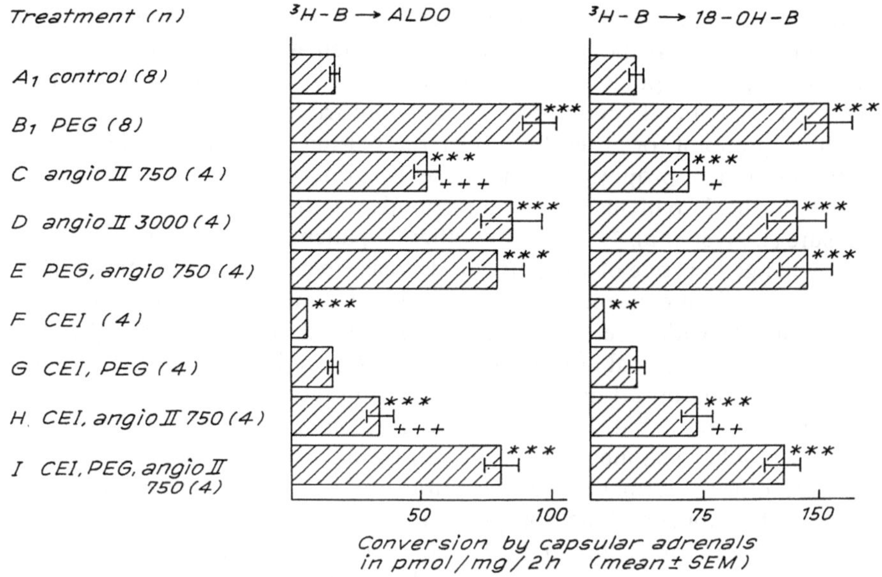

Fig. 34. Effects of different treatments (1-day) of potassium-deficient rats on the capsular adrenal conversions of tritiated corticosterone (*^{3}H-B;* 300 nmol/flask) to aldosterone *(ALDO)* and 18-hydroxycorticosterone *(18-OH-B)*. Statistical significance (*t*-tests) of differences from control animals *(A_1)*: *** $P<0.001$, ** $P<0.01$. Statistical significance of differences from group *B_1*: $^{+++}P<0.001$, $^{++}P<0.01$, $^{+}P<0.05$. *PEG*, subcutaneous polyethylene glycol; *angio II*, angiotensin II, s.c. infused at rates of 750 or 3000 $ng \cdot kg^{-1} \cdot min^{-1}$; *CEI*, converting-enzyme inhibition with captopril. (From J. Müller et al. 1982)

5.4.3 Beta-Adrenergic Blocking Agents

Treatment with propranolol during 1 week of dietary sodium restriction completely blocked any increase in plasma renin activity in six of 11 human subjects (nine normal volunteers and two patients with mild essential hypertension), according to Lowenstein and Steele (1977). In the other five subjects there was a blunted increase in plasma renin activity. Nevertheless, there was a significant increase in plasma aldosterone in both groups. In the individuals in whom plasma renin and angiotensin II levels remained unchanged, the aldosterone response could not be attributed to changes in plasma sodium or potassium concentration or to increased ACTH secretion. However, according to Plouin et al. (1977), treatment with propranolol or acebutolol of six patients with mild essential hypertension on a low sodium intake significantly lowered the plasma renin activity as well as the plasma concentration and the secretion rate of aldosterone to the sodium-replete range. In eight patients on a normal sodium intake, beta-blockade suppressed the plasma renin activity but not the aldosterone secretion. When normal human subjects on a low sodium intake were treated with oxprenolol for 1 day, there was a suppression of plasma renin activity to sodium-replete levels associated with a small decrease in urinary aldosterone excretion (Pedrinelli et al. 1982). In 22 patients with essential hypertension on a liberal sodium intake, treatment with alprenolol for 1 month resulted in moderate but statistically significant drops in plasma aldosterone (−26%) and renin (−33%) concentrations (blood samples taken after 1 h in the supine position), which were accompanied by a small increase in plasma potassium (E.B. Pedersen and Kornerup 1976a). When 50 patients with essential hypertension on an unrestricted sodium intake were treated with propranolol for at least 3 weeks, the plasma renin activity decreased in all instances, but a proportional decrease in urinary aldosterone excretion occurred only in the 21 patients who also responded with an unequivocal drop in blood pressure (Drayer et al. 1978). In the other 29 patients, aldosterone excretion remained unchanged in spite of a significant decrease in plasma renin activity. The mean plasma potassium concentration increased by 0.3 mmol/l in both groups of patients. In patients with essential hypertension treated with a diuretic (hydrochlorothiazide), the additional administration of propranolol lowered the plasma renin activity (−25%) but did not affect the secretion rate and plasma concentration of aldosterone (Pratt et al. 1980). However, the combined therapy reduced the aldosterone metabolic clearance rate by an average of 21%. The acute administration of propranolol to normal subjects significantly decreased the plasma renin activity in both the supine and the upright posture but had no effect on the plasma aldosterone concentration (Ulmann et al. 1973). In other studies, beta-adrenergic blockers suppressed the increase in plasma aldosterone in response to postural changes or to physical exercise. Propranolol treatment of patients with essential hypertension for 1 month significantly decreased plasma renin activity but not the plasma aldosterone concentration when measured in the supine posture (Golub et al. 1981). However, both parameters were significantly lowered by propranolol therapy when measured after 3 h of upright posture. Responses in plasma aldosterone to graded infusions of angiotensin II or ACTH remained unchanged. Treatment of hypertensive patients with propranolol completely suppressed the response in plasma renin activity and partially blocked the response in plasma aldosterone to head-up tilt (Morganti et al. 1979). The rise in plasma aldosterone during tilt was completely abolished by the combined treatment with propranolol and dexamethasone. In normal human subjects, submaximal exercise elicited marked increases in plasma renin and aldosterone concentrations (Bonelli et al. 1977). After 2 weeks of propranolol treatment, the response in plasma renin was completely blocked and the response in plasma aldosterone partially blocked. Similar effects were induced by 4 weeks of treatment with metoprol in a group of hypertensive patients (Lijnen et al. 1979). On the other hand, short-term propranolol treatment of six normal human subjects on a low sodium intake significantly augmented the rise in plasma aldosterone elicited by heat exposure while blunting the rise in plasma renin activity (Brandenberger et al. 1981).

When patients with essential hypertension were treated with labetalol, an alpha- and beta-adrenergic blocking agent, a decrease in upright plasma renin activity was accompanied by an unchanged aldosterone and an elevated plasma potassium concentration (Weidmann et al. 1978a). Upon combined therapy with labetalol and a diuretic (chlorthalidone) upright plasma aldosterone remained unchanged, whereas there was a rise in plasma renin activity and a fall in the plasma potassium concentration.

Treatment of rats on a normal sodium intake with timolol maleate for 6 days induced a marked fall in plasma renin activity (−64%) and a smaller decrease in plasma aldosterone (−22%), according to Rebuffat et al. (1979). These changes were accompanied by moderate (−22%−−44%) but statistically significant decreases in the following morphometric parameters of the zona glomerulosa: (a) volume of the zona glomerulosa, (b) volume of parenchymal cells, (c) volume of nuclei, (d) volume of mitochondrial compartment, (e) surface of mitochondrial cristae, and (f) surface of smooth endoplasmic reticulum. Similar changes in the plasma hormone concentrations and in the ultrastructure of the zona glomerulosa were induced by 12 days of timolol maleate administration to rats treated simultaneously with dexamethasone and ACTH (Mazzocchi et al. 1982). Sodium loading (2.5% NaCl solution as drinking fluid) had similar effects on plasma aldosterone and adrenocortical morphology but did not suppress the plasma renin activity. Most of the effects of the beta-blocking agent and of sodium loading were additive.

5.4.4 Inhibitors of Prostaglandin Synthesis

Inhibitors of prostaglandin synthesis have been used in several studies on the role of the renin-angiotensin system in the physiological control of aldosterone. However, it must be kept in mind that prostaglandins may influence aldosterone biosynthesis in a number of different ways. There is good evidence that prostaglandins are involved in mediating within the juxta-glomerular apparatus the response in renin release to stimulation by catecholamines or diuretics (Weber et al. 1976; Romero et al. 1976; Yun et al. 1977; Oates et al. 1979; Suzuki et al. 1981a, 1981b). The direct stimulatory and inhibitory effects of prostaglandins on steroidogenesis in zona glomerulosa cells and the possible role of prostaglandins as regulators (or mediators of other stimulators) of aldosterone production have been discussed in Sect. 4.6 (see p. 68). In addition, it must be considered that in the intact organism prostaglandins may also indirectly affect aldosterone biosynthesis through their effects on sodium and water homeostasis and on blood pressure (J. B. Lee et al. 1976; M. G. Dunn and Hood 1977).

In normal subjects kept on a high sodium intake, short-term treatment with indomethacin did not alter supine plasma aldosterone concentration or its response to angiotensin II infusion (Golub et al. 1976). In five patients with previous episodes of malignant hypertension, treatment with indomethacin significantly lowered supine plasma renin activity, supine plasma aldosterone concentration, and urinary sodium excretion, but it did not affect the response in plasma aldosterone to an angiotensin II infusion (J. C. Frölich et al. 1976). In normal human subjects, treatment with indomethacin or ibuprofen for 4 days led to significant drops in supine and upright plasma renin activity and plasma aldosterone concentration (Speckart et al. 1978). Indomethacin induced similar effects in six patients with renin-dependent renovascular hypertension. Treatment of normal subjects with acetyl salicylic acid was followed by a significant decrease in supine plasma aldosterone but not in supine plasma renin activity (Norbiato et al. 1978). However, the drug significantly reduced both parameters when measured in the upright posture. Another prostaglandin synthesis inhibitor, diclofenac sodium, led to significant decreases in plasma aldosterone and in plasma renin activity in the supine as well as in the upright position. The suppressive effects of diclofenac sodium on aldosterone and renin were also seen in subjects on a moderately low sodium intake (15–30 mmol/l) but were absent in subjects on a more severe sodium restriction (15 mmol/day). Indomethacin treatment of normal sodium-replete human subjects led to significant decreases in the upright plasma concentrations of aldosterone, 18-hydroxycorticosterone, corticosterone and 18-OH-DOC (Witzgall and Hassan-Ali 1981). Acute administration of captopril to indomethacin-treated subjects elicited significant rises in plasma corticosterone and 18-OH-DOC while not further affecting plasma aldosterone and 18-hydroxycorticosterone. Treatment of normal human subjects with indomethacin markedly suppressed the responses in plasma aldosterone and in plasma renin activity to furosemide administration (Tan and Mulrow 1977). The marked increases in plasma renin activity and in plasma and urinary aldosterone induced by 2-day spironolactone administration in normal human volunteers were completely absent in arthritic patients chronically treated with indomethacin, although the mineralocorticoid antagonist elicited the same urinary sodium loss in both groups (Glasson et al. 1979). According to these investigators, treatment of normal subjects with indomethacin during the last 5 days of a 10-day period of sodium restriction significantly lowered plasma renin activity but did not suppress a high rate of urinary aldosterone excretion. When indomethacin was given at the beginning of sodium restriction, two different types of response were seen. Plasma renin activity and urinary aldosterone excretion significantly rose in half of the subjects, but remained unchanged in the other half. Irrespective of these hormonal changes, urinary sodium excretion was reduced to the same extent in both groups. These findings stressed the importance of the temporal relation between prostaglandin synthesis inhibition and sodium restriction on the final effect on renin and aldosterone secretion and demonstrated that, under suppression of endogenous prostaglandin production, the organism can adapt normally to salt restriction without any elevation of renin and aldosterone. In another study, indomethacin treatment lowered the plasma renin activity to sodium-replete levels in sodium-restricted human subjects but had only an insignificant suppressive effect on urinary aldosterone excretion (Pedrinelli et al. 1982). However, the urinary aldosterone excretion rate was reduced to sodium-replete values by a combined treatment with indomethacin and a beta-adrenergic blocker (oxprenolol).

Treatment of hypophysectomized rats with indomethacin for 5 days did not significantly alter aldosterone secretion rate, either during unrestricted sodium intake or under stimulation by dietary sodium restriction (Spät und Fejes-Toth 1978). Indomethacin completely blocked an increase in plasma aldosterone concentration in response to furosemide administration in rats, while only partially suppressing a rise in plasma renin activity (Spät et al. 1977c, 1979c). However, in rats that were acutely sodium depleted by peritoneal dialysis, indomethacin treatment significantly attenuated a rise in plasma renin activity but did not affect a marked rise in the plasma aldosterone concentration (Spät et al. 1979b).

5.5 Role of the Kidneys

5.5.1 Renovascular Hypertension

Renovascular hypertension, i.e., reversible elevation of arterial blood pressure due to a spontaneous or experimentally induced partial obstruction of a renal artery, is in many instances thought to be caused by an excessive secretion of renin by the ischemic kidney and by elevated levels of angiotensin II. In other instances, in which renovascular hypertension results mainly from impeded renal sodium excretion, "normal" peripheral plasma concentrations of renin and angiotensin II could still be considered inappropriately high in relation to the body sodium status (Langford 1963; Menard et al. 1967; Tobian et al. 1969; Brunner et al. 1971; Swales et al. 1971). It is also possible that, in the malignant phase of renovascular hypertension, the high blood pressure induces urinary sodium loss, which in turn elicits a physiological increase in renin secretion.

Hyperaldosteronism secondary to renovascular hypertension was first described by Wrong (1957). Subsequently, Dollery and co-workers (1959) observed two patients with malignant hypertension due to unilateral renal artery stenosis with strikingly decreased serum potassium levels, who became normokalemic upon removal of the diseased kidney. Two similar cases were described by Laidlaw et al. (1960); one had a normal, the other an elevated urinary aldosterone excretion rate. These investigators also found increased aldosterone excretion in two of five patients with renovascular hypertension and normal serum potassium levels. Increases in the aldosterone secretion or excretion rate or elevated plasma aldosterone levels have since been reported in a large number of patients with hypertension caused by unilateral renal artery stenosis (Gowenlock and Wrong 1962; Cope et al. 1962; Sambhi et al. 1963; Wrong 1964; Genest et al. 1964; Barraclough et al. 1965; Laragh et al. 1966; Kaufmann et al. 1967; Weidmann and Siegenthaler 1967; Schönbeck et al. 1968; Veyrat et al. 1968; George et al. 1968; Lommer et al. 1968, 1972; Biglieri et al. 1968; Vecsei et al. 1969; Streeten et al. 1969; Kotchen et al. 1970; Grim and Keitzer 1974; Hansson et al. 1981). The overall incidence of hyperaldosteronism among patients with renovascular hypertension is difficult to evaluate, but it may be around 50%. In individual reports it varies from 25% (Weidmann and Siegenthaler 1967) to 90% (Kaufmann et al. 1967). Urinary aldosterone excretion was found to be increased in patients with renovascular hypertension during high (200 mmol/day) but not during low (10 mmol/day) sodium intake (Streeten et al. 1969). In a number of patients who excreted aldosterone at a normal rate, plasma aldosterone concentration rose to abnormally high levels in the evening, particularly when the patients were in the upright posture (Grim and Keitzer 1974). Hyperaldosteronism appears to occur more frequently in those patients with renovascular hypertension who are eventually cured or improved by surgical treatment than in patients who are not responsive to surgery (Hansson et al. 1981). Increased peripheral plasma concentrations of renin were found with a similar frequency among patients with renal artery stenosis (J.J. Brown et al. 1966a). According to Laragh et al. (1966), values of plasma renin activity were generally proportional to normal or increased aldosterone secretion rates. A good correlation between increases in the aldosterone secretion rate and increases in plasma renin activity was also found by Barraclough et al. (1965) and Veyrat et al.

(1968). On the other hand, only two of eight patients of Weidmann and Siegenthaler (1967) with a pathologically high peripheral plasma renin activity also had a pathologically increased aldosterone secretion rate. No significant correlation between plasma renin activity and the aldosterone secretion rate was observed in ten patients with renovascular hypertension by Schönbeck et al. (1968). However, no instances of renovascular hypertension have been reported in which hypersecretion of aldosterone was associated with normal or low plasma renin activity. This is consistent with the obvious assumption that in renovascular hypertension hyperaldosteronism is due to increased activity of the renin-angiotensin system. However, the causal relation between increased renin activity and increased aldosterone secretion may be more complex and involve alterations in sodium balance. Acute renal ischemia in a 22-year-old man led not only to malignant hypertension with extremely high plasma levels of renin, hyperaldosteronism with hypokalemia, and metabolic alkalosis, but also to severe sodium and water depletion and a reduced plasma volume; surgical treatment of the renal artery stenosis corrected all these abnormalities (Barraclough 1965). An inverse correlation between plasma renin and sodium concentrations in patients with renal hypertension has been observed by J.J. Brown et al. (1965a). In four of six patients with hyperaldosteronism due to unilateral renal artery stenosis, serum sodium concentrations were abnormally low, and in all patients total blood volume, plasma volume, and red cell volume were in the low normal range or decreased (Slaton and Biglieri 1965). Available evidence does not allow firm conclusions as to what degree the elevated plasma renin level is a cause or a consequence of the disturbance of sodium balance in renovascular hypertension. However, a low serum sodium concentration, a low total body sodium, or both may substantially contribute to increased aldosterone secretion in some patients with renovascular hypertension. This is consistent with the finding by Kem et al. (1971) that saline infusion suppressed plasma aldosterone levels to a normal extent in patients with renovascular hypertension. Elevated urinary aldosterone rates were also normally suppressed in patients with renovascular hypertension by 3 days of treatment with deoxycorticosterone acetate (Biglieri et al. 1972).

Animal studies on the influence of experimental renovascular hypertension on aldosterone secretion have generally confirmed the observations made in human renovascular hypertension cited above. In all models of experimental renovascular hypertension, aldosterone production was found to be independent of the blood pressure and was increased only under experimental conditions, which also induced a marked increase in the peripheral plasma renin level or in the renin content of the ischemic kidney.

Carpenter et al. (1961) induced renal hypertension in dogs by constricting both renal arteries. According to the tightness of the second clamp, two different types of hypertension developed. In dogs with a benign form of hypertension, aldosterone secretion remained within normal limits, and the renin content of the kidneys was only slightly increased. In dogs with a malignant form of hypertension, aldosterone secretion rose markedly, and there was a tenfold increase of the kidney renin contents. In dogs in whom hypertension was induced by unilateral renal artery constriction and contralateral nephrectomy, the aldosterone secretion rate fell during the first week after application of the clamp (C.S. McCaa et al. 1965). After the blood pressure had reached its maximum value, aldosterone secretion rose again but did not reach preoperative values. In short-term experiments, constriction of one or both renal arteries or renal veins in intact as well as in unilaterally nephrectomized dogs was promptly followed by a sharp increase in aldosterone secretion without concomitant changes in cortisol and corticosterone outputs (A. F. Muller 1963; A. F. Muller et al. 1964). In hypophysectomized dogs, bilateral renal artery constriction led to marked increases in the secretion rates of aldosterone, corticosterone, and cortisol (Slater et al. 1965). Constriction of the remaining renal artery in unilaterally nephrectomized dogs raised plasma aldosterone only transiently for a day or two (Liard et al. 1974). The increase in plasma aldosterone was accompanied by a sharp - but also transient - rise in plasma renin activity. After release of the renal artery constriction there was a significant drop in plasma renin activity to subnormal values, whereas plasma aldosterone remained unchanged.

In sheep, constriction of one renal artery with the opposite kidney left intact resulted in only a temporary increase in the peripheral plasma concentrations of aldosterone, cortisol, and renin (Blair-West et al. 1968c). After a few days all these parameters returned to normal, but blood pressure remained elevated. Some animals with severe hypertension became sodium-deficient due to sodium loss by the unclamped kidney. In these sheep, renin and aldosterone levels did not return to normal but rose progressively in proportion to the sodium deficit. When the remaining renal artery in unilaterally nephrectomized sheep was constricted, hypertension developed, but sodium balance was maintained: plasma renin and aldosterone rose initially but became normal within 3 days.

In rabbits, two-kidney renovascular hypertension (unilateral renal artery constriction, opposite kidney left intact) led to a significant increase in the mean plasma renin activity, when measured after 30 days (Lohmeier and Davis 1976). However, plasma renin activity was not elevated in all hypertensive animals but only in the ones with negative sodium balance. Increases in the aldosterone secretion rate were found to be proportionally related to increases in plasma renin activity.

Spät et al. (1966b) constricted the aorta of rats above or below the renal arteries and found that suprarenal constriction resulted in increased, infrarenal constriction in decreased aldosterone and corticosterone production by incubated adrenal tissue. Adrenals of rats with acute renal hypertension, induced by renal capsulation, converted more [^{14}C]-labeled progesterone to aldosterone and 18-hydroxycorticosterone than adrenals or control animals (Vecsei et al. 1966b). Adrenals of rats with chronic renal hypertension due to unilateral renal artery constriction showed a normal rate of incorporation of ^{14}C-progesterone into aldosterone but a decreased incorporation into 18-hydroxycorticosterone. Chronic experimental hypertension after unilateral renal artery constriction led to a significantly (+200%) increased aldosterone secretion but to unchanged corticosterone secretion, according to Singer et al. (1963a) and Kinson et al. (1967). If one renal artery was clamped and the contralateral kidney was removed, aldosterone secretion remained unchanged despite a marked increase in blood pressure. These findings are consistent with the results of our own study (J. Müller and Gross 1969) on aldosterone biosynthesis in adrenal tissue of rats with different forms of experimental renovascular hypertension. Adrenals of rats with one renal artery constricted and the other kidney left intact produced 200% more aldosterone from endogenous precursors and converted 50% more tritiated pregnenolone, progesterone, and corticosterone to aldosterone than adrenals of control animals when incubated under basal conditions (Fig. 35). The difference in aldosterone production was less marked when serotonin, KCl, or ACTH were added to the incubation medium. The production of corticosterone and deoxycorticosterone by quartered adrenals of renal hypertensive rats was almost normal when incubated with or without aldosterone-stimulating substances. This contrasted with the decreased production of corticosterone and deoxycorticosterone observed in quartered or capsular adrenals of sodium-restricted rats (J. Müller and Huber 1969; J. Müller 1970b). If rats were given a 1% NaCl solution instead of water as drinking fluid, the rise of aldosterone production was not prevented (Fig. 36). On the other hand, the increase in aldosterone production was partially or totally inhibited when, simultaneously with renal artery constriction, the contralateral kidney was removed or when the ureter of the ischemic or the contralateral kidney was ligated. Gross et al. (1965, 1968) have shown that in experimental renovascular hypertension a rise in the renin content of the ischemic kidney and in peripheral plasma renin-like activity is dependent on an intact, functioning contralateral kidney. Normal plasma renin activity and normal aldosterone production by incubated quartered adrenals were also observed in unilaterally nephrectomized rats with experimental renal hypertension due to constriction of the remaining kidney (Karppanen et al. 1975). However, by putting a clip of 0.2 mm internal diameter on the remaining renal artery of unilaterally nephrectomized rats, R. E. McCaa et al. (1979) induced malignant hypertension which was associated with moderately increased plasma levels of renin activity and aldosterone. In these animals, plasma aldosterone levels returned to normal under treatment with the converting-enzyme inhibitor captopril. In rats with a benign one-kidney Goldblatt hypertension (clip diameter 0.25 mm), plasma renin activity was suppressed and plasma aldosterone

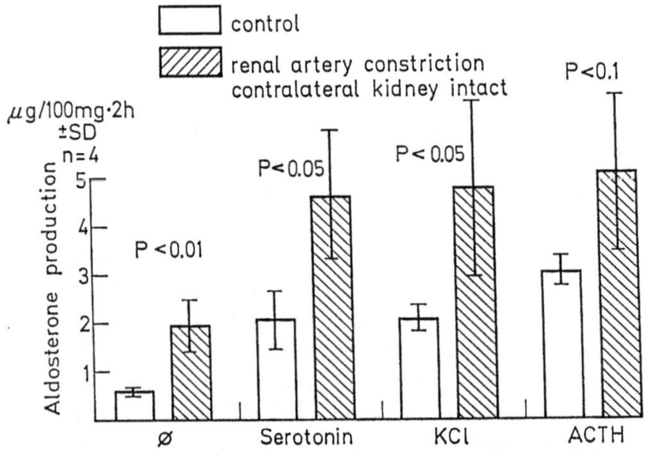

Fig. 35. Aldosterone production in vitro by quartered adrenals of normal rats *(white columns)* and rats with experimental renal hypertension due to clamping of one renal artery *(hatched columns)*. Mean values of two experiments ($n = 4$) ± standard deviation. *P* values of significance calculated by *t*-tests. The following substances were added to the incubation medium: serotonin creatinine sulfate ($1.7 \times 10^{-5} M$), KCl (final K^+ concentration 8.5 mmol/l), and ACTH (5 IU/flask). (Data from J. Müller and Gross 1969)

concentration was normal. In these rats, captopril treatment had a moderate and slow blood pressure-lowering effect but did not suppress the plasma aldosterone concentration. In rats with two-kidney Goldblatt hypertension, increases in aldosterone secretion depended on the duration and severity of the hypertension. In hypophysectomized rats, aldosterone secretion increased by almost 100% within 1 h of unilateral renal artery constriction (Spät et al. 1976). According to Leenen et al. (1975), there was a transient elevation in plasma renin activity and in aldosterone production by incubated adrenals on the 2nd day following the constriction of a renal artery with a 0.2 mm diameter clip. However, a permanent increase in aldosterone production became manifest only on day 14, 5 days after the beginning of a continuing rise in plasma renin activity. In rats with moderate renovascular hypertension (clip diameter 0.25 mm) no endocrine changes were observed. Dietz et al. (1975) found an increased secretion and plasma concentration of aldosterone after 9 days of renal artery constriction. After 21 days, both parameters were slightly elevated in animals with benign hypertension and markedly elevated in animals with malignant hypertension. At the same time, there was a close correlation between the plasma concentrations of aldosterone and angiotensin II in malignantly hypertensive rats. On the other hand, DeForrest et al. (1979) found markedly elevated plasma aldosterone levels - particularly when measured at 4 p.m. - not only in severely but also in moderately hypertensive rats 4 and 10 weeks after unilateral renal artery constriction by a 0.2-mm clip. Again,

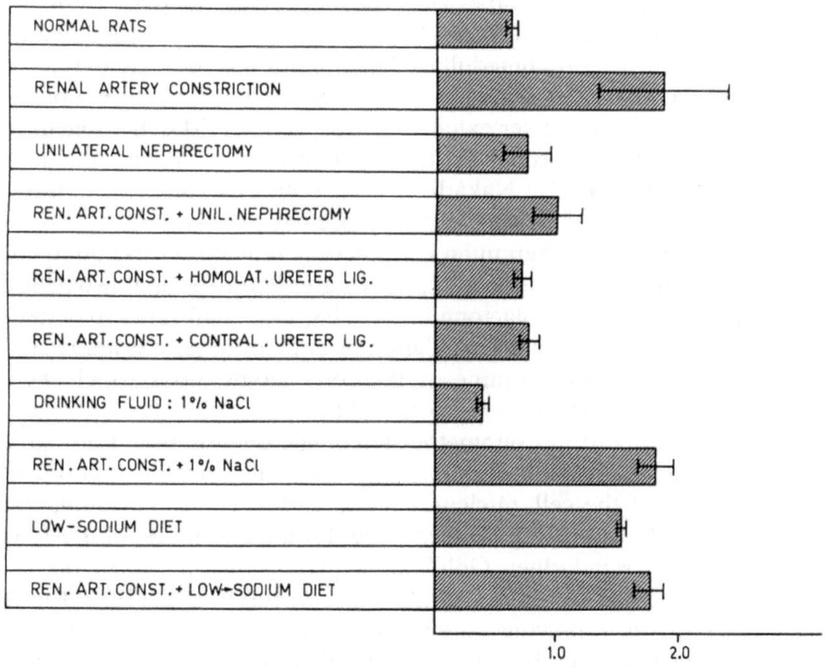

Fig. 36. Aldosterone production by quartered adrenal glands of various experimental groups of rats incubated under basal conditions. Mean values of two flasks ± range. (Data from J. Müller and Gross 1969)

R. E. McCaa et al. (1979) observed markedly elevated plasma aldosterone levels in rats with two-kidney Goldblatt malignant hypertension (clip diameter 0.2 mm) but only moderately elevated levels in rats with benign hypertension (clip diameter 0.25 mm). Plasma renin activity was elevated in proportion to raised plasma aldosterone. In both groups of animals, plasma aldosterone returned to normal under captopril treatment. The close correlation between aldosterone production and the activity of the renin-angiotensin system are at least compatible with the assumption that increases in aldosterone biosynthesis seen in renovascular hypertension are mediated by angiotensin II. However, there is also evidence of chronic increases in ACTH secretion, which could have contributed to the overall stimulation of aldosterone secretion in several instances of experimental renovascular hypertension. Thus, significantly elevated plasma concentrations of corticosterone and increased total corticosteroid production by incubated adrenal tissue in rats with severe or malignant two-kidney Goldblatt hypertension have been reported by Leenen et al. (1975) and by Dietz et al. (1975). The level to which plasma corticosterone was raised in the course of malignant renovascular hypertension depended on the sex and strain of the rats (J. Möhring et al. 1976). Also, significant increases in relative and absolute adrenal weight have been observed in rats with renovascular or renal encapsulation-induced hypertension (Fisher and Horvath 1971; Dietz et al. 1975; Kasemsri and Nickerson 1976).

An ultrastructural study of the adrenal zona glomerulosa of two children with renovascular hypertension and high plasma renin activity revealed an increased number of membrane-bound dense bodies and - in some cells - increased smooth endoplasmic reticulum (Hashida and Yunis 1972). In one patient, "corticomedullary cells" were found in the corticomedullary junction, in the zona glomerulosa, and in capsular vascular spaces. An increase in zona glomerulosa glucose-6-phosphate dehydrogenase and - to a lesser extent - 3β-hydroxysteroid dehydrogenase activity was found in adrenal cortical tissue of patients with renovascular and renal parenchymal hypertension by Nakada et al. (1970). The zona glomerulosa width was significantly increased in rats with renovascular (two-kidney) hypertension (Fischer and Horvath 1971). Agranular endoplasmic reticulum was abundant. Mitochondrial cristae showed a tendency for concentric lamination, approaching in appearance the so-called spironolactone bodies. Experimental renal hypertension, induced in young rats by latex encapsulation of both kidneys, was associated with increases in the width and volume of the zona glomerulosa which were directly proportional to the elevation in systolic blood pressure (Kasemsri and Nickerson 1976). According to morphometric-stereologic quantitation, both zona glomerulosa and zona fasciculata cells of these animals showed significant increases in the volume of the cell, nucleus, smooth endoplasmic reticulum, and lipid droplets. Using similar stereological techniques for a quantitation of the zona glomerulosa of rats with two-kidney Goldblatt hypertension (increased plasma renin activity and plasma aldosterone concentration), Rebuffat et al. (1979) found significant increases in the following parameters: volume of zona glomerulosa, volume of parenchymal cells, number of parenchymal cells, volume of nuclei, volume of mitochondrial compartment, surface of mitochondrial cristae, surface of smooth endoplasmic reticulum, and volume density of granules. However, there was a significant decrease in the volume of the lipid compartment. The zona fasciculata was not studied in these experiments.

5.5.2 Renal Denervation, Kidney Transplantation, and Unilateral Nephrectomy

The secretion of renin by the juxtaglomerular cells depends to some extent on the sympathetic innervation of the kidneys (Vander 1967). In dogs, bilateral denervation of the kidneys was followed by a delayed but undiminished rise in plasma renin activity in response to sodium restriction (Brubacher and Vander 1968). According to Mogil et al. (1969), an increase in plasma renin activity in response to sodium depletion (diuretic and low-sodium diet) was completely absent during the first 2 months after renal sympathectomy of dogs but became normal again after 3 months. This was not confirmed by Gotshall et al. (1973), who found that sodium depletion a few days after bilateral renal denervation induced a normal rise in plasma renin activity. In unilaterally nephrectomized sheep, 1-2 weeks after the denervation of the remaining kidney, sodium depletion by saliva sequestration elicited a normal increase in the blood aldosterone concentration, although there was a delayed and diminished rise in blood angiotensin II; plasma renin concentration remained within the sodium-replete range, and plasma potassium was unchanged (Blair-West et al. 1972). In rats, bilateral renal denervation had no effect on the aldosterone secretion rate during normal sodium intake, but it partially inhibited the rise in aldosterone secretion induced by 1 week of dietary sodium restriction (Kinson and Singer 1969). However, 2 weeks of sodium restriction stimulated aldosterone secretion to the same extent in intact rats and in rats with denervated kidneys.

A number of investigators have regarded human kidney transplants as models of renal denervation. This assumption is probably sound during the first weeks after transplantation. However, most of the following studies were carried out many months or years after transplantation. By that time, some of the transplants might well have become re-innervated. Thus, Greene et al. (1968) compared three patients who had received renal homotransplants 1 year or more previously with their kidney donors. There were no differences between the two groups in their response in plasma renin activity and urinary aldosterone excretion to dietary sodium restriction and to upright posture. Normal responses in urinary or plasma aldosterone and plasma renin activity to high or low sodium intake were also observed in patients with kidney transplants by Cooke et al. (1973b), Bennett et al. (1974), Keogh et al. (1976), E.B. Pedersen and Kornerup (1976b), and Sufrin et al. (1978). A normal increase in plasma aldosterone and plasma renin activity in response to upright posture in renal transplant recipients was found by Cooke et al. (1973a) and R.E. McCaa et al. (1973c) but not by E.B. Pedersen and Kornerup (1976). R.E. McCaa et al. (1973c) also reported normal responses in plasma aldosterone to stimulation by ACTH (in the absence of a responses in plasma cortisol, most likely because of prednisone medication) and by angiotensin II in a group of kidney allograft recipients. Beckerhoff et al. (1974) measured plasma aldosterone and plasma renin activity in 19 kidney allograft recipients repeatedly during the first days and weeks after transplantation. Both parameters were elevated in most patients during the first days. The values were significantly higher in patients with late onset of graft function. There was a good correlation between plasma renin activity and plasma aldosterone in most instances, even in previously anephric patients whose plasma renin activity had been unmeasurably low. However, a dissociation between the two parameters was observed in several cases. Sodium concentration was low (mean 131 mmol/l) and serum potassium high (mean 4.7 mmol/l) on day 1. Both parameters became normal within a period of days or weeks. Within a month after transplantation, normal plasma aldosterone concentrations were seen in all and normal plasma renin activity in most patients. There was a significant increase in the mean value of both parameters in response to 2 h of ambulation after overnight recumbency. In nonsteroid-treated kidney transplant recipients, there was a normal circadian rhythm of plasma aldosterone, which closely paralleled the circadian rhythm of plasma cortisol, whereas plasma renin activity stayed constant throughout the day (Ølgaard et al. 1977a). In steroid-treated patients with kidney allografts, peak aldosterone values were seen at 8 p.m.

Plasma renin activity was higher in patients with post-transplant hypertension than in normotensive kidney transplant recipients (Bennett et al. 1974). However, urinary aldosterone excretion was the same in both groups, and both parameters were equally suppressed by a high sodium intake. In contrast, Kornerup and Pedersen (1977) found increased plasma renin concentrations and reduced renin suppression by a high sodium intake only in hypertensive patients with an angiographically verified renal artery stenosis. In most of these patients, plasma aldosterone was normal and responded normally to a high sodium intake. According to Keogh et al. (1976) and to Sufrin et al. (1978), basal plasma renin activity was slightly higher in hypertensive than in normotensive post-transplant patients, whereas plasma aldosterone concentrations were equal. The aldosterone secretion rate was elevated in all post-transplant patients but to a greater extent in hypertensive ones. This parameter did not increase in response to sodium depletion in either group of patients in spite of normal increases in plasma aldosterone and plasma renin activity.

In a patient with hyperkalemia and renal salt loss following kidney transplantation, the urinary aldosterone excretion rate was found to be inappropriately low (H.J. Kramer et al. 1972). Renin was not measured, but no increase in the granulation of the juxtaglomerular apparatus was observed at autopsy. In two other patients, described by Roll et al. (1979), hyperkalemia following renal transplantation was associated with abnormally low values of plasma renin activity and plasma aldosterone. In one case, plasma aldosterone did not respond to an angiotensin II infusion. Six weeks after the transplantation there was a spontaneous recovery, and aldosterone and renin secretions became normal again.

In a group of apparently healthy young men who had had one kidney removed 1.4–13.2 years previously, R.D. Gordon et al. (1975) found a marked decrease in recumbent plasma renin activity (−86%). There was also a moderate but statistically significant decrease in the urinary aldosterone excretion rate (−36%). Plasma volume was normal, but diastolic blood pressure was elevated in four of 11 patients.

5.5.3 Renal Failure and Hemodialysis

In normokalemic patients with chronic renal failure, plasma aldosterone and 18-hydroxycorticosterone concentrations and plasma renin activity were found to be normal or moderately elevated (Weidmann et al. 1975a, Saruta et al. 1981).

Sodium restriction or upright posture elicited normal rises in these parameters, and there were also normal increases in plasma aldosterone in response to furosemide, angiotensin II, and ACTH. In hyperkalemic patients with chronic renal failure and similar creatinine clearance rates, plasma aldosterone and plasma renin activity were decreased and showed a diminished rise in response to the various stimuli. In this latter group of patients, plasma 18-hydroxycorticosterone was diminished in proportion to the decrease in plasma aldosterone, whereas the plasma concentrations of corticosterone, deoxycorticosterone, and cortisol were normal. The hyperkalemia of these patients was probably due to insufficient aldosterone secretion. Apparently, chronic decreases in renin secretion had led to a diminished responsiveness of the zona glomerulosa to high extracellular potassium. On the other hand, only one of three hyperkalemic patients with chronic renal failure not treated by hemodialysis showed a decreased plasma aldosterone with a diminished response to ACTH, according to G.H.Williams et al. (1973). In the other two cases, plasma aldosterone was normal and elevated respectively, and showed a marked increase in response to ACTH.

In 27 patients with terminal renal failure, plasma aldosterone ranged from subnormal to excessively elevated values and correlated significantly with plasma renin activity as well as with plasma potassium concentrations (Weidmann et al. 1973a). Plasma aldosterone and plasma renin activity were significantly higher in patients with uncontrollable hypertension than in patients with normal blood pressure or with hypertension controllable by chronic dialysis treatment. This finding was in agreement with the observations by J.J.Brown et al. (1971). In three of four patients with terminal renal failure and uncontrollable hypertension, plasma aldosterone fell from abnormal high values into the lower normal range within 24 h after bilateral nephrectomy (Weidmann et al. 1973a). In the 4th patient, plasma aldosterone was still elevated 1 day after nephrectomy, although the plasma renin activity had decreased to a hardly detectable low level. According to G.H.Williams et al. (1973), plasma aldosterone was low and did not increase in response to upright posture or an ACTH infusion in most of their patients with chronic renal failure kept on hemodialysis treatment. This suppression of aldosterone secretion was thought to be possibly due to long-term intermittent heparin administration. However, this was not supported by the observations of Weidmann et al. (1975b), who found an intact responsiveness of plasma aldosterone to postural change and ACTH in a number of non-nephrectomized patients with normal or high plasma renin activity who had been treated by hemodialysis for up to 90 months. In five patients with abnormally low plasma renin activity and plasma aldosterone, Krause et al. (1980) found the posthemodialysis body weight to be considerably above the estimated dry weight. After removal of the excess fluid by chronic hemodialysis treatment, both plasma renin activity and plasma aldosterone rose to normal or elevated values. Accordingly, hyporeninemic hypoaldosteronism in dialysis patients is not always irreversible; it can be corrected in some cases by long-term fluid depletion. Woods et al. (1974) observed a significant but subnormal rise of plasma aldosterone in response to angiotensin II infusion in non-nephrectomized patients with terminal renal failure. An intact responsiveness of plasma aldosterone to angiotensin II and ACTH in non-nephrectomized patients on regular hemodialysis has also been reported by Ølgaard and co-workers (Ølgaard and Madsen 1977a; Ølgaard et al. 1977b). In such patients, a lowering of the plasma potassium concentration, achieved with a combined infusion of insulin and glucose, was accompanied by a parallel decrease in plasma aldosterone (Ølgaard 1975).

Generally, acute changes in aldosterone secretion induced by hemodialysis in non-nephrectomized patients seem to result from the competitive effects of potassium loss and volume depletion. According to Schurr et al. (1976) and Henrich et al. (1977), plasma aldosterone was significantly increased after isokalemic dialysis and significantly decreased after hypokalemic dialysis. Ølgaard and Madsen (1977b) also observed that hypokalemic dialysis of non-nephrectomized patients regularly resulted in a decrease in plasma aldosterone, which was, however, not proportional to the decrease in plasma potassium. In the same patients, ultrafiltration dialysis (isokalemic) raised the plasma aldosterone, whereas conventional hemodialysis (hypokalemic) lowered it (Farinelli et al. 1980). During hypokalemic hemodialysis there was a clear-cut dissociation between the change in plasma aldosterone and the change in plasma renin activity. In patients with terminal renal failure who had controllable hypertension, isokalemic dialysis resulted variably in decreased, unchanged, or increased plasma aldosterone, whereas isonatremic dialysis either did not affect plasma aldosterone or decreased it (W. Vetter et al. 1977). Van Stone et al. (1982) reported that hypernatremic dialysis with fluid removal led to a decrease in plasma aldosterone in spite of a marked increase in plasma renin activity. Decreases in plasma aldosterone accompanying increases in plasma renin activity were also induced by hyponatremic or isonatremic dialysis without weight loss.

5.5.4 Bilateral Nephrectomy

Anephric human subjects and bilaterally nephrectomized animals have been extensively used as models for testing the role of the renin-angiotensin system in the physiological control of aldosterone secretion. In classical experimental endocrinology, removal of an endocrine gland is an important step in assessing the biological function of its hormones. However, bilateral nephrectomy involves much more than just the elimination of the biological source of a hormone. In particular, the abolishment of the vital excretory function of the kidneys creates a complex and artificial physiological situation. Uremia with hyperkalemia can itself be a potent stimulus of aldosterone secretion. Besides producing renin, the kidneys might secrete other yet unknown hormones involved in aldosterone regulation. The kidneys may also play a critical role in the metabolism or excretion of extrarenal aldosterone regulators. Finally, bilateral nephrectomy does not completely eliminate the renin-angiotensin system from the mammalian organism.

Isoenzymes of renin which hydrolyze angiotensinogen to form angiotensin I are found in considerable quantities in the extrarenal tissues of most mammals, including man (for reviews see Ganten et al. 1976a, 1976b). Sources of "isorenins" or "tissue angiotensinogenases" are – among others – the brain, the uterus, blood vessel walls, the salivary glands, the lungs, the adrenal glands, tumors, and tissue cultures. It can be assumed that extrarenal isorenins function mainly as tissue hormones and are generally not secreted into the circulation. Nevertheless, variable small amounts of isorenin enter the blood stream by leakage or by an active secretory process and make up the measurable renin activity found in some anephric human beings and experimental animals. The release of isorenin from extrarenal tissue into the blood may set the minimal level of activity of the renin-angiotensin system achievable in an individual organism (Yu et al. 1972). Whereas measurable normal or subnormal plasma renin concentrations were found in all three and seven patients respectively, who had been anephric for periods of 3 weeks to 17 months by Hannon et al. (1969) and by McKenzie and Montgomerie (1969), no renin was detected by Berman et al. (1972) in 61 serum samples taken from ten anephric subjects in the supine and upright posture and before and after severe dehydration. Capelli et al. (1968) had previously reported normal levels of plasma renin activity in four (all female) anephric patients and unmeasurably low levels in two others. In a male patient with a normal plasma angiotensin II concentration 9 months after bilateral nephrectomy, the plasma renin concentration was subnormal, while the plasma renin substrate concentration was elevated (Yu et al. 1972). In this man, plasma renin was affected neither by angiotensin II infusion nor by sodium loading or depletion.

Isorenin has been detected in adrenal tissue of man (Ganten et al. 1976a), cattle (Ryan 1972), dogs (Ganten et al. 1976b), rabbits (Ryan 1967), and rats (Ganten et al. 1974) as well as in human adrenocortical tumors and nontumorous adrenocortical tissue (Ganten et al. 1976a; Naruse et al. 1973). In the context of this monograph, it is of particular potential interest that there is some evidence indicating that adrenal isorenin appears to be subject to physiological regulation and may be involved in local control of aldosterone biosynthesis. Thus, dietary sodium restriction and, to a lesser extent, potassium restriction of rats significantly raised the isorenin content of adrenal tissue (Ganten et al. 1974). Adrenal isorenin concentration was also elevated 16 h after bilateral nephrectomy of rats bearing a transplanted ectopic pituitary tumor, in whom plasma corticosterone was elevated, adrenal weight was markedly increased, and plasma renin concentration was unmeasurably low (Ganten et al. 1974). "Renin activity" was also found in the rat adrenal gland, with a preferential localization in the capsular portion, by Aguilera et al. (1980a) and by Doi et al. (1983, 1984). It was significantly increased by dietary sodium restriction and significantly suppressed by sodium loading. Bilateral nephrectomy led to a marked and selective increase in the capsular adrenal isorenin concentration (Doi et al. 1983, 1984). Most likely, this was due to hyperkalemia, since potassium loading of intact rats had a similar effect. The angiotensin II antagonist saralasin (see p. 30) significantly lowered the aldosterone output when added without angiotensin II to incubated human adrenal tissue (Ganten et al. 1976b), rat adrenal tissue (Ganten et al. 1976a), and rat adrenal glomerulosa cells (G. H. Williams et al. 1974). This indicated an effect of locally formed angiotensin II under baseline conditions on incubation. On the other hand, converting-enzyme inhibitors did not suppress the aldosterone output by rabbit or rat zona glomerulosa cells, unless they were incubated with added angiotensin I or des-asp^1-angiotensin I (Ackerly et al. 1977; Mendelsohn and Kachel 1980).

Some investigators have reported normal basal plasma aldosterone concentrations in anephric patients (R. E. McCaa et al. 1972; Mitra et al. 1972). Others have measured subnormal values in most patients and unmeasurably low values in some of them (Bayard et al. 1971; G. H. Williams et al. 1973; Weidmann et al. 1973b; T. J. Goodwin et al. 1974; Sealey et al. 1978). Low plasma aldosterone concentrations in bilaterally nephrectomized people result from the combination of a very low production rate (Pratt et al. 1978) and an increased metabolic clearance rate of the hormone (J. Möhring et al. 1968; G. H. Williams et al. 1973; Corvol et al. 1974). The discrepancy in basal values of plasma aldosterone reported by different investigators may account for similar discrepancies in observed responses of this parameter to various stimulatory maneuvers, because there seems to be a close correlation between basal plasma aldosterone and its responsiveness to stimuli (Weidmann et al. 1973b; Deheneffe et al. 1975; Deheneffe and Castermans 1976; Ølgaard et al. 1977b).

A significant increase in plasma aldosterone in response to 2 h of quiet standing was observed in eight nephrectomized, normally hydrated patients by Mitra et al. (1972). However, this was associated with a nonsignificant increase in plasma cortisol. Tuma et al. (1981) induced a significant rise in plasma aldosterone in four anephric patients by 1 h of ambulation; plasma cortisol remained unchanged. In contrast, postural change had no effect on plasma aldosterone in the patients of Bayard et al. (1971), R. E. McCaa et al. (1973c), and G. H. Williams et al. (1973). Weidman et al. (1973b) observed a rise in plasma aldosterone in response to upright posture in six of 12 anephric patients. In the responsive patients, there was proportional increase in the aldosterone secretion rate and only a small (but statistically significant) decrease in the aldosterone metabolic clearance rate during upright posture.

Plasma aldosterone rose significantly in anephric patients in response to hemodialysis, according to R. E. McCaa et al. (1972). This was associated with decreases in plasma sodium and potassium concentrations (with an increase of the plasma sodium/plasma potassium ratio) and with an average weight loss of 1.5 kg. Plasma cortisol remained unchanged. The rise in plasma aldosterone was due to an increase in the aldosterone blood production rate, whereas the metabolic clearance rate of the steroid did not change (Read et al. 1973). The increase in aldosterone secretion was not caused by volume reduction, since removal of fluid volume by ultrafiltration was ineffective (R. E. McCaa et al. 1973b). In contrast, isovolemic sodium depletion without a change in plasma potassium elicited increases in plasma aldosterone which were proportional to the decreases in the plasma sodium concentration (R. E. McCaa et al. 1973b). W. Vetter et al. (1975) observed an increase in plasma aldosterone during hemodialysis in four of five anephric patients, while in all patients the plasma concentrations of cortisol, sodium and potassium decreased. The same group of investigators had previously reported secretory episodes of aldosterone in the night following hemodialysis of anephric patients (W. Vetter et al. 1974). These episodic rises in plasma aldosterone were not accompanied by parallel increases in plasma cortisol and were not suppressed by dexamethasone treatment. Other investigators found either no consistent change (G. H. Williams et al. 1973; T. J. Goodwin et al. 1974; Sealey et al. 1978; Cooke et al. 1983) or a decrease in plasma aldosterone (Weidmann et al. 1973b; Birkhäuser et al. 1974; Ølgaard and Madsen 1977b) in response to hemodialysis in a majority of their anephric patients. According to Cooke et al. (1983) hemodialysis with volume depletion of recently nephrectomized patients induced an increase in plasma aldosterone which was associated with a comparable rise in plasma cortisol and ACTH concentrations. In chronically nephrectomized subjects, a similar response in ACTH and cortisol secretion elicited by volume depletion was no longer accompanied by an increase in aldosterone secretion. The response in plasma aldosterone to extracellular fluid volume reduction or sodium depletion in anephric patients was found by T.J. Goodwin et al. (1974) to be generally poor. Moreover, central hypervolemia induced by immersion in water to the neck did not suppress plasma aldosterone in 12 anephric patients (M. Epstein et al. 1978). In most nephrectomized patients, the response in plasma aldosterone to ACTH or angiotensin II infusions was absent or markedly diminished (G. H. Williams et al. 1973; Weidmann et al. 1973; R. E. McCaa et al. 1973c; Woods et al. 1974; T. J. Goodwin et al. 1974; Ølgaard and Madsen 1977a; Ølgaard et al. 1977b; Tuma et al. 1981). According to Deheneffe et al. (1976), the aldosterone responsiveness to angiotensin II is also dependent on the sodium status in anephric man. Thus, angiotensin II infusions raised the plasma aldosterone concentration only in two of six anephric subjects when they were on their usual sodium intake. However, after 3 days of dietary sodium restriction combined with weight-reducing hemodialysis, angiotensin II also stimulated the aldosterone secretion in the four previously unresponsive subjects, while the aldosterone response was enhanced in the two previously responsive subjects. In all six subjects, plasma renin and angiotensin II were detectable, but did not change with sodium restriction and weight-reducing hemodialysis.

The plasma potassium concentration appears to be an important regulatory factor of aldosterone secretion in most anephric persons. Thus, the increases in plasma potassium occurring between dialysis sessions were generally found to be accompanied by parallel increases in plasma aldosterone (Weidmann et al. 1973b; Cooke et al. 1973b; Ølgaard and Madsen 1977b). However, in a number of anephric patients with very low plasma aldosterone concentrations there was no apparent relationship between the two parameters (Weidmann et al. 1973b; T. J. Goodwin et al. 1974). Cooke et al. (1973c) produced a significant decrease in the plasma potassium concentration in seven anephric patients by glucose and insulin administration. This resulted in a significant and sustained drop in the plasma aldosterone concentration. However, just the opposite effect was observed in similar experiments by Ølgaard (1975), who found plasma aldosterone to rise simultanously with an induced decrease in the plasma potassium concentration in seven anephric subjects.

In hypophysectomized dogs with hyperaldosteronism due to thoracic inferior vena cava constriction, bilateral nephrectomy resulted in a marked fall in aldosterone secretion (J. O. Davis et al. 1961b). In nephrectomized dogs with intact pituitary glands, aldosterone and cortisol secretions increased and decreased normally in response to caval restriction and release (Gann and Travis 1964). Nephrectomy led to transient low rates of aldosterone secretion in otherwise intact dogs and prevented an increase of aldosterone secretion in response to carotid artery constriction in most animals. In sodium-depleted hypophysectomized dogs, removal of the kidneys led to a striking decrease in the secretion rates of aldosterone, corticosterone, and cortisol, according to J. O. Davis et al. (1961b, 1966) but not according to Slater et al. (1965). Hemorrhage markedly stimulated aldosterone secretion in hypophysectomized dogs but not in hypophysectomized, nephrectomized dogs (Ganong and Mulrow 1961, 1962; J. O. Davis et al. 1961a; Slater et al. 1965). Bilateral nephrectomy combined with decapitation of dogs led within an hour to a marked decrease in the aldosterone secretion rate (R. E. McCaa et al. 1973a). At that time the aldosterone secretion rate showed no response to hemorrhage. Within 8 h there was a continuous increase in the serum potassium concentration from 3.2 to 5.6 mmol/l, which was associated by an increase in the aldosterone secretion rate and plasma concentration to control levels. Under these conditions, hemorrhage elicited a marked increase in aldosterone secretion and a further rise in serum potassium. Aldosterone secretion did not increase following hemorrhage when the increase in serum potassium concentration was prevented by hemodialysis. Bilateral nephrectomy markedly diminished the rise in aldosterone secretion in dogs in response to a lethal dose of *Escherichia coli* endotoxin (White et al. 1967). During anaphylactic shock induced by horse serum, the aldosterone secretion increased from 1.3 to 8.4 ng/kg/min in intact dogs and from 0.2 to 2.4 ng/kg/min in hypophysectomized, nephrectomized dogs, according to Aikawa et al. 1981 b. In the latter group of animals there was only a minor increase in plasma potassium concentration, and the authors concluded that aldosterone secretion had been stimulated either directly by histamine or by an as yet unknown mechanism.

Thoracic inferior vena cava constriction significantly stimulated aldosterone secretion in midcollicular-decerebrated, hypophysectomized, pinealectomized sheep, but had no effect in animals which had also been bilaterally nephrectomized (Blair-West et al. 1963). In sodium-replete uninephrectomized, mid-collicular-decerebrated sheep, removal of the second kidney was followed by a marked decrease in aldosterone output (Blair-West et al. 1963, 1964). In sodium-deficient sheep, the second nephrectomy did not significantly affect aldosterone secretion, which continued at an elevated rate for several hours. On the other hand, midcollicular decerebration combined with hypophysectomy and pinealectomy of sodium-deficient bilaterally nephrectomized sheep was followed by a substantial fall in aldosterone secretion. However, after all these ablations, aldosterone secretion remained elevated above the level found in normal sodium-replete sheep for periods of 3–7 h. Prolonged studies carried out in conscious sodium-deficient sheep showed that bilateral nephrectomy following hypophysectomy eventually resulted in a reduction of the aldosterone secretion rate to values seen in sodium-replete animals, but in a majority of experiments this occurred 10–12 h after nephrectomy (Blair-West et al. 1968 b). The sustained elevated aldosterone secretion rate was not due to hyperkalemia, which was prevented by appropriate dietary measures. In sodium-replete hypophysectomized sheep, the aldosterone secretion rate did not change for several hours after bilateral nephrectomy; it then rose coincidently with a rise in the plasma potassium concentration. In these animals, the aldosterone secretion rate was stimulated by a renin infusion to values observed in sodium-deficient sheep, but returned to normal within 2 h after cessation of the infusion. In sheep in which a sodium deficit had been produced by loss of saliva through a chronic parotid fistula, blood aldosterone concentration fell to near sodium-replete values within 24 h of bilateral nephrectomy (Blair-West et al. 1973 a). At this time the angiotensin II/aldosterone dose-response curve was extremely steep; i.e., normal plasma angiotensin II concentrations raised the plasma aldosterone to maximum levels. In a majority of the animals the plasma potassium concentration was moderately elevated. On the other hand, when the sodium deficit was produced by parotid cannulation and the sheep were treated with dexamethasone, plasma potassium and the response of aldosterone to angiotensin II were normal 24 h after bilateral nephrectomy (Coghlan et al. 1979). In these animals, the aldosterone response to angiotensin II was markedly diminished 48 h after bilateral nephrectomy.

In sodium-replete rats, acute bilateral nephrectomy was followed by a 30% decrease in aldosterone and corticosterone secretion (Eilers and Peterson 1964). When nephrectomy was combined with hypophysectomy, the decrease in aldosterone secretion was approximately 50%. Nephrectomy prevented the stimulation of aldosterone secretion by inferior vena cava constriction. Sodium-deficient hypophysectomized rats secreted aldosterone at a rate twice that of sodium-replete intact animals; bilateral nephrectomy induced only a 25% decrease in aldosterone secretion within 2 h. According to Marieb and Mulrow (1965), combined hypophysectomy and bilateral nephrectomy lowered the aldosterone secretion rate by 66% in sodium-replete rats; infusion of angiotensin II restored aldosterone to the preoperative level. Rats that had been kept on a sodium-deficient diet for 10 days secreted aldosterone at a markedly elevated rate 6-18 h after bilateral nephrectomy and 2-3 h after nephrectomy combined with hypophysectomy (Palmore et al. 1969). Sustained hypersecretion of aldosterone was not dependent on a low sodium or a high potassium concentration in the plasma. If sodium-replete rats were first nephrectomized and then depleted of sodium by peritoneal dialysis, aldosterone secretion increased by 100%, corticosterone secretion decreased markedly, and zona glomerulosa width increased slightly. When nephrectomy was followed by peritoneal sodium loading, the aldosterone secretion rate and zona glomerulosa width remained unchanged, but the corticosterone secretion rate decreased to an even greater extent than it did in nephrectomized sodium-depleted rats. Elema et al. (1968a) observed that lipid depletion and increases in enzyme activities (isocitrate dehydrogenase, glucose-6-phosphate dehydrogenase, NADP-tetrazolium reductase, 3β-hydroxysteroid dehydrogenase) induced in the zona glomerulosa by sodium depletion of intact rats was also induced by bilateral nephrectomy. In nephrectomized rats, these parameters were not further affected by subsequent sodium loading or sodium depletion. In rats that had been kept on a sodium-deficient diet for 7-10 days, bilateral nephrectomy failed to decrease aldosterone secretion rate within 90 min (Spielman and Davis 1974). Plasma aldosterone and corticosterone concentrations increased to the same high levels in response to hypovolemia induced by subcutaneous polyethylene glycol in intact and in acutely nephrectomized rats (Stricker et al. 1979). However, nephrectomy alone raised the plasma concentrations of both steroids to a considerable extent. Moreover, even in intact rats, polyethylene glycol-induced sequestration of extracellular fluid was followed by a significant increase in the plasma potassium concentration. Thus, aldosterone secretion may have been stimulated in nephrectomized animals by both ACTH and extracellular potassium. Spät et al. (1966a) had previously observed that aldosterone production by rat adrenals in vitro was significantly increased 2 h after bilateral nephrectomy. Bilateral nephrectomy prevented a further increase in aldosterone production due to suprarenal constriction of the aorta or a decrease due to infrarenal constriction (Spät et al. 1966b). Infusions of potassium chloride to bilaterally nephrectomized rats resulted in an additional enhancement of aldosterone production in vitro (Sturcz et al. 1967). Adrenals taken from uremic rats 40-48 h after bilateral nephrectomy produced normal amounts of aldosterone and decreased amounts of corticosterone and converted ^{14}C-labeled progesterone and tritiated deoxycorticosterone at an increased rate to aldosterone, 18-hydroxycorticosterone, and 18-OH-DOC but at a normal rate to corticosterone (Steinacker et al. 1968). These findings were at least partially consistent with the results of our incubation studies with adrenal tissue taken form uremic rats 40 h after nephrectomy (J. Müller and Huber 1969). Adrenal tissue of uremic rats produced significantly increased amounts of aldosterone when incubated in the presence of stimulators, such as serotonin, potassium chloride, or ACTH, or of precursor steroids (progesterone, deoxycorticosterone, corticosterone) added in substrate amounts to the incubation medium (see Figs. 16, 23). Under stimulation by serotonin or potassium ions, adrenals of uremic rats produced less corticosterone and deoxycorticosterone than adrenals of control animals. These changes were most likely due to a marked hyperkalemia, because they were almost exactly opposite of the alterations of aldosterone biosynthesis induced by potassium deficiency. Bilateral nephrectomy of previously potassium-depleted rats was followed by a delayed increase in the plasma potassium concentration not yet manifest after 24 h, i.e., at a time when the serum creatinine level was markedly elevated (Haldy and Müller 1981 a). In potassium-deficient nephrectomized rats, experimental edema induced by subcutaneous polyethylene glycol or formalin did not affect the capsular adrenal conversions of tritiated corticosterone and deoxycorticosterone to aldosterone and 18-hydroxycorticosterone (Fig. 27). There was a moderate increase in the plasma aldosterone concentration in polyethylene glycol-treated nephrectomized rats, which was, however, accompanied by a similar increase in the plasma corticosterone concentration. When uremia was induced in potassium-depleted rats by bladder resection, with the kidneys left intact, polyethylene glycol treatment still significantly stimulated aldosterone biosynthesis. In nephrectomized rats without edema, angiotensin II infused at a rate of 750 ng/kg/min stimulated the final steps of aldosterone to the same extent as it did in intact rats (J. Müller et al. 1982). The same dose of angiotensin II had no effect on the capsular adrenal conversions of deoxycorticosterone or corticosterone to aldosterone and 18-hydroxycorticosterone when infused into bilaterally nephrectomized rats with polyethylene glycol-induced edema (Fig. 37). This indicated that the role of the kidneys in mediating the stimulation of aldosterone biosynthesis by sodium sequestration was twofold. First, they acted through the secretion of renin. In addition, a second, yet unknown kidney factor was necessary for a full response of the zona glomerulosa to the stimulatory action of angiotensin II. The dopaminergic antagonist metoclopramide induced the same increase in plasma aldosterone in intact and in bilaterally nephrectomized rats (Sowers et al. 1980 a). In rats studied 30 h after nephrectomy, plasma renin activity was still measurable, and plasma potassium was increased. However, these parameters did not respond to metoclopramide administration. According to these results, dopaminergic modulation of aldosterone secretion occurred independent of the renin-angiotensin system. Bilateral nephrectomy of rats caused a marked increase in the number of adrenal angiotensin II receptors within 48 h (Douglas 1981). Since this effect was absent when nephrectomized rats were kept normokalemic by treatment with a cation exchange resin, it was apparently modulated by the plasma potassium concentration independently of changes in angiotensin II blood levels.

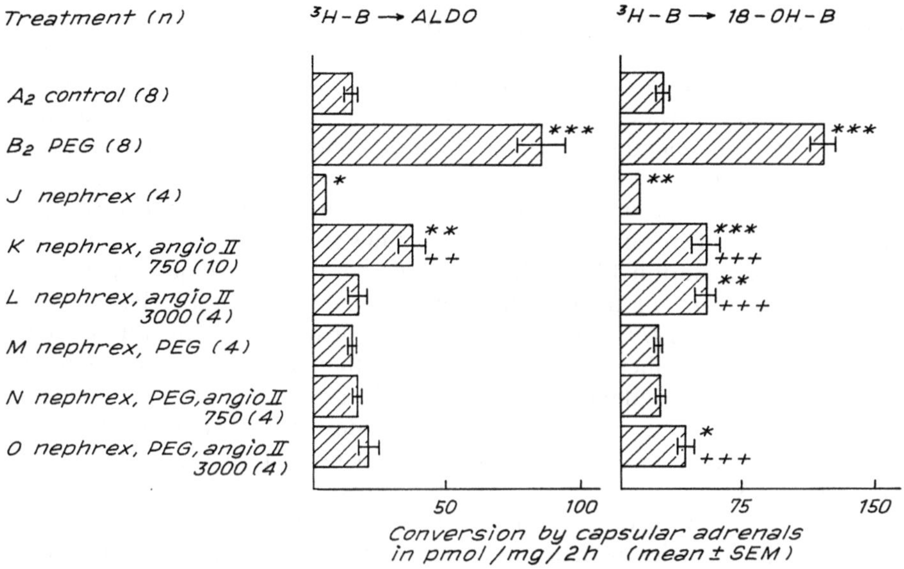

Fig. 37. Effects of different treatments (1-day) of potassium-deficient rats on capsular adrenal conversions of tritiated corticosterone (*^{3}H-B;* 300 nmol/flask) to aldosterone *(ALDO)* and 18-hydroxycorticosterone *(18-OH-B)*. Statistical significance (*t*-tests) of differences from control animals *(A_2)*: $^{***}P<0.001$, $^{**}P<0.01$, $^{*}P<0.05$. Statistical significance of differences from group *B_2*: $^{+++}P<0.001$, $^{++}P<0.01$. *PEG,* s.c. polyethylene glycol; *nephrex,* bilateral nephrectomy; *angio II,* angiotensin II, s.c. infused at rates of 750 or 3000 $ng \cdot kg^{-1} \cdot min^{-1}$. (From J. Müller et al. 1982)

5.5.5 Primary Reninism

A rare cause of hypersecretion of renin is a juxtaglomerular cell tumor of the kidney, which occurs in children and young adults and causes hypertension, urinary potassium loss, hypokalemia, and metabolic alkalosis (P. W. Robertson et al. 1967; Kihara et al. 1968; Eddy and Sanchez 1971; Bonnin et al. 1972; Conn et al. 1972; Schambelan et al. 1973; J. J. Brown et al. 1973). These usually benign tumors have a diameter of 0.8–4 cm, are highly vascularized, and contain large quantities of renin. In most instances reported plasma renin activity was markedly elevated and generally responded to postural changes but only exceptionally to sodium restriction (Conn et al. 1972; J. J. Brown et al. 1973b). Urinary aldosterone excretion, the aldosterone secretion rate, and plasma aldosterone levels were found to be elevated in proportion to the increases in plasma renin activity. Removal of the affected kidney was followed by a prompt disappearance of hypertension and all endocrine and metabolic abnormalities. A similar syndrome, characterized by malignant hypertension, hypokalemia, and hyponatremia, was caused by a renin-secreting nephroblastoma (Wilms' tumor) in a 2-year-old child (Mitchell et al. 1970). Ectopic secretion of renin by an oat cell carcinoma of the lungs in a 38-year-old man led to mild general edema, hypokalemia, and metabolic alkalosis (Hauger-Klevene 1970). Plasma renin activity was markedly elevated but was not affected by changes in posture or sodium intake. The aldosterone secretion rate was abnormally high and increased further in response to sodium restriction.

5.5.6 Hyporeninemic Hypoaldosteronism

Whereas primary mineralocorticoid deficiency, due to Addison's disease or to an inborn error of aldosterone biosynthesis, is generally associated with elevated plasma levels of renin and angiotensin II, the most common form of acquired isolated aldosterone deficiency is characterized by a suppressed activity of the renin-angiotensin system and is often referred to as "hyporeninemic hypoaldosteronism" (for reviews see DeFronzo et al. 1980; Veldhuis and Melby 1981; Vaamonde et al. 1981; Tan and Burton 1981). The syndrome occurs most frequently in middle-aged persons, many of them with advanced diabetes mellitus (Perez et al. 1977b; de Chatel et al. 1977; Christlieb et al. 1978) or with a chronic tubulointerstitial nephropathy (of diabetic or other etiology) with moderate azotemia (Schambelan et al. 1972, 1980; Weidmann et al. 1973c, 1975a; Saruta et al. 1981). According to available evidence, hyporeninemic hypoaldosteronism appears to be a rather rare complication of diabetes mellitus. It was not detected in patients with uncomplicated juvenile-type diabetes mellitus (Moss et al. 1978). Ferriss et al. (1982) found no evidence of overt hypoaldosteronism in 101 consecutive unselected outpatient diabetics with or without complications. Many patients with hyporeninemic hypoaldosteronism are hypertensive. Others suffer from autonomic insufficiency (R.D. Gordon et al. 1967a; Slaton and Biglieri 1967). Hyporeninemic hypoaldosteronism is not a regular feature of autonomic insufficiency, however. Thus, two patients with severe idiopathic orthostatic hypotension were found to show normal responses in the plasma renin activity and in the aldosterone secretion rate to upright posture and to sodium restriction respectively (Bliddal and Nielssen 1970). Salt wasting is not common in adult hyporeninemic hypoaldosteronism. The condition generally manifests as unexplained hyperkalemia with hyperchloremic metabolic acidosis, which in a majority of patients does not lead to cardiac or muscular dysfunctions. Hyperkalemia is not always present. Several cases of hypoaldosteronism with normal serum potassium concentrations were detected by Weidmann et al. (1980) among patients with diabetes mellitus or renal diseases. Treatment with a mineralocorticoid such as 9α-fluorocortisol or deoxycorticosterone acetate usually corrects the hyperkalemia and ameliorates the metabolic acidosis (Perez et al. 1974; Szylman et al. 1976; Sebastian et al. 1977). Only exceptional patients with hyporeninemic hypoaldosteronism also display a tubular resistance to mineralocorticoids (Vaamonde et al. 1981; Pelleya et al. 1983).

Hyporeninemic hypoaldosteronism can develop in patients with primary aldosteronism after the surgical removal of an aldosterone-producing adrenocortical adenoma (Biglieri et al. 1966a) and in patients with 17α-hydroxylation deficiency under suppressive therapy with dexamethasone (Biglieri et al. 1966b). The syndrome can be a consequence of previous excessive ingestion of sodium bicarbonate (Lowder and Brown 1975; Oster et al. 1976) or licorice (M.T. Epstein et al. 1977). It can also be caused by indomethacin (Tan et al. 1979) or ibuprofen (Tan and Burton 1981) medication or by chronic lead intoxication (Gonzalez et al. 1979). However, in most instances the etiology of hyporeninemic hypoaldosteronism is uncertain and probably multifactorial. Possible pathogenetic factors include age (Weidmann et al. 1975c, 1977; Noth et al. 1977b; Hegstad et al. 1983), increased total body sodium and expanded extracellular fluid volume (Oh et al.

1974; de Châtel et al. 1977), excessive sodium chloride intake (Rado et al. 1979), impaired conversion of prorenin to renin (DeLeiva et al. 1976; Tan et al. 1980), defective control of renin release by the autonomic nervous system (Tuck et al. 1979), and decreased prostaglandin E_2 production (Tan et al. 1980).

In patients with hyporeninemic hypoaldosteronism, aldosterone secretion obviously does not adequately respond to chronically elevated levels of extracellular potassium. In most patients, ACTH or angiotensin II infusions also elicited subnormal increases in plasma or urinary aldosterone or failed to stimulate aldosterone secretion (Schambelan et al. 1972; Weidmann et al. 1973c, 1975a; Kuhlmann et al. 1978; Tuck and Mayes 1980; Saruta et al. 1981). In all the patients studied by the investigators cited, ACTH stimulated cortisol secretion to a normal extent. The responses of plasma renin activity and aldosterone secretion to upright posture, sodium restriction, and short-term treatment with a diuretic were also diminished or absent in patients with hyporeninemic hypoaldosteronism (Schambelan et al. 1972; J.J. Brown et al. 1973a; Weidmann et al. 1973c, 1975a; Perez et al. 1977b, 1977c; Kuhlmann et al. 1978; Tuck and Mayes 1980; Saruta et al. 1981). However, prolonged sodium depletion and treatment with a diuretic stimulated both parameters in at least some patients (Oh et al. 1974; Perez et al. 1977c). Hahn et al. (1980) observed a normal increase in plasma renin activity in response to a prostaglandin A_1 infusion in two of nine patients with hyporeninemic hypoaldosteronism with normal prostaglandin E excretion.

The available evidence pointing to a selective defect of a particular step in the aldosterone biosynthetic pathway in patients with hyporeninemic hypoaldosteronism is conflicting and not conclusive. Thus, in a patient with hyporeninemic hypoaldosteronism and progressive renal failure, angiotensin II infusion produced an excessive rise in plasma pregnenolone, indicating an impaired conversion of pregnenolone to progesterone in the zona glomerulosa (Lebel and Grose 1976). The patterns of urinary corticosteroid metabolites in two patients with diabetes mellitus and acquired hypoaldosteronism - one with normal and one with decreased plasma renin activity - were suggestive of corticosterone methyl oxidase deficiency type 1 and type 2 respectively (DeLeiva et al. 1976). In five patients with diabetes and hyporeninemic hypoaldosteronism, Tuck and Mayes (1980) found normal plasma levels of 18-hydroxycorticosterone, which responded normally to sodium restriction as well as to ACTH or angiotensin II infusions. This indicated a partial blockade of the conversion of 18-hydroxycorticosterone to aldosterone. In contrast, Schambelan et al. (1973) found decreased urinary excretion of the major metabolite of 18-hydroxycorticosterone in three patients with hyporeninemic hypoaldosteronism. Similarly, the plasma levels of 18-hydroxycorticosterone were decreased in proportion to the plasma levels of aldosterone in the patients with chronic renal failure and hyporeninemic hypoaldosteronism studied by Saruta et al. (1981).

5.6 Role of the Pituitary Gland

It is generally assumed that the pituitary gland plays neither an essential nor a predominant role in the physiological control of aldosterone secretion. However, it is also undisputed that the pituitary gland can markedly influence aldosterone secretion in many direct or indirect ways, in particular through ACTH, but perhaps also through growth hormone or prolactin, through hormones which are exclusively secreted under pituitary control such as glucocorticoids and thyroid hormones or through the effects of antidiuretic hormone on salt and water homeostasis. Moreover, there is clinical and experimental evidence pointing to the possible involvement of ACTH by-products or unknown pituitary factors in the control of aldosterone.

5.6.1 Hypophysectomy and Hypopituitarism

In man and in experimental animals, such as dogs, sheep, and rats, aldosterone secretion continues in the complete absence of the pituitary gland. A great part of our knowledge on the physiological mechanisms regulating aldosterone secretion was obtained in experiments with animals which were hypophysectomized in order to abolish the influence of ACTH, a hormone known to stimulate aldosterone production (see Sect. 4.4.1). However, many investigators have found that hypophysectomy or spontaneous hypopituitarism leads to an impairment of aldosterone secretion under basal conditions or under various physiological stimuli. The extent of this impairment appears to depend greatly on the species, the replacement therapy with hormones, and the experimental protocol used.

In man, hypophysectomy did not acutely affect urinary aldosterone excretion (Jessiman et al. 1959), or it was followed by a temporary increase of this parameter, perhaps in response to transient diabetes insipidus (Ross et al. 1960). The morphology of the zona glomerulosa was normal in human adrenal glands removed 9 months after hypophysectomy from patients with breast cancer, whereas the inner zones were diffusely atrophic (Jessiman et al. 1959). Ross et al. (1960) found a decreased urinary aldosterone excretion rate with a reduced response to a low-sodium diet in patients with long-standing hypopituitarism. According to Lieberman and Luetscher (1960), basal aldosterone excretion was normal in hypopituitary patients, but it did not adequately increase during sodium restriction. Hauger-Klevene and Cole (1970) observed a decreased aldosterone secretion rate with a diminished response to sodium restriction in hypopituitary patients with chromophobe adenomas. In contrast, Raiti et al. (1968) reported normal aldosterone secretion rates with normal responses to sodium restriction in four hypopituitary children. The basal aldosterone secretion rate was also normal in 17 adult hypopituitary patients studied by G.H. Williams et al. (1971). However, a low-sodium diet elicited a delayed or insufficient increase of this parameter in 11 of them. After a week of dietary sodium restriction, 15 patients failed to show a significant increase in aldosterone secretion on the first day of an ACTH infusion. In the same study, six patients with secondary adrenocortical failure due to prolonged treatment with pharmacological doses of glucocorticoids showed normal aldosterone responses to the two stimuli. Evidence for a complex involvement of the pituitary gland in the regulation of aldosterone secretion was also found in patients with panhypopituitarism by R.E. McCaa and coworkers (1978b, 1981). Plasma aldosterone was normal under basal conditions but failed to increase in response to an intravenous ACTH injection, to an angiotensin II infusion, or to low sodium intake. After 12 days of sodium restriction, ACTH elicited a normal response in aldosterone secretion, whereas angiotensin II was still ineffective. Seifert and Oelkers (1981) found basal plasma aldosterone concentrations to be moderately decreased in eight hypopituitary patients. Plasma aldosterone rose in response to angiotensin II infusion as well as to sodium depletion (furosemide and low-sodium diet), although to a lesser extent than in normal subjects. Sodium depletion also significantly enhanced the adrenocortical sensitivity to angiotensin II in hypopituitary patients. In most of the 12 patients with long-standing pituitary insufficiency studied by Birkhäuser et al. (1981), plasma aldosterone was subnormal under basal conditions and showed a subnormal response to angiotensin II infusion and upright posture but a normal response to ACTH infusion. According to Lopez et al. (1980), plasma aldosterone was normal and responded normally to an angiotensin II infusion in six hypopituitary patients, even when cortisol replacement was withheld on the test day. However, in these patients a KCl infusion did not raise plasma aldosterone unless cortisol was given on the same day. The information available on the function of the renin-angiotensin system in hypopituitary patients is conflicting. Thus, a normal basal plasma renin activity with a normal response to sodium depletion has been reported by G.H. Williams et al. (1971), Murakami et al. (1972), and R.E. McCaa et al. (1978b, 1981). According to Hauger-Klevene and Cole (1970), plasma renin activity was normal in supine patients but did not increase during upright posture. Plasma renin activity was abnormally low and showed a diminished reponse to upright posture and furosemide, respectively, in the patients of Lefebvre et al. (1979) and of Bakiri et al. (1983a) and in some of the patients of Birkhäuser et al. (1981). The plasma angiotensin II concentration was elevated during normal sodium intake but showed a subnormal increase in response to sodium depletion in the patients of Seifert and Oelkers (1981). Chupin et al. (1981) found a decreased plasma renin activity only in hypopituitary patients who had not yet been treated with cortisol. However, plasma aldosterone was also subnormal in cortisol-treated patients with normal plasma renin activity. In individual hypopituitary patients a decreased plasma renin activity was restored to normal by long-term substitution therapy with glucocorticoids (Major et al. 1978; Merriam and Baer 1980). On the other hand, cortisol treatment alone raised the plasma renin substrate level but had no effect on plasma renin activity and lowered plasma aldosterone in the 20 patients with Sheehan's syndrome studied by Bakiri et al. (1983, 1986). Subnormal plasma aldosterone and normal renin activity with a nearly normal responsiveness to orthostasis and furosemide were observed in these patients, however, following combined therapy with cortisol and thyroxine.

Hypophysectomy resulted in an acute decrease of aldosterone secretion of approximately 90% in conscious or anesthetized dogs with hyperaldosteronism due to constriction of the thoracic inferior vena cava (J.O. Davis et al. 1960a, 1960b). However, in some dogs, aldosterone secretion remained above normal for several days. It was assumed by J.O. Davis (1961) that the effect of hypophysectomy was due to abolished ACTH secretion, since it was prevented by ACTH infusions. Mulrow and Ganong (1961) found that the increment in aldosterone secretion induced by hemorrhage was markedly reduced but not abolished in dogs 1 h after hypophysectomy; ACTH infusion stimulated aldosterone secretion to values seen after hemorrhage in intact dogs. Hypophysectomy abolished the increase in aldosterone secretion in response to bilateral carotid artery constriction in dogs (Biglieri and Ganong 1961). Most likely, the stimulation had been due to

ACTH, since carotid constriction also induced an increase in 17-hydroxycorticoid secretion in intact dogs and failed to stimulate aldosterone secretion in hypophysectomized dogs infused with ACTH. In dogs with hyperaldosteronism induced by infrarenal aortocaval fistula, hypophysectomy led to a greater fractional decrease in aldosterone secretion than did subsequent bilateral nephrectomy (J. O. Davis et al. 1964). In hypophysectomized dogs, the aldosterone secretion rate was approximately one third that of intact dogs with normal sodium intake as well as with sodium depletion (Binnion et al. 1965). In dogs hypophysectomized 4 weeks previously, neither ACTH nor angiotensin II stimulated 17-hydroxycorticoid secretion, and both agents had a markedly diminished effect on aldosterone output (Ganong et al. 1967b). In contrast, dogs with intact pituitaries but treated with large doses of a glucocorticoid for 4 weeks showed a normal response in aldosterone secretion to stimulation by these two agents, although the responses in 17-hydroxycorticoid production were also completely absent. It was concluded that a pituitary factor other than ACTH was responsible for the decreased responsiveness in aldosterone secretion in chronically hypophysectomized animals. According to R. E. McCaa et al. (1974), acute sodium depletion by isovolemic hemodialysis significantly raised the plasma aldosterone concentration in nephrectomized dexamethasone-treated dogs but not in nephrectomized hypophysectomized or nephrectomized decapitated dogs, even when they were maintained on a constant infusion of ACTH or angiotensin II. These experiments indicated the involvement of an unidentified pituitary factor in the acute regulation of aldosterone secretion during altered sodium balance.

In sodium-replete sheep, hypophysectomy combined with pinealectomy and superior collicular decerebration significantly reduced the secretion of cortisol, corticosterone, and aldosterone; in sodium-deficient sheep, however, these operations had in most instances no effect on aldosterone secretion, which continued at its elevated rate for severel hours (Blair-West et al. 1963). In hypophysectomized sheep that were maintained on cortisone or ACTH treatment for several weeks, sodium deficiency stimulated aldosterone secretion to the same extent as it did in intact animals.

In sodium-replete rats, hypophysectomy resulted in a 70% decrease in aldosterone secretion, according to Singer and Stack-Dunne (1955). Acute treatment with ACTH restored aldosterone secretion to normal as long as 7 days after hypophysectomy. In hypophysectomized rats, sodium deficiency induced only a small increase in aldosterone secretion, which remained below the level observed in intact sodium-replete rats. According to Eilers and Peterson (1964), acute hypophysectomy led to a 30% decrease in aldosterone secretion in sodium-replete rats. In rats that were kept for 2-4 weeks on a sodium-deficient diet prior to hypophysectomy, the aldosterone secretion rate was twice of that of intact animals on a normal sodium intake. Palmore and Mulrow (1967) observed that in chronically hypophysectomized rats the response in aldosterone secretion to sodium deficiency was completely abolished; it was partially restored by injections of "whole rat pituitary glands" (preparation not specified by the authors) but not by treatment with ACTH, growth hormone, or ACTH and thyroxine. These findings were partially confirmed by T. C. Lee et al. (1968) and by T. C. Lee and de Wied (1968), who studied the influence of hypophysectomy and sodium deficiency on aldosterone production by rat adrenals in vitro. They found that hypophysectomy prevented the increase in aldosterone production induced in intact rats by sodium deficiency; adrenal responsiveness was not restored by treatment with ACTH alone, but by treatment with porcine pituitary powder or with bovine growth hormone and ACTH. According to Palmore et al. (1970), treatment with a combination of growth hormone and ACTH or with rat pituitary extract completely restored the high aldosterone secretion rate in rats that were sodium depleted before hypophysectomy. Following acute hypophysectomy of sodium-depleted rats, aldosterone secretion decreased by 38% (Spielman and Davis 1974). In these animals, treatment with the converting-enzyme inhibitor teprotide caused an additional significant drop in aldosterone secretion. Such an effect of teprotide was not observed in intact sodium-depleted rats or in sodium-depleted animals in which ACTH secretion had been suppressed by treatment with dexamethasone alone or in combination with morphine. In sodium-deficient rats, hypophysectomy significantly lowered the aldosterone output from endogenous precursors by incubated quartered whole or capsular adrenals within 2 or 6 days but did not affect the conversions of exogenous corticosterone to aldosterone and 18-hydroxycorticosterone by capsular adrenal tissue or mitochondria (J.E. Boyd et al. 1972; K. Baumann and Müller 1974). In potassium-deficient rats, hypophysectomy markedly stimulated the capsular adrenal conversions of corticosterone to aldosterone and to 18-hydroxycorticosterone (K. Baumann and Müller 1974; Fig. 38); this effect was prevented by treatment with maintenance doses of ACTH. Hypophysectomy did not impair the marked stimulation of the final steps of aldosterone biosynthesis by potassium repletion of potassium-deficient rats. Zona glomerulosa cells of hypophysectomized rats kept on a normal diet produced normal amounts of aldosterone but diminished amounts of corticosterone under basal conditions of incubation (Kenyon et al. 1978a). The conversion of added corticosterone to aldosterone was decreased, but not to a statistically significant extent. Plasma aldosterone concentration was the same in hypophysectomized as in intact rats kept on a normal sodium and potassium intake and responded with the same increase to a 6-day angiotensin II infusion (Aguilera et al. 1980b). In capsular adrenals of hypophysectomized rats, mitochondrial pregnenolone production and 11β-hydroxylase activity, as well as microsomal cytochrome P-450 content and 21-hydroxylase activity were approximately 50% lower than in capsular adrenals of intact rats. In contrast, the mitochondrial conversion of corticosterone to aldosterone was 48% higher. Treatment with angiotensin II significantly stimulated all these parameters of early and late steps of the aldosterone biosynthetic pathway and raised the capsular adrenal concentration of angiotensin II receptors in hypophysectomized animals. In hypophysectomized rats, plasma renin concentrations were only slightly lower than in intact rats and increased normally in response to sodium restriction (F.J. Goodwin et al. 1970). However, hypophysectomy produced a significant drop in the plasma renin substrate concentration. Although the zona glomerulosa was wider in hypophysectomized than in intact rats (Deane et al. 1948; Palmore and Mulrow 1967; Nickerson and Brownie 1975), this was due rather to shrinkage of the inner zones than to an increase in zona glomerulosa volume. According to Nussdorfer et al. (1973), hypophysectomy of rats leads to a decreased volume of the zona glomerulosa, which may, however, be only transient (Nickerson and Brownie 1975). Sodium depletion or potassium loading also led to hypertrophy of the zona glomerulosa in hypophysectomized rats (Deane et al.

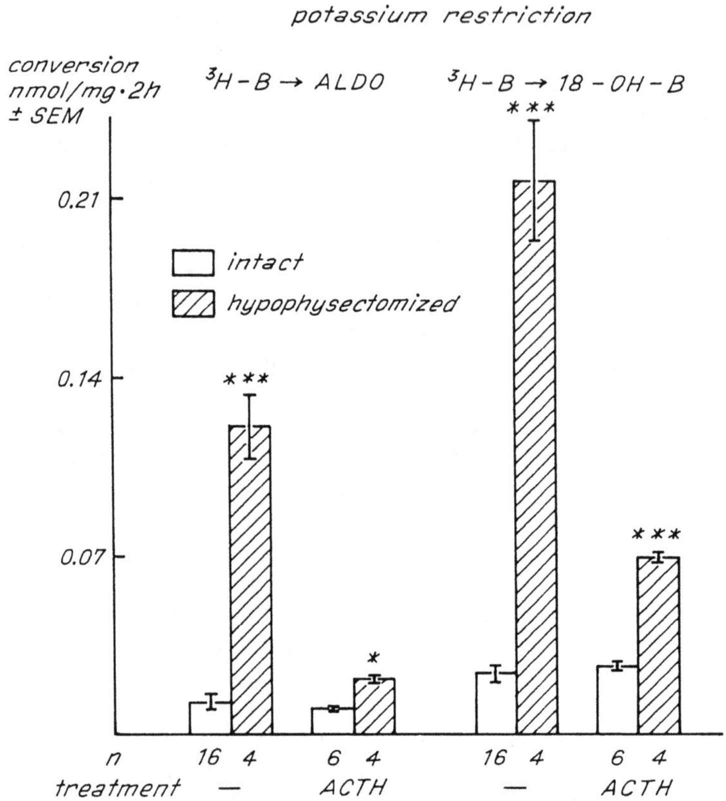

Fig. 38. Effects of hypophysectomy, treatment with maintenance doses of ACTH, or both (6 days) on the conversions of tritiated corticosterone (*^{3}H-B;* 300 nmol/flask) to aldosterone *(ALDO)* and 18-hydroxycorticosterone *(18-OH-B)* by capsular adrenals of rats kept on a potassium-deficient diet for 2 weeks. Statistical significance (*t*-tests) of the difference from the respective control value (untreated rats): *** $P<0.001$, * $P<0.05$. (From K. Baumann and Müller 1974)

1948; Lever 1956; R.A. Miller 1965; Palmore and Mulrow 1967). According to measurements with stereologic techniques, hypophysectomy of rats led within 11 days to marked and statistically significant decreases in the following parameters of zona glomerulosa cells: volume of cells, volume of nuclei, volume of the mitochondrial fraction, surface of the mitochondrial cristae, membrane space, and surface of the smooth endoplasmic reticulum (Nussdorfer et al. 1973). It also induced an increased volume of the lipid compartment. All these ultrastructural changes were completely reversed by treatment of the hypophysectomized animals for 5 days with either ACTH or a combination of cyclic AMP and cyclic GMP (Nussdorfer et al. 1973; Mazzocchi et al. 1974).

In a primitive mammal, the American opossum, hypophysectomy was followed by a significant decrease in aldosterone secretion (Johnston et al. 1965b). There is only limited information available on the effect of hypophysectomy on aldosterone secretion in submammalian vertebrates. Hypophysectomy significantly lowered the concentration of aldosterone in adrenal vein plasma from chickens (A.A. Taylor et al. 1970). On the other hand, the adrenal aldosterone content of ducks, pigeons, and quails was not affected by hypophysectomy (Baylé et al. 1971). In the bullfrog, hypophysectomy led to a marked fall in aldosterone secretion, which was, however, statistically significant in spring but not in winter (A. A. Taylor et al. 1972b). Adenohypophysectomy of toads *(Xenopus laevis)* resulted in a marked suppression of the conversions of tritiated pregnenolone to corticosterone, aldosterone and 18-hydroxycorticosterone by incubated adrenal tissue (Chan and Edwards 1970). In frogs *(Rana ridibunda)*, hypophysectomy equally reduced corticosterone and aldosterone outputs by superfused interrenal tissue (Leboulenger et al. 1980).

5.6.2 Adrenocorticotropin (ACTH)

In short-term experiments, ACTH has consistently been found to be a very potent stimulator of aldosterone biosynthesis in vitro and in vivo within physiological concentrations (see Sect. 4.4.1.). However, the physiological role of ACTH in the long-term regulation of aldosterone secretion in man and in experimental animals is very difficult to define, since the available evidence is often controversial. Thus, synchronous fluctuations in the plasma concentrations of cortisol and aldosterone observed in normal, supine, sodium-replete human subjects give the impression that, under basal conditions, aldosterone secretion is mainly under ACTH control (Katz et al. 1972a, 1975a; Kowarski et al. 1975; Armbruster et al. 1975). However, similar concentrations and diurnal variations of plasma aldosterone were observed in untreated and in dexamethasone-treated subjects (Armbruster et al. 1975; Weinberger et al. 1975). Also, "ACTH-related aldosterone secretion" accounted for only about 20% of the total daily aldosterone secretion rate in six normal women, as calculated from the gradients in plasma aldosterone and cortisol concentrations between adrenal venous and peripheral blood (Huq et al. 1976). Although physiological and clinical evidence strongly indicates that the functional capacity of the zona glomerulosa remains completely intact for prolonged periods of time in the total absence of ACTH, this has in many instances not yet been unequivocally established and is not entirely supported by morphological evidence. It is also uncertain whether increased plasma levels of ACTH substantially contribute to chronic increases in aldosterone secretion under certain pathological conditions.

Increased production of aldosterone and corticosterone in dogs with an experimental constriction of the inferior vena cava has been partially ascribed to increased ACTH secretion (J. O. Davis et al. 1960a, 1960b). In these animals, high doses of cortisone, as well as hypophysectomy, resulted in marked decreases in aldosterone secretion. Since hypovolemia is known to lead to increased ACTH secretion, it is likely that ACTH contributes to the hypersecretion of aldosterone in experimental situations associated with hypovolemia, such as hemorrhage or sodium and water sequestration by subcutaneous polyethylene glycol or formalin (see p. 122). The increases in plasma corticosterone and in adrenal weight observed by some investigators in rats with experimental renovascular hypertension may also be mentioned in this context (see. p. 162).

Excessive long-term stimulation of aldosterone production by normal levels of ACTH appears to be the crucial pathogenetic factor in so-called glucocorticoid-suppressible hyperaldosteronism. This disorder is found in a small subgroup of patients with nontumorous primary aldosteronism. It is characterized by a familial occurrence (Sutherland et al. 1966; New and Peterson 1967; Giebink et al. 1973; Grim and Weinberger 1980; Oberfield et al. 1981, Gill and Bartter 1981; Ganguly et al. 1981a) and is inherited by autosomal dominant transmission (New et al. 1980). In glucocorticoid-suppressible hyperaldosteronism - but not in other forms of primary aldosteronism - aldosterone secretion, blood pressure, and plasma potassium concentration became completely and permanently normal under treatment with small doses of dexamethasone or prednisone (Sutherland et al. 1966; New and Peterson 1967; Salti et al. 1969a; Ganguly 1982). Continued treatment with a glucocorticoid also led to a rise of plasma renin activity to the normal or supranormal range (Oberfield et al. 1982). In patients with this disorder, small doses of exogenous ACTH elicited an excessive rise in plasma aldosterone (Ganguly et al. 1983a). Prolonged administration of pharmacological doses of ACTH induced a sustained elevation of aldosterone secretion and plasma concentration (New et al. 1976; Rauh et al. 1978; Gill and Bartter 1982). Patients with glucocorticoid-suppressible hyperaldosteronism exhibited an abnormal postural response of aldosterone secretion, with a marked decrease of plasma aldosterone after 2-4 h of standing (Ganguly et al. 1981b). After treatment with dexamethasone there was a normal increase in plasma aldosterone in response to upright posture. Metoclopramide acutely raised plasma aldosterone in untreated patients but was completely ineffective in dexamethasone-treated patients (Ganguly et al. 1983a). In only two reported instances were patients with glucocorticoid-suppressible hyperaldosteronism partially adrenalectomized. In the case of Sutherland et al. (1966), multiple adenocortical nodules, 0.1-0.4 cm in diameter and composed of clear cells, were found mainly in the zona fasciculata. There was diffuse bilateral cortical hyperplasia, especially of the zona fasciculata and reticularis, in the adrenals removed from the patient of Miura et al. (1968), in whom glucocorticoid-suppressible hyperaldosteronism was combined with a partial 17α-hydroxylation defect. The primary defect causing glucocorticoid-suppressible hyperaldosteronism is as yet completely unknown. In principle, the disorder could be due to either a pituitary or an adrenocortical dysfunction (Ganguly 1982). Mulrow (1981) speculated that aldosterone secretion might be controlled by an abnormal pituitary hormone which was secreted simultaneously with ACTH and was also suppressed by dexamethasone. Such a hormone might derive from the precursor molecule of ACTH, pro-opiomelanocorticotropin, and might be related to either β-lipotropin or the amino-terminal peptide containing the γ-MSH sequence. No direct evidence supporting this hypothesis has as yet been presented. Güllner et al. (1983) found normal plasma concentrations of immunoreactive ACTH, β-endorphin/β-LPH, and γ_3-MSH in a woman with glucocorticoid-suppressible hyperaldosteronism. On the other hand, there is some evidence pointing to a primary adrenocortical defect. Thus, Ulick and co-workers (Ulick and Chu 1982; Ulick et al. 1983a) and Gomez-Sanchez et al. (1984) found substantial amounts of 18-hydroxycortisol and 18-oxocortisol in the urine of patients with glucocorticoid-suppressible hyperaldosteronism. This observation may indicate that aldosterone and cortisol were biosynthesized in the same cells. Such abnormal cells might be located in the zona fasciculata and be predominantly controlled by ACTH. The inappropriate secretion of aldosterone by these cells in response to normal levels of ACTH would be expected to suppress the normal zona glomerulosa cells by negative feedback mechanisms, e.g., by suppression of the renin-angiotensin system or by hypokalemia. This assumption is consistent with the observation by Oberfield et al. (1981) that in dexamethasone-treated patients there was an appropriate increase in aldosterone in response to sodium restriction, which was paralleled by increases in plasma renin activity and potassium concentration.

In some, but not all patients with other forms of primary aldosteronism, i.e., aldosterone-producing adrenocortical adenoma or idiopathic hyperaldosteronism with bilateral adrenal hyperplasia, aldosterone secretion was found to be more dependent on ACTH than in normal persons, although not to the extent seen in patients with glucocorticoid-suppressible hyperaldosteronism. In these patients, plasma aldosterone and plasma cortisol concentrations were significantly correlated in their spontaneous fluctuations (Kem et al. 1973, 1976; Ganguly et al. 1973, 1983b; H. Vetter et al. 1974; H. Vetter and Vetter 1975a; Wenting et al. 1978). The aldosterone response to ACTH infusions was exaggerated (Kem et al. 1978). Treatment with dexamethasone for 1 or 2 days significantly lowered plasma aldosterone concentrations, sometimes even to subnormal values (Kem et al. 1976; Ganguly et al. 1977, 1983b; Wenting et al. 1978). However, the suppressive effect of glucocorticoids was transient and was no longer apparent after a few days or weeks of continued treatment.

When presumably normal adrenal glands are chronically exposed to elevated levels of endogenous ACTH or to high doses of exogenous ACTH, aldosterone secretion either remains normal or decreases. Thus, normal or decreased aldosterone secretion rates were observed in patients with Cushing's syndrome due to increased ACTH secretion by the pituitary gland or by a nonendocrine tumor (Biglieri et al. 1968; Mantero et al. 1978; Ganguly et al. 1983c; Baba et al. 1983). Administration of high or low doses of ACTH to normal human subjects on a liberal or restricted sodium intake or to patients with essential hypertension or primary aldosteronism caused only transient increases in urinary aldosterone excretion, the aldosterone secretion rate, or plasma aldosterone concentration, which were followed by decreases to below control values (Liddle et al. 1956; A. F. Muller et al. 1956b; Tucci et al. 1967a; Newton and Laragh 1968a; Biglieri et al. 1969; Benraad and Kloppenborg 1970; Beitins et al. 1972a[2]; Bayard et al. 1975; Kraiem et al. 1979; Rauh et al. 1978, 1979; Mantero et al. 1980; Fuchs-Hammoser et al. 1980; Gaillard et al. 1983). Similar effects of long-term ACTH treatment on plasma levels or adrenal secretion rates of aldosterone were also observed in experiments with dogs (Ganong 1982), sheep (Scoggins et al. 1974; McDougall et al. 1980), rats (Komor and Müller 1979; Aguilera et al. 1981b; Vazir et al. 1981), and rabbits (Llano et al. 1982). Moreover, Kubo et al. (1974) found decreased plasma aldosterone levels in rats bearing a mammotropic (ACTH-, growth hormone-, and prolactin-secreting) MtT-F4 tumor. In contrast, only a few investigators observed a continued stimulation of aldosterone secretion under prolonged ACTH administration. In normal human subjects, ACTH induced a progressive increase in the urinary excretion of tetrahydroaldosterone (Pratt et al. 1976). In a

[2] These investigators observed a continuous elevation of plasma aldosterone concentrations during prolonged ACTH administration to patients with non-salt-wasting congenital adrenal hyperplasia, which contrasted with the transient effect of ACTH on plasma aldosterone in normal subjects.

study with rats, urinary excretion rates of tetrahydroaldosterone and 18-hydroxy-corticosterone remained markedly elevated during treatment with ACTH for 10 days, whereas urinary excretion of aldosterone returned to control values after an initial increase (Haack et al. 1978). According to Payet and Lehoux (1982), there was a persistent elevation of plasma aldosterone in rats treated with ACTH for 2 weeks, which contrasted with a diminished aldosterone output by zona glomerulosa cells in vitro.

Treatment of rats, rabbits or sheep with ACTH in vivo consistently resulted in a marked suppression of aldosterone biosynthesis by adrenal cells or tissue in vitro

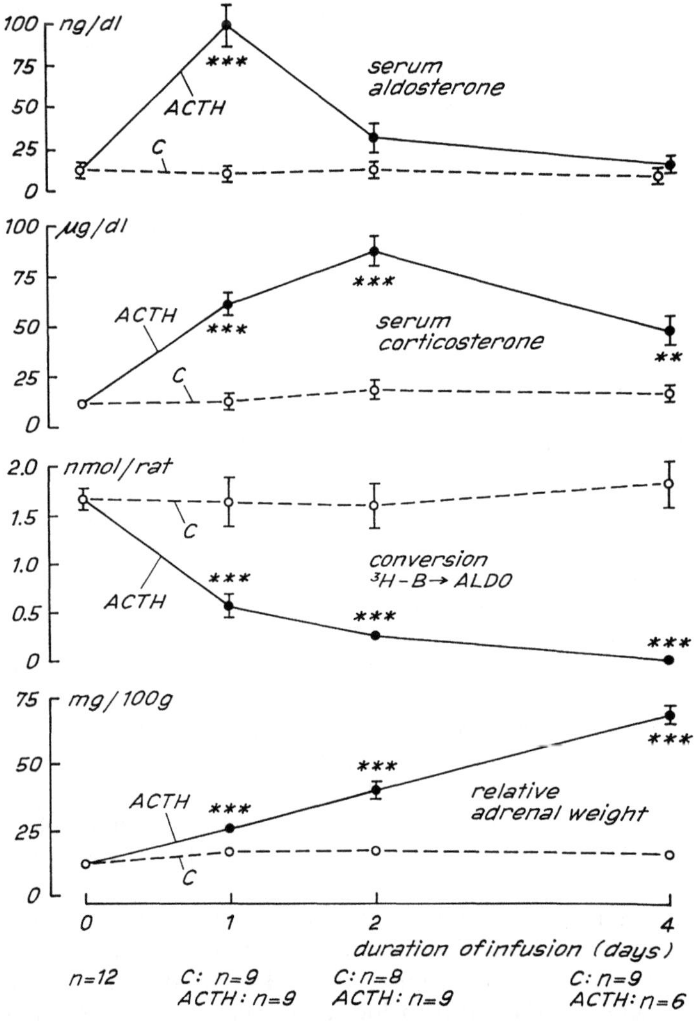

Fig. 39. Effect of an i.v. infusion of ACTH (50 µg/day, *solid lines* and *full dots*) in 5% glucose into rats on the serum concentrations of aldosterone and corticosterone, the relative adrenal weight, and the conversion of tritiated corticosterone *(^{3}H-B)* to aldosterone *(ALDO)* by quartered adrenals. Control animals received an i.v. infusion of 5% glucose *(C, broken lines, open dots)*. Mean values ± SEM. Significance (*t*-tests) of differences between experimental values and corresponding control values: *** $P<0.001$, ** $P<0.01$. (From Komor and Müller 1979)

(Vecsei et al. 1966b; Kolena 1968; J. Müller 1970a, 1978; K. Baumann and Müller 1974; Komor and Müller 1979; McDougall et al. 1980; Aguilera et al. 1981b; Vazir et al. 1981; Llano et al. 1982). Zona glomerulosa cells from ACTH-treated animals produced less aldosterone from endogenous precursors - under baseline conditions or in the presence of stimulators - and converted less added corticosterone, deoxycorticosterone, progesterone, or pregnenolone to aldosterone and 18-hydroxycorticosterone than zona glomerulosa cells from control animals. Aldosterone biosynthesis was similarly impaired in adrenal tissue from rats bearing an ACTH-secreting mammotropic pituitary tumor (MtT-F4; de Nicola et al. 1973). Prolonged ACTH treatment also significantly suppressed the in vivo conversion of radioactive corticosterone to aldosterone in rats (Vecsei and Kessler 1970, 1971). According to this evidence, ACTH has a marked long-term suppressive effect on the activity of the enzyme(s) involved in the conversion of corticosterone to aldosterone. Generally, this suppressive effect became manifest within a day or two. In rats receiving a continuous intravenous infusion of ACTH, the conversion of corticosterone to aldosterone decreased to less than half of its rate within 24 h; at that time, the plasma aldosterone concentration was still markedly elevated (Komor and Müller 1979; Fig. 39). The available evidence about a possible effect of ACTH on zona glomerulosa 11β-hydroxylase activity is conflicting. On the one hand, the capsular adrenal output of deoxycorticosterone in response to added serotonin was markedly diminished in rats treated with ACTH for 4 days (J. Müller 1978). After 2 weeks of ACTH treatment, capsular adrenal tissue was no longer responsive to serotonin (Fig. 40). Capsular adrenals of ACTH-treated rats showed a strikingly increased conversion of added deoxycorticosterone to corticosterone and 18-OH-DOC (J. Müller 1978; Fig. 41) and produced more corticos-

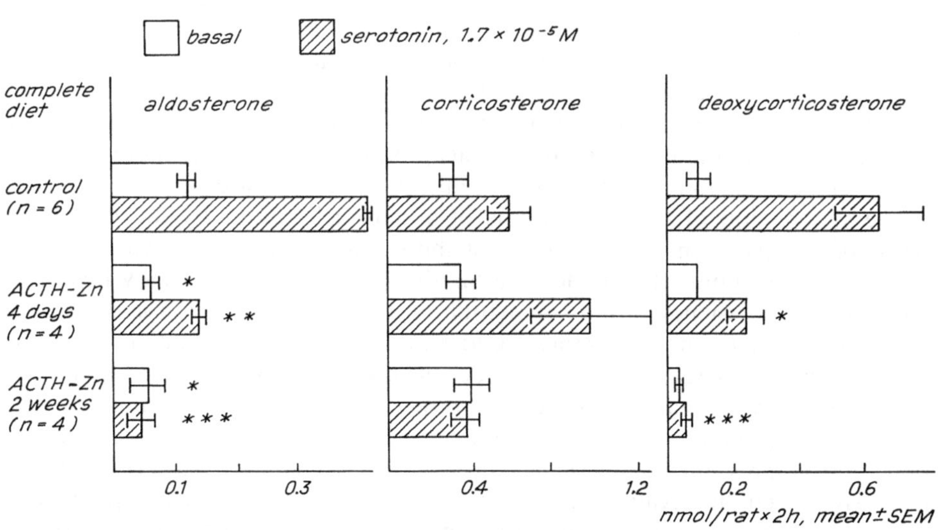

Fig. 40. Effect of treatment with ACTH of rats (complete diet) on the output of endogenous corticosteroids by capsular adrenals incubated with or without serotonin. Statistical significance (*t*-tests) of differences from the respective control values: *** $P<0.001$, ** $P<0.01$, * $P<0.05$. Long-acting ACTH preparation consisted of a peptide-zinc complex (Synacthen Depot). (From J. Müller 1978).

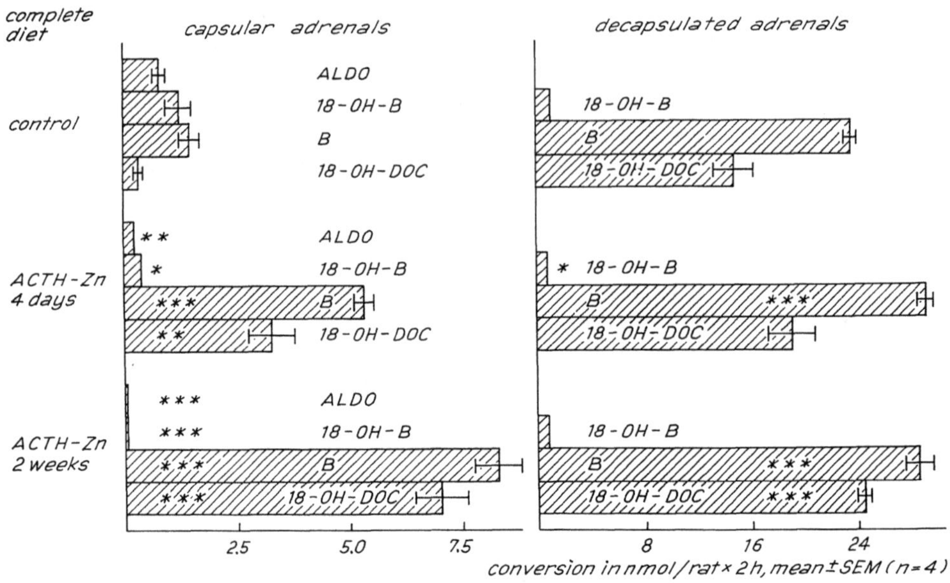

Fig. 41. Effect of treatment with ACTH of rats (complete diet) on the conversions of tritiated deoxycorticosterone (*³H-DOC;* 300 nmol/flask) to aldosterone *(ALDO),* 18-hydroxycorticosterone *(18-OH-B),* corticosterone *(B),* and 18-hydroxy-11-deoxycorticosterone *(18-OH-DOC)* by adrenocortical tissue. Statistical significance (*t*-tests) of the differences from the respective control values: *** $P<0.001$, ** $P<0.01$, * $P<0.05$. (From J. Müller 1978)

terone and 18-OH-DOC from endogenous precursors (Vazir et al. 1981). On the other hand, Aguilera et al. (1981b) found an increased deoxycorticosterone output by zona glomerulosa cells from ACTH-treated rats (6 days) during incubation with ACTH, cyclic AMP, angiotensin II, or a high potassium concentration. They also observed a decreased conversion of deoxycorticosterone to corticosterone by capsular adrenal mitochondria from ACTH-treated rats. In addition, capsular adrenal mitochondria of ACTH-treated rats produced significantly more pregnenolone from endogenous cholesterol than mitochondria of untreated rats.

The suppression of the biosynthesis of a steroid hormone by the prolonged action of a tropic hormone with a potent short-term stimulatory activity is a phenomenon rather unusual in endocrine physiology. According to many investigators, both the transiency of the stimulation of aldosterone secretion and the suppression of aldosterone biosynthesis under long-term ACTH treatment could be due to overriding negative feedback mechanisms. It is to be expected that inappropriately elevated plasma concentrations of aldosterone and other mineralocorticoids will eventually inhibit aldosterone production through either sodium retention and suppression of renin secretion or potassium loss. ACTH treatment regularly caused an elevation in plasma sodium and a drop in plasma potassium, but these changes were often very small, although statistically significant. Reported alterations of plasma renin activity in response to prolonged ACTH treatment are conflicting. Some investigators observed no consistent changes of this parameter (Newton and Laragh 1968a; Pratt et al. 1976). Others found a marked decrease, but only in subjects kept on a normal sodium intake (Bayard et al. 1975; Rauh et al. 1978; Gaillard et al. 1983). However, according to Benraad and Kloppenborg (1970), prolonged treatment with ACTH (8-21 days) resulted in a substantial suppression of plasma renin activity (upright or recumbent) in patients kept on a low-sodium diet. On the other hand, prolonged treatment with a low dose of ACTH had a transient *stimulatory* effect on plasma renin activity in man - irrespective of sodium intake - with an increase on the first 2 days and a return to control values in the further course of the experiment (Fuchs-Hammoser et al. 1980). Similar changes in plasma renin activity were induced by the prolonged administration of a synthetic substituted α-1-18 ACTH preparation (Mantero et al. 1980). In these two studies, changes in plasma renin activity were parallel to changes in plasma aldosterone. In dogs, ACTH treatment resulted in a significant drop in plasma renin activity, even during sodium restriction (Ganong et al. 1972). In the rat, ACTH induced a short-term transient rise in plasma renin activity which could be blocked by dexamethasone administration (Hauger-Klevene 1970b). Continued ACTH treatment for 7 days significantly lowered the plasma renin activity and reduced by half the angiotensin II receptor concentration in zona glomerulosa cells (Aguilera et al. 1981b) of rats kept on an unrestricted sodium intake. When rats on a sodium- and potassium-deficient diet were treated with a high dose of ACTH, plasma renin activity was unchanged after 24 h but decreased after 48 h, whereas the capsular adrenal conversion of corticosterone to aldosterone decreased after 24 h (J. Müller, unpublished observations).

If the ACTH-induced suppression of aldosterone production is due to normal feedback mechanisms, ACTH treatment and mineralocorticoid treatment should result in similar alterations of aldosterone biosynthesis. However, at least according to our own observations, there were some material differences between ACTH-induced and mineralocorticoid-induced changes in zona glomerulosa function in rats. Capsular adrenals of mineralocorticoid-treated rats responded to serotonin added in vitro with a markedly increased output of deoxycorticosterone (Hunziker and Müller 1977; Fig. 45), whereas the serotonin-elicited deoxycorticosterone output was decreased or absent in capsular adrenals of ACTH-treated animals (J. Müller 1978; Fig. 40). The conversions of tritiated deoxycorticosterone to corticosterone and 18-OH-DOC were unaltered or decreased in capsular adrenals of mineralocorticoid-treated rats but strikingly increased in capsular adrenals of ACTH-treated rats. Finally, mineralocorticoids did not affect aldosterone biosynthesis in rats kept on a sodium-deficient diet, whereas ACTH acted independently of the sodium intake. ACTH treatment for 5 days also caused a marked decrease in plasma aldosterone in sheep kept on a sodium-free diet (Humphrey et al. 1983). According to most investigators - with the exception of Bayard et al. (1975), sodium restriction increased the stimulatory effect of ACTH on aldosterone secretion in man but did not prevent its transiency. Nevertheless, sodium restriction during ACTH treatment of man and dogs prevented a complete suppression of the response in plasma aldosterone to short-term stimulation by ACTH or angiotensin II (Gaillard et al. 1983; Ganong 1972). When ACTH was given in combination with the diuretic triamterene for 4 days, urinary aldosterone excretion reached its peak value on day 2 and was still significantly elevated on day 4 (Gaillard et al. 1983). In contrast, the combined treatment with ACTH and spironolactone prevented neither sodium retention nor a return of aldosterone excretion to the control value on the 4th day.

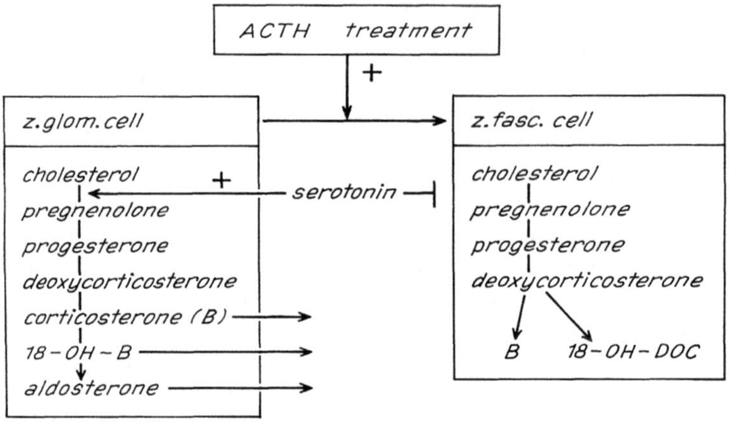

Fig. 42. Alterations of zona glomerulosa cell functions under prolonged treatment of rats with ACTH

Capsular adrenals of ACTH-treated rats became similar to normal zona fasciculata tissue in their lack of steroidogenic response to serotonin and their preferential conversion of deoxycorticosterone to corticosterone and 18-OH-DOC (J. Müller 1978; Fig. 42). This could have been due alternatively to (a) a diminished function of zona glomerulosa cells accompanied by an increased activity of the zona fasciculata cells contaminating the capsular portion of the adrenal glands, (b) a disappearance of zona glomerulosa cells and replacement with zona fasciculata cells, or (c) a conversion of zona glomerulosa cells to cells with functional properties of zona fasciculata cells. The available morphological evidence in support of any of these possibilities is conflicting. In the sheep, ACTH treatment resulted within 2-3 days in a progressive decrease in the zona glomerulosa cell number and in a disruption of normal zona glomerulosa structure (McDougall et al. 1980). Most of the remaining cells in the subcapsular area were of the intermediate type, with mixed lamellar and vesicular mitochondrial cristae. Treatment of mice with ACTH for 6 days led to a significant decrease of zona glomerulosa width (Shire and Stewart 1982). In rats bearing an ACTH-, growth hormone-, and prolactin-secreting tumor for 6 weeks, the zona glomerulosa was reduced to a few clusters of cells which were structurally identical with normal glomerulosa cells and had tubular mitochondrial cristae (Nickerson et al. 1970). As previously mentioned, treatment of hypophysectomized rats with ACTH for 5 days normalized the decreased volume of the zona glomerulosa, its cells, and its main subcellular structures (Nussdorfer et al. 1973). This was associated with a change of mitochondrial cristae from a lamellar to a (normal) tubulovesicular type. When the same investigators treated intact rats with ACTH for 2 weeks, they observed a linear increase in the volume of the zona glomerulosa cells and their organelles (Nussdorfer et al. 1977). This was not associated with any ultrastructural change indicative of a loss of zona glomerulosa characteristics. However, a change from tubular to vesicular mitochondrial cristae was induced by ACTH administration for 8 days to rats that were sodium loaded and treated with a beta-adrenergic blocking agent (timolol maleate; Nussdorfer et al. 1982). In these animals, an increase in zona glomerulosa volume was associated with a decrease in plasma aldosterone. Treatment of rats with a long-acting ACTH preparation for 2 days brought about a striking increase in the frequency of mitosis in zona glomerulosa cells but only a minor enlargement of the zona glomerulosa surface area (Payet et al. 1980a). After 2 weeks of treatment, the zona glomerulosa cells became fasciculata-like with a columnar arrangement (Payet and Lehoux 1982).

Important information about a possible mechanism of the ACTH-induced suppression of aldosterone biosynthesis has been obtained in studies with zona glomerulosa cell or tissue cultures, in which the influence of extra-adrenal regulatory systems was completely excluded. Addition of ACTH or of dibutyryl cyclic AMP to the medium in which adult rat adrenal zona glomerulosa cells were cultured in monolayers initially stimulated aldosterone production but later suppressed it to below control levels (Hornsby et al. 1973, 1974). In contrast, ACTH elicited a continuous elevation in corticosterone output. Addition of ACTH to the culture medium also brought about a change from tubular to vesicular mitochondrial cristae (Hornsby et al. 1974). A similar ultrastructural transformation was induced by ACTH in cultures of mixed fetal rat adrenocortical cells or tissue; however, it was associated with a continuously elevated aldosterone output (Kahri et al. 1970; Manuelidis and Mulrow 1973, 1974; Salmenperä and Kahri 1976). When monobutyryl cyclic AMP was added to a high-potassium (18 mmol/l) medium in which adult rat adrenal zona glomerulosa cells were kept in monolayer culture, the metabolism of tritiated deoxycorticosterone was converted from a "zona glomerulosa-like" (main products: 18-hydroxycorticosterone and aldosterone) to a "zona fasciculata-like" (main products: corticosterone and 18-OH-DOC) pattern (Hornsby and O'Hare 1977). This effect of cyclic AMP could be imitated by the addition of cortisol, corticosterone, or 18-OH-DOC - but not of aldosterone - to the culture medium. Thus, the long-term suppressive action of ACTH on aldosterone biosynthesis in cell cultures could be due to the accumulation of corticosteroids in the medium. In cultured bovine adrenal zona glomerulosa cells, a rapid decline in the conversion of deoxycorticosterone to aldosterone and 18-hydroxycorticosterone could be prevented by the addition of metyrapone (but not of aminoglutethimide or cyanoketone) to the medium (Crivello et al. 1982). However, metyrapone was not sufficient to maintain aldosterone biosynthesis by cells cultured in a medium that also contained ACTH or angiotensin II. The long-term suppressive action of these two peptides on aldosterone biosynthesis could be prevented only when the radioprotectant dimethyl sulfoxide and the antioxidant H_2SeO_3 were added together with metyrapone to the culture medium. In cultures of bovine adrenal zona fasciculata cells, 11β-hydroxylase activity was substantially decreased by the addition of various steroids to the medium for 24 h (Hornsby 1980). The suppressive compounds included 11β-hydroxysteroids such as cortisol, corticosterone, prednisolone, and 11β-hydroxyandrostenedione, as well as 11-deoxysteroids such as deoxycorticosterone, 11-deoxycortisol, testosterone, and androstenedione. 11-Ketosteroids and 11α-hydroxysteroids were ineffective. Some of the same steroids, in particular cortisol, androstenedione, and 11β-hydroxyandrostenedione, as well as 18-OH-DOC, brought about a marked suppression of the final steps in aldosterone biosynthesis when added for 4 h to the culture medium of bovine adrenal zona glomerulosa cells (Crivello et al. 1983). In both preparations androstenedione was the most active suppressor. Cortisol suppressed 11β-hydroxylation and the final steps of aldosterone biosynthesis in similar concentrations, but the cortisol-induced decrease in corticosterone methyl oxidase activity occurred more rapidly (half-life of 4 h) than the decrease in zona glomerulosa (13 h) and zona fasciculata (9 h) 11β-hydroxylase activity. In both preparations, the steroid-induced decays in the activity of steroidogenic enzymes could be prevented or reduced by the simultaneous addition of antioxidants (butylated hydroxyanisole, H_2SeO_3, and dimethylsulfoxide). According to a theory stated by Crivello et al. (1983) and by Hornsby and Crivello (1983a, 1983b), the interaction of cortisol or other steroids with zona glomerulosa cytochrome P-$450_{11\beta}$ could cause the production of oxygen-derived free radicals, resulting in lipid peroxidation. The accumulation of lipid hydroperoxides in zona glomerulosa mitochondria could then cause the destruction of both cytochrome P-$450_{11\beta}$ and - at a more rapid rate - of cytochrome P-450_{CMO}. Other effects of ACTH on zona glomerulosa cells observed in experiments with cell cultures cannot be explained by such a mechanism but are consistent with the assumption of an ACTH-induced transformation to zona fasciculata cells. Thus, ACTH induced 17α-hydroxylase activity in cultured bovine adrenal zona glomerulosa cells (Crivello and Gill 1983). Moreover, the addition of ACTH to the medium of cultured bovine zona glomerulosa cells resulted within 1 day in a marked decrease in the number of angiotensin II receptor binding sites without affecting their binding affinity (Andoka et al. 1984). This effect of the peptide was probably mediated by cyclic AMP, since it could also be induced by the addition of forskolin, cholera toxin, or 8-bromo-cyclic AMP to the culture medium.

For a correct interpretation of plasma concentrations, secretion rates, and urinary excretion rates of aldosterone and its metabolites measured in human subjects or experimental animals treated with ACTH, it has to be considered that these parameters not only are representative of aldosterone production but also reflect the distribution and metabolism of the hormone. In particular, evidence has been presented by several investigators that in man, ACTH administration leads to increases in the metabolic clearance rate of plasma aldosterone, whereas the metabolic clearance rate of whole blood aldosterone remains unchanged (Pratt et al. 1976; Messerli et al. 1976; Zisper et al. 1976). This is mainly due to a redistribution of aldosterone from plasma to red blood cells in the presence of high concentrations of cortisol, apparently because aldosterone is displaced by cortisol from plasma carrier proteins, in particular, from cortisol-binding globulin (Messerli et al. 1976; Zipser et al. 1976, 1980; Chavarri et al. 1977a; Zager et al. 1979).

5.6.3 ACTH-Related Pituitary Peptides

Although the direct short-term steroidogenic effects of proopiomelanocortin-derived peptides have been extensively investigated (see Sect. 4.4.2), there is as yet very little information available about a possible role of these substances in the long-term physiological control of aldosterone biosynthesis. Treatment of sodium-deprived rats with α-melanocyte-stimulating hormone (α-MSH; 60 µg/day s.c.) for 2 days following hypophysectomy did not prevent marked decreases in adrenal weight and in the aldosterone production by incubated adrenal tissue (Page et al. 1974). Under combined treatment with α-MSH and growth hormone, adrenal weight was kept normal and aldosterone production remained significantly higher than in untreated hypophysectomized rats. However, it decreased below the level seen in intact rats or in hypophysectomized rats treated with the combination of ACTH and growth hormone. Treatment of intact rats with subcutaneous injections of α-MSH for 5 days resulted in a significant hypertrophy of the zona glomerulosa and an increase (+60%) in plasma aldosterone (Robba et al. 1986). The zona fasciculata morphology and the plasma corticosterone concentration were not affected. A significant increase in plasma aldosterone, together with a marked lipid depletion of the zona glomerulosa cells, was observed 1 hour after the intraperitoneal injection of the peptide. In hypophysectomized, sodium-deficient rats, a 6-day s.c. infusion of α-MSH raised plasma aldosterone to the level seen in intact sodium-deficient rats without affecting the plasma corticosterone (Shenker et al. 1985a). ACTH had no effect on plasma aldosterone in these animals. In intact rats, plasma aldosterone did not alter in response to an α-MSH infusion. In intact rats, the plasma α-MSH concentration was not affected by substantial alterations of sodium intake (Whitehouse et al. 1982). Irrespective of sodium intake, 1 week of bromocriptine administration to rats significantly lowered the plasma concentrations of both α-MSH and aldosterone but increased the sensitivity of zona glomerulosa cells to the steroidogenic action of α-MSH added in vitro (Whitehouse et al. 1982; Vinson et al. 1983). In contrast, treatment of rats with metoclopramide for a week significantly raised the plasma levels of α-MSH and aldosterone but completely eliminated any steroidogenic response to α-MSH of incubated zona glomerulosa cells (Whitehouse et al. 1982). According to these observations, the increase in the sensitivity of zona glomerulosa cells to α-MSH induced by sodium restriction could not have been mediated by increased plasma levels of the peptide.

Elevated plasma concentrations of immunoreactive γ-MSH (Griffing et al. 1985a) and of β-endorphin (Griffing et al. 1985b) were found in patients with idiopathic hyperaldosteronism but not in patients with primary aldosteronism due to an adrenocortical adenoma or with essential hypertension.

5.6.4 Glucocorticoids

In many studies on the physiological regulation of aldosterone secretion, glucocorticoids (mainly dexamethasone) have been given to human subjects or experimental animals in order to selectively suppress endogenous ACTH secretion and to avoid a possible stimulation of aldosterone production by elevated levels of circulating ACTH, particularly during periods of experimental stress. However, some of the observations made in these and other experiments have indicated that glucocorticoids may influence aldosterone secretion by other mechanisms as well.

Administration of dexamethasone (2 mg/day) for 2 days to patients with essential hypertension (with a normal or elevated aldosterone secretion rate) led in 16 of 17 instances to a moderate decrease (average −34%) in the aldosterone secretion rate (Spark et al. 1968). Under continued dexamethasone intake, the aldosterone secretion rate was stimulated by an angiotensin II infusion but not to the level seen before glucocorticoid treatment. Also, in a normal woman kept on a sodium-deficient diet, dexamethasone had a marked inhibitory effect on aldosterone secretion. However, treatment of 18 male hypertensive patients with dexamethasone (1.5 mg/day) for 8 weeks did not significantly alter plasma aldosterone concentration, although it significantly lowered diastolic blood pressure and plasma cortisol (Hamilton et al. 1979). In normal human subjects on a low sodium intake (40 mmol/day), short-term treatment with dexamethasone (3 mg) or cortisol (70 mg) for 9 h significantly lowered the basal plasma aldosterone concentration but did not change the responsiveness of this parameter to infused angiotensin II (Nicholls et al. 1980). Administration of dexamethasone (2 mg/day) to normal volunteers on a fixed sodium intake of 120 mmol/day significantly suppressed plasma aldosterone and 18-hydroxycorticosterone concentrations within 5 days (Sowers et al. 1983a). Also, in a study by Ferrier et al. (1983), there was a significant drop in plasma aldosterone in response to 2 days of dexamethasone treatment (2 mg/day) of normal human volunteers on a liberal sodium intake. In subjects treated with the diuretic indapamide, an elevated plasma aldosterone concentration returned to normal following a week of combined treatment with the diuretic and dexamethasone. Treatment with cortisol (80 mg/day) and - to a lesser extent - treatment with dexamethasone (2 mg/day) significantly lowered urinary aldosterone excretion in normal subjects and in some patients with essential hypertension kept on a low-sodium diet (Newton and Laragh 1968b). Glucocorticoids were less effective in suppressing aldosterone excretion in subjects on a normal sodium intake. In glucocorticoid-treated subjects kept on a sodium-deficient diet, infusion of ACTH for 1 day failed to stimulate aldosterone excretion. This observation indicated that glucocorticoids inhibited aldosterone secretion by a mechanism different from that suppressing ACTH secretion.

Treatment of monkeys with nivazol[3] for a week had no effect on plasma aldosterone while markedly lowering the plasma cortisol concentration (Schane et al. 1984). This corticosteroid analogue displayed only ACTH-suppressive but not peripheral glucocorticoid activity in primates. Fekete and Szeberényi (1968) observed that a number of glucocorticoids differed considerably in their suppressive effect on aldosterone production by adrenal tissue (incubated with or without substrate amounts of progesterone) when given to rats for 6 days in doses which were equivalent in suppressing corticosterone output. Administration of cortisol to rats 16 h before they were killed lowered the production of corticosterone but not the production of aldosterone and 18-hydroxycorticosterone by incubated adrenal tissue (Vinson and Whitehouse 1969). Prolonged treatment of rats on a normal or on a sodium-deficient diet with dexamethasone or corticosterone for 2 weeks resulted in a significant suppression of late steps in aldosterone biosynthesis by incubated quartered adrenals or capsular hemiadrenals (J. Müller 1970a, 1978; Fig. 43), an effect not elicited by hypophysectomy (J. E. Boyd et al. 1972; K. Baumann and Müller 1974). As previously mentioned (see p. 80 and p. 189), some glucocorticoids (in particular cortisol and corticosterone) have been found to have direct short-term or long-term inhibitory effects on aldosterone biosynthesis in adrenocortical mitochondria, adrenal homogenate, quartered adrenals, and zona glomerulosa cell cultures, but only when added in very high concentrations (Kahnt and Neher 1966a; Burrow et al. 1966; Burrow 1969; Hornsby et al. 1977; Crivello et al. 1983). Apparently, these aldosterone-suppressing effects were independent of glucocorticoid activity, since they were not displayed by such potent

[3] Nivazol = Win 27914 = (17α)2'-(4-fluorophenyl)2'-H-pregna-2,4-dien-20-yno[3,2c]pyrazol-17-ol. This compound has peripheral glucocorticoid effects in rats comparable to the ones induced by methylprednisolone (but not in monkeys).

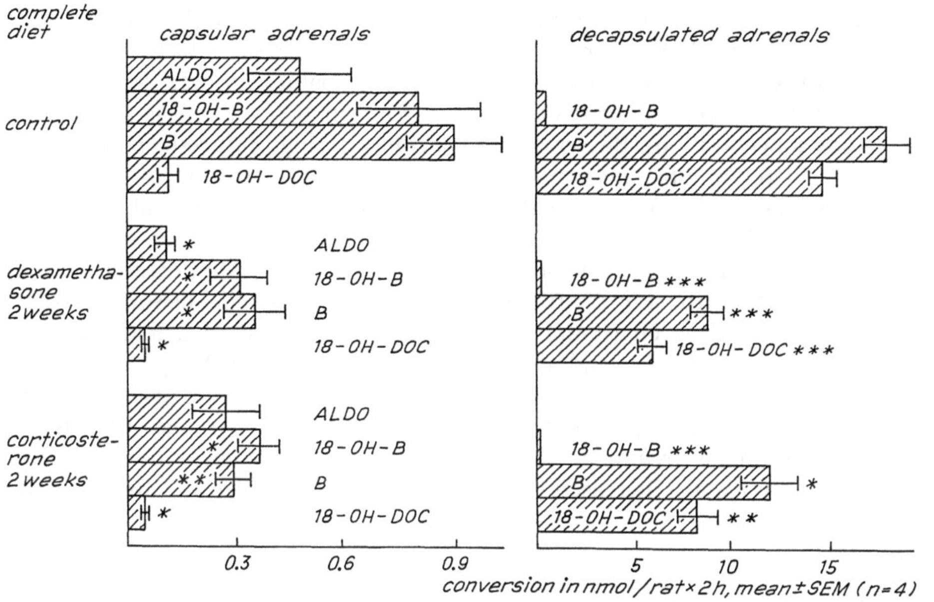

Fig. 43. Effect of 2 weeks of treatment of rats (complete diet) with dexamethasone or corticosterone on the conversions of tritiated deoxycorticosterone (*^{3}H-DOC;* 300 nmol/flask) to aldosterone *(ALDO)*, 18-hydroxycorticosterone *(18-OH-B)*, corticosterone *(B)*, and 18-hydroxy-11-deoxycorticosterone *(18-OH-DOC)* by adrenocortical tissue. Mean values of two experiments. Statistical significance (*t*-tests) of differences from the respective control values: ***$P<0.001$, **$P<0.01$, *$P<0.05$. (From J. Müller 1978)

synthetic glucocorticoids as prednisolone and dexamethasone. Perhaps a long-term direct action on the zona glomerulosa may contribute to the overall aldosterone-suppressing effect of some glucocorticoids. In addition, glucocorticoid-induced changes in mineral and water homeostasis might influence aldosterone biosynthesis by extra-adrenal control systems. In particular, glucocorticoids can cause potassium losses by an action which is independent of mineralocorticoid activity (Bartter and Fourman 1957; Liddle 1959; Bia et al. 1982).

On the other hand, glucocorticoid-induced suppression of aldosterone biosynthesis is not mediated by a decreased activity of the renin-angiotensin system. Glucocorticoid treatment of normal human subjects or patients with essential hypertension did not lead to any significant change in plasma renin activity (Newton and Laragh 1968b; Kelsch et al. 1971; Hamilton et al. 1979; Sowers et al. 1983a; Ferrier et al. 1983) or in the plasma angiotensin II concentration (Nicholls et al. 1980). In other instances, glucocorticoid treatment of man even induced significant increases in plasma renin activity (Krakoff 1973; Weinberger et al. 1975). This was at least partly due to a substantial increase in the plasma renin substrate concentration (Krakoff 1973). Also in patients with Cushing's syndrome, Krakoff (1973) observed a significant increase in the plasma renin substrate concentration but not in plasma renin activity. Plasma renin activity was also normal in five patients with Cushing's syndrome investigated by Ganguly et al. (1983c). However, according to Mantero et al. (1978), plasma renin activity was subnormal in more than half of 16 patients with pituitary-dependent Cushing's syndrome but elevated in two patients with a cortisol-producing adrenocortical adenoma. In patients with primary adrenocortical failure there was a substantial decrease in the plasma renin substrate concentration (Stockigt et al. 1979). This parameter was significantly correlated with the plasma cortisol concentration but not with plasma renin or aldosterone concentrations. Bilateral adrenalectomy of dogs induced a marked fall in the plasma renin substrate concentration, which was corrected by treatment with dexamethasone but not by treatment with aldosterone (Reid et al. 1973). In adrenalectomized rats, deoxycorticosterone acetate and aldosterone were more active in raising plasma renin substrate concentration and in suppressing plasma renin activity than cortisol or corticosterone were (Nasjletti and Masson 1969). In nephrectomized and adrenalectomized rats, aldosterone and 21-dehydrocortisol were as potent as cortisol in inducing plasma renin substrate (Voigt and Köster 1980). In intact rats, short-term (15 min to 5 h) treatment with cortisol or dexamethasone significantly suppressed the plasma renin activity (Hauger-Klevene et al. 1969). Prolonged treatment (2 weeks) with methylprednisolone, however, resulted in significantly increased plasma levels of both renin and its substrate (Krakoff et al. 1975).

5.6.5 Growth Hormone

Although the maintenance of a high aldosterone production by adrenal glands of sodium-deficient hypophysectomized rats has been shown to depend on the administration of growth hormone (see Sect. 5.6.1, p. 178; T. C. Lee and De Wied 1968; Palmore et al. 1970), the bulk of the available clinical and experimental evidence does not indicate that growth hormone plays an important role in the regulation of aldosterone biosynthesis in man.

Acute elevation of the plasma growth-hormone level due to the intravenous injection of the hormone itself or to the stimulation of endogenous hormone secretion by an injection of α-MSH did not affect plasma aldosterone concentrations in normal human subjects (Birkhäuser et al. 1975). Short-term (2–3 days) treatment with human growth hormone of patients with hypopituitarism or nonendocrine diseases significantly stimulated urinary aldosterone excretion in five of seven cases (Beck et al. 1960). Administration of a single dose of human growth hormone to normal human subjects or to patients with primary aldosteronism resulted in moderate sodium retention but either did not affect urinary aldosterone excretion or decreased it (Biglieri et al. 1961). In untreated hypopituitary dwarfs, aldosterone secretion rates were normal and did not consistently change upon prolonged treatment with human growth hormone (Finkelstein et al. 1965). In other patients with isolated growth hormone deficiency, plasma aldosterone concentration and plasma renin activity were normal and rose to a normal extent in response to furosemide or to 10 days of dietary sodium restriction (M. H. Connors et al. 1977; R. E. McCaa et al. 1978b). An injection of ACTH elicited normal increases in plasma aldosterone and cortisol concentrations (R. E. McCaa et al. 1978b). Basal plasma aldosterone and renin activity did not change in the course of 8 or 12 months of replacement therapy with growth hormone. However, the response in plasma aldosterone to furosemide became significantly greater under treatment (M. H. Connors et al. 1977). In patients with panhypopituitarism, long-term therapy with growth hormone did not restore a normal plasma aldosterone response to ACTH or to sodium restriction (R. E. McCaa et al. 1978b).

The occurrence of primary aldosteronism due to an adrenocortical adenoma in a patient with acromegaly was first described by Dluhy and Williams (1969). A remarkably high incidence of primary aldosteronism among hypertensive patients with acromegaly has been observed by some investigators (Strauch et al. 1972; Mantero et al. 1979), but not by others (Werning et al. 1970; Cain et al. 1972; Ogihara et al. 1979; Moore et al. 1980). Perhaps the association of acromegaly with primary aldosteronism represents a special form of endocrine polyadenomatosis. However, abnormalities in aldosterone secretion and in the renin-angiotensin system have also been observed in acromegalic patients without primary aldosteronism. Some of them might be due to an increase in exchangeable sodium, a parameter which was found to be positively correlated with the plasma growth-hormone concentration (Snow et al. 1977). The rise in aldosterone secretion rate in response to sodium restriction was abnormally low in normotensive as well as in hypertensive acromegalics, whereas a blunted increase in plasma renin activity was observed only in the hypertensive patients (Cain et al. 1972). Urinary excretion of tetrahydroaldosterone was abnormally high in eight of 15 patients with acromegaly studied by Modigliani et al. (1972). In eight of 16 instances, this parameter did not adequately decrease upon sodium loading or treatment with deoxycorticosterone acetate. In the patients of Mantero et al. (1979), urinary aldosterone excretion and plasma renin activity were neither significantly suppressed by treatment with 9α-fluorocortisol nor significantly raised by treatment with furosemide. Abnormally low plasma aldosterone concentrations and plasma renin activities were measured in sodium-replete supine acromegalics by Ogihara et al. (1979). Subnormal plasma aldosterone concentrations but normal levels of plasma renin activity and angiotensin II under sodium restriction characterized the patients of Moore et al. (1980). In comparison with sodium-restricted normal subjects, sodium-restricted acromegalics showed an increased blood pressure response but a decreased plasma aldosterone response to graded angiotensin II infusions. Moreover, infusions of the angiotensin antagonist saralasin lowered neither their blood pressure nor their plasma aldosterone concentration.

High doses of human growth hormone directly stimulated the output of aldosterone, corticosterone, and 17-hydroxycorticosteroids by isolated perfused dog adrenal glands, but this effect may have been due to contamination with ACTH (Cushman et al. 1966).

According to whole-body autoradiography of immature hypophysectomized rats performed 6 min after the injection of radioiodine-labeled human growth hormone, radioactivity was accumulated preferentially in the kidneys, liver, submandibular glands, and adrenals (Mayberry et al. 1971). Within the adrenal gland, radioactivity was greatest in the zona glomerulosa. Administration of pig, monkey, or human growth hormone to hypophysectomized rats in vivo led to increased aldosterone production by adrenal tissue in vitro without affecting the output of corticosterone (O. J. Lucis and Venning 1960; Venning and Lucis 1962). Growth hormone did not directly stimulate aldosterone production in vitro (O. J. Lucis and Venning 1960). However, the biosynthesis of aldosterone was selectively increased when plasma from hypophysectomized growth hormone-treated rats was added to the incubation medium. The long-term aldosterone-stimulating effect of growth hormone in hypophysectomized, sodium-deficient, ACTH-treated rats described by Palkovits et al. (1970, 1971 a) was apparently not mediated by the kidneys, since it was also manifest in adrenal tissue of rats that had been bilaterally nephrectomized 6 h before they were killed (Palkovits et al. 1971 b). Sodium restriction of rats for 3 weeks markedly raised the aldosterone production by incubated adrenal tissue but did not lead to any change in the plasma concentration or pituitary content of growth hormone (Relkin and Adachi 1973).

5.6.6 Prolactin

Although prolactin plays a prominent role in the osmoregulation of fish (Ensor and Ball 1972), it does not appear to be materially involved in sodium and water homeostasis of man (G. Baumann and Loriaux 1976; G. Baumann et al. 1977). Nevertheless, the possible function of prolactin as an adrenoglomerulotropic factor has at least been seriously considered by a number of investigators either studying the mechanism of dopaminergic modulation of aldosterone secretion (see Sect. 5.7, p. 200) or looking for pituitary aldosterone regulators other than ACTH. Most of the available evidence, however, does not indicate that prolactin is of physiological or pathophysiological importance in the control of aldosterone secretion in man.

Rubin et al. (1978) observed that spontaneous fluctuations in the plasma aldosterone concentration in normal human subjects during sleep were synchronous with fluctuations in the plasma prolactin concentration. However, according to Holland et al. (1977), a positive correlation between plasma prolactin and aldosterone concentrations was found in only one of three normal subjects on a normal sodium intake and in two of four patients with primary aldosteronism under dexamethasone treatment. Acute increases in the plasma prolactin level induced in normal subjects or patients with primary hypothyroidism by the injection of TRH had no effect on plasma aldosterone concentrations (G. Baumann and Loriaux 1976; Re et al. 1979). Changes in plasma aldosterone observed during hemodialysis of anephric patients or patients with terminal renal failure were not related to changes in plasma prolactin (Ølgaard et al. 1977 c; Sciarra et al. 1977). In such patients, treatment with bromocriptine substantially lowered the plasma prolactin concentrations but did not affect plasma aldosterone (Ølgaard et al. 1977 c). Plasma prolactin concentration did not change in response to sodium restriction, sodium loading, or upright posture in either normal human subjects or patients with primary aldosteronism (Holland et al. 1977). Holland and co-workers observed one case with a prolactin-producing pituitary microadenoma among 33 patients with primary aldosteronism. Removal of the prolactinoma did not affect the primary aldosteronism (bilateral adrenal hyperplasia). In other patients with primary aldosteronism, plasma prolactin concentrations were normal (Holland et al. 1977; Düsing et al. 1979).

In rats, sodium restriction induced significant rises in the pituitary content and plasma concentration of prolactin (Relkin and Adachi 1973). Prolonged treatment of rats with sheep prolactin induced an elevation of the plasma aldosterone concentration which reached statistical significance within 3 days and was associated with a progressive hypertrophy of zona glomerulosa cells and their organelles (Mazzocchi et al. 1986).

5.6.7 Thyroid Hormones

The available evidence does not indicate that thyroid hormones directly affect aldosterone biosynthesis in vitro (Müller and Ziegler 1968). However, thyroid hormones would be expected to influence aldosterone secretion in the intact organism in many ways through their known actions on corticosteroid metabolism, sodium and potassium homeostasis, and the sympathetic nervous system.

According to Peterson (1958), the biological half-life of infused labeled or unlabeled cortisol in plasma was increased in patients with myxedema and decreased in patients with thyrotoxicosis, whereas plasma cortisol levels were normal. Similarly, the metabolic clearance rate of radiolabeled aldosterone from plasma was low in hypothyroid and high in hyperthyroid patients (Luetscher et al. 1963). Mean plasma aldosterone concentrations, as calculated from metabolic clearance rates and secretion rates, were normal in both groups of patients and did not change upon treatment. This latter observation was not confirmed by investigators who directly measured plasma aldosterone concentrations by radioimmunoassay. In patients with myxedema, both the plasma aldosterone concentration and the urinary aldosterone excretion rate were low and rose progressively in response to treatment with increasing doses of thyroxine (Marks et al. 1978). Abnormally low levels of plasma aldosterone and plasma renin activity were observed in hypothyroid patients by Ogihara et al. (1973) and by Saruta et al. (1980). Both parameters rose subnormally in response to upright posture or to furosemide. Increases in plasma aldosterone elicited by an ACTH injection, an angiotensin II infusion, or a potassium infusion were also subnormal. Plasma sodium and potassium concentrations were normal in hypothyroid patients, but urinary potassium excretion in response to furosemide administration was significantly lower than in normal subjects (Saruta et al. 1980). In patients with hyperthyroidism, plasma renin activity and aldosterone concentration were significantly elevated both in recumbent and in upright posture (Ogihara et al. 1973). One week of treatment with propranolol significantly lowered these parameters, but not into the normal range. In hyperthyroid patients kept on a low-sodium diet (10 mmol Na^+ and 100 mmol K^+ per day), the plasma concentration and the secretion rate of aldosterone were significantly lower, but the plasma renin activity and the metabolic clearance rate of aldosterone were significantly higher than in normal sodium-restricted subjects (Cain et al. 1973). Although the mean serum potassium concentration was the same in hyperthyroid and in euthyroid man, the lower aldosterone secretion of thyrotoxic patients may have been due to potassium depletion. Under a high-potassium (200 mmol per day), low-sodium (10 mmol per day) regime, aldosterone secretion rates of hyperthyroid patients were not significantly different from those of normal subjects.

Hypothyroidism induced in rats by the long-term administration of propylthiouracil substantially decreased the aldosterone secretion rate (Fregly et al. 1965). Propylthiouracil treatment significantly lowered serum aldosterone and plasma renin concentrations in sodium-restricted as well as in sodium-replete rats (Montiel et al. 1983). Administration of L-triiodothyronine markedly raised plasma renin and serum aldosterone concentrations in rats on a normal sodium intake but not in rats on a low-sodium diet. Administration of propylthiouracil to rats on a low-sodium diet significantly lowered the plasma concentration as well as the urinary excretion of potassium. On the other hand, treatment with L-triiodothyronine significantly suppressed the urinary sodium excretion in sodium-replete as well as in sodium-restricted rats. In thyroidectomized rats the plasma renin concentration was normal, but plasma renin activity and renin substrate concentration were subnormal, and the renal renin content was increased (Bouhnik et al. 1981). An injection of isoproterenol raised the plasma renin concentration in intact rats but was ineffective in thyroidectomized animals. Thyroidectomy of rats led to a 30% decrease in the conversion of corticosterone to aldosterone and to 18-hydroxycorticosterone by adrenocortical mitochondria (Benelli et al. 1982b). However, this effect was not restricted to aldosterone biosynthesis, since the mitochondrial conversion of cholesterol to pregnenolone and the 11β-hydroxylation of deoxycorticosterone as well as the microsomal 21-hydroxylation of progesterone were found to be similarly suppressed in adrenals of thyroidectomized rats (Benelli et al. 1982a, 1982b).

5.6.8 Vasopressin

The available experimental evidence indicates that vasopressin has little direct influence on aldosterone biosynthesis (see p. 58), but it appears to be an active growth factor for the zona glomerulosa of the rat adrenal cortex. Addition of vasopressin - alone or in combination with ACTH - to the medium in which rat adrenal zona glomerulosa cells were kept in primary culture significantly stimulated the incorporation of radiolabeled thymidine into DNA (Payet et al. 1984). The presence of vasopressin with or without ACTH in the culture medium also markedly stimulated the aldosterone output, whereas ACTH alone had only a minor effect on aldosterone output and mainly stimulated the corticosterone output. It had previously been shown by Isler (1973) that treatment of intact or hypophysectomized rats with injections of posterior pituitary powder significantly increased the mitotic activity of the adrenal zona glomerulosa. According to a later study with intact rats (Payet and Isler 1976), this effect on mitosis was statistically significant only in female animals. However, the posterior pituitary extract induced a significant increase in the area and thickness of the zona glomerulosa in both sexes. The effect of posterior pituitary powder on the mitotic activity of the zona glomerulosa in female rats could be reproduced by treatment with synthetic vasopressin and, to a lesser extent, with synthetic oxytocin or porcine renin (but not with angiotensin II, ACTH, growth hormone, α-MSH, β-MSH, serotonin, or 6-methoxytetrahydroharman). In hypophysectomized female rats, 2 days of treatment with vasopressin alone stimulated the mitotic activity of the zona glomerulosa to a markedly greater extent than treatment with ACTH or with a combination of vasopressin and ACTH (Payet and Lehoux 1980). However, a significant long-term stimulatory effect of vasopressin on plasma aldosterone and corticosterone concentrations was observed only in intact, and not in hypophysectomized rats.

Besides having some direct effects on the zona glomerulosa, vasopressin would be expected to indirectly influence aldosterone secretion in the mammalian organism through its action on water, sodium, and potassium homeostasis and on renin secretion. However, reported effects of administered or excessively secreted vasopressin or of diabetes insipidus on aldosterone secretion in man or experimental animals are somewhat conflicting.

Treatment of normal human subjects with vasopressin alone did not affect the urinary aldosterone excretion rate (A. F. Muller 1963). However, the combination of vasopressin with water loading significantly suppressed this parameter of aldosterone secretion. Treatment of sodium-restricted human subjects with vasopressin resulted in a decrease in urinary aldosterone only in the individuals in which it also caused a weight gain (Cox et al. 1961, 1962). In contrast, a drop in aldosterone excretion was caused without a weight gain by administration of pitressin tannate to normal subjects on a constant sodium and water intake (Franken et al. 1961). Again, prolonged treatment of normal human subjects with vasopressin or oxytocin for 5-10 days did not lead to any consistent changes in plasma renin activity or urinary aldosterone excretion, according to F. J. Goodwin et al. (1970b). Other investigators, however, observed a significant drop in plasma renin activity in response to short-term or long-term administration of vasopressin to normal human subjects or to patients with heart or liver disease on maintenance diuretic therapy (Fichman et al. 1974; Khokhär et al. 1976; Hesse and Nielsen 1977). Abnormally low values of plasma renin activity were also found in patients with the syndrome of inappropriate secretion of antidiuretic hormone (R. D. Gordon and Pawsey 1971; Fichman et al. 1974; Fyhrquist et al. 1977). Fichman et al. (1974) observed a striking discrepancy between suppressed values of plasma renin activity and normal values of plasma aldosterone concentration in vasopressin-treated normal subjects as well as in patients with inappropriate secretion of endogenous antidiuretic hormone. In these subjects, plasma aldosterone, but not plasma renin activity, rose in a normal fashion in response to upright posture, sodium restriction, or both. The two parameters increased equally upon fluid restriction. A marked suppression and stimulation of the urinary aldosterone excretion rate had previously been induced by fluid loading and deprivation, respectively, in a patient with inappropriate secretion of antidiuretic hormone by Knochel et al. (1964). In another patient with this syndrome, plasma and urinary aldosterone were markedly elevated with an unrestricted water intake, in spite of suppressed plasma renin activity (Fyhrquist et al. 1977). Plasma aldosterone became normal with a water intake of 400 ml per day and a sodium intake of 100 mmol per day.

In two patients with untreated diabetes insipidus, markedly elevated levels of plasma renin activity contrasted with normal urinary aldosterone excretion rates (Werning et al. 1969). In a child with diabetes insipidus and adipsia an extremely high level of plasma renin was associated with a normal plasma aldosterone level during dehydration induced by withholding of fluid and by vasopressin therapy (Schalekamp et al. 1976). Under these conditions, the serum sodium rose to levels as high as 188 mmol/l, whereas the serum potassium and plasma cortisol concentration remained normal. When the serum osmolality was normalized by rehydration and treatment with vasopressin, a substantial drop in plasma renin was not accompanied by any change in plasma aldosterone. In this patient, the ratio of plasma aldosterone/renin concentrations was inversely correlated to the serum sodium level, which indicated that the aldosterone-stimulating activity of the endogenous angiotensin II was blocked by the high extracellular sodium concentration.

According to experiments in which the hormone was infused into the renal artery of nonfiltering kidney preparations in dogs (Shade et al. 1973) or added to the medium with which isolated rat kidneys were perfused (Vandogen 1975), vasopressin suppressed renin secretion by a direct effect on the kidneys, even in the absence of a functioning macula densa, most likely by acting on the renin-producing cells. A constant intravenous infusion of pitressin for 4 days to dogs with a water intake limited to 700 ml/day did not produce any changes in plasma renin activity or plasma aldosterone concentration (D. B. Young and McCaa 1981 b).

Treatment of normal male rats with pitressin tannate significantly lowered the plasma renin activity, according to Gutman and Benzakein (1971). These investigators found the plasma renin activity to be significantly elevated in male but not in female rats with hereditary diabetes insipidus (Brattleboro strain). According to B. Möhring et al. (1974), rats with hereditary diabetes insipidus were deficient in potassium. Moreover, they appeared to have a partial secondary adrenocortical insufficiency, as manifested by a decreased adrenal weight and a low plasma corticosterone concentration. The plasma aldosterone concentrations were also subnormal, possibly due to hypokalemia, whereas the plasma angiotensin II concentrations were elevated. Laulin et al. (1984) also observed a decreased aldosterone plasma concentration in male Brattleboro rats but a normal aldosterone output by incubated quartered adrenals and a normal adrenal weight. Zona glomerulosa cells from Brattleboro rats produced normal amounts of aldosterone but increased amounts of corticosterone under basal conditions of incubation as well as under stimulation by potassium ions and ACTH (Kenyon et al. 1978 b). They converted added corticosterone to aldosterone at a normal rate. Under the same conditions of incubation, adrenal glomerulosa cells from sodium-restricted rats with diabetes insipidus produced more aldosterone and corticosterone than cells from sodium-restricted normal animals. On the other hand, zona fasciculata cells of Brattleboro rats were less responsive to ACTH than cells of control rats. According to Vinson et al. (1978 b), incubated capsular adrenal tissue of Brattleboro rats with diabetes insipidus (homozygous) produced less aldosterone, deoxycorticosterone, and 18-OH-DOC than capsular adrenal tissue of animals without diabetes insipidus (heterozygous). Decapsulated glands of rats with diabetes insipidus also produced decreased amounts of deoxycorticosterone and 18-OH-DOC. There were no differences between rats with diabetes insipidus and rats without diabetes insipidus in their adrenal production of corticosterone and 18-hydroxycorticosterone.

5.7 Dopaminergic Regulation of Aldosterone Secretion

5.7.1 General Considerations

According to a widely accepted hypothesis, first stated by Carey et al. (1979), aldosterone secretion in man and certain species of experimental animals is under maximal chronic dopaminergic inhibition. This negative tonic dopaminergic influence can be abolished by dopamine antagonists or overridden by stimulators such as angiotensin II. The hypothesis is based mainly on experiments with the drug metoclopramide, which is a potent dopaminergic antagonist acting on D-2 receptors but also has many other pharmacological properties. The aldosterone-stimulating activity of metoclopramide greatly varies among different animal species. It is uncertain whether metoclopramide acts mainly by blocking dopaminergic influences at the adrenocortical level or rather by affecting the secretion of an extra-adrenal aldosterone stimulator or inhibitor. Experiments with other dopamine antagonists and dopaminergic agonists have partially supported the existence of a negative dopaminergic modulation of aldosterone secretion but have yielded conflicting information about its physiological role and importance. Altogether, the available experimental evidence, which will be summarized on the following pages, is strongly suggestive of a dopaminergic regulation of aldosterone secretion but does not allow the construction of an internally consistent model of its mechanism (for reviews see D.J. Campbell et al. 1981a; Ganguly 1984; Aguilera et al. 1984).

5.7.2 Experiments with Metoclopramide

A low dose of metoclopramide, 5 mg given as an intravenous bolus injection, led to a rise in plasma prolactin and to a drop in plasma cortisol after 30 min but did not affect plasma aldosterone, plasma renin activity, or serum sodium and potassium concentrations in normal male volunteers (Ogihara et al. 1977). At a higher dose, 10 mg, intravenous injections of the drug reproducibly raised plasma aldosterone concentrations in normal human subjects (Norbiato et al. 1977; Carey et al. 1979; Noth et al. 1980; Edwards et al. 1980b; Pratt et al. 1981; Sowers et al. 1981a, 1981b, 1982a, 1982b, 1983a; R.D. Brown et al. 1982; Ganguly et al. 1983a; Nagahama et al. 1983a; N.L. Gilchrist et al. 1984). This stimulatory dose of the drug did not alter the metabolic clearance rate of aldosterone (R.D. Brown et al. 1981). Oral administration of metoclopramide (30 mg/day) led to a transient increase in urinary aldosterone excretion and in the aldosterone secretion rates (R.D. Brown et al. 1981). Both parameters significantly rose on the 1st day but returned to normal on the 4th day of treatment. In all of the studies cited above, metoclopramide administration did not alter plasma sodium and potassium concentrations, and plasma renin activity remained unchanged in most instances. However, a small and delayed increase of this parameter in response to intravenous metoclopramide in normal human subjects was observed by several investigators (R.D. Brown et al. 1979; Norbiato et al. 1979; Sowers et al. 1981a, 1982a). A more extensive rise in plasma renin activity was elicited by metoclopramide in women with idiopathic edema, according to Norbiato et al. (1979) but not according to Sowers et al. (1982c, 1982d). Metoclopramide raised plasma aldosterone in three of six anephric patients (Pratt et al. 1981). Moreover, the aldosterone-stimulating effect of metoclopramide in normal human subjects was not blunted by the simultaneous infusion of the angiotensin antagonist saralasin (Sowers et al. 1981b). When given at a dose of 10 mg, metoclopramide selectively elevated the plasma concentrations of aldosterone and 18-hydroxycorticosterone, without altering the plasma concentrations of other corticosteroids such as cortisol, corticosterone, 18-OH-DOC, and deoxycorticosterone (Sowers et al. 1982b, Sowers and Beck 1983a; Nagahama et al. 1983). However, Nishida et al. (1983a) observed a late and transient increase in plasma cortisol in response to 10 mg of metoclopramide and a more rapid and extensive increase in both plasma cortisol and ACTH concentrations in response to 20 mg of metoclopramide. These effects could be eliminated by pretreatment with dexamethasone. According to the same investigators, pretreatment with dexamethasone had no effect on the rise in plasma aldosterone induced by 10 mg of metoclopramide, but it markedly blunted the aldosterone-stimulating effect of 20 mg of metoclopramide (Nishida et al. 1983b). Apparently, a high dose of metoclopramide can also stimulate aldosterone secretion through an ACTH-mediated effect. On the other hand, even prolonged treatment with dexamethasone lowered only the baseline plasma concentrations of aldosterone and 18-hydroxycorticosterone, but not the peak values of these two parameters observed after a bolus injection of 10 mg of metoclopramide (Sowers et al. 1983a). Metoclopramide significantly raised plasma aldosterone as well in all of six hypopituitary patients studied by Pratt et al. (1981). Treatment of normal subjects with the inhibitor of prostaglandin biosynthesis indomethacin during a low sodium intake did not alter the response in plasma aldosterone to metoclopramide, although under these conditions metoclopramide induced a small decrease in plasma renin activity (Golub et al. 1982). An infusion of ketanserin, a selective blocker of serotonin type-2 receptors, did not significantly affect the rise in plasma aldosterone elicited by the injection of metoclopramide but enhanced the response in plasma renin activity (Degli Uberti et al. 1983a). However, pretreatment of normal subjects with the antiserotoninergic drug metergoline for half a day significantly reduced the plasma aldosterone response to an injection of metoclopramide but only slightly blunted the response in plasma prolactin (Opocher et al. 1983). In normal human subjects, the response in plasma aldosterone to metoclopramide was not impaired by an infusion of the ganglionic blocker trimethaphan (T. A. Wilson et al. 1983a). This observation indicated that in man – in contrast to sheep – the aldosterone response to metoclopramide is not mediated by the autonomic nervous system. On its own, trimethaphan led to a small but significant rise in both aldosterone and cortisol plasma concentrations, most likely by stimulating ACTH secretion. According to Bevilacqua et al. (1984), metoclopramide injection into normal human subjects was followed by the transient appearance of an unknown aldosterone stimulator in the serum. This factor selectively stimulated aldosterone production by incubated calf adrenal zona glomerulosa cells and had no effect on cortisol output. Its action was not inhibited by the angiotensin antagonist saralasin.

The increase in plasma aldosterone secretion elicited in man by metoclopramide is influenced by the sodium intake. The maximal increments in plasma aldosterone and 18-hydroxycorticosterone concentrations induced by metoclopramide injections or infusions were substantially greater in subjects on a low (10 mmol/day) than in subjects on a high (200 mmol/day) sodium intake (Sowers et al. 1982a; Sowers and Stern 1982; M. B. Gordon et al. 1983a). The increased response could have been mediated by the renin-angiotensin system, since metoclopramide also significantly raised plasma renin activity and angiotensin II concentration, but only in subjects on the low and not in subjects on the high sodium intake (M. B. Gordon et al. 1983a). On the other hand, metoclopramide can also abolish or prevent the suppression of the angiotensin II sensitivity of the zona glomerulosa by a high sodium intake. After a 2-h infusion of saline into normal human subjects, a subsequent 1-h infusion of angiotensin II no longer raised the plasma aldosterone level (I. Kojima and Ogata 1982). When metoclopramide was added to the saline infusion, the aldosterone response to angiotensin II was completely preserved. An infusion of metoclopramide also enhanced the response in plasma aldosterone to graded doses of angiotensin II in human subjects who had been kept on a high sodium intake (200 mmol/day; M.B.Gordon et al. 1983b). Thus, under metoclopramide administration, the response curves of plasma aldosterone elicited by increasing doses of angiotensin II in these subjects became identical with the curves obtained in subjects on a low-sodium diet (10 mmol/day). However, metoclopramide did not affect the increased response in blood pressure to angiotensin II due to a high sodium intake.

In most patients with tumorous or nontumorous primary aldosteronism, metoclopramide elicited an excessive rise in plasma aldosterone and 18-hydroxycorticosterone concentrations, even during treatment with dexamethasone (Gniadek et al. 1982; Mizuno et al. 1983; Holland et al. 1983; Ganguly et al. 1983a; Witzgall et al. 1983). In some patients there was also an increased response in plasma prolactin. However, metoclopramide did not stimulate aldosterone secretion in dexamethasone-treated patients with glucocorticoid-suppressible hyperaldosteronism (Ganguly et al. 1983a). In untreated patients with this disorder the plasma aldosterone response to metoclopramide was normal.

At somewhat higher doses than in humans, metoclopramide also raised aldosterone and 18-hydroxycorticosterone plasma concentrations in rhesus monkeys without affecting the levels of other corticosteroids, plasma renin activity, or serum sodium and potassium concentrations (Sowers et al. 1981c, 1982e; Sowers 1982; R.D.Brown et al. 1982).

In sodium-replete and sodium-deficient sheep, a bolus injection followed by a systemic intravenous infusion of metoclopramide elicited a transient rise in plasma aldosterone but did not significantly affect plasma cortisol (McDougall et al. 1981). Intravenous administration of the drug also stimulated aldosterone secretion by an autotransplanted adrenal gland. However, metoclopramide did not influence aldosterone secretion when injected and infused into the adrenal artery of an autotransplanted gland in sodium-replete or sodium-deficient sheep. According to these observations, the drug stimulated aldosterone secretion by an extra-adrenal action. According to another study in sheep, an intravenous bolus injection of metoclopramide induced significant increases in the plasma concentrations of aldosterone, 18-hydroxycorticosterone, corticosterone, and cortisol (Sowers et al. 1983b). The secretory response of each of the four corticosteroids was completely abolished by 4 days of pretreatment with a high dose of dexamethasone. While these results indicated that metoclopramide acted on the adrenal cortex through ACTH secretion, another study suggested that metoclopramide stimulated aldosterone secretion in sheep through a mechanism involving the autonomic nervous system. Thus, metoclopramide did not further stimulate aldosterone secretion in lambs when injected during a constant infusion of the ganglionic blocker trimethaphan (T.A.Wilson et al. 1983b).

An intravenous bolus injection of 200 μg/kg metoclopramide, i.e., a dose that significantly stimulated aldosterone secretion in monkeys, had no effect on plasma aldosterone and cortisol concentrations in rabbits and dogs, according to Sowers et al. (1981c). However, Zanella and Bravo (1982) observed a significant increase of plasma aldosterone in rabbits in response to the intra-arterial injection of a higher dose (1 mg/kg) of metoclopramide. The rise in plasma aldosterone was accompanied by a parallel rise in plasma renin activity and a small but significant decrease in the plasma potassium concentration. Prolonged treatment of rabbits with metoclopramide for 2 weeks resulted in a 90% increase in the plasma aldosterone concentration and in a 60% increase in the adrenal weight but did not alter the plasma corticosterone concentration, plasma renin activity, or plasma potassium concentration. Adrenocortical zona glomerulosa cells of metoclopramide-treated rabbits produced normal amounts of aldosterone under basal conditions of incubation but were more sensitive to the stimulatory action of angiotensin II. Their sensitivity toward ACTH was normal.

An intra-arterial injection of metoclopramide (200 μg/kg) elicited a marked increase in plasma aldosterone, with peak values after 15 min, in intact rats as well as in rats that had been bilaterally nephrectomized 30 h before (Sowers et al. 1980a). In intact rats, there was also a significant rise in plasma renin activity. The aldosterone response to metoclopramide was affected neither by the converting-enzyme inhibitor captopril nor by the angiotensin antagonist saralasin (Sowers et al. 1980b). In contrast, D.J. Campbell et al. (1981b) observed no increase in plasma aldosterone in response to a similar injection of metoclopramide into untreated and dexamethasone-treated rats. According to these investigators, intraperitoneal administration of metoclopramide to rats in doses of up to 100 mg/kg had no effect on urinary aldosterone excretion, either. However, in two other studies, marked increases in plasma aldosterone were elicited by the intraperitoneal injection of smaller amounts of metoclopramide to rats, i.e., 5 and 10 mg/kg (Nagahama et al. 1983b), 0.1 and 1 mg/kg (Jungmann et al. 1983). Higher doses (20 or 50 mg/kg) of the drug also significantly raised plasma renin activity, but had no effect on plasma corticosterone or on sodium and potassium concentrations (Nagahama et al. 1983b). The increase of plasma aldosterone in response to 10 mg of metoclopramide was greater in rats kept on a sodium-deficient diet than in rats on a high sodium intake (Nagahama et al. 1983b). A 2-day intravenous infusion of metoclopramide (5 μg/kg/min) into rats on a normal sodium intake raised the plasma aldosterone concentration to a moderate extent (Aguilera and Catt 1984). In rats on a high-sodium diet, neither an intravenous infusion of metoclopramide nor an intraperitoneal infusion of angiotensin II had any effect on the plasma aldosterone concentration, the concentration of capsular adrenal angiotensin II receptor binding sites, or the capsular adrenal mitochondrial conversion of corticosterone to aldosterone. However, when angiotensin II and metoclopramide were infused simultaneously, there were significant increases in each of these three parameters of zona glomerulosa function. Intravenous infusion of metoclopramide for 48 h did not significantly alter the plasma renin activity in rats on a normal sodium intake or in rats on a high-sodium diet. Prolonged treatment of rats on a normal sodium intake for a week with metoclopramide (4 mg/kg/day; route of administration not specified by the authors) did not alter plasma aldosterone concentration, but it completely abolished the steroidogenic response of zona glomerulosa cells to α-MSH in vitro (Whitehouse et al. 1982). The treatment of rats on a low-sodium diet with metoclopramide for a week did not affect the response of zona glomerulosa cells in corticosterone and aldosterone output to ACTH added in vitro (Vinson et al. 1983). In intact rats, an intracerebroventricular injection of metoclopramide caused a significant elevation of plasma renin activity and plasma aldosterone concentration within 30 min, without any alteration in plasma corticosterone and potassium concentrations (Kawabe et al. 1983). The responses in plasma renin activity and in plasma aldosterone to intracerebroventricular metoclopramide were completely absent in nephrectomized animals, which suggested that the drug acted on aldosterone secretion by a central nervous stimulation of renin release.

5.7.3 Other Dopamine Antagonists

Sulpiride (another D-2 dopamine antagonist, structurally related to metoclopramide) not only stimulated prolactin secretion but also elicited an acute rise in plasma aldosterone, when intramuscularly administered in a dose of 100 mg to normal women (Costa et al. 1980). The drug had no effect on plasma renin activity or on plasma cortisol and potassium concentrations. A lower dose of sulpiride (50 mg i.m.) did not affect plasma aldosterone in normal subjects, even when they were on a low-sodium diet (Mori et al. 1980). *Domperidone* (a dopamine antagonist that does not cross the blood-brain barrier) markedly raised plasma prolactin and also significantly elevated plasma thyrotropin (TSH) concentration, but it had no effect on plasma renin activity or on the plasma concentrations of aldosterone, 18-hydroxycorticosterone, cortisol, or corticosterone when given as an intravenous bolus injection of 10 mg to normal human volunteers (Sowers et al. 1982f; Sowers and Beck 1983a). Similarly, the acute oral intake of 20 mg of *zetidoline* (an antipsychotic drug with antidopaminergic properties) significantly elevated the plasma prolactin concentration but induced no significant changes in the blood levels of aldosterone, cortisol, or renin (Barbieri et al. 1984). In patients with schizophrenia, the intravenous injection of *chlorpromazine* (a central and peripheral dopamine antagonist as well as an α-adrenergic receptor blocking agent) induced a delayed rise in plasma aldosterone, which became significant within an hour and was preceded by an increase in plasma renin activity (Robertson and Michelakis 1975).

In rhesus monkeys, a bolus injection of domperidone which elicited a 16-fold increase in the plasma prolactin concentration had no effect on the plasma levels of aldosterone, 18-hydroxycorticosterone, cortisol, corticosterone, 18-OH-DOC, or deoxycorticosterone or on plasma renin activity (Sowers 1982; Sowers et al. 1982g).

A number of different dopamine antagonists, i.e., metoclopramide, sulpiride, pimozide, domperidone, haloperidol, and chlorpromazine, did not alter urinary aldosterone excretion when administered intraperitoneally into rats, according to D.J. Campbell et al. (1981b). By contrast, Jungmann et al. (1983) elicited significant rises in plasma aldosterone by the intraperitoneal injection of *haloperidol* in doses from 0.1 to 5 mg/kg into normally fed rats. The highest dose of haloperidol also brought about a significant increase in plasma renin activity. In rats pretreated with furosemide for 12 h, the drug no longer stimulated aldosterone secretion but still raised the plasma renin activity. A high dose of haloperidol (5 mg/kg) induced marked increases in both parameters in rats that had been kept on a sodium-deficient diet for 2 weeks.

5.7.4 Dopamine and L-Dopa

An intravenous infusion of dopamine (2–4 μg/kg/min) significantly lowered the plasma prolactin concentration but did not affect the basal aldosterone concentration in normal human subjects (Carey et al. 1980; Noth et al. 1980; R. D. Brown et al. 1982; Sowers and Beck 1983a). However, it significantly blunted or totally inhibited the increases in the plasma concentrations of aldosterone and 18-hydroxycorticosterone elicited by a bolus injection of metoclopramide. In subjects on a normal sodium intake, a dopamine infusion did not alter the response in plasma aldosterone to graded infusions of angiotensin II or ACTH, according to Carey (1982). However, according to I. Kojima and Ogata (1982), there was an increase in blood pressure but not in plasma aldosterone in response to a concomitant 1-h infusion of angiotensin II (5 ng/kg/min) and dopamine (3 μg/kg/min). There was also no response in plasma aldosterone to a 1-h infusion of angiotensin II that had been preceded by a 1-h infusion of dopamine. During a dopamine infusion (4 μg/kg/min) the increases in plasma aldosterone and in plasma renin activity elicited by 65° head-up tilt remained normal (Lightman 1981). In dexamethasone-treated normal human subjects on a sodium-deficient diet, a dopamine infusion (4 μg/kg/min) did not affect baseline plasma aldosterone concentration, but it significantly diminished the response of this parameter to graded doses of angiotensin II (Drake and Carey 1984; Drake et al. 1984). In patients with essential hypertension and a normal plasma renin activity as well as in patients with an aldosterone-producing adenoma, dopamine infusions did not affect plasma aldosterone neither under basal conditions nor under stimulation with ACTH (Holland et al. 1983). In contrast, they caused a significant fall in basal plasma aldosterone in patients with primary aldosteronism due to bilateral zona glomerulosa hyperplasia or with low-renin essential hypertension. However, even in these patients, dopamine did not blunt the plasma aldosterone response to ACTH.

Oral administration of 500 mg of L-dopa (3-hydroxytyrosine; a biological precursor of dopamine that passes the blood-brain barrier) to normal human subjects resulted in moderate drops in plasma aldosterone and plasma renin activity, which became statistically significant after 90 and 120 min respectively (Barbieri et al. 1981). A lower dose of L-dopa (100 mg), given together with carbidopa (an inhibitor of extracerebral dopa decarboxylase) to subjects pretreated with carbidopa caused a similar but more rapid decrease in the plasma aldosterone concentration and again a delayed decrease in plasma renin activity. These observations indicated that the changes in aldosterone and renin secretion were due to increased catecholamine levels in the central nervous system. In untreated normal human subjects on a normal sodium intake, the oral intake of L-dopa (500 mg) resulted in significant drops in plasma and urinary aldosterone, as well as in plasma cortisol, but not in any alterations of the plasma renin activity or of the serum potassium concentration (Frisina et al. 1983). In dexamethasone-treated subjects, L-dopa lowered plasma and urinary aldosterone to the same extent as in untreated subjects without having any further effect on plasma cortisol. According to Lightman (1981), an infusion of L-dopa did not significantly affect the rise in plasma aldosterone elicited by head-up tilt in normal subjects, even when they had been pretreated with carbidopa. However, a dopamine infusion without carbidopa pretreatment significantly diminished the rise in plasma renin activity, and the combination of L-dopa with carbidopa significantly raised the plasma cortisol concentration. In healthy pregnant women (24–36 weeks of gestation) the oral intake of L-dopa (1 g) was followed by significant decreases in plasma aldosterone concentration and in plasma renin activity, but it did not impair a marked increase in plasma aldosterone in response to an angiotensin II infusion (Kaulhausen et al. 1982).

Barbeau and collaborators (1969) observed significantly decreased levels of plasma renin activity in patients with Parkinson's disease, which became unmeasurably low upon treatment with L-dopa. However, according to Michelakis and Robertson (1970), decreased levels of plasma renin activity, measured in the upright position in patients with Parkinson's disease, became normal under L-dopa therapy. In most patients with Parkinson's disease treated with L-dopa, plasma aldosterone, urinary aldosterone, aldosterone secretion, and plasma renin activity were normal and increased normally in response to upright posture and sodium restriction (Sullivan et al. 1973; Katz and Hoehn 1977).

In rhesus monkeys, an infusion of dopamine (4 μg/kg/min) significantly lowered plasma prolactin concentrations but did not alter plasma corticosteroid concentrations (Sowers 1982; Sowers et al. 1982e). However, it markedly blunted the rises in plasma aldosterone and 18-hydroxycorticosterone levels elicited by a bolus injection of metoclopramide. In a later study by the same senior investigator (Sowers et al. 1984), the infusion of a lower dose of dopamine (2 μg/kg/min) decreased the plasma concentration of 18-hydroxycorticosterone (but not of aldosterone or other corticosteroids) in untreated as well as in dexamethasone-treated rhesus monkeys.

In sodium-depleted sheep with a cervical adrenal autotransplant, neither a systemic intravenous nor a local adrenoarterial infusion of dopamine had any effect on the aldosterone secretion rate (McDougall et al. 1981). In sodium-replete sheep, neither systemic nor local adrenal infusion of dopamine blunted the stimulatory effect of locally infused angiotensin II on aldosterone secretion (Lun et al. 1983).

An intravenous infusion of L-dopa (1 mg/kg/min) to rabbits prevented the rise in plasma aldosterone and blunted the increase in plasma renin activity in response to an intra-arterial bolus injection of metoclopramide (Zanella and Bravo 1982).

A simultaneous infusion of dopamine (2 μg/min) completely prevented the long-term stimulatory effect of the combined infusions of metoclopramide and angiotensin II upon the plasma aldosterone concentration, the capsular adrenal angiotensin II receptor site concentration, and the capsular adrenal mitochondrial conversion of corticosterone to aldosterone in rats maintained on a high sodium intake (Aguilera and Catt 1984). The intra-arterial administration of L-dopa (30 mg/kg) to rats significantly lowered plasma aldosterone concentration and plasma renin activity on its own and diminished the response of these two parameters to a subsequent infusion of metoclopramide (Sowers et al. 1980a). However, L-dopa was ineffective when given 30 min after the onset of a metoclopramide infusion. An intracerebroventricular injection of dopamine resulted within 30 min in a significant suppression of plasma renin activity and plasma aldosterone in intact conscious rats (Kawabe et al. 1983). In bilaterally nephrectomized rats, intracerebroventricularly administered dopamine had no effect on plasma renin activity and slightly raised plasma aldosterone, although not to a statistically significant extent.

5.7.5 Bromocriptine

Bromocriptine (2-bromoergocryptine), a derivative of the ergotoxin group of ergot alkaloids, is a dopaminergic agonist acting on peripheral and central D-1 and (preferably) D-2 receptors, with a particular specificity for the pituitary prolactin-producing cells. Bromocriptine treatment had no effect on the basal plasma concentrations of aldosterone, cortisol, and deoxycorticosterone or on urinary aldosterone excretion in normal sodium-replete human subjects (Del Pozo et al. 1977; Birkhäuser et al. 1979; Carey et al. 1979, 1980; Whitfield et al. 1980; Sowers and Beck 1983b). However, it significantly raised plasma corticosterone and it markedly suppressed plasma 18-hydroxycorticosterone levels (Sowers and Beck 1983a, 1983b). According to degli Uberti et al. (1979), bromocriptine treatment did not alter the diurnal rhythms of plasma aldosterone and cortisol concentrations, but shifted the acrophase of plasma renin activity from 6 a.m. to 10 a.m. In spite of a substantial reduction of mean plasma renin activity in all five subjects treated with bromocriptine, a statistically significant decrease in the mean plasma aldosterone concentration was observed in only two of them. Bromocriptine medication did not affect the increase in plasma aldosterone in response to upright posture, according to del Pozo et al. (1977) and to Birkhäuser et al. (1979). However, according to Whitfield et al. (1980), plasma aldosterone did not increase in response to 10 min of upright posture in bromocriptine-treated persons, but it did increase in response to an isometric handgrip. Bromocriptine administration did not suppress the rise in plasma aldosterone elicited by a bolus injection of metoclopramide, whereas it completely abolished the rise in serum prolactin (Carey et al. 1980). There are conflicting reports about a possible impairment of the zona glomerulosa responses to angiotensin II and ACTH by bromocriptine treatment. Carey et al. (1979) did not observe any difference between untreated and bromocriptine-treated normal subjects in their plasma aldosterone responses to graded doses of angiotensin II. In contrast, bromocriptine administration markedly blunted the increments in plasma aldosterone and 18-hydroxycorticosterone concentrations elicited by small doses (0.5–2 ng/kg/min) of angiotensin II in the subjects studied by Whitfield et al. (1980) and by Sowers and Beck (1983b). In another study (Birkhäuser et al. 1979), plasma aldosterone responses to a higher dose of angiotensin II (7 ng/kg/min) were markedly diminished or even absent in bromocriptine-treated subjects. Normal responses in aldosterone secretion to graded infusion of ACTH were found in bromocriptine-treated subjects by Whitfield et al. (1980) and by Sowers and Beck (1983b). On the other hand, a bolus injection of ACTH (0.5 mg) raised plasma aldosterone to a significantly lesser extent in bromocriptine-treated than in untreated persons (Birkhäuser et al. 1979). Treatment of three normal subjects with bromocriptine for 1 or 2 weeks completely eliminated the increase in plasma aldosterone induced by an intravenous injection of furosemide (Edwards et al. 1975). In two other subjects, a single dose of bromocriptine taken 4 h before an intravenous injection of furosemide raised the basal plasma aldosterone concentration but prevented a further stimulation of aldosterone secretion by the diuretic. Responses in plasma renin activity to furosemide injections were not affected by bromocriptine. Normal increases in the plasma concentrations of aldosterone, 18-hydroxycorticosterone, renin, and angiotensin II were elicited in bromocriptine-treated subjects by 4 days of dietary sodium deprivation (10–14 mmol/day; Semple and Mason 1978). In another, similar study, bromocriptine treatment did not alter the gradual increases in plasma aldosterone concentration and in urinary aldosterone excretion observed in normal human subjects during 1 week of dietary sodium restriction (Carey and van Loon 1982).

In patients with idiopathic hyperaldosteronism, a single dose of bromocriptine markedly reduced the response in plasma aldosterone to 4 h of upright standing (Mantero et al. 1977). Plasma aldosterone concentrations (supine and upright) were not changed by bromocriptine intake in patients with an aldosterone-producing adenoma. In patients with both types of primary aldosteronism, plasma levels of conjugated and free dopamine were found to be elevated and to rise further during dexamethasone suppression (Kuchel et al. 1979, 1980). Opposing fluctuations in plasma aldosterone and in plasma dopamine in these patients suggested that surges in aldosterone secretion could have been due to periodically decreased dopaminergic inhibition.

In patients with idiopathic edema, plasma 18-hydroxycorticosterone concentrations were abnormally high under basal conditions and rose excessively in response to upright standing and isometric handgrip (Sowers et al. 1982d). Plasma aldosterone was normal under basal conditions but also responded excessively to the two stimuli. Bromocriptine treatment brought back to normal the plasma concentration of 18-hydroxycorticosterone as well as the aldosterone and 18-hydroxycorticosterone responses to orthostasis and isometric exercise in these patients. It had previously been shown by Kuchel et al. (1977) that urinary dopamine excretion was abnormally low in patients with idiopathic edema.

In primary tissue cultures of human adrenocortical aldosterone-producing adenomas, bromocriptine acted as an agonist-antagonist of angiotensin II (Armato and Mantero 1984). When added alone, it stimulated the incorporation of labeled uridine into RNA. However, when added together with angiotensin II, bromocriptine blocked the stimulation of uridine incorporation as well as the trophic effect of the peptide on the cytoplasmic compartment.

In rats on a low, normal, or high sodium intake, 7 days of treatment with bromocriptine (3 mg/kg/day) led to moderate but statistically significant decreases in plasma aldosterone (Whitehouse et al. 1982). Bromocriptine treatment in vivo also increased the sensitivity of zona glomerulosa cells to the steroidogenic action of α-MSH in vitro but did not alter the sensitivity of zona fasciculata-reticularis cells to ACTH. According to a later report by the same investigators, bromocriptine treatment of rats altered the response in corticosterone output by zona glomerulosa cells to increasing doses of α-MSH but had no effect on the zona glomerulosa cell response in corticosterone output to ACTH or on the stimulation of aldosterone output by α-MSH or ACTH (Vinson et al. 1983).

5.8 Role of the Central Nervous System

5.8.1 General Considerations

It is generally assumed that the central nervous system plays a minor role in the physiological control of aldosterone secretion and that its only important influence on the zona glomerulosa of the adrenal cortex is mediated by pituitary ACTH secretion (Mulrow 1966; J.O.Davis 1967). This assumption is mainly based on the finding by J.O.Davis et al. (1961a) that decapitation of dogs did not alter aldosterone secretion and did not prevent a marked increase in aldosterone output in response to hemorrhage. Although the presence and integrity of the central nervous system is not essential for aldosterone secretion under basal conditions or under various stimuli, there is some evidence indicating that certain areas of the brain can modify aldosterone secretion independently of ACTH secretion, either through the renin-angiotensin system or by unknown mechanisms. For instance, some experimental evidence reviewed in the preceding section (5.7) points to an involvement of the central nervous system in the dopaminergic modulation of aldosterone secretion.

Some observations indicating a role of the central nervous system in aldosterone control have been made in sheep with cervical adrenal autotransplants (Blair-West et al. 1974). In particular, the rapid drop of the aldosterone secretion rate in response to satiation of sodium appetite by sodium-depleted sheep most likely reflected an alimentary-gustatory reflex mechanism (see p.125; Denton et al. 1977). The central nervous system must also have been involved in the marked increases in the aldosterone secretion rate which were elicited in a mildly sodium-deficient sheep by environmental events, such as the preparation of sodium solutions within sight and hearing of the animal or the staring into the animal's eyes by six observers (Denton 1982). In man, acute psychological strains such as the hypnotic suggestion of running a race or mental arithmetic caused considerable increases in plasma renin activity and angiotensin II concentration, which were, however, accompanied by significant drops in the plasma aldosterone concentration (Kosunen et al. 1977; Kosunen 1977). Marked increases in plasma renin activity were elicited by psychosocial stimuli (exposure to a novel environment, presence of a hungry cat) in rats as well (Clamage et al. 1975). Plasma aldosterone concentrations were not measured in these experiments.

Based on indirect evidence, Gann and Travis (1964) postulated an unknown control mechanism, independent of the release of ACTH or the renin-angiotensin system but involving the central nervous system and an aldosterone inhibitor, to explain the biphasic effect of carotid artery constriction on aldosterone secretion in dogs.

5.8.2 Electrical and Humoral Stimulation of the Central Nervous System

Electrical stimulation by means of chronically implanted electrodes in the lateral hypothalamus and certain localized sites in the cingulate area and lower medial parts of the frontal lobes of rhesus monkey brains resulted in considerable increases in plasma aldosterone concentration and in plasma renin activity (Frankel et al. 1976). Plasma cortisol also rose considerably after hypothalamic stimulation but much less after stimulation of the extrahypothalamic sites. The increases in plasma aldosterone and renin activity observed after extrahypothalamic stimulation were completely suppressed by prior administration of the β-adrenergic blocker propranolol, whereas α-adrenergic blockade by phentolamine was ineffective (Jenkins et al. 1976). Propranolol only partially reduced the response in aldosterone secretion to hypothalamic stimulation but completely inhibited the response in plasma renin activity. Dexamethasone treatment also partially inhibited the aldosterone response to hypothalamic stimulation, completely suppressed the cortisol response, and had no effect on the response in plasma renin activity. Accordingly, the increase in aldosterone secretion elicited by stimulation of the lateral hypothalamus was mediated partly by the renin-angiotensin system and partly by ACTH secretion. On the other hand, extrahypothalamic stimulation raised aldosterone secretion exclusively through the renin-angiotensin system.

Ventriculocisternal perfusion of dogs with a hyposmotic, low-sodium fluid caused a marked decrease in renal sodium excretion and raised the plasma renin activity (Mouw and Vander 1970). Aldosterone secretion was not measured, but the rapid onset and decay of antinatriuresis and a simultaneous reduction in potassium excretion made it unlikely that the renal effect was mediated by aldosterone (Mouw and Vander 1971). In a previous study, cerebroventricular perfusion of dogs had resulted in increases in adrenal venous output of aldosterone, which were, however, paralleled by simultaneous increases in cortisol secretion and depended more on the time of perfusion than on the sodium and potassium content of the perfusion fluid (Norman 1961). Intraventricular infusion of a low-sodium iso-osmotic fluid into sodium-replete rats led to a significant rise in plasma renin activity but did not affect plasma aldosterone (Gomez-Sanchez et al. 1977). In sodium-restricted rats, the intraventricular infusion of either the same fluid or a hyperosmolar, high-sodium fluid resulted in elevated plasma corticosterone but in unchanged plasma aldosterone and renin activity. As previously mentioned (p. 125), cerebroventricular perfusion with a high-sodium (170 mmol/l) fluid led within 5 h to a significant drop in plasma aldosterone in sodium-deficient sheep without any concomitant changes in plasma sodium and potassium concentrations, blood angiotensin II, or plasma renin concentration and despite a significant rise in plasma cortisol and corticosterone concentrations (Abraham et al. 1976). In sodium-replete sheep, the intraventricular infusion of mannitol (0.7 mol/l) elicited marked increases in plasma aldosterone concentrations, which preceded a rise in plasma renin concentration by 2 h and were not accompanied by any changes in the plasma cortisol or potassium concentrations (Coghlan et al. 1980).

In sodium-deficient dogs, the intraventricular infusion of angiotensin II resulted in significant decreases in plasma aldosterone, renin activity, and potassium concentration but did not affect the plasma cortisol concentration (Brooks and Malvin 1981). Intraventricular infusions of the converting enzyme inhibitor teprotide or of the angiotensin antagonist saralasin led to marked elevations in the plasma aldosterone concentrations which were not accompanied by any changes in plasma renin activity, cortisol, or potassium concentrations. According to Eguchi and Bravo (1984), intracerebroventricular infusion of angiotensin II lowered plasma renin activity, cortisol, and potassium concentration in both sodium-replete and sodium-depleted dogs. However, a suppression of plasma aldosterone was observed only in sodium-depleted dogs or in sodium-replete dogs that had been pretreated with dexamethasone. The drop in plasma aldosterone induced by intraventricular angiotensin II in sodium-deficient dogs could be prevented by a simultaneous intravenous infusion of angiotensin II. In sodium-replete sheep, the intracerebroventricular infusion of angiotensin II raised plasma aldosterone, together with plasma cortisol (Nicholls et al. 1983). This response was abolished by pretreatment with dexamethasone. Central infusion of angiotensin II did not alter the plasma angiotensin II concentration in these animals.

Direct or indirect experimental evidence pointing to a CNS site of action of dopaminergic agonists or antagonists in their effect on aldosterone secretion has been cited in Sect. 5.7. According to indirect pharmacological evidence obtained in dogs, renal renin secretion can be increased by stimulation of central nervous serotonergic pathways (Zimmermann and Ganong 1980). Plasma aldosterone was not measured in these experiments. In contrast, the oral administration of 5-hydroxytryptophan to human volunteers elicited increases in plasma aldosterone concentration that were apparently due to a central serotonergic stimulation and were not accompanied by increases in plasma renin activity, potassium, or ACTH (Shenker et al. 1985b, 1985c). The effect of the serotonin precursor on aldosterone secretion was enhanced by the peripheral inhibition of its decarboxylation by pretreatment with carboxydopa (Shenker et al. 1985b). Also, blockade of peripheral serotonin$_2$ receptors with ketanserin or methysergide did not inhibit the aldosterone-stimulating action of 5-hydroxytryptophan in man (Shenker et al. 1985c).

5.8.3 Brain Lesions

A normally synchronized circadian rhythm of plasma cortisol was observed in *comatose* patients a few days after *severe brain injury* who were under treatment with high doses of dexamethasone (Agrimonti et al. 1981). In contrast, the circadian pattern of plasma aldosterone appeared to be disrupted, with irregular fluctuations during the entire day. According to Krieger and Krieger (1964), four of seven patients with *pretectal lesions* did not increase urinary aldosterone excretion in response to hemorrhage. However, none of these subjects exhibited clinical or laboratory evidence of hypoaldosteronism.

In dogs, *hypothalamic lesions* had no effect on aldosterone secretion, provided that the median eminence was left intact (J.O. Davis et al. 1959). Bilateral injury of the *median eminence* was followed by adrenal atrophy and decreased secretion rates of 17-hydroxycorticoids and aldosterone (J.O. Davis et al. 1959, Ganong et al. 1959). *Complete midbrain transection* did not affect the aldosterone secretion rate of dogs or its stimulation by hemorrhage (J.O. Davis et al. 1961c). In dogs with secondary hyperaldosteronism due to thoracic inferior vena cava constriction, midbrain transection did not lower the elevated secretion rate of aldosterone if the high venous pressure was maintained. *Midcollicular brain section with removal of the forebrain* in hypophysectomized, nephrectomized dogs led to significant increases in aldosterone, cortisol, and corticosterone outputs (Barbour et al. 1965). Whereas the secretion rate of the two glucocorticoids remained below the level observed in intact animals, aldosterone was secreted at a rate comparable to the one seen in intact or hypophysectomized dogs after hemorrhage, caval constriction, or sodium depletion.

In cats, discrete lesions in the *dorsal central region of the rostral midbrain* (involving the posterior commissure and the periaqueductal gray substance) markedly decreased aldosterone secretion during normal or low sodium intake (A. N. Taylor and Farrell 1962), whereas larger lesions (involving the habenulae, stria medullaris, and the pineal gland) obscured this effect at least partially.

In sodium-deficient sheep, *midcollicular decerebration* impaired the rapid decrease in aldosterone secretion in response to a systemic NaCl infusion (Blair-West et al. 1960, 1963).

In rats, electrocoagulation of the *subcommissural organ* resulted in decreases in aldosterone production in vitro as well as in the width and nuclear volume of the zona glomerulosa (Palkovits et al. 1965). Intraperitoneal injection of saline extracts of the subcommissural organ induced a transient increase of the nuclear volume in zona glomerulosa cells (Palkovits and Födvari 1963). The effects of small *bilateral hypothalamic lesions* in sodium-deficient rats on aldosterone and corticosteroid ("corticosterone") production by incubated adrenal tissue were studied by Palkovits et al. (1974). Lesions located in the median eminence or in the ventromedial nucleus lowered both aldosterone and corticosteroid productions. Lesions in the nucleus paraventricularis, the nucleus supraopticus, or the supraopticoneurohypophyseal tract selectively decreased aldosterone production. Treatment with growth hormone restored aldosterone production by adrenals of rats with ventromedial nuclear lesions but had no effect in rats with lesions of the supraopticoneurohypophyseal tract. In the latter group of rats, aldosterone production was restored by treatment with vasopressin.

5.8.4 Hypothalamic Factors

The direct effect of *somatostatin* on zona glomerulosa cells has been described earlier (see p. 75). There is limited information available about a possible effect of this hypothalamic, intestinal, and pancreatic hormone on aldosterone secretion in vivo. In normal human subjects, an intravenous somatostatin infusion did not affect plasma renin activity under basal conditions but inhibited an increase of this parameter in response to the diuretic furosemide (Rosenthal et al. 1976) or to the beta-adrenergic receptor-stimulating substance orciprenaline (Rosenthal et al. 1977). A 90-min infusion of somatostatin into furosemide-treated normal human subjects led to a 30% drop in plasma renin activity but not in plasma aldosterone (Gomez-Pan et al. 1976). In patients with unilateral renovascular hypertension, somatostatin significantly lowered renin activity in peripheral as well as in renal venous blood but did not alter the serum aldosterone concentration (Alart et al. 1978). According to Millar et al. (1983) and to C. R. Jones et al. (1984), an infusion of somatostatin significantly attenuated the increase in plasma aldosterone in response to 80° or 60° head-up tilt or to graded infusions of angiotensin II in normal human subjects but had no effect on basal plasma aldosterone, the increase in plasma renin activity in response to tilt, or the increase in blood pressure in response to angiotensin II. On the other hand, an infusion of somatostatin-28 enhanced the increases in serum aldosterone and in serum potassium elicited by the infusion of arginine-HCl in normal human subjects (Barron et al. 1984). The rise in serum aldosterone in response to the amino acid was most likely due to hyperkalemia. Somatostatin blocked an arginine-induced insulin release and thus prevented an insulin-dependent cellular uptake of potassium. No somatostatin receptors were found in membrane preparations of human, simian, or canine adrenal zona glomerulosa (Aguilera and Catt 1983). In contrast, specific high-affinity somatostatin receptors were discovered in the zona glomerulosa of rat, rabbit, and bovine adrenals. In rats, capsular adrenals showed the highest uptake of radioiodinated somatostatin among several tissues tested after intravenous injection of the labeled peptide (Aguilera et al. 1982). Treatment of rats with subcutaneous injections of somatostatin for 5 days resulted in significant decreases in the plasma aldosterone concentrations and in the following morphometric parameters of adrenal zona glomerulosa cells: volume of cells, volume of nuclei, volume of mitochondrial compartment, surface of mitochondrial cristae, and surface of SER membranes (Rebuffat et al. 1984). Similar effects were induced by treatment with the converting-enzyme inhibitor captopril; the inhibitory actions of captopril and somatostatin were not additive (Mazzocchi et al. 1985). Somatostatin completely blocked the adrenoglomerulotropic effects of angiotensin II in rats. Somatostatin administration did not affect morphometric parameters of zona fasciculata cells, plasma corticosterone concentration, or plasma renin activity (Rebuffat et al. 1984). A small stimulatory effect of angiotensin II on morphometric parameters of zona fasciculata cells and on the plasma corticosterone concentration was not affected by somatostatin administration (Mazzocchi et al. 1985).

An intravenous bolus injection of ovine *corticotropin-releasing factor* (CRF; 100 μg) into normal human subjects did not affect plasma aldosterone, while markedly raising plasma ACTH and cortisol concentrations, according to O. A. Müller et al. (1982). In other studies, a higher dose of CRF (200 μg) elicited rises in plasma aldosterone which paralleled rises in plasma cortisol and ACTH concentrations (Hermus et al. 1984; Conaglen et al. 1984). Nakahara et al. (1983) observed similar small increases in plasma aldosterone after the intravenous injection of 100 and 500 μg of CRF. In sheep, an intravenous injection of CRF led to a small rise in plasma aldosterone which was accompanied by a drop in plasma angiotensin II (Donald et al. 1983). In rats, the intravenous injection of ovine CRF led to marked elevations in the plasma concentrations of ACTH, corticosterone, aldosterone, and 18-hydroxycorticosterone (Maser-Gluth et al. 1984). The maximal elevations were observed 30 min after the injection of CRF. After 6 h, the plasma corticosterone and 18-hydroxycorticosterone concentrations were lower in CRF-treated animals than in control animals, whereas the plasma aldosterone level was still slightly elevated in the CRF-treated rats.

5.8.5 Pineal Gland

Kinson et al. (1967) observed that pinealectomy of normal rats resulted in a marked increase in aldosterone secretion and in a smaller increase in corticosterone secretion. In pinealectomized rats, unilateral renal artery constriction stimulated aldosterone secretion more markedly than it did in intact rats and resulted in an increased secretion rate of corticosterone (which was not observed in rats with an intact pineal gland). Sodium restriction induced the same fractional increase in aldosterone production in intact and in pinealectomized rats (Kinson and Singer 1967). The simulatory effect of pinealectomy on the adrenal cortex was a transient one (Kinson et al. 1968). Six months after the operation, aldosterone and corticosterone secretion rates had returned to normal levels. In another study, pinealectomy of rats resulted within 5 weeks in significant increases in blood pressure and plasma renin activity (Karppanen et al. 1975). However, aldosterone production in vitro by quartered adrenals incubated without stimulators or with ACTH, dibutyryl cyclic AMP, or serotonin was not affected by pinealectomy. According to Acuña et al. (1982, 1984) pinealectomy of female rats (male rats had been used in the studies cited above) led to a decrease in plasma renin activity and an increase in plasma corticosterone but left plasma aldosterone levels unchanged. No changes in plasma renin activity and corticosterone were observed after combined pinealectomy and ovariectomy.

In intact dogs, the administration of ubiquinone, a constituent of the lipid fraction of the pineal gland, led to a considerable decrease in aldosterone secretion without having a significant effect on cortisol secretion (Fabre et al. 1965). A hexane extract of bovine pineal gland appeared to inhibit 11β-hydroxylation in vitro (Lommer 1966). It decreased the conversion of ^{14}C-labeled progesterone to aldosterone, corticosterone, cortisol, and cortisone and enhanced its incorporation into deoxycorticosterone and 11-deoxycortisol by bovine adrenal slices. The effects of serotonin and melatonin, two other constituents of pineal gland extracts, on aldosterone biosynthesis in vitro are described in Sect. 4.5. Treatment of potassium-deficient rats with 5-hydroxytryptophan (a precursor of serotonin) or with reserpine (a drug which releases stored serotonin from the blood platelets) for 2 days resulted in significant increases in the capsular adrenal conversions of tritiated corticosterone to aldosterone and 18-hydroxycorticosterone (J. Müller and Baumann 1974). However, these observations did not unequivocally demonstrate a long-term stimulatory effect of serotonin on late steps in aldosterone biosynthesis. Treatment with 5-hydroxytryptophan induced hemorrhagic infarctions of the kidneys and a marked elevation in plasma renin activity, and serpasil-treated rats consumed hardly any food or water. Thus, the observed stimulation of aldosterone biosynthesis may have been due to activation of the renin-angiotensin system in the former group and to dehydration in the latter. Nevertheless, both agents also stimulated aldosterone biosynthesis in bilaterally nephrectomized potassium-depleted rats, although to a lesser extent (J. Müller, unpublished observations). An intraperitoneal injection of 6-methoxytetrahydro-β-carboline (=pinoline, a compound which is found in high concentrations in mammalian and avian pineal glands) in normal rats induced an acute increase in the plasma aldosterone concentration, which was preceded by a rise in plasma renin activity (Airaksinen et al. 1984). There was also a significant increase in plasma β-endorphin/β-lipotropin immunoreactivity. Plasma ACTH and corticosterone concentrations were not measured in these experiments.

5.8.6 Spinal Cord Transection

In four patients with complete cervical spinal cord transections, resting plasma renin activities were elevated, whereas plasma aldosterone concentrations were within the normal range (Mathias et al. 1975). Head-up tilting (45°) elicited a blunted or absent increase in plasma norepinephrine concentrations but an enhanced elevation in plasma renin activity and a delayed but marked rise in plasma aldosterone concentrations. James et al. (1976) observed normal diurnal variations in plasma aldosterone, corticosterone, cortisol, and renin activity in two other patients with cervical spinal cord transection.

5.8.7 Manic-depressive Psychosis

In bipolar manic-depressive patients, a decrease in the production rate of aldosterone accompanied by urinary sodium loss was observed during the switch from mania to depression (Allsopp et al. 1972). In a long-term study of four patients with manic-depressive psychosis, Hullin et al. (1977) saw no obvious relationship between mood and plasma renin activity. Nevertheless, resting plasma renin activity was high and showed a blunted response to upright posture. Aldosterone production rates were inappropriately low in relation to plasma renin activity. This might have been due to an unknown circulating inhibitor of aldosterone biosynthesis. Sera from manic-depressive patients caused a significant inhibition of aldosterone production in vitro when added to dispersed rat adrenal zona glomerulosa cells incubated with ACTH, angiotensin II, or serotonin but did not influence unstimulated aldosterone production (O'Brian et al. 1979). According to a later report by the same group of investigators (Hullin et al. 1981), a high proportion of sera from patients with a bipolar manic-depressive disorder inhibited aldosterone production. The inhibitory activity was greatest in depression and lowest during periods of normal affect. In one patient studied through 11 affective cycles, the inhibitory activity increased during transitions from mania to depression.

The single-dose overnight dexamethasone suppression test with measurement of plasma cortisol (Nugent et al. 1965; Pavlatos et al. 1965) is not only a valuable tool for the diagnosis of Cushing's syndrome; it has also been used for the detection and management of endogenous depression (for literature see Health and Public Policy Committee 1984). According to Holsboer et al. (1983), morning plasma aldosterone concentrations after dexamethasone intake were increased in normal control subjects but decreased in patients with endogenous depression. However, Targum et al. (1984) did not observe any consistent change in plasma aldosterone in response to dexamethasone intake in either normal persons or depressed patients. No abnormality in plasma aldosterone was seen in seven depressed patients in whom plasma cortisol was not suppressed by dexamethasone.

5.9 Normal Life

5.9.1 Fetal Development

The human adrenal cortex becomes capable of aldosterone biosynthesis after 16 weeks of gestation, according to Dufau and Villee (1969), who incubated fetal adrenal tissue with labeled progesterone and observed a measurable conversion of this precursor to aldosterone (but not to 18-hydroxycorticosterone). The biosynthesis of aldosterone by human fetuses perfused with corticosterone at mid-term (20 weeks of gestation) had previously been reported by Pasqualini et al. (1966). Bayard et al. (1970) gave constant infusions of tritiated aldosterone to mothers at the time of elective cesarean section at 38-40 weeks of gestation and observed that there was a significant secretion of aldosterone by the fetus in spite of maternal aldosterone crossing the placenta. The aldosterone concentration was 2-12 times higher in the fetal than in the maternal plasma. Plasma aldosterone was also higher in cord blood than in maternal blood according to Beitins et al. (1972b), Katz et al. (1974), Sippell et al. (1978), and Hubl et al. (1979), whereas approximately equal aldosterone concentrations were measured in maternal and cord plasma by Godard et al. (1976). Plasma renin concentrations were significantly higher in cord than in maternal blood, but this was offset by considerably higher concentrations of renin substrate in maternal plasma, resulting in similar renin activities and angiotensin II concentrations in maternal and cord plasma (Katz et al. 1974; Godard et al. 1976). Sodium depletion of the mother led to substantial increases in the aldosterone concentrations in cord plasma as well as in the plasma of the newborn (Beitins et al. 1972b). Sippell et al. (1981) measured a series of corticosteroids in human amniotic fluid. With the exception of cortisone, they all increased between the 14th and 38th week of gestation. In particular, aldosterone concentration in the amniotic fluid rose from a mean value of 4 ng/dl at 14-16 weeks to 53 ng/dl at 36-38 weeks and declined again to 27 ng/dl toward term.

The ontogeny of aldosterone biosynthesis and its physiological regulation has been extensively studied in sheep. The mean gestation time in this species is 147 days. During incubations with ACTH, sheep 40-day fetal adrenals produced more aldosterone and cortisol per unit body weight than term adrenals (Wintour et al. 1975). ACTH elicited the highest increase in aldosterone output in vitro in adrenals of fetuses with an age of 80-89 days. The response of fetal adrenals in cortisol and aldosterone outputs declined in the period from day 90 to day 120. Blood levels of aldosterone and cortisol were also significantly lower in 90- to 120-day-old fetuses than in younger and older ones. Plasma aldosterone and plasma potassium were significantly higher in 126- to 145-day fetuses than in 100- to 125-day fetuses, whereas the difference in plasma renin activity was not statistically significant (Robillard et al. 1980). However, plasma aldosterone concentrations correlated better with plasma renin activities than with plasma potassium concentrations. A high potassium concentration in the incubation medium elicited a response in aldosterone output only in adrenals of fetal sheep with a gestational age of more than 120 days (Wintour et al. 1975). When added to adrenals of fetuses of less than 100 days, angiotensin II significantly stimulated the outputs of aldosterone, cortisol, and corticosterone (Wintour et al. 1977). A much smaller, nonsignificant steroidogenic response to angiotensin II was observed in adrenal tissue of fetuses older than 100 days but younger than 140 days. At term, again, angiotensin II significantly stimulated aldosterone and cortisol outputs and was a more potent stimulator of aldosterone production than ACTH. Lowering the sodium concentration in the incubation medium from 140 to 130 mmol/l had no effect on aldosterone output by fetal adrenals in vitro at either a low (4 mmol/l) or a high (8 mmol/l) potassium concentration. Sodium depletion of pregnant ewes did not induce any measurable changes in the corticosteroid pattern produced by incubated fetal adrenals in vitro. In chronically catheterized ovine fetuses, intravenously infused ACTH did not affect peripheral blood corticosteroid concentrations between 90 and 120 days of gestation, but it significantly raised the blood levels of aldosterone, corticosterone, and cortisol in 120- to 142-day-old fetuses (Wintour et al. 1977). However, according to a later report by the same investigators (E. H. Brown et al. 1978), ACTH infusions had no consistent effect on the fetal peripheral blood aldosterone concentration at any stage of gestation but significantly raised blood corticosterone levels in all fetuses and blood cortisol levels in fetuses of 135 days or older. Increments in the plasma potassium concentration which caused marked elevations in the plasma aldosterone concentrations of pregnant or nonpregnant ewes did not raise blood aldosterone concentrations in fetal sheep (Wintour et al. 1979). In chronically catheterized fetal lambs between 131 and 143 days of gestation, hypoxemia elicited a marked rise in plasma aldosterone, which was associated with a rise in the plasma potassium and angiotensin II concentrations (Robillard et al. 1984). This increase in the aldosterone level was most likely mediated by ACTH, since it was suppressed by treatment with dexamethasone but not by treatment with the converting-enzyme inhibitor captopril. According to aldosterone blood clearance rates measured during constant infusions of tritiated aldosterone into pregnant sheep or into chronically cannulated fetal sheep, 4% of the maternal production rate was transferred to the fetus, whereas 29% of the fetal production rate was transferred to the maternal compartment (Wintour et al. 1980). In sodium-deplete sheep, more than 80% of the aldosterone in fetal blood was of fetal origin if the actual fetal concentration of aldosterone was above 1.5 ng/dl.

Cultured adrenals of fetal calves (3-5.5 months of gestation) converted ^{14}C-labeled acetate to cortisol, corticosterone, and 11-dehydrocorticosterone but not to aldosterone (Chouraqui and Weniger 1970). In contrast, cultured fetal rabbit (26 days of gestation) and mouse (13.5-19 days of gestation) adrenals incorporated this precursor into aldosterone. Adrenal and plasma aldosterone of fetal guinea pigs increased severalfold between gestation days 62 and 67, to reach a peak value on the first day after birth (Giry and Delost 1977). A marked increase in plasma aldosterone during the last 2 days of gestation with a peak value on the day after birth was also observed in fetal mice (Dalle et al. 1978). This was paralleled by a decrease in the fetal plasma corticosterone concentration. Sodium restriction of pregnant rats during the last week of gestation significantly raised the aldosterone output without altering the corticosterone output of incubated fetal (term) adrenals but did not alter the aldosterone production by maternal adrenals (G. Schneider and Mulrow 1973). The conversion of radiolabeled progesterone to 18-hydroxycorticosterone and aldosterone by adrenal glands of rat fetuses from normal intact mothers became measurable at the 21st gestation day (Klepac et al. 1977).

Adrenals of 15-day-old chick embryos converted tritiated progesterone to aldosterone and 18-hydroxycorticosterone (Pedernera and Lantos 1973). According to Lehoux (1974), aldosterone was formed from tritiated progesterone by whole adrenocortical homogenates of 18-day-old (but not of 12-day-old) chick embryos and from tritiated corticosterone by adrenocortical mitochondria of 14-day-old chick embryos.

5.9.2 Infancy and Childhood

In normal human newborns, both the secretion rate and the plasma concentrations of aldosterone are markedly elevated during the first days of life (Weldon et al. 1967; Beitins et al. 1972b; Dillon et al. 1976; Sippell et al. 1978, 1980; Hubl et al. 1978; Sparano et al. 1978). Plasma aldosterone concentrations were also found to be high in preterm newborns with and without respiratory distress and were not related to gestational age (Siegel et al. 1974). When sodium was given in larger quantities to preterm infants with respiratory distress, plasma aldosterone became inversely correlated with sodium intake and sodium excretion (Siegel et al. 1974). Both plasma renin activity and plasma potassium concentrations were also elevated in newborns but were not directly correlated with plasma aldosterone concentrations (Dillon et al. 1976; Godard et al. 1976; Siegler et al. 1977; Sparano et al. 1978). Plasma cortisol, cortisone, and corticosterone concentrations were not elevated in newborns (Sippell et al. 1978). Plasma aldosterone was substantially higher in newborns of sodium-deficient mothers than in newborns of sodium-replete mothers (Beitins et al. 1972b).

According to Kowarski et al. (1974), the mean plasma aldosterone concentration was higher in 1-week-old to 1-year-old children than in older children or newborns. This was only partly confirmed by the other investigators cited above, who found plasma aldosterone to be highest in newborns and observed a continuous decrease of this parameter throughout infancy and childhood until puberty. A continuous decrease in the mean plasma aldosterone concentration throughout childhood and a negative correlation between age and plasma aldosterone were also observed by P. Stark et al. (1975), Dillon and Ryness (1975), Van Acker et al. (1979), Zadik and Kowarski (1980), Vincent et al. (1980), Goldfarb et al. (1981), and Fiselier et al. (1983). All these investigators also observed a continuous decrease in plasma renin activity from very high values in newborns to adult values in prepubertal children. The close parallelism of the two parameters resulted in a relatively constant ratio of plasma aldosterone/plasma renin activity throughout childhood (Zadik and Kowarski 1980). The aldosterone secretion rate of children was negatively correlated with sodium intake (Loras et al. 1967; Bryan 1971). Goldfarb et al. (1981) found a significant negative correlation between plasma renin activity and urinary sodium excretion in children. In 3- to 12-month-old children, there was a high correlation between plasma aldosterone and plasma renin activity and between plasma aldosterone and plasma potassium concentrations (Siegler et al. 1977). Considering children of all age groups, plasma aldosterone was significantly correlated with the plasma renin activity but not with the plasma potassium concentration (P. Stark et al. 1975). After correction for age, the correlation between plasma aldosterone and plasma angiotensin II was still significant (Fiselier et al. 1983). Plasma aldosterone and plasma renin activity significantly increased in response to tilting and oral administration of furosemide in all children (Van Acker et al. 1979). The fractional increase of these two parameters was greatest in older children in whom the lowest baseline values were found.

Plasma aldosterone concentration of newborn calves was relatively high immediately after birth (120 pg/ml) and decreased sharply to 25 pg/ml within 12 h (Safwate et al. 1980). After a small rise on day 3, plasma aldosterone dropped into the adult range (13 pg/ml) by day 7. These changes in plasma aldosterone were not accompanied by any changes in the plasma sodium and potassium concentrations. In newborn lambs, peaks in plasma aldosterone were observed at birth and on the 4th postnatal day (Moncaup et al. 1980). At an age of 5–10 days, a potassium chloride infusion elicited a significant rise in plasma aldosterone in lambs (Siegel 1979). Also in newborn lambs (8–21 days), an infusion of the converting-enzyme inhibitor captopril suppressed the plasma aldosterone to 50% of the baseline concentration (Robillard et al. 1983). Such an effect of captopril was not observed in chronically catheterized sheep fetuses. Thus, potassium and the renin-angiotensin system become operative regulators of aldosterone secretion in the sheep after birth. In newborn guinea pigs, plasma aldosterone was very high at birth and increased further during the first day of life (Malinowska and Nathanielzs 1974; Giry and Delost 1977). It decreased again on days 2–5, to reach adult levels by 14–25 days. Plasma protein binding and the metabolic clearance rate of aldosterone remained unchanged in neonatal guinea pigs from birth to day 8, and the values were similar to those found in adults (Giry and Delost 1980). This indicated that the high plasma aldosterone level seen on the first day was representative of a high secretion rate. High concentrations in plasma aldosterone, reaching a peak 12 h after birth, were also observed in newborn mice (Dalle et al. 1978). In normal Wistar rats, the plasma aldosterone concentration was relatively low on day 5 after birth and remained relatively constant up to day 30 (Mullins et al. 1982). In a normotensive Wistar-Kyoto strain, plasma aldosterone was similarly low on days 5 and 10 but significantly higher after 15 days. In spontaneously hypertensive rats, plasma aldosterone was significantly elevated on postnatal days 5, 10, and 20. An interesting difference between immature and adult rats in the adaptation of urinary aldosterone excretion to varying sodium intakes was observed by Aviv et al. (1982). In rats kept on a low sodium intake during the 3rd–7th week of life, urinary aldosterone excretion became markedly elevated. When these animals received a high sodium intake during weeks 8 and 9, their aldosterone excretion decreased but remained significantly higher than the aldosterone excretion of rats that had continuously received the high sodium diet from the 3rd week. In contrast, urinary aldosterone excretion rates returned to control values after 2 weeks (13 and 14) of high sodium intake in rats that had been sodium deprived during weeks 8–12.

5.9.3 Age

The aldosterone secretion rate was 56% lower and the aldosterone metabolic clearance rate 21% lower in a group of elderly persons (ages 67–88) than in a group of younger subjects (ages 18–35), according to Flood et al. (1967). A decrease in the aldosterone secretion rate with increasing age was observed in hypertensive patients by Tuck et al. (1973). Plasma renin activity and plasma aldosterone concentrations were lower in a group of elderly subjects (age 62–70) than in a group of young subjects (age 20–30; Weidmann et al. 1975c). The differences were very small and not statistically significant while the subjects were supine and on a normal sodium intake (120 mmol/day), but they became more pronounced and statistically significant when the subjects were upright and during sodium depletion. Body sodium and fluid balance were equal in the two age-groups. Plasma aldosterone and cortisol responses to ACTH were also the same in the two groups. Crane and Harris (1976), who studied groups of at least 20 normal subjects from each decade from age 20 to 80, observed age-related decreases in plasma renin activity (supine and upright) and in urinary aldosterone excretion under normal as well as under low sodium intake. Similar decreases in plasma renin activity, but not in urinary aldosterone excretion, were observed in middle-aged (47–61) normotensive and hypertensive persons by Noth et al. (1977b). According to a later report by the same investigators (Tan et al. 1978), plasma aldosterone significantly decreased with increasing age. Age-dependent decreases in upright plasma aldosterone and in urinary aldosterone excretion were also observed by Salvetti et al. (1977) and by Hegstad et al. (1983). The urinary excretion rate and plasma concentration of norepinephrine were found to be positively correlated with age and to be increased in elderly subjects (Weidmann et al. 1977, 1978b). However, the possible causal relationship between aldosterone secretion and urinary or plasma norepinephrine is complex, since both plasma aldosterone and plasma norepinephrine were negatively correlated with urinary sodium excretion (Weidmann et al. 1978b). In a group of elderly normal subjects plasma aldosterone was positively correlated with urinary norepinephrine excretion (Matsuda 1984). Mean plasma and blood volumes and exchangeable sodium were the same in young, middle-aged, and elderly subjects (Weidmann et al. 1977). However, elderly persons showed a greater weight loss on a low-sodium diet (Takeda et al. 1980). The response in plasma aldosterone to angiotensin II was proportional to the baseline plasma aldosterone concentration and was thus smaller in elderly subjects than in young and middle-aged ones (Takeda et al. 1980). On the other hand, elderly subjects showed a greater blood pressure response to the peptide.

5.9.4 Diurnal Rhythms

Circadian variations in aldosterone secretion were first described by investigators measuring urinary aldosterone excretion rates (Venning et al. 1956b; Luetscher and Lieberman 1958; A. F. Muller et al. 1958b; Bartter et al. 1962; Wolfe et al. 1966). They observed a higher aldosterone excretion during the day than during the night and a higher aldosterone excretion in the morning than in the afternoon in most normal human subjects. This diurnal variation could be abolished by keeping the subjects recumbent or inactive during the daytime (A. F. Muller et al. 1958b; Wolfe et al. 1966). When normal subjects assumed the upright posture at 8 p.m. rather than at 8 a.m., their aldosterone excretion became higher at night than it was during the day (Wolfe et al. 1966). Diurnal variation of aldosterone excretion was absent in patients with pituitary insufficiency unless they were treated with cortisone (A. F. Muller et al. 1958b). On the other hand, the aldosterone excretory rhythm of normal persons was not affected by treatment with dexamethasone at a dosage sufficient to suppress urinary excretion of 17-hydroxycorticoids (Wolfe et al. 1966). Under most of these experimental conditions, changes in plasma renin activity tended to parallel those in urinary aldosterone excretion (Wolfe et al. 1966). Thus, according to these early studies, posture is the dominant factor in determining circadian rhythmicity of aldosterone secretion in normal upright man, and the effect of posture is mediated mainly by the renin-angiotensin system.

The development of radioimmunoassays for plasma aldosterone led to the discovery of other diurnal rhythms of aldosterone secretion, which are independent of posture and are also manifest during recumbency. These diurnal rhythms are often synchronous with the diurnal rhythms in plasma cortisol but are generally not dependent on ACTH secretion. In supine normal subjects on a normal sodium intake, Michelakis and Horton (1967) found plasma aldosterone and plasma renin activity to be higher at 6 a.m. and at midnight than at noon and at 6 p.m. Most other studies on diurnal variations of plasma aldosterone in supine normal man on a normal or low sodium intake have confirmed the occurrence of a peak in the early morning but have shown low concentrations at midnight (G.H. Williams et al. 1972b; Katz et al. 1972a, 1975a; Kowarski et al. 1975; Armbruster et al. 1975; James et al. 1976; Lightman et al. 1981; Kawasaki et al. 1983). However, a rise in plasma aldosterone occurring at the onset of sleep has been observed in sodium-replete and in sodium-deficient subjects by Breuer et al. (1974) and by Vagnucci et al. (1974) respectively. Aldosterone, cortisol, and renin are secreted in an episodic manner (Katz et al. 1972a; Hellman et al. 1970; Junnila et al. 1975). In sodium-replete untreated supine subjects, some but not all of the secretory bursts of aldosterone were found to be coincident with secretory bursts of cortisol. In most of the normal, sodium-replete, untreated subjects studied by Katz et al. (1975a), Weinberger et al. (1975), Armbruster et al. (1975), Kowarski et al. (1975), Sulon et al. (1978), and Lightman et al. (1981), there was a significant correlation between the diurnal concentrations of aldosterone and cortisol. In some of the subjects, plasma aldosterone was also significantly correlated with plasma renin activity. According to Cugini et al. (1981), there was a significant lag of the acrophase of plasma cortisol behind the acrophase of plasma aldosterone in sodium-replete as well as in sodium-deficient persons. On the other hand, there was no significant difference between the acrophases of plasma aldosterone and plasma renin activity. However, Lightman et al. (1981) did not observe a diurnal rhythm of plasma renin activity or a correlation between plasma aldosterone and plasma renin activity in sodium-replete recumbent subjects. According to these investigators, plasma renin activity declined sharply during periods of REM sleep. In contrast, Rubin et al. (1978) had observed the highest plasma aldosterone concentrations during the REM phases of sleep. In sodium-depleted subjects, plasma aldosterone was still correlated with plasma cortisol, according to Katz et al. (1975a) but not according to Armbruster et al. (1975) and Weinberger et al. (1975). Significant correlations between plasma aldosterone and renin activity were seen in the sodium-depleted subjects studied by Vagnucci et al. (1974), Katz et al. (1975a), and Armbruster et al. (1975), but in only one of the three sodium-depleted subjects studied by Weinberger et al. (1975). Under suppression of ACTH secretion by treatment with dexamethasone, secretory bursts of aldosterone were diminished neither in size nor in frequency. However, there are conflicting reports on the effect of glucocorticoid treatment on aldosterone diurnal rhythms. According to Katz et al. (1975a), Armbruster et al. (1975), Weinberger et al. (1975), and James et al. (1976), a normal rhythm with peak values in the early morning was also seen in dexamethasone-treated subjects. However, in the subjects studied by Takeda et al. (1984), aldosterone rhythmicity was completely abolished by dexamethasone treatment. Ølgaard et al. (1977a) observed an inverse circadian rhythm of plasma aldosterone and of plasma renin activity with peak values in the evening in steroid-treated patients with kidney transplants or nonrenal diseases. Generally, plasma aldosterone correlated better with plasma renin activity in steroid-treated than in untreated persons. Treatment with the dopaminergic agonist bromocriptine did not affect the diurnal rhythm of plasma aldosterone, although it induced a significant shift in the acrophase of plasma renin activity (degli Umberti et al. 1979; Takeda et al. 1984). The diurnal rhythm of plasma aldosterone disappeared when normal supine subjects took their meals (and thus sodium) during the night instead of during the day (Kunita et al. 1976). In anephric subjects, plasma aldosterone remained relatively constant throughout the day (irrespective of posture and sleep), in spite of a normal diurnal rhythm in plasma cortisol (Cooke et al. 1979). The diurnal rhythm of plasma aldosterone in sodium-replete, supine normal subjects was not altered under converting-enzyme inhibition with captopril (Takeda et al. 1984). Treatment with the beta-adrenergic blocker propranolol completely abolished the diurnal rhythms of plasma aldosterone, cortisol, and renin activity in normal supine human subjects (Cugini et al. 1977).

There are only a few reports on diurnal rhythms of plasma aldosterone in animals. In rams, circadian variations of plasma aldosterone were synchronous with those of plasma cortisol, with the highest levels in the morning and the lowest levels at midnight (Mesbah and Brudieux 1982).

In sodium-replete rats, circadian variations of plasma aldosterone were parallel to circadian variations in plasma corticosterone, with peak concentrations in the evening, at the beginning of darkness, i.e., at 6 p.m. in the study of Hilfenhaus (1976) and at 4 p.m. in the study of Gomez-Sanchez et al. (1976). The diurnal rhythm of plasma renin activity was also synchronous with those of the two plasma steroids, according to Hilfenhaus (1976) but not according to Gomez-Sanchez et al. (1976; peak value of plasma renin activity at noon). Inversion of the light-dark circle induced a complete inversion of the circadian rhythms of the three endocrine parameters (Hilfenhaus 1976). According to both groups of investigators, plasma renin activity lost its diurnal rhythmicity under dietary sodium restriction, whereas the normal diurnal rhythm of plasma corticosterone was maintained. Only Gomez-Sanchez et al. (1976) and not Hilfenhaus (1976) observed a shift of the peak of plasma aldosterone from 4 p.m. to 8 p.m. in sodium-restricted rats. The diurnal rhythms of plasma aldosterone and corticosterone concentrations were no longer apparent in rats treated with propranolol or dexamethasone (Hilfenhaus 1974, 1976). Normal circadian rhythmicity in plasma aldosterone was maintained in rats with renovascular hypertension (two-kidney, one clip) in which the circadian rhythm of plasma renin activity had been abolished (De Forrest et al. 1979).

A diurnal rhythm of plasma corticosteroid concentrations was also observed in frogs living in a natural environment (Leboulenger et al. 1982). High levels of aldosterone and corticosterone were measured during the night and low levels during daylight. Variations in aldosterone and corticosterone concentrations were synchronous, but the amplitude of corticosterone fluctuations was 3.5 times greater than that of aldosterone fluctuations.

5.9.5 Posture

A stimulatory effect of upright posture on urinary aldosterone excretion in normal human subjects was first observed by A.F. Muller et al. (1958b) and by Gowenlock et al. (1959). Aldosterone excretion was also markedly stimulated by passive 82° head-up tilt (E.L. Cohen et al. 1967) but not by sitting (Gowenlock et al. 1959). Later, substantial increases in aldosterone plasma concentrations and blood production rates in response to standing or sitting were reported by Balikian et al. (1968). Standing - but not sitting - also caused a statistically significant decrease in the metabolic clearance rate of aldosterone, which was probably due to a decreased hepatic blood flow (Bougas et al. 1964). However, the increases in blood production rate (eight fold) contributed substantially more to the rise in plasma aldosterone concentration than did the decreases in metabolic clearance rate (−23%; Balikian et al. 1968).

The active assumption of an upright posture or passive tilt also induced marked increases in plasma renin (J.J. Brown et al. 1966b; E.L. Cohen et al. 1967). A close correlation between posturally induced alterations in plasma renin or angiotensin II and urinary or plasma aldosterone has been observed by a great number of investigators (among others: Luetscher et al. 1967; R.D. Gordon et al. 1967a; Michelakis and Horton 1970; G.H. Williams et al. 1972b; Tuck et al. 1975; Reck et al. 1975; Sassard et al. 1976; Kosunen and Pakarinen 1978; Messerli et al. 1978). Most of their results are consistent with the view that postural effects on aldosterone secretion are predominantly mediated by the renin-angiotensin system.

Evidence for an important role of the sympathetic nervous system in the regulation of renin and aldosterone secretion during postural changes was first presented by R.D. Gordon et al. (1967a). These investigators elicited marked increases in plasma renin activity and urinary aldosterone excretion in supine human subjects with a 4-h infusion of a mixture (10:1) of norepinephrine and epinephrine. They also observed that upright posture failed to raise plasma renin activity and urinary aldosterone in a patient with autonomic insufficiency, unless he received a catecholamine infusion. A 1-h infusion of a norepinephrine-epinephrine 10:1 mixture also caused marked rises in plasma renin activity as well as in plasma aldosterone in normal supine subjects, according to Michelakis and Horton (1970). However, a short-term infusion of a pressor dose of norepinephrine induced only a minor (although statistically significant) increase in plasma renin activity, which was not accompanied by a significant change in plasma aldosterone (Beretta-Piccoli et al. 1980b). Reports on the effect of posture on urinary and plasma catecholamines are conflicting. Upright posture stimulated the urinary excretion of norepinephrine, according to Vagnucci et al. (1969), but not according to Campese et al. (1980). After 10 min of standing there were significant increases in the plasma concentrations of norepinephrine and epinephrine in the subjects studied by Campese et al. (1980). Under the same conditions, Sowers et al. (1982h) observed a rise in plasma norepinephrine but not in plasma epinephrine.

The posture-induced stimulation of renin and aldosterone secretion seems to depend on blood volume redistribution within the organism. Thus, urinary aldosterone did not increase in response to standing in subjects immersed to the neck in water (Gowenlock et al. 1959), and plasma renin activity was not stimulated by standing in subjects whose lower abdomen, hips, and legs had been firmly wrapped in bandages (R.D.Gordon et al. 1966). Generally, sodium deficiency enhanced the posture-induced increments in plasma aldosterone and plasma renin activity (Katz et al. 1972b, 1972c; Tuck et al. 1975), and the responsiveness of both parameters was closely correlated to their basal values in normal subjects as well as in patients with renal diseases (Messerli et al. 1978). On the other hand, prior stimulation of aldosterone by standing markedly diminished the response in plasma aldosterone to subsequent ACTH or angiotensin II infusions (Katz et al. 1972c). According to Tuck et al. (1975), changes in plasma aldosterone followed changes in plasma renin activity and angiotensin II with a delay of 20-30 min after transition from supine to upright posture, as well as after the resumption of the supine position. In contrast, plasma aldosterone increased almost synchronously with plasma renin activity during orthostatism and remained elevated within an hour of return to the supine posture in spite of a marked fall in plasma renin activity in the subjects studied by Sassar et al. (1976). Finally, Reck et al. (1975) observed a significant rise in plasma renin activity within 15 min of orthostasis in all their subjects, but a similarly rapid increase in plasma aldosterone in only some of them, i.e., in those who also showed a simultaneous increase in plasma potassium or plasma cortisol. In the other subjects, aldosterone rose after a delay of 30-60 min. Consistent increases in plasma potassium and decreases in plasma sodium in response to ambulation or head-up tilt were observed in normal subjects by Messerli et al. (1978) and by Sonkodi et al. (1981). An increase in plasma potassium was also seen by Sassard et al. (1976). However, in many instances, these changes were very small. Neither an increase in plasma potassium nor increased secretion of ACTH (according to unchanged plasma cortisol) could have accounted for the small but statistically significant rise in plasma aldosterone in anephric patients (Messerli et al. 1978; see Sect. 5.5.4, p. 166, for other observations in bilaterally nephrectomized subjects).

After 3 days of continuous bed rest there was an exaggerated rise in plasma aldosterone in response to 4 h of ambulation (Chavarri et al. 1977b). In these subjects, plasma renin activity was already substantially increased in the supine posture and did not rise further after standing. After longer periods of bed rest (5-14 days), plasma renin activity rose excessively in response to standing or head-up tilt, whereas the increases in plasma aldosterone were within normal limits (Melada et al. 1975).

5.9.6 Exercise

In normal human subjects, 4 h of exercise during recumbency had little effect on plasma renin activity and urinary aldosterone secretion (E.L.Cohen et al. 1967). In upright man, however, exercise of varying intensity and duration consistently led to significant increases in aldosterone secretion and in the activity of the renin-angiotensin system. Thus, aldosterone secretion was stimulated by exercise on a bicycle ergometer (Maher et al. 1975; Bonelli et al. 1977; Lijnen et al. 1979; Fagard et al. 1984), isometric handgrip (Sowers et al. 1982h), short-distance running (Kosunen and Pakarinen 1976), marathon running (Newmark et al. 1976), and cross-country skiing (Sundsfjord et al. 1973). Generally, the observed rises in plasma aldosterone were accompanied by simultaneous and commensurate increases in plasma renin activity and in plasma norepinephrine concentration. This suggested that exercise stimulated the renin-angiotensin system through sympathetic activation and that increased aldosterone secretion was due mainly to elevated blood levels of angiotensin II. However, other regulatory factors seemed to be involved as well. Thus, exercise-induced elevations of plasma cortisol indicated a considerable contribution of ACTH to the overall stimulation of aldosterone secretion (Sundsfjord 1973; Newmark et al. 1976). Treatment with beta-adrenergic blocking agents suppressed the exercise-induced increase in aldosterone secretion less extensively than the activation of the renin-angiotensin system (Bonelli et al. 1977; Lijnen et al. 1979). On the other hand, treatment with the dopaminergic agonist bromocriptine (Sowers et al. 1982h) or with the serotonin antagonist ketanserin (Fagard et al. 1984) blunted the exercise-induced rise in plasma aldosterone but not the rise in plasma renin activity. Moreover, the hypnotic suggestion of running caused an increase in the plasma angiotensin II concentration (comparable to the one seen in the actual running exercise) without any stimulation of aldosterone secretion (Kosunen et al. 1977). As will be shown later (see below), major determining factors for the increase in aldosterone secretion during exercise are the environmental temperature, the loss of water and electrolytes by sweating, and their replacement.

In a group of normal human subjects, prolonged exercise of 5 days of hill walking led to marked increases in the plasma aldosterone concentration and in plasma renin activity, which were associated with a positive sodium balance, water retention, and plasma volume expansion (Milledge et al. 1982).

In thoroughbred horses, galloping at maximal pace for 2.5 km induced a sixfold increase in plasma aldosterone, which was accompanied by small rises in plasma renin activity (+29%) and in plasma cortisol (+41%; Guthrie et al. 1982). Most likely, an increase in plasma potassium from 3.9 to 4.4 mmol/l substantially contributed to the stimulation of aldosterone secretion. ACTH probably did not contribute to the increase in aldosterone output, since exogenous ACTH lowered plasma aldosterone while raising plasma cortisol and corticosterone in these animals.

5.9.7 Heat and Cold

Acute heat stress (1 h at 46°-51° C) led to marked increases in plasma aldosterone and in plasma renin activity in normal human subjects kept on a high- or a low-sodium diet (Bailey et al. 1972). Hematocrit was not altered. A 20-min exposure to 85°-90° C in a Finnish sauna also induced substantial rises in plasma renin activity and angiotensin II concentration, as well as a more extensive but somewhat delayed increase in plasma aldosterone; these were not accompanied by measurable changes in hematocrit or in the serum sodium and protein concentrations (Kosunen et al. 1976). According to these observations, the stimulation of the renin-angiotensin system and of aldosterone secretion by heat need not be due to water and sodium losses.

Heat potentiated the effects of exercise on plasma renin activity and aldosterone. A 30-min period of light exercise, which did not affect plasma renin activity and had only a minor aldosterone-stimulating effect at 25° C in normal human subjects, markedly raised both plasma renin activity and aldosterone concentration when performed at 50° C (Finberg and Berlyne 1976). Light exercise in a thermoneutral environment had no effect on plasma renin activity and aldosterone concentration but led to significant increases in both parameters in a hot, wet and particularly in a hot, dry environment (Francesconi et al. 1983). The observed endocrine responses to exercise in the heat depended on water and salt replacement and on acclimation. Increments in both plasma renin activity and aldosterone concentration were greater in hypohydrated subjects than in euhydrated subjects (Francesconi et al. 1983). Two hours of exercise in the heat elicited greater increases in plasma aldosterone and in plasma renin activity in subjects receiving only water supplementation than in subjects also receiving an electrolyte solution (Francis and MacGregor 1978). According to Finberg and Berlyne (1977), the responses in plasma renin activity and aldosterone concentration to exercise in the heat were markedly lower in summer than in winter (natural acclimatization) and were also blunted by a week of daily repeated heat exposure in winter (artificial acclimation). However, according to Francesconi et al. (1983), 10 days of acclimation led to diminished responses in plasma renin activity but did not alter the responses in plasma aldosterone, in either euhydrated or hypohydrated subjects. Irrespective of the responses in plasma renin activity and plasma aldosterone to exercise in the heat, plasma cortisol did not increase (Finberg and Berlyne 1977). Treatment with propranolol significantly decreased the response of the heart rate to exercise in the heat but had no consistent effect on the increases in plasma renin activity and aldosterone concentration (Finberg and Berlyne 1976). The observation that plasma aldosterone rose markedly in a propranolol-treated subject in whom plasma renin activity did not increase indicated that angiotensin II was not the only aldosterone stimulator operative during exercise in the heat.

Exposure of cattle to heat (35° C) resulted in a rapid increase in the plasma antidiuretic hormone (ADH) concentration and in a slow reduction of plasma aldosterone, which became significant after 24 h (El-Nouty et al. 1980). Dehydration under heat induced a slow increase in plasma aldosterone concentration but not into the normal range.

Ambient temperature influenced the response in plasma aldosterone to dehydration of the camel (Finberg et al. 1978). During mild spring weather (23.5° C), complete water deprivation for 10 days brought about an increase in the plasma sodium concentration from 138 to 147 mmol/l and resulted in marked and significant increases in plasma aldosterone concentration and renin activity. Severe dehydration in hot summer weather (36.9° C) raised the plasma sodium from 146 to 157 mmol/l and led to a greater increase in plasma renin activity but to a blunted increase in plasma aldosterone. At both temperatures, rehydration induced a marked further increase in plasma aldosterone but a return of plasma renin activity to control values.

Neither plasma aldosterone nor plasma renin activity was significantly affected by exposure of rats kept on food with a normal sodium content to prolonged heat (35.5° C for 7 days), in spite of a significant rise in the plasma potassium concentration (Francesconi and Hubbard 1985). In rats kept on a sodium-deficient diet, plasma aldosterone was markedly elevated under basal conditions while plasma renin activity was normal. During exposure to heat there were substantial and sustained rises in both parameters, which became manifest after 6 (aldosterone) and 24 (renin) h respectively.

Exposure of normal human subjects wearing summer clothes to an ambient temperature of 4° C for 1 h did not affect plasma aldosterone despite increases in blood pressure and plasma levels of norepinephrine, and cortisol and a fall in plasma renin activity (Hiramatsu et al. 1984). On the other hand, immersion of the hands in cold water (0° C) for 10 min elicited a marked rise in plasma aldosterone, which was preceded by a rise in plasma norepinephrine. It was accompanied by a parallel increase in plasma cortisol but was not associated with any changes in plasma renin activity or angiotensin II concentration. Most likely, the cold-induced increase in aldosterone secretion was mediated by ACTH, because it was abolished by a single dose of 2 mg dexamethasone given 8 h before the test.

5.9.8 Altitude

The effects of high altitude - real, or simulated in a decompression chamber - on aldosterone secretion in normal man have been extensively studied by several groups of investigators. Although some of the results are conflicting, a majority of investigators observed a decrease in urinary excretion of aldosterone-18-glucuronide, in the aldosterone secretion rate, or in plasma aldosterone concentrations in response to a few days of exposure to the atmospheric conditions of high altitude (Ayres et al. 1961; Slater et al. 1969a, 1969b; Hogan et al. 1973, Maher et al. 1975; Pines et al. 1977; Sutton et al. 1977; Keynes et al. 1982; Milledge et al. 1983a; Maresh et al. 1984). Hypobaria alone without hypoxia did not affect aldosterone excretion, either under basal conditions or under stimulation by sodium restriction (M. Epstein and Saruta 1972, 1973). On the other hand, plasma aldosterone was lowered only by hypobaric but not by isobaric hypoxia (Heyes et al. 1982). Time of adaptation seems to be of critical importance. Thus, urinary aldosterone excretion was lowest on the 2nd day of a 3-day exposure to a simulated altitude of 4000 m and rose again on the 3rd day (Hogan et al. 1973). Plasma aldosterone levels at rest and after moderate exercise, as well as urinary aldosterone excretion, were markedly suppressed on day 1 at an altitude of 4300 m but returned to the values seen at sea level on day 11 (Maher et al. 1975). In a group of subjects, plasma aldosterone remained unchanged within a few days of exposure to an altitude of 3300 m (Frayser et al. 1975). In the same individuals, a higher altitude, 5700 m, elicited temporary increases in plasma aldosterone, renin activity, and cortisol. After 2-4 weeks at altitudes of 4500 and 6300 m, respectively, plasma aldosterone concentration and plasma renin activity had returned to within ±10% of sea-level values (Milledge et al. 1983a, 1983b). Exposure to moderate altitude (2000 m) resulted in a slow increase in plasma aldosterone, which became significant on day 10 (Humpeler et al. 1980). It was accompanied by increases in plasma cortisol and potassium concentrations and by a nonsignificant drop in plasma renin activity. According to Jung et al. (1971), the response in aldosterone secretion to high altitude (3800 m) was age dependent. Thus, a significant drop in plasma aldosterone was seen in four middle-aged subjects (age above 40), whereas in four younger subjects (age below 40) plasma aldosterone either remained unchanged or showed a tendency to rise.

Reports on effects of high altitude on the renin-angiotensin system are equivocal. Decreases in plasma renin activity in proportion to the decreases in aldosterone secretion were observed by Hogan et al. (1973) and Maher et al. (1975). Keynes et al. (1982) also found a decrease in plasma renin activity, but its onset was delayed, and it was less extensive than the drop in plasma aldosterone. Insignificant changes in plasma renin activity were observed by Slater et al. (1969b); Sutton et al. (1977) and Heyes et al. (1982). Acute exposure to hypobaric hypoxia for 1 h resulted in a rise in plasma renin activity which was accompanied by a drop in blood angiotensin converting-enzyme activity (Milledge and Catley 1982). At an altitude of 4500 m, Milledge et al. (1983a) observed a temporary drop in blood angiotensin converting-enzyme activity which paralleled a drop in plasma aldosterone. This phenomenon, however, was not responsible for a blunted response in plasma aldosterone to exercise (in spite of an exaggerated response in plasma renin activity), because the same discrepancy between plasma aldosterone and plasma renin activity responses to exercise was also observed after a few weeks of adaptation to an altitude of 6300 m, when converting-enzyme activity had returned to normal (Milledge et al. 1983b).

Suppression of aldosterone secretion by high altitude is probably not mediated by hypokalemia or decreased ACTH levels. Thus, in the study of Slater et al. (1969a) a decreased aldosterone secretion rate during exposure to high altitude was accompanied by potassium retention and elevated plasma potassium levels. Decreased urinary potassium excretion at high altitude was also observed by Hogan et al. (1973) and Sutton et al. (1977). However, according to Keynes et al. (1982), there were no significant changes in urinary potassium excretion or in plasma potassium concentrations during a period of high altitude. A decreased aldosterone secretion rate due to high altitude contrasted with increased cortisol secretion rates (Slater et al. 1969b). Plasma cortisol levels and urinary cortisol excretion remained normal at high altitude in spite of decreases in plasma aldosterone (Maher et al. 1975; Keynes et al. 1982).

Chronic hypobaric hypoxia of rats in a decompression chamber over a period of 7 months led to significant increases in plasma renin activity but did not significantly alter plasma aldosterone concentrations (I.H. Martin et al. 1982).

5.9.9 Fasting and Refeeding

In lean and obese human subjects, fasting induces a period of marked natriuresis and negative sodium balance, which lasts about a week and is followed by a period of restored sodium conservation if fasting is continued (Bloom and Mitchell 1960; Bloom et al. 1966; Veverbrants and Arky 1969). In contrast, refeeding is characterized by antinatriuresis and sodium retention. During the natriuretic phase of fasting, urinary excretion of aldosterone-18-glucuronide, the aldosterone secretion rate, and plasma aldosterone concentrations were substantially increased (Hansen et al. 1967; Smith et al. 1969; Kolanowski et al. 1970; Chinn et al. 1970; Boulter et al. 1974). According to two studies (Chinn et al. 1970; Boulter et al. 1974), this increase in aldosterone secretion was not mediated by the renin-angiotensin system, although it was most likely a normal response to the stimulus of sodium deficiency. In the study by Chinn et al. (1970), a period of sodium restriction (5 days) preceded the period of total fasting (12 days). During the first days of fasting, plasma aldosterone rose markedly, whereas the plasma renin concentration decreased in eight of ten subjects and remained unchanged in the other two despite urinary sodium loss. Both plasma aldosterone and renin concentrations increased during later stages of fasting and dropped together in response to resumed food and sodium intake. The subjects studied by Boulter et al. (1974) were kept on a constant water and electrolyte intake (50 mmol Na^+ per day) during fasting and refeeding. During the natriuretic phase of fasting the aldosterone secretion rate rose, while plasma renin activity decreased. The opposite changes in these two parameters were seen during the refeeding period, when a decrease in the aldosterone secretion rate contrasted with a rise in plasma renin activity.

Total fasting of potassium-deficient rats for 2 days did not affect the final steps of aldosterone biosynthesis, whereas treatment of such animals with a sodium- and potassium-deficient diet resulted in a three fold increase in the conversion of corticosterone to aldosterone (J. Müller and Baumann 1974; Haldy and Müller 1981b). Urinary sodium excretion was the same under both experimental conditions. Plasma renin activity increased with the diet but was significantly suppressed by fasting (Haldy and Müller, unpublished observations).

5.10 Estrogens and Progestogens

5.10.1 Exogenous Estrogens

Natural and synthetic estrogens probably do not directly influence aldosterone biosynthesis at physiological concentrations (see Sect. 4.10.2). Nevertheless, according to known side effects of high doses of endogenous or exogenous estrogens, they would be expected to modify aldosterone secretion and its regulation in different indirect ways, in particular through

1. Stimulation of the synthesis of renin substrate (Helmer and Griffith 1952)
2. Stimulation of the synthesis of transcortin and perhaps other plasma aldosterone carrier proteins (Sandberg and Slaunwhite 1959)
3. Alteration of aldosterone metabolism (Layne et al. 1962)
4. Salt and water retention (Preedy and Aitken 1956a, b)

Peterson et al. (1960) observed that treatment of three normal human subjects (two men, one woman) with *ethinyl estradiol* (0.5 mg/day) resulted in marked elevations of the plasma cortisol concentrations but did not alter urinary aldosterone excretion. According to Laidlaw et al. (1962), long- or short-term treatment with diethylstilbestrol or ethinyl estradiol (various doses) led to moderate increases in aldosterone excretion. However, aldosterone excretion was elevated to a pathological level in only one of eight subjects. These investigators assumed that the increase in urinary aldosterone(-18-glucuronide) excretion reflected altered metabolism rather than increased secretion rates, since ethinyl estradiol (0.3 mg/day) also augmented aldosterone excretion in a patient with Addison's disease who was treated with aldosterone infusions. Layne et al. (1962) observed that treatment of five premenopausal women with ethinyl estradiol 3-methyl ether (0.3 mg/day) for 20 days resulted in a statistically significant increase of the mean aldosterone secretion rate from 76 to 151 μg/day, in a decrease of the fractional metabolism of tritiated aldosterone to glucuronidase-hydrolyzable urinary metabolites, and in a markedly increased binding of aldosterone to plasma globulins from 16.4% to 40%. When the same estrogen was given at a lower dosage (0.1 mg/day), it had no effect on the secretion rate and plasma protein binding of aldosterone but influenced aldosterone metabolism in the same way as it did at the higher dosage. Crane and Harris (1969 a) treated 13 normal subjects with ethinyl estradiol (0.5 mg/day) for 3 weeks and observed significant increases of the mean aldosterone excretion rate from 8.5 to 15 μg/day and from 29 to 61 μg/day on unrestricted and on low sodium intake respectively. The increases in aldosterone excretion were accompanied by striking elevations in plasma renin activity. Apparently, increases in plasma renin activity induced by ethinyl estradiol are due mainly to increased plasma renin substrate concentrations. According to Oelkers et al. (1976), treatment of men and women with a low dose of ethinyl estradiol (1 μg per kg body wt. per day) for 14 days led to significant increases in the plasma renin substrate concentration and in plasma transcortin levels but not in plasma renin activity, plasma angiotensin II concentration, or urinary aldosterone excretion. The increases in plasma renin substrate concentration were greater in women than in men. Treatment of ovariectomized women with high doses of ethinyl estradiol (0.4 mg/day) led within 24 h to a substantial rise of the plasma renin substrate concentration (Kaulhausen et al. 1980). The increases in plasma renin activity were less extensive and became statistically significant only after 2 days. Increases in plasma aldosterone were observed in only two of four women. On the other hand, treatment of women with an intramuscular injection of *estradiol benzoate* (10 mg) affected the renin-angiotensin system in a more complex manner (Kaulhausen and Weyand 1982). It not only induced renin substrate synthesis but also elicited a marked increase in plasma renin concentration which preceded the rise in plasma renin substrate concentration by 35 h. In extent and time course, increases in plasma aldosterone paralleled the increases in plasma renin concentration. Long-term treatment of postmenopausal women with *estriol succinate* did not lead to any consistent alteration in plasma renin activity or in urinary aldosterone excretion (Erkkola et al. 1978).

5.10.2 Exogenous Progestogens

Progesterone, the major natural progestogen, has a potent natriuretic activity, which appears to be due mainly to a competitive inhibition of aldosterone action on the distal renal tubule. Progesterone was found to be an inhibitor of mineralocorticoid effects on the toad bladder and to displace aldosterone from its receptors in this tissue (for reviews see Sharp and Leaf 1966). Similarly, it displaced aldosterone from mineralocorticoid and from glucocorticoid binding sites in human and rat renal tissue (Fuller and Funder 1976; Wambach and Higgins 1978) and competitively blocked the metabolic and hypertensive effects of aldosterone and deoxycorticosterone acetate in vivo (Wambach and Higgins 1978, 1979). Treatment of normal human subjects with progesterone (50–100 mg/day) resulted in marked but transient excessive urinary sodium and chloride excretion (Landau et al. 1955). In mineralocorticoid-treated but not in untreated addisonian patients, progesterone induced a pronounced sodium loss. In short-term experiments carried out in two patients with Addison's disease, 150 mg of progesterone completely blocked the sodium-retaining effect of 10 μg, but not of 40 or 50 μg of aldosterone (Landau and Lugibihl 1958). Some of the short-term sodium diuretic effects of progesterone may, however, be independent of its antimineralocorticoid activity, since progesterone failed to acutely stimulate sodium excretion in subjects chronically treated with deoxycorticosterone acetate (Oparil et al. 1975).

Laidlaw et al. (1962) treated six subjects with progesterone (50–200 mg/day, i.m.) and observed an increase in urinary aldosterone excretion in all instances. However, this increase was moderate in two elderly men (62 and 64 years) who also showed no reversal of progesterone-induced natriuresis. In four younger subjects, sodium diuresis was reversed within 3 days of continuous progesterone administration, and urinary aldosterone excretion rose to values above the normal range. In two subjects the aldosterone secretion rate was also determined; it rose in proportion to rises in aldosterone excretion. By contrast, the increases in aldosterone excretion induced by progesterone (100–200 mg/day, i.v.) treatment in nine women were very small, and abnormally high values were observed in only three instances (G. Stark and Kossmann 1963). In this study, sodium diuresis was either absent or moderate. Layne et al. (1962) induced significant increases in the aldosterone secretion rate in five women by oral treatment with progesterone (300 mg/day); this medication had no effect on aldosterone metabolism or on cortisol secretion rate. In four young men, treatment with progesterone suppositories (500 mg/day) during 5 days of normal sodium intake led to marked and statistically significant increases in urinary aldosterone excretion and plasma renin activity, which were positively correlated with plasma progesterone concentrations (Sundsfjord 1971). A high sodium intake (12 g NaCl/day) during continued progesterone administration suppressed urinary aldosterone and plasma renin activity to very low values. A stimulatory effect of progesterone treatment on urinary aldosterone excretion and plasma renin activity in young men was also observed by Oelkers et al. (1974b). The increases in plasma renin activity were accompanied by proportional increases in plasma angiotensin II concentrations, whereas the plasma renin substrate concentrations remained unchanged. Administration of progesterone (200 mg/day) significantly enhanced the aldosterone secretion rate in women with normal pregnancy as well as in patients with a pregnancy terminated by intrauterine fetal death (Watanabe et al. 1965). In several instances, aldosterone secretion did not return to pretreatment values or increased even further upon cessation of progesterone administration.

Treatment of rats with progesterone (10 mg/day, s.c.) for a few days resulted in a 100% increase in the aldosterone secretion rate and in a smaller but significant decrease in corticosterone production (Singer et al. 1963b; Telegdy and Lissák 1965). In hypophysectomized rats, exogenous progesterone stimulated aldosterone secretion to the same extent that it did in intact animals but had no effect on corticosterone output (Singer et al. 1963b).

The natriuretic activity of progesterone appears to be independent of its progestational activity, since synthetic steroids with potent progestational activity were found either to have no effect on sodium excretion or to lead to sodium retention (Landau et al. 1958; Chagoya et al. 1961). On the other hand, the aldosterone antagonist spironolactone was progestationally active in rabbits and rhesus monkeys (Schane and Potts 1978). Treatment with *norethynodrel* (17-ethinyl-17β-hydroxy-5(10)-estren-3-one) induced a small but statistically significant elevation of the aldosterone secretion rate in premenopausal women (Layne et al. 1962). Administration of *medroxyprogesteroneacetate* (6α-methyl-17α-hydroxy-4-pregnene-3,20-dione acetate; 10 or 20 mg/day) led to a marked reduction of the aldosterone excretion rate in five normal subjects on an unrestricted sodium intake but did not alter the response in aldosterone secretion to dietary sodium restriction (Crane and Harris 1969a). This indicated that the suppression of aldosterone secretion was due to a sodium-retaining activity of the drug. However, suppression of aldosterone excretion was not accompanied by a decrease in plasma renin activity. The influence of four other synthetic progestogens on electrolyte balance, urinary aldosterone excretion, and the renin-angiotensin system in normal young men was investigated by Oelkers et al. (1974b). *Megestrol acetate* (17α-hydroxy-6-methyl-pregna-4,6-diene-3,20-dione acetate) led to a negative potassium balance, did not affect sodium balance, and induced transient increases in aldosterone excretion and in plasma renin activity. Under treatment with *dydrogesterone* (9β, 10α-pregna-4,6-diene-3,20-dione), transient increases in aldosterone excretion (significant) and in plasma renin activity and angiotensin II concentration (not significant) on days 3 and 4 were accompanied by transient sodium retention and were followed by urinary potassium loss. *Norethisterone acetate* (19-nor-17α-ethinyltestosterone acetate) did not significantly affect sodium and potassium balance but led to a progressive increase in aldosterone excretion, which was accompanied by an increase in plasma renin activity and plasma renin substrate concentration. The last finding was indicative of estrogenic properties of the steroid. Finally, treatment with *D-norgestrel* (D-13-ethyl-17α-ethinyl-17β-hydroxy-4-gonen-3-one) induced delayed sodium and potassium losses, which were accompanied by decreased plasma renin activity and angiotensin II concentration and had been preceded by a transient increase in urinary aldosterone excretion. Prolonged treatment of women with *chlormadinone acetate* (6-chloro-t-dehydro-17α-hydroxyprogesterone acetate) for 6 months did not result in any alteration of plasma aldosterone, plasma renin activity, or plasma renin substrate concentration (Katz and Beck 1974).

5.10.3 Oral Contraceptive Medication

Since oral contraceptives generally contain an estrogenic and a progestational component, they would be expected to exert a combination of estrogen-like and progesterone-like effects on aldosterone secretion and its regulation. However, the dose of estrogen in many of these preparations is lower than the minimal dose needed to induce a predictable effect on aldosterone secretion and metabolism. The progestational component of oral contraceptives is mainly of the type that either does not influence sodium excretion or has a transient sodium-retaining activity. It must also be considered that most of the studies with exogenous estrogens and progestogens described in the preceding sections lasted only a few days or weeks, whereas oral contraceptives are taken for months and years.

The aldosterone secretion rate increased considerably (+167%) in five women after treatment with Enovid (10 mg/day; 98.5% 17-ethinyl-17β-hydroxy-5(10)-estren-3-one, 1.5% ethinylestradiol 3-methyl ether) for at least two cycles (Layne et al. 1962). In these women, aldosterone metabolism was altered in the same manner as it was in women treated with ethinylestradiol 3-methyl ether (0.3 mg/day, see p. 230), and the binding of tritiated aldosterone to plasma globulins was also markedly increased (Meyer et al. 1961). Urinary aldosterone excretion rates were found to be elevated in most of the normotensive and hypertensive women on oral contraceptive medication (various preparations) studied by Crane and Harris (1969b), Laragh et al. (1967), and Weinberger et al. (1969). Increases in aldosterone excretion due to oral contraceptive treatment became more pronounced during sodium restriction or treatment with a diuretic. Generally, increases in urinary aldosterone excretion were associated with increases in plasma renin activity. Three of five women who were treated with Enovid (5 mg/day) for one cycle showed a marked increase of the aldosterone secretion rate on the 7th day of the Enovid cycle compared with their control cycle (Gray et al. 1968). On the 19th and 25th days of the Enovid cycle, aldosterone secretion rates were in most instances smaller than they were on the respective days of the control cycle. In ten healthy young women, oral contraception with two different preparations led to a significant increase in the mean plasma aldosterone concentration from 83 to 183 pg/ml after a week and to 229 pg/ml after 3 months (Beckerhoff et al. 1973). The rise in plasma aldosterone was paralleled by increases in plasma renin activity and in the plasma renin substrate concentration. On the other hand, there was a significant drop in the plasma renin concentration. These observations were only partially confirmed by other investigators. In women treated with four different combination oral contraceptives, plasma aldosterone concentrations were normal in spite of marked increases in the plasma renin substrate concentration and smaller increases in plasma renin activity (Katz and Beck 1974). Normal values of plasma aldosterone and of plasma renin activity were measured in eight oral-contraceptive users by Weinberger et al. (1977). Both parameters increased normally in response to upright posture and were normally suppressed by a saline infusion. In a group of hypertensive oral-contraceptive users, plasma aldosterone concentrations were the same as in a matched group of untreated women with essential hypertension (McAreavey et al. 1983). Plasma renin substrate concentrations and plasma angiotensin II levels were significantly higher in the contraceptive users than in the women with essential hypertension, whereas there was no difference between the two groups in their plasma renin concentrations. These latter findings support the hypothesis stated by Saruta et al. (1970) that hypertension induced by oral contraceptives could be due to diminished feedback suppression of renin release by angiotensin II. According to Katz et al. (1975b), there were considerable day-to-day variations in the plasma levels of aldosterone, renin activity, and renin substrate throughout the cycle in four women taking oral contraceptives. The first peaks in plasma aldosterone and plasma renin activity appeared on the 4th or 5th day of taking "the pill". In only one of these four women did an increase in plasma renin substrate precede the peaks in plasma aldosterone and plasma renin activity. This indicated that factors other than the estrogen-induced increased renin substrate production were responsible for rises in plasma aldosterone and renin activity during oral contraceptive treatment. Carr et al. (1979) observed that plasma ACTH concentrations were markedly lowered by oral contraceptive medication. This finding was consistent with previously described decreases in the cortisol secretion rate due to treatment with estrogens or oral contraceptives (Peterson et al. 1960; Layne et al. 1962). Perhaps suppression of ACTH secretion could explain the normal aldosterone concentrations found by McAreavey et al. (1983) in hypertensive oral-contraceptive users with elevated plasma angiotensin II concentrations.

5.10.4 Menstrual Cycle

The mean aldosterone secretion rate of 13 normal women was 140 μg/day during the follicular phase of the menstrual cycle (Gray et al. 1968). This corresponded well to the mean aldosterone secretion rate of normal men (135 μg/day; Watanabe et al. 1963). During the luteal phase, there was a significant rise in aldosterone secretion rate to a mean value of 235 μg/day. A similar increase was observed in two women on a high salt intake (260 mmol Na^+ and 240 mmol K^+ per day). The rise in aldosterone secretion was associated with an increase in urinary pregnanediol excretion but was not accompanied by any consistent alteration in sodium balance. Sundsfjord and Aakvaag (1972) observed a 100% increase in urinary aldosterone excretion as well as in plasma renin activity in the luteal phase of the cycle of normal young women, provided that there also was a clear-cut increase in the plasma progesterone concentration, i.e., that corpus luteum function was normal. Neither urinary aldosterone nor plasma renin activity increased during the second half of cycles with luteal failure (no increase in plasma progesterone) in spite of significant increases in plasma and urinary estrogens. This indicated that progesterone rather than estrogens was responsible for the rises in aldosterone and renin secretion in the luteal phase of the normal cycle. Marked rises of plasma aldosterone concentrations and of plasma renin activity during the luteal phase of ovulatory cycles were also reported by Michelakis et al. (1975). Both parameters remained relatively constant during an anovulatory cycle. According to Katz et al. (1975b), plasma aldosterone concentrations rose during the luteal phase of the cycle in only two of four women, whereas the plasma renin activity increased significantly in all of them. In contrast, Ruilope et al. (1977) observed significant increases in plasma aldosterone during the luteal phase of normal women (free diet, upright posture, untreated or treated with furosemide i.v.) but insignificant increases in plasma renin activity. Plasma aldosterone concentrations of women in the luteal phase were markedly higher than those of women in the follicular phase but not significantly different from those of normal men, according to Schöneshöfer and Wagner (1977). These investigators found plasma aldosterone concentrations to be significantly lower in "follicular" women than in men. Moreover, plasma concentrations of corticosterone, deoxycorticosterone, and 17-hydroxyprogesterone were also significantly lower in "follicular" women than in "luteal" women or in men. Katz et al. (1972b) also compared plasma aldosterone concentrations of women in the follicular phase of the cycle with those of men and found them to be equal under liberal sodium intake, irrespective of posture. Under sodium restriction, the plasma aldosterone levels were higher in women than in men after 2 h of standing. Looking for day-to-day variations in plasma aldosterone and renin activity throughout the menstrual cycle, Sundsfjord and Aakvaag (1973) discovered a distinct aldosterone peak on the day before ovulation, which was not accompanied by any rise in plasma renin activity. This preovulatory peak in plasma aldosterone was also observed by M. Frölich et al. (1976) but not by Katz and Romfh (1972). It appeared several days before the rise in plasma progesterone, but it might have been elicited by increasing plasma levels of estrogens or of 17-hydroxyprogesterone. However, the mechanism of aldosterone stimulation in this situation is completely unknown.

The elevated levels of plasma renin activity observed in the luteal phase of the menstrual cycle seem to be due to increased renin release. According to J.J. Brown et al. (1964b), the plasma renin concentration was significantly elevated in the 3rd or 4th week of the cycle. Plasma renin substrate concentration remained relatively constant throughout ovulatory and anovulatory cycles (Sundsfjord and Aakvaag 1972; Katz et al. 1975b).

In guinea pigs, whole blood concentrations of aldosterone and cortisol were significantly higher in females than in males, but there was no significant difference of these steroid levels at diestrus compared with those at estrus (Whipp et al. 1976). In rats, blood concentrations of aldosterone and corticosterone were significantly higher at estrus than at diestrus (Whipp et al. 1978). However, plasma renin concentrations and plasma renin substrate concentrations remained equal at the different stages of the estrous cycle.

5.10.5 Pregnancy

From theoretical considerations, the following factors would be expected to influence aldosterone secretion during pregnancy:

1. Increased sodium and potassium requirements for the growth of the fetus, placenta, and uterus, for an expansion of maternal blood and interstitial fluid volumes, and for the production and maintenance of amniotic fluid
2. Increased levels of circulating progesterone, which is a potent mineralocorticoid antagonist (see p. 231) but which can also be converted to the active mineralocorticoid deoxycorticosterone (Casey and MacDonald 1982)
3. Increased levels of circulating estrogens with multiple potential effects on the secretion, plasma protein binding, and metabolism of aldosterone (see p. 230)

An increased secretion rate and elevated plasma levels of aldosterone are commonly found in the third trimester of normal human pregnancy. Urinary excretion of aldosterone-18-glucuronide is also increased, but this is partially due to an alteration in aldosterone metabolism (K.M. Jones et al. 1959; J.F. Tait and Little 1968). A mean aldosterone secretion rate of 582 μg/day (nonpregnant women 192 μg/day) was found in six women in the 32nd–38th weeks of pregnancy (K.M. Jones et al. 1959). Similar values were observed by G. Stark (1966), and more pronounced increases in aldosterone secretion were reported by Watanabe et al. (1963). These investigators found elevated aldosterone secretion rates in women who were in only the 15th and 20th weeks of pregnancy. Between the 28th and 35th week of pregnancy aldosterone secretion rates were relatively constant in individual subjects; the mean levels were around 1000 μg/day. Between the 36th week and term there was a further increase to a mean value of 1600 μg/day. Since the metabolic clearance rate of aldosterone was found to be the same in pregnant and nonpregnant women (J.F. Tait et al. 1962), increased aldosterone secretion in pregnancy does not reflect increased catabolism of the steroid.

Plasma aldosterone concentrations are also generally found to be elevated during the normal human pregnancy. However, there appear to be great interindividual variations of this parameter. In a single case, plasma aldosterone was elevated a few days after conception (M. Frölich et al. 1976), whereas in other women it stayed within the nonpregnant range from early pregnancy through to the puerperium (Weir et al. 1975). In general, substantial and statistically significant increases in this parameter were observed at midterm or during the last trimester of pregnancy (Katz and Beck 1874; Ledoux et al. 1975; Weir et al. 1975; Smeaton et al. 1977; Wintour et al. 1978; Nolten et al. 1978; Hubl et al. 1979; M. Wilson et al. 1980). According to equilibrium dialysis experiments, the plasma protein binding of aldosterone was not increased in the last trimester of pregnancy (Meyer et al. 1961). J. F. Dunn et al. (1981) computed from the known plasma concentrations and binding affinities for albumin, transcortin (corticosteroid-binding globulin), and testosterone-binding globulin of 21 endogenous steroid hormones an expected fraction of unbound plasma aldosterone of 37% in men and nonpregnant women and of 32% in pregnant women (third trimester). An increased binding to transcortin (from 21% to 41%) was partly compensated by a decreased binding to albumin (from 42% to 25%). The nature of aldosterone carrier proteins in human plasma is controversial. Zipser et al. (1980) found that transcortin binding of aldosterone accounted for all the nonalbumin protein binding of this steroid. On the other hand, evidence for the existence of one or two specific aldosterone carrier proteins has been presented by Richardson et al. (1977) and by Katayama and Yamaji (1982).

The human fetal adrenal produces aldosterone, and aldosterone levels were found to be higher in the cord blood than in the maternal blood at the time of birth (see p. 217). Nevertheless, most of the aldosterone circulating in pregnant women has its origin in the maternal adrenals. Aldosterone was not detected in the urine of a pregnant woman with Addison's disease (Baulieu et al. 1957). Extremely low values of urinary aldosterone excretion were measured in two bilaterally adrenalectomized women during pregnancy (Laidlaw et al. 1958). Increased aldosterone secretion depends on a living fetus. After intrauterine fetal death, aldosterone secretion dropped to levels found in nonpregnant women (Watanabe et al. 1965).

Laidlaw et al. (1958) treated two adrenalectomized women with a constant maintenance dose of cortisol and 9α-fluorocortisol throughout pregnancy and concluded that "the high output of aldosterone during pregnancy was not a requisite for healthy gestation and served no apparent purpose". However, according to Venning et al. (1959), increased aldosterone secretion may contribute to the substantial sodium retention necessary for uterine and fetal development and for the expansion of maternal blood volume during normal pregnancy. Enhanced aldosterone production may also be necessary to overcome the aldosterone-antagonistic activity of the high levels of circulating progesterone in pregnant women. Hypokalemia, metabolic alkalosis, and hypertension markedly improved during pregnancy in a woman with primary aldosteronism, despite a large increase in aldosterone secretion (Biglieri and Slaton 1967). A close correlation between aldosterone secretion rates and urinary pregnanediol excretion (but also with urinary free cortisol excretion) was observed in pregnant women by K. M. Jones et al. (1959). Plasma concentrations of aldosterone and progesterone in pregnant women were significantly correlated, according to Ledoux et al. (1975) but not according to Smeaton et al. (1977).

The role of the renin-angiotensin system in causing a high aldosterone secretion rate in pregnancy is uncertain. Among the components of the system, the plasma renin substrate concentration has been most consistently found to be markedly elevated in pregnant women, to increase with progressive gestation time, and to be positively correlated with plasma aldosterone (Gould et al. 1966; Helmer and Judson 1967; Skinner et al. 1972; Weir et al. 1975; M. Wilson et al. 1980; Hsueh et al. 1982). However, similar elevations of this parameter induced by oral contraceptive medication were accompanied by unchanged aldosterone concentrations (Katz and Beck 1974; Weinberger et al. 1977). Moreover, plasma aldosterone concentrations had returned to nonpregnant values by the 3rd day post partum, when plasma renin substrate concentrations were still substantially elevated (Jadoul et al. 1982). Plasma angiotensin II concentrations in pregnant women were not correlated with plasma aldosterone or with plasma renin substrate levels (Weir et al. 1975). According to J. J. Brown et al. (1963), Gould et al. (1966), and Helmer and Judson (1967), plasma renin concentration was elevated in approximately half of all the pregnant women investigated, whereas the other half had plasma renin concentrations in the upper range of normal nonpregnant subjects. Other investigators found elevated plasma renin concentrations in all (Skinner et al. 1972) or most (Weir et al. 1975) pregnant women. There was no apparent correlation between this parameter and the duration of pregnancy. Some of the highest renin levels were measured in the first trimester.

Skinner et al. (1972) discovered a physiologically inactive component of the plasma renin in maternal blood, which was activated by treatment with acid. The concentration of this component was highest in early pregnancy, and the investigators proposed that it might be secreted by the fetal chorion. High concentrations of renin were found in human amniotic fluid by J. J. Brown et al. (1964c). According to Hsueh et al. (1982), most of the renin in the amniotic fluid was inactive and had the same molecular weight as the inactive renin circulating in the maternal blood. Considerable amounts of renin were also found in the uterus and placenta of rabbits (Gross et al. 1964). Renin isolated from rabbit uterus had the same molecular weight and pH optimum as the renin isolated from kidneys (R. C. Anderson et al. 1968). On the other hand, plasma renin concentration was significantly lower in human uterinc venous blood than in peripheral maternal blood during cesarean section at term (Skinner et al. 1972). Hsueh et al. (1982) observed that active and inactive renin increased similarly in response to sodium restriction in pregnant women and concluded that the kidneys were an important source of both forms of renin during pregnancy, but that other sources were not excluded. Abnormally low values of plasma renin activity had previously been found in a pregnant woman with primary aldosteronism (R. D. Gordon et al. 1976b). Bay and Ferris (1979) measured substantially increased plasma levels and urinary excretion rates of prostaglandin E in pregnant women and proposed that increased renin secretion in pregnancy could be secondary to increased prostaglandin synthesis.

Plasma sodium concentration was below the normal nonpregnant range in most women observed throughout gestation by Weir et al. (1975). However, since plasma sodium and aldosterone concentration were positively correlated, hyponatremia was probably not a major stimulus of aldosterone secretion. On the other hand, aldosterone and renin secretion are inversely related to sodium intake in normal pregnancy. Urinary excretion of aldosterone-18-glucuronide, the aldosterone secretion rate, and plasma aldosterone concentrations, as well as plasma renin concentration, rose to very high values in response to dietary sodium restriction or treatment of pregnant women with a diuretic (Watanabe et al. 1963; G. Stark 1966; Boonshaft et al. 1968; Becker et al. 1978; Bay and Ferris 1979). A high sodium intake or the infusion of saline elicited significant decreases in urinary and plasma aldosterone as well as in plasma renin activity in pregnant women; nevertheless, these parameters also remained elevated under suppression in comparison with nonpregnant women (Stark and Lehmann-Achilles 1964; Weinberger et al. 1977; Becker et al. 1978; Bay and Ferris 1979). In pregnant women, a change from a lateral recumbent to a supine position led to an elevation in systolic and diastolic blood pressure but not to any significant alteration in the plasma aldosterone concentration (Becker et al. 1978; Öney et al. 1981). A small but statistically significant decrease in plasma renin activity in response to this postural change was observed by Öney et al. (1981) but not by Becker et al. (1978). Significant rises in both parameters were elicited in pregnant women by a change from the supine to the upright posture (Weinberger et al. 1977). The oral intake of L-dopa by pregnant women caused decreases in plasma aldosterone and renin activity but did not affect the rise in plasma aldosterone in response to an angiotensin II infusion (Kaulhausen et al. 1982).

The aldosterone secretion rate, urinary excretion of aldosterone-18-glucuronide, plasma aldosterone, and plasma renin concentrations were significantly lower in women with *pre-eclampsia* than in normal pregnant women (Watanabe et al. 1965; G. Stark 1966; J. J. Brown et al. 1965b, 1966c; Weir et al. 1973; Hubl et al. 1979; E. B. Pedersen et al. 1982). These changes might have been secondary to pathological sodium retention in this condition. Kumar et al. (1959) found an increased intracellular sodium concentration in pre-eclamptic women, whereas this parameter decreased in normal late pregnancy.

High plasma concentrations of *deoxycorticosterone* are regularly observed late in human pregnancy and are not affected by a suppression of maternal pituitary ACTH secretion with dexamethasone (R. D. Brown et al. 1972b; Nolten et al. 1978). A considerable part of the deoxycorticosterone circulating in pregnant women was found to be produced in the maternal compartment by extra-adrenal 21-hydroxylation of plasma progesterone (Winkel et al. 1980a). Steroid 21-hydroxylase activity has been found in adult and fetal human kidneys, as well as in a number of other tissues (Winkel et al. 1980b; Casey and MacDonald 1982). Plasma concentrations of deoxycorticosterone in women with toxemia of pregnancy were not significantly different from those seen in women with normal pregnancies during the third trimester (R. D. Brown et al. 1972b).

High secretion rates and elevated plasma concentrations of aldosterone, characteristic of human pregnancy, are not common to pregnancy in all other mammalian species. Thus, in pregnant rhesus monkeys plasma aldosterone was not significantly elevated; plasma renin activity and renin substrate concentration also remained in the nonpregnant range (Wintour et al. 1974).

In pregnant (Australian) merino sheep, peripheral blood concentration and the secretion rate of aldosterone remained in the nonpregnant range under basal experimental conditions but increased more extensively than in nonpregnant sheep in response to stress, sodium deficiency, and intravenous infusions of angiotensin II or ACTH (Wintour et al. 1976). In contrast, plasma aldosterone rose markedly in parallel with a rise in plasma progesterone 6 weeks before parturition in pregnant (Algerian) Tadmit ewes (Boulfekhar and Brudieux 1980). Plasma aldosterone reached maximal levels (+200%) 17 days before term, decreased during the next 10 days, and increased again during the last week of pregnancy.

In dogs, the mean values of the aldosterone secretion rate and of plasma renin activity were significantly elevated in the second half of pregnancy (Robb et al. 1970). However, in several individual dogs one or both of these parameters remained normal. Plasma renin substrate concentration was consistently elevated in seven dogs during pregnancy. Removal rates of plasma aldosterone were the same in pregnant and in nonpregnant dogs (J.A.Johnson et al. 1974).

In guinea pigs (with a gestation time of 68 days), plasma aldosterone concentrations were significantly elevated on the 40th day of pregnancy, but not different from nonpregnant values on the 20th and the 60th day (Whipp et al. 1976). According to Giry and Delost (1977), plasma aldosterone was at a nonpregnant level on day 62 of gestation, increased to approximately twice this value by day 66, and stayed at this level until 24 h after birth.

In rats, plasma renin substrate was found to be significantly decreased during pregnancy (Carretero and Gross 1967). Nevertheless, significantly elevated blood aldosterone levels were measured in rats in the second half of pregnancy (gestation time 21-22 days) by Whipp et al. (1978). Sodium loading significantly suppressed blood aldosterone in 18- to 20-day pregnant but not in nonpregnant rats. Plasma renin concentration increased significantly by day 11 of gestation, whereas plasma renin substrate concentration dropped significantly by day 11 and even further by days 18-20. According to S.Churchill et al. (1981), plasma aldosterone was significantly lower in full-term pregnant than in nonpregnant rats when the animals received water as drinking fluid, but slightly higher in pregnant than in nonpregnant rats when water had been replaced by saline. Independent of sodium intake, plasma corticosterone was markedly lower in pregnant than in nonpregnant rats. In totally adrenalectomized rats, plasma aldosterone and corticosterone became measurable during pregnancy, reaching levels only 50% lower than in intact animals at term. The plasma levels of both corticosteroids were unmeasurably low again within 2 days after birth. Adrenals of rats killed 4 h after delivery produced more aldosterone in vitro than adrenals of nonpregnant rats (G.Schneider and Mulrow 1973). Sodium restriction during pregnancy did not further increase the aldosterone output of maternal adrenals.

In mice, maternal plasma aldosterone concentrations during the last 4 days of pregnancy were higher than in nonpregnant mice but lower than in the fetus (Dalle et al. 1978).

5.11 Aspects of Negative Feedback Regulation

5.11.1 Mineralocorticoids

Autonomous overproduction of aldosterone by adrenocortical adenomas not only causes the syndrome of primary aldosteronism but also suppresses the secretion of aldosterone by the adjacent and the contralateral adrenal cortex. Aldosterone concentrations in adrenal venous plasma taken from the contralateral side of an aldosterone-producing tumor were generally within the range of peripheral plasma aldosterone concentrations (Melby et al. 1967; Egdahl et al. 1968; Horton and Finck 1972; Nicolis et al. 1972). Marked decreases in the urinary excretion of aldosterone-18-glucuronide were found to persist in some patients for several weeks and months after the surgical removal of an aldosterone-producing adenoma (Conn et al. 1964; Biglieri et al. 1966a). During this period, aldosterone secretion did not adequately increase in response to stimulation by ACTH, angiotensin II, or dietary sodium restriction. Sodium restriction resulted in severe urinary sodium losses and hyperkalemia. In patients without preoperative treatment, plasma renin activity also remained suppressed during the first week after the removal of the aldosterone-producing tumor but recovered earlier than aldosterone secretion did (Conn et al. 1964; Morimoto et al. 1970). The recovery of renin secretion, but not the recovery of aldosterone secretion, was hastened by prolonged preoperative treatment of the patients with spironolactone (Morimoto et al. 1970). Postoperative selective hypoaldosteronism was also not prevented by prolonged (6-72 months) preoperative combined treatment with spironolactone and chlorothiazide, which caused a substantial elevation in plasma renin activity (Bravo et al. 1975). Within 1-3 months after unilateral adrenalectomy, sodium restriction evoked a blunted response in urinary aldosterone excretion in spite of a normal response in plasma renin activity. The response in aldosterone excretion became normal 6-12 months after the operation. Thus, at least in these pretreated patients, renin deficiency was not the principal cause of hypoaldosteronism.

Administration of deoxycorticosterone acetate or 9α-fluorocortisol to normal human subjects or to patients with essential hypertension brought about marked decreases in the urinary excretion rates of aldosterone metabolites as well as in plasma aldosterone concentrations (Biglieri et al. 1967, 1970; Horton 1970; Cangiano et al. 1971; Shade and Grim 1975; Lund and Nielsen 1980). Substantial decreases in the secretion rate, urinary excretion, and plasma concentrations of aldosterone were also induced by treatment with deoxycorticosterone in rats (Singer and Stack-Dunne 1955; Eilers and Peterson 1964; W. B. Campbell and Pettinger 1975; Hilfenhaus 1975) and in sheep (Fan et al. 1975).

Nontumorous adrenal tissue adjacent to tumors removed from patients with primary aldosteronism contained either no aldosterone or only traces of it (Louis and Conn 1961). Such a tissue converted tritiated pregnenolone to 18-hydroxycorticosterone to only a very small extent; conversion to aldosterone was unmeasurably low (Dufau et al. 1968). According to Marusic and Mulrow (1969), the conversion of ^{14}C-labeled corticosterone to aldosterone by a homogenate and by mitochondria of an aldosterone-producing tumor was three times higher than by material from the adjacent adrenal tissue. On the other hand, the conversion of corticosterone to 18-hydroxycorticosterone was somewhat higher in the adjacent adrenal than in the adenoma. In aldosterone-producing adenomas, as well as in adjacent adrenal tissue, the angiotensin II receptor concentration was three times lower than in a normal surgically removed human adrenal gland (G. Brown et al. 1980). Whereas cells from aldosterone-producing adenomas generally responded to added angiotensin II or ACTH with an increased aldosterone output, cells from the adjacent adrenal were in most instances not responsive to these two stimuli.

Treatment of rats with 9α-fluorocortisol for 2 weeks resulted in a 70%–90% decrease in aldosterone production in vitro by quartered adrenal glands incubated with or without aldosterone-stimulating substances but had no effect on corticosterone or deoxycorticosterone outputs (J. Müller 1970a). In rats kept on a normal sodium intake, aldosterone production by incubated adrenal tissue in vitro was also significantly diminished after 5 days of treatment with aldosterone acetate or deoxycorticosterone acetate (Debreceni and Csete 1973b). According to a later study performed in our laboratory (Hunziker and Müller 1977), treatment of rats with either 9α-fluorocortisol or deoxycorticosterone trimethyl acetate for 2 weeks markedly lowered the conversions of tritiated corticosterone and deoxycorticosterone to 18-hydroxycorticosterone and aldosterone (Fig. 44) and suppressed – less extensively – the conversions of added deoxycorticosterone and 11-deoxycortisol to 11β-hydroxylated derivatives by capsular adrenal tissue. It did not influence the rates of 18-hydroxylation and 11β-hydroxylation of these precursors by the decapsulated portions of the glands. When incubated with serotonin, capsular adrenals of mineralocorticoid-treated rats produced less aldosterone but more deoxycorticosterone than capsular adrenals of control rats (Fig. 45).

Aldosterone biosynthesis from endogenous or added precursors by adrenal tissue of rats which had been kept on a sodium-deficient diet during treatment with a mineralocorticoid was not different from aldosterone biosynthesis by adrenals of

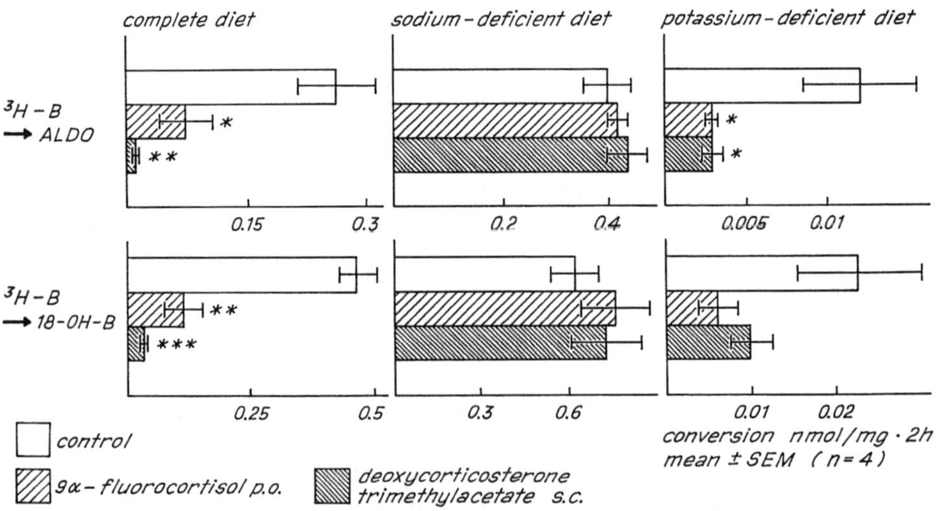

Fig. 44. Effects of 2 weeks of treatment of rats kept on different diets with 9α-fluorocortisol or deoxycorticosterone trimethyl acetate on the conversion of tritiated corticosterone (*^{3}H-B;* 300 nmol/flask) to aldosterone *(ALDO)* and 18-hydroxycorticosterone *(18-OH-B)* by capsular adrenals. Statistical significance (*t*-tests) of the difference from the respective control value: *** $P<0.001$, ** $P<0.01$, * $P<0.05$. (From Hunziker and Müller 1977)

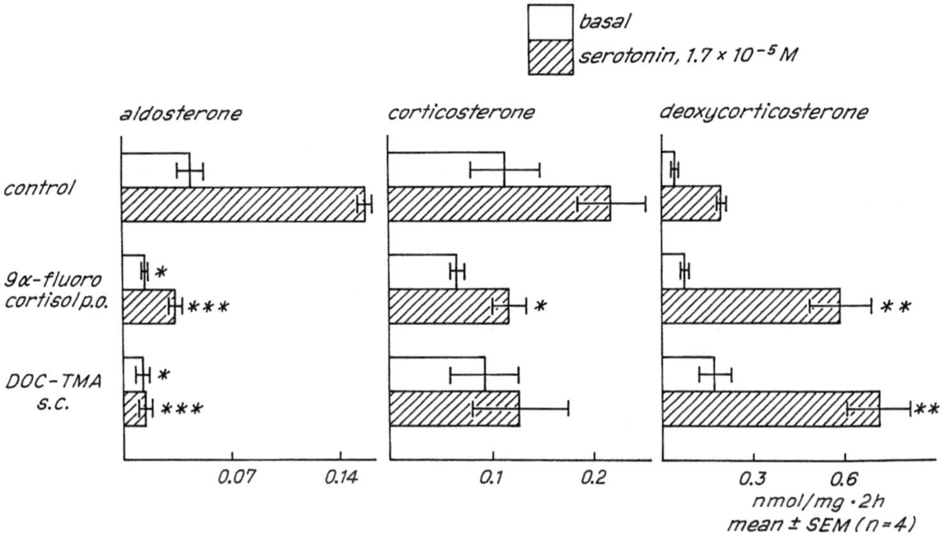

Fig. 45. Effects of 2 weeks of treatment of rats (complete diet) with 9α-fluorocortisol or deoxycorticosterone trimethyl acetate *(DOC-TMA)* on the output of endogenous corticosteroids by capsular adrenals incubated with or without serotonin. Statistical significance (*t*-tests) of the difference from the respective control value: *** $P<0.001$, ** $P<0.01$, * $P<0.05$. (From Hunziker and Müller 1977)

untreated, sodium-restricted rats under most conditions of incubation (J. Müller 1970a; Debreceni and Csete 1973b; Hunziker and Müller 1977; Fig. 44). In sheep with moderate sodium deficiency, aldosterone secretion was not suppressed by adrenal arterial or systemic infusion of aldosterone (Blair-West et al. 1962b, 1965b). These findings indicate that decreased aldosterone secretion is not due to a direct effect of plasma mineralocorticoids on the adrenal cortex or a receptor organ involved in negative feedback control, but is rather a consequence of a mineralocorticosteroid-induced alteration in sodium balance or sodium distribution. However, suppression of aldosterone secretion is not proportional to sodium ret-

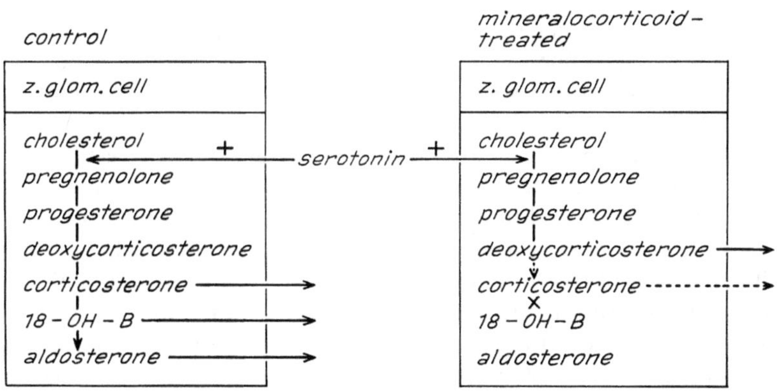

Fig. 46. Alterations of zona glomerulosa cell functions under prolonged treatment of rats with a mineralocorticoid

ention. Thus, the same decreases in aldosterone excretion were induced by a 3-day treatment with deoxycorticosterone acetate in healthy men and in patients with essential hypertension (Biglieri et al. 1967). In the first group, the cumulative sodium retention ranged between 250 and 750 mmol, weight gain between 2 and 4 kg; in the second group sodium retention was less than 150 mmol, weight gain less than 0.5 kg. No significant changes in serum electrolyte concentrations, blood pressure, or hematocrit levels were observed in either group. In rats, suppression of aldosterone biosynthesis continued progressively during the 2nd week of treatment with deoxycorticosterone trimethyl acetate (Hunziker and Müller 1977), i.e., at a time when the animals would have been expected to be escaping from the sodium-retaining action of the mineralocorticoid but still be losing potassium (J. Möhring and Möhring 1972). However, at the end of 2 weeks of treatment with deoxycorticosterone trimethyl acetate, serum potassium concentration was unchanged (Hunziker and Müller 1977). The potassium loss must therefore have been small or compensated by a shift from the intracellular to the extracellular compartment.

Decreases in aldosterone production due to exogenous or endogenous mineralocorticoids could be mediated by the renin-angiotensin system. Decreased or low levels of plasma renin activity or plasma renin concentration were found in patients with primary aldosteronism (Conn et al. 1964; J. J. Brown et al. 1968b). In normal human subjects on high or low sodium intake, treatment with small amounts of deoxycorticosterone acetate significantly suppressed plasma renin activity and plasma aldosterone concentration (Shade and Grim 1975). Significant suppression of renin secretion was induced in dogs by treatment with aldosterone (Gelhoud and Vander 1967) and with deoxycorticosterone acetate (Robb et al. 1969); this suppression was prevented by sodium deficiency. Plasma renin activity, renin concentration, and plasma angiotensin II concentration were also significantly suppressed by treatment of rats with deoxycorticosterone acetate, provided that the animals received sodium in their diet and were not anesthetized for blood sampling (F. J. Goodwin et al. 1969; Pettinger et al. 1971; W. B. Campbell and Pettinger 1973; Ayachi et al. 1976; Haack et al. 1977). In rats kept on a sodium-deficient diet, the kidney renin content was suppressed by daily injections of deoxycorticosterone acetate but not by daily injections of deoxycorticosterone trimethyl acetate (Christlieb et al. 1972). A suppressive effect of aldosterone on renin release was also demonstrated in an isolated perfused canine kidney preparation (Greco and Murphy 1971). On the other hand, aldosterone did not directly affect the release of renin by human kidney slices during a 30-min period of incubation (Rosset and Veyrat 1971).

Prolonged administration of deoxycorticosterone acetate to rats resulted in marked narrowing and lipid depletion of the zona glomerulosa (Deane et al. 1948). This effect was also observed in hypophysectomized rats (Greep and Deane 1947) but not in animals kept on a sodium-deficient diet (Deane et al. 1948). A broad but small-cell and lipid-free zona glomerulosa characterizes the adrenal cortex adjacent to aldosterone-producing adenomas removed from patients with primary aldosteronism (Tonutti and Bayer 1964). Mitochondria of many, but not all, glomerulosa cells in the adrenal cortices of rats treated with aldosterone or deoxycorticosterone acetate were larger and more spherical than those of control animals (Fisher and Horvath 1971). Cristae were large and extended across the diameter of the mitochondrion in parallel arrays. In addition, there was a relative paucity of agranular endoplasmic reticulum and Golgi structures in the cytoplasm of glomerulosa cells of mineralocorticoid-treated rats.

5.11.2 Licorice and Carbenoxolone

Licorice is widely consumed as a sweet and is also a component of nonalcoholic beverages and cough medicines. Its active component is glycyrrhizic acid, which is not a steroid but a structurally related substance. Carbenoxolone sodium is a derivative of glycyrrhizic acid, i.e., 18β-glycyrrhizic acid sodium hemisuccinate, and is used in the treatment of gastric and duodenal ulcers. In intact human subjects as well as in adrenalectomized patients and experimental animals, licorice extracts display a potent mineralocorticoid-like activity (Molhuysen et al. 1950; Fletcher et al. 1967; Baron et al. 1969; Werning et al. 1972; Armanini et al. 1982). According to a study by Armanini et al. (1982), the overall effect of carbenoxolone on electrolyte metabolism in the intact organism is due partly to an intrinsic mineralocorticoid activity and partly to an amplification of aldosterone actions. In addition, a study on the effect of carbenoxolone on the plasma concentration and uptake by kidney cell fractions of tritiated aldosterone in adrenalectomized rats indicated that the drug also affected the normal distribution and metabolism of the hormone (Humphrey et al. 1979).

Long-term abuse of licorice leads to the syndrome of pseudoaldosteronism, characterized by hypertension, hypokalemia, metabolic alkalosis, abnormally low plasma renin activity, and pathologically decreased urinary excretion rate and plasma concentration of aldosterone (Molhuysen et al. 1950; Jenny et al. 1961; Salassa et al. 1962; Conn et al. 1968; Trono et al. 1983). Even moderate licorice consumption or treatment with normal therapeutic doses of carbenoxolone led in many instances to edema, hypokalemia, or hypertension and consistently caused suppression of plasma renin activity, plasma angiotensin II and aldosterone concentrations, and urinary aldosterone-18-glucuronide excretion (Turpie and Johnson 1965; Baron et al. 1969; Werning et al. 1971; Baas et al. 1976; M.T. Epstein et al. 1977).

5.11.3 Pseudohyperaldosteronism (Liddle's Syndrome)

A familial disorder characterized by hypertension, hypokalemia, metabolic alkalosis, and very low aldosterone secretion was first described by Liddle et al. (1964) and was thought to be due to a renal disorder of electrolyte handling. Later studies revealed that patients with Liddle's syndrome also showed an abnormality in the sodium transport across erythrocyte membranes (Helbock and Reynolds 1970; Gardner et al. 1971; Hyman et al. 1979; Wang et al. 1981). Characteristically, hypertension and hypokalemia did not respond to treatment with spironolactone but became normal under treatment with triamterene, alone or combined with dietary sodium restriction (Liddle et al. 1964; Aarskog et al. 1967; Wang et al. 1981). In untreated patients plasma renin activity was very low (Gardner et al. 1971; Hyman et al. 1979; Wang et al. 1981). In the patient described by Wang et al. (1981), urinary and plasma aldosterone rose markedly in response to ACTH administration for 3 days but not to triamterene medication for 12 days, although the latter treatment caused substantial rises in plasma renin activity and in serum potassium. According to Aarskog et al. (1967), aldosterone secretion of their patient became normal only after she had been treated with triamterene and sodium restriction for 9 months.

5.11.4 Acquired Forms of Primary Hypoaldosteronism

In patients with *Addison's disease* with severe deficiency in both aldosterone and cortisol secretion, plasma renin concentration was grossly elevated, whereas plasma renin substrate concentration was substantially diminished (Stockigt et al. 1979). Renin substrate deficiency was closely related to the degree of cortisol deficiency. In an addisonian patient with severe glucocorticoid deficiency but intact aldosterone secretion, renin substrate concentration was also very low in spite of a normal renin concentration. In this patient, treatment with dexamethasone resulted in water diuresis and weight loss, which was associated with a transient increase in plasma renin concentration and urinary aldosterone excretion and led to a slow increase in plasma renin substrate concentration.

Bilateral adrenalectomy of rats was followed by a marked increase in plasma renin activity and a marked decrease in the plasma renin substrate concentration (Nasjletti and Masson 1969). When water was replaced by a 1% NaCl solution as drinking fluid, the drop in the plasma renin substrate level was completely prevented, but the rise in plasma renin activity was only partially impeded. Treatment with deoxycorticosterone acetate effectively prevented both alterations, whereas treatment with aldosterone, cortisol, or corticosterone was sufficient to maintain renin substrate concentration but had a relatively weak suppressive effect on plasma renin activity. In adrenalectomized, untreated rats, plasma renin concentration was significantly suppressed by plasma volume expansion due to an albumin infusion but not by an NaCl infusion (Welch et al. 1983). Treatment with dexamethasone lowered the pre-infusion plasma renin concentration of adrenalectomized rats to the normal range and restored the suppressive effect of NaCl infusion on the plasma renin concentration.

Whereas hyporeninemic hypoaldosteronism is the most common form of acquired selective mineralocorticoid deficiency in man (see p. 174), there are also cases of *hyperreninemic* forms of *acquired selective hypoaldosteronism*. The combination of selective hypoaldosteronism with elevated plasma renin activity has been observed in two patients with long-standing type-II diabetes mellitus (Morimoto et al. 1979; F. A. Williams et al. 1983), in a patient with hemochromatosis (Thomas 1984), but also in a 57-year-old woman without previous diseases (Kokko 1985). In these patients, plasma aldosterone and 18-hydroxycorticosterone concentrations were abnormally low and did not adequately increase in response to ACTH, angiotensin II, upright posture, or sodium restriction, whereas plasma cortisol, corticosterone, and deoxycorticosterone concentrations were normal. Plasma renin activity and plasma potassium concentrations were elevated. A relatively frequent form of acquired hyperreninemic hypoaldosteronism was discovered by Zipser et al. (1981) among severely ill patients in an intensive-care unit. In 18 of 28 patients with persistent hypotension, plasma aldosterone was inappropriately low in spite of elevated plasma renin activity, elevated plasma cortisol, and normal plasma potassium concentrations. All these patients suffered from major infections. The rises in plasma aldosterone in response to exogenous angiotensin II or ACTH were markedly diminished. Plasma 18-hydroxycorticosterone concentrations were higher than in normal control persons (Zipser et al. 1981). Nevertheless, they were inappropriately low in comparison with the 18-hydroxycorticosterone concentrations found in intensive-care unit patients with similarly increased plasma cortisol and renin levels but with intact aldosterone secretion (Stern et al. 1983). By the same criteria, plasma corticosterone concentrations were also inappropriately low in critically ill patients with hyperreninemic hypoaldosteronism. These observations indicated a generalized impairment of zona glomerulosa function, with early and late steps of aldosterone biosynthesis similarly affected. Although the pathogenesis of the condition is unknown, the frequency of persistent hypotension in these patients might suggest that impaired adrenocortical circulation could be one of the causes (Davenport and Zipser 1983). Morphological evidence of zona glomerulosa necrosis was found in some patients (Zipser et al. 1981).

An experimental model of acquired hyperreninemic hypoaldosteronism was found in the hypercalcemic tumor-bearing Fischer rat by Sowers and co-workers (Sowers et al. 1981d; Sowers and Barrett 1983). Implantation of tumorous Leydig's cells into these animals led within a few weeks to marked hypercalcemia, hypophosphatemia, and slightly impaired renal function associated with substantial elevations in active and inactive plasma renin and abnormally low plasma aldosterone concentrations. The increases in plasma aldosterone in response to graded infusions of angiotensin II, ACTH, or KCl were severely blunted in tumor-bearing animals (Sowers and Barrett 1983). However, the impairment of adrenocortical function was not entirely limited to the zona glomerulosa, since the responses in plasma corticosterone to immobilization stress and ACTH administration were also blunted (Sowers et al. 1981d).

5.11.5 Congenital Deficiencies of Steroidogenic Enzymes

In principle, aldosterone secretion can be affected by a congenital defect of a steroidogenic enzyme in four different ways:

1. Direct impairment of aldosterone biosynthesis due to deficiency of a zona glomerulosa enzyme
2. Excessive production of a mineralocorticoid by the zona fasciculata (e.g., deoxycorticosterone)
3. Excessive production of a steroid with antimineralocorticoid activity by the zona fasciculata (e.g., progesterone)
4. Chronically elevated plasma ACTH concentrations

Two inborn errors of steroid metabolism which exclusively affect aldosterone biosynthesis in the zona glomerulosa of the human adrenal cortex are *corticosterone methyloxidase deficiency type 1* and *corticosterone methyloxidase deficiency type 2,* which have already been referred to in this monograph (see pp. 11, 114). The latter of these two rare disorders appears to be the more common one and has been detected relatively frequently among members of isolated families of Iranian Jews with a high incidence of consanguinity (Rösler et al. 1977a; Rösler 1984). There was a great variation of clinical manifestations which dependend not only on quantitative differences in hormone secretion but also on changing patterns of aldosterone requirement throughout life, ranging from severe life-threatening sodium loss and dehydration during the first year of life to postural hypotension and growth retardation in later childhood and to asymptomatic states in adults detectable only by biochemical screening. The disorder is recognized in the combination of low or normal levels of plasma aldosterone and urinary aldosterone metabolites with grossly elevated plasma concentrations of 18-hydroxycorticosterone and excessively increased urinary excretion of the 18-hydroxycorticosterone metabolite 18-hydroxytetrahydro-11-dehydrocorticosterone (Ulick et al. 1964b; Rappaport et al. 1968; David et al. 1968; Ulick 1976; Rösler et al. 1977a; Veldhuis et al. 1980; Drop et al. 1982). According to markedly increased plasma levels and urinary excretion rates of metabolites, the secretion of corticosterone, deoxycorticosterone, and 18-OH-DOC was also abnormally high in patients with untreated corticosterone methyloxidase deficiency type 2 (Ulick et al. 1964b; Rappaport et al. 1968; David et al. 1968; Veldhuis et al. 1980; Drop et al. 1982). The suppression of these steroids to normal levels upon treatment with a mineralocorticoid and salt supplementation - but not by treatment with a glucocorticoid - indicated that they were produced in the zona glomerulosa. Cortisol secretion was found to be normal in most patients with the disorder. However, David et al. (1972) observed that dexamethasone only partially suppressed the cortisol production rate during periods of sodium depletion and that treatment with deoxycorticosterone acetate and sodium chloride transiently lowered the urinary excretion of cortisol metabolites to subnormal levels. These findings suggested an influence of sodium balance on cortisol secretion by a mechanism not dependent upon ACTH secretion. Whereas plasma renin activity was markedly elevated in most untreated patients (Rappaport et al. 1968; Rösler et al. 1977a; Veldhuis et al. 1980; Drop et al. 1982), pathologically increased plasma potassium and decreased plasma sodium concentrations were usually observed only in patients with severe sodium deficiency. Plasma aldosterone did not rise in response to angiotensin II infusions, even when baseline values were within the normal range (Rösler et al. 1977b). In the same patients, plasma aldosterone responded normally to KCl or ACTH infusions.

The most common inborn error of corticosteroid biosynthesis is *21-hydroxylase deficiency,* which exists in two different clinical variants, i.e. *simple virilizing* 21-hydroxylase deficiency and *salt-wasting* 21-hydroxylase deficiency. Older theories about the biochemical basis of these two forms of the disease, in particular a two-enzyme theory assuming a different 21-hydroxylase for 17-hydroxy and 17-deoxy steroids, have been replaced by a new theory based on the concept that the adrenal zona fasciculata and zona glomerulosa function as two separate glands (for a review see New and Levine 1984). According to this theory, the simple virilizing form is characterized by an enzyme deficiency which is restricted to the inner zones of the adrenal cortex and leaves 21-hydroxylation intact in the zona glomerulosa. In the salt-wasting form, the biosynthetic defect affects steroidogenesis to a similar extent in the zona glomerulosa and the zona fasciculata.

In untreated patients with the simple virilizing form of 21-hydroxylase deficiency, urinary excretion of aldosterone-18-glucuronide, the aldosterone secretion rate, or plasma aldosterone concentrations were often found to be elevated during periods of liberal sodium intake (Prader et al. 1955; Blizzard et al. 1959; Kowarski et al. 1965; Degenhart et al. 1965; Bartter et al. 1968; Kuhnle et al. 1981; Kater and Biglieri 1983). Generally, parameters of increased aldosterone secretion became normal under suppressive therapy with glucocorticoids. According to Bartter et al. (1968), increased aldosterone secretion in patients with 21-hydroxylase deficiency could be either directly caused by elevated plasma ACTH concentrations or the result of a regulatory mechanism compensating for the antimineralocorticoid activity of excessively produced ACTH-dependent steroids. The former possibility is supported by the observation that prolonged ACTH administration leads to a *persistent* stimulation of aldosterone secretion in patients with classical simple virilizing 21-hydroxylase deficiency (Bartter et al. 1968; Beitins et al. 1972a; Kater et al. 1985a), which contrasts with the *transient* stimulation of aldosterone secretion in normal subjects (see p. 183). Increased production rates and elevated plasma concentrations of progesterone and 17α-hydroxyprogesterone are characteristic of untreated 21-hydroxylase deficiency, and both steroids have been reported to exert antimineralocorticoid activity (Landau and Lugibihl 1958; Jacobs et al. 1961; Wambach and Higgins 1978). Also consistent with the antimineralocorticoid theory are the markedly increased levels of plasma renin activity observed in many patients with untreated simple virilizing 21-hydroxylase activity (Godard et al. 1968; Strickland and Kotchen 1972; Grant et al. 1977; Kuhnle et al. 1981). However, plasma renin activity has also been found to be elevated in patients receiving adequate glucocorticoid treatment. Normalization of plasma renin activity of such patients by sodium repletion with or without additional mineralocorticoid therapy was accompanied by a further suppression of plasma ACTH concentration, indicating an interrelationship between the renin-angiotensin system and ACTH secretion (Rösler et al. 1977c). Moreover, infusion of angiotensin II into cortisone-treated women with simple virilizing 21-hydroxylase deficiency raised not only plasma aldosterone but also plasma 17α-hydroxyprogesterone without affecting plasma ACTH levels (Schaison et al. 1980). This suggested that the peptide directly stimulated steroidogenesis in the zona fasciculata without the interaction of ACTH.

Evidence for an aldosterone biosynthetic defect in 21-hydroxylase deficiency has been presented by Ulick et al. (1980), who found an abnormally high ratio of tetrahydro-18-hydroxy-11-dehydrocorticosterone (TH-18-OH-A) over tetrahydroaldosterone in the urine of all their patients with the salt-wasting form and of most of their patients with the simple virilizing form of 21-hydroxylase deficiency, provided that they were older than 1 month. The same pattern of urinary steroids is typical for patients with corticosterone methyloxidase deficiency type 2. The increased excretion rates of TH-18-OH-A were suppressed by treatment with a mineralocorticoid but not by treatment with a glucocorticoid. This indicated that the excessive 18-hydroxycorticosterone was produced by the zona glomerulosa. However, Kater and co-workers (Kater and Biglieri 1983; Kater et al. 1985b) found plasma 18-hydroxycorticosterone levels to be elevated only in proportion to elevated plasma aldosterone levels, and they observed a normal ratio of plasma 18-hydroxycorticosterone to aldosterone concentrations in all their patients with simple virilizing 21-hydroxylase deficiency. Moreover, Kuhnle et al. (1981) reported that urinary excretion rates of free 18-hydroxycorticosterone were in proportion to aldosterone-18-glucuronide excretion rates in four patients with the simple virilizing form as well as in four patients with the salt-wasting form of

21-hydroxylase deficiency. Thus, the observation of inappropriately increased TH-18-OH-A excretion in this disorder remains as yet unconfirmed and unexplained. Biglieri et al. (1981) measured increased plasma concentrations of deoxycorticosterone in patients with the simple virilizing form of 21-hydroxylase deficiency and concluded that this steroid was produced by the zona glomerulosa. However, according to Kuhnle et al. (1981), urinary and plasma deoxycorticosterone levels were low in dexamethasone-treated patients and, in most instances, did not increase in response to sodium restriction. Most likely, the steroid resulted mainly from the peripheral conversion of progesterone produced by the zona fasciculata. Normal *extra-adrenal* 21-hydroxylase activity in a woman with simple virilizing adrenal 21-hydroxylase deficiency was demonstrated by Winkel et al. (1983).

As previously mentioned, the salt-wasting form of congenital adrenal hyperplasia has been ascribed by some investigators to a more extensive 21-hydroxylase deficiency, involving not only the inner zones but also the zona glomerulosa of the adrenal cortex (New and Levine 1984). This hypothesis is supported by the finding of decreased aldosterone secretion rates, decreased urinary excretion rates of aldosterone-18-glucuronide and free 18-hydroxycorticosterone, and decreased plasma aldosterone concentrations contrasting with high levels of plasma renin activity in such patients (Degenhart et al. 1965; Kowarski et al. 1965; Godard et al. 1968; Imai et al. 1968; Rösler et al. 1977c; Kuhnle et al. 1981). However, several observations are in direct conflict with the zona glomerulosa 21-hydroxylase deficiency hypothesis. On the one hand, not all salt losers have been found to be unequivocally aldosterone deficient, but rather mineralocorticoid resistant (Keenan et al. 1979). Normal or even elevated values of plasma aldosterone, urinary excretion rates of aldosterone-18-glucuronide or tetrahydroaldosterone, or the aldosterone secretion rate were measured after withdrawal of mineralocorticoid therapy in children with previous episodes of salt wasting (Limal et al. 1977; Horner et al. 1979; Keenan et al. 1979). On the other hand, in cases with proven inadequate aldosterone secretion, a particular block of zona glomerulosa steroidogenesis at the stage of 21-hydroxylation of progesterone has (to my knowledge) never been documented. Thus, under treatment with dexamethasone and sodium restriction, plasma progesterone concentrations were not higher in patients with salt wasting than in normal children or in patients with the simple virilizing form of 21-hydroxylase deficiency (Kuhnle et al. 1981). Moreover, Osterwalder (1971) reported the complete absence of a zona glomerulosa in six autopsied cases of treated or untreated salt-wasting congenital adrenal hyperplasia, whereas the zona glomerulosa was hypertrophic in children with the simple virilizing form of the disorder. Thus, deficient aldosterone secretion in salt losers could be due to a general atrophy of the zona glomerulosa rather than to just a block in aldosterone biosynthesis.

Congenital *11β-hydroxylase deficiency* is often clinically characterized by a combination of virilization with high blood pressure. Hypertension is generally ascribed to excessive production of the mineralocorticoid deoxycorticosterone (New and Levine 1984). However, in several patients with 11β-hydroxylase deficiency, only the 11β-hydroxylation of 11-deoxycortisol was impaired, whereas the conversion of deoxycorticosterone to corticosterone was normal (Zachmann et al. 1971, 1983). Moreover, hypertension is not a regular feature of 11β-hydroxylase deficiency and has been found to be often poorly correlated with deoxycorticosterone secretion (Zachmann et al. 1983). Overproduction of deoxycorticosterone in patients with 11β-hydroxylase deficiency may even be associated with neonatal salt wasting (Holcombe et al. 1980; Zachmann et al. 1983). Abnormally low aldosterone secretion rates, urinary excretion rates of aldosterone metabolites, and plasma aldosterone concentrations were consistently measured in untreated patients with the hypertensive form of the disease (New et al. 1966; Kowarski et al. 1968; New and Seaman 1970; Sizonenko et al. 1972; Kater and Biglieri 1983). Plasma renin activity (Imai et al. 1968; Sizonenko et al. 1972) and urinary and plasma 18-hydroxycorticosterone levels were also subnormal (Levine et al. 1980; Kater and Biglieri 1983). In untreated patients, aldosterone excretion and plasma renin activity did not increase in response to sodium restriction (Sizonenko et al. 1972). On the other hand, sodium restriction elicited significant increases in the urinary excretion of aldosterone metabolites and in plasma aldosterone in dexamethasone-treated and in cortisone-treated patients (Kowarski et al. 1968; New and Seaman 1972; Sizonenko et al. 1972; Levine et al. 1980; Zadik et al. 1984). However, these increases were in many instances subnormal and not sufficient to prevent urinary sodium losses (Kowarski et al. 1968; Holcombe et al. 1980; Zadik et al. 1984). In these patients, plasma renin activity rose to higher levels than in sodium-restricted normal subjects (Zadik et al. 1984). Plasma 18-OH-DOC (Kater and Biglieri 1983) and urinary excretion of 18-OH-DOC (Levine et al. 1980) were abnormally low in untreated patients with 11β-hydroxylase deficiency. The latter parameter did not rise in response to sodium restriction in dexamethasone-treated patients. According to these findings, 18-OH-DOC is secreted mainly by the zona fasciculata of the human adrenal cortex, and its biosynthesis from deoxycorticosterone is dependent on the catalytic activity of 11β-hydroxylase. In a patient studied by New and Seaman (1970), sodium restriction during dexamethasone treatment elicited a marked increase in the secretion rate of deoxycorticosterone. However, in four patients of Levine et al. (1980) and in 11 patients of Zadik et al. (1984) urinary excretion of free deoxycorticosterone and plasma deoxycorticosterone concentrations, respectively, did not increase in response to a low-salt diet during treatment with dexamethasone. Thus, in most patients with 11β-hydroxylase deficiency, the biosynthetic defect appears to be limited to the inner zones of the adrenal cortex and not to involve the zona glomerulosa. The mechanism of impaired aldosterone secretion in some patients receiving glucocorticoid treatment remains unknown. Insufficient suppression of ACTH and thus also of deoxycorticosterone secretion might be an important pathogenetic factor.

Congenital *17α-hydroxylase insufficiency* is clinically characterized by the combination of androgen and estrogen deficiency - causing primary amenorrhea in women (Biglieri et al. 1966b) and pseudohermaphroditism in men (New 1970) - with hypermineralocorticoidism (hypertension and hypokalemia). Hypermineralocorticoidism is generally thought to be due to excessive production of deoxycorticosterone and 18-OH-DOC by the zona fasciculata (Kater et al. 1983). In untreated patients, the secretion rate and plasma concentrations of aldosterone, as well as plasma renin activity, have generally been found to be subnormal (Biglieri et al. 1966b; Goldsmith et al. 1967; de Lange et al. 1977; Rovner et al. 1979; Kater et al. 1982; Kater and Biglieri 1983), but normal values of these parameters have been observed in a few instances (Saruta et al. 1980b; Tvedegaard et al. 1981). Since decreased aldosterone secretion was combined with a markedly increased secretion of 18-hydroxycorticosterone, and since aldosterone secretion remained low under long-term dexamethasone therapy, Biglieri et al. (1966) considered the possibility of a second biosynthetic defect in the aldosterone pathway, i.e., corticosterone methyl oxidase deficiency type 2. However, the increased secretion of 18-hydroxycorticosterone in 17α-hydroxylase deficiency appears to be due to excessive production of this steroid by the zona fasciculata rather than by the zona glomerulosa, according to indirect evidence presented by Kater et al. (1982). Thus, plasma concentrations of 18-hydroxycorticosterone became normal under treatment with dexamethasone. Moreover, the plasma level of 18-hydroxycorticosterone closely paralleled the levels of other corticosteroids known to be produced mainly by the zona fasciculata (corticosterone, deoxycorticosterone, 18-OH-DOC) but diverged from plasma renin activity in its circadian rhythm and its drop in response to upright posture. A deficiency of "18-hydroxylase" was postulated to be the cause of hypoaldosteronism in a patient with 17α-hydroxylase deficiency, in whom urinary excretion rates of 18-OH-DOC and of 18-hydroxycorticosterone were pathologically decreased before treatment and in whom aldosterone secretion remained subnormal within 10 months of dexamethasone treatment (Waldhäusl et al. 1978). In two patients studied by Saruta et al. (1980b), plasma 18-hydroxycorticosterone concentrations were found to be normal or only slightly elevated. However, in one of these patients plasma aldosterone rose to the low-normal range after 6 months of treatment with dexamethasone. In the patients studied by Kater et al. (1982), plasma aldosterone levels did not significantly rise during short-term glucocorticoid treatment, in spite of a marked increase in plasma renin activity. However, plasma aldosterone levels became very high in the first 2 weeks after cessation of treatment, when plasma renin activity was still elevated. After 3-4 weeks without treatment, plasma renin activity was subnormal and plasma aldosterone had returned into the low normal range. Incubated slices of an adrenal gland surgically removed from a patient with 17α-hydroxylase deficiency converted significant amounts of tritiated 18-hydroxycorticosterone to aldosterone, but only in the pressence of angiotensin II (5×10^{-10} mol/l; Rovner et al. 1979). Addition of angiotensin II to the incubation medium also stimulated the conversion of progesterone to aldosterone in this tissue by a factor of 6. Thus, at least in the majority of cases, hypoaldosteronism that manifests in treated patients with 17α-hydroxylase deficiency is not due to an inborn error of aldosterone biosynthesis.

5.11.6 Pseudohypoaldosteronism

Familial renal hyperkalemia and tubular acidosis resistant to mineralocorticoids occur in two clinical forms, type-1 and type-2 pseudohypoaldosteronism (Morris and Sebastian 1983). Type-1 pseudohypoaldosteronism is characterized by salt wasting and hypotension, whereas in type-2 pseudohypoaldosteronism renal sodium conservation is normal and blood pressure is normal or elevated.

A complete or partial deficiency of mineralocorticoid receptors has been found in three patients with *type-1 pseudohypoaldosteronism* and may be the primary lesion of this disorder (Armanini et al. 1985). In previously published cases there was a complete lack of response by the kidneys, salivary or sweat glands, colon mucosa, or all these target organs to high doses of mineralocorticoids (aldosterone, deoxycorticosterone acetate, 9α-fluorocortisol; for a review of case reports published from 1958 to 1977 see Oberfield et al. 1979). Irrespective of sodium intake, urinary and plasma aldosterone levels were abnormally high. Increased aldosterone secretion was associated with elevated plasma renin activity, hyperkalemia, hyponatremia, and in several instances with abnormally high levels of plasma cortisol and urinary 17-hydroxycorticoids as well (Rösler et al. 1977b; Rosenberg et al. 1980). In spite of high basal values, plasma and urinary aldosterone levels rose unequivocally in response to infusions of angiotensin II or KCl as well as to ACTH administration (Rösler et al. 1977b). The urinary excretion rate of TH-18-OH-A and the plasma concentrations of 18-hydroxycorticosterone were also found to be increased, but in proportion to the increased aldosterone secretion (Rosenberg et al. 1980; Rösler 1984). According to these observations, aldosterone biosynthesis is normal and is regulated by normal physiological control mechanisms in type-1 pseudohypoaldosteronism.

According to Schambelan et al. (1981), *type-2 pseudohypoaldosteronism* is characterized by a pathologically increased distal nephron chloride reabsorption leading to insufficient potassium and hydrogen secretion and increased sodium reabsorption and causing hyperkalemia, metabolic acidosis, hypervolemia, hypertension, and suppression of renin secretion. In this disorder, plasma and urinary aldosterone levels were either in the high-normal or in the moderately elevated range despite a very low plasma renin activity. On the other hand, aldosterone secretion was considered to be inappropriately low with regard to the hyperkalemia.

5.11.7 Spirolactones

Most likely, the major pharmacological activity of spirolactones is due to their ability to displace aldosterone and other mineralocorticoids competitively from their receptor binding sites in target organs. However, according to short-term experiments with adrenocortical tissue or cells in vitro, these agents are also potent inhibitors of aldosterone biosynthesis (see Sect. 4.8.5 of this monograph). Moreover, spirolactones have been found to interfere with the hepatic metabolism of aldosterone (Tsai and Morris 1978; Tsai et al. 1980). These multiple sites of interaction between spirolactones and aldosterone biodynamics may explain at least some of the conflicting results reported by investigators who have studied the long-term in vivo effects of spirolactones on aldosterone secretion in man or animals under physiological or pathological conditions.

Abshagen et al. (1978) treated young men on a constant sodium and potassium intake (each 75 mmol/day) for 7 days either with spironolactone (300 mg/day) or with triamterene (200 mg/day). Both drugs induced a similar urinary sodium loss and elicited a similar marked and progressive rise in plasma aldosterone concentrations. However, plasma renin activity and angiotensin II concentration increased to a considerably greater extent under spironolactone than under triamterene medication. Under spirolactone treatment there were also significant rises in the plasma concentrations of deoxycorticosterone, corticosterone, and 18-OH-DOC. These observations indicated that a partial block of aldosterone biosynthesis by spironolactone was compensated by a greater stimulation of the renin-angiotensin system. Neither drug had any significant effect on the plasma concentrations of progesterone, 17α-hydroxyprogesterone, 11-deoxycortisol, or cortisol. In a study by Gaillard et al. (1980), normal young men were treated with spironolactone (400 mg/day) for 3 days during a constant sodium intake of 130 mmol/day. The treatment resulted in a fourfold increase in the aldosterone secretion rate and in proportional increases in the urinary excretion rates of aldosterone-18-glucuronide and tetrahydroaldosterone-3-glucuronide. During spironolactone administration there were parallel progressive rises in plasma aldosterone and renin activity, whereas plasma cortisol and corticosterone concentrations remained constant. Spironolactone medication did not affect ACTH-induced increases in urinary aldosterone excretion. Also, short-term administration of the drug did not lower the plasma aldosterone concentration in subjects receiving an angiotensin II infusion. According to this evidence, aldosterone biosynthesis was not impaired by the drug. However, Tuck et al. (1981) reached a different conclusion from plasma corticosteroid concentrations measured in normal subjects after 5 and 10 days of spironolactone administration (400 mg/day). During the first 5 days, plasma aldosterone, 18-hydroxycorticosterone, and renin activity rose to similar extents. From day 5 to day 10, plasma aldosterone returned to the pretreatment level, plasma 18-hydroxycorticosterone continued to increase, and plasma renin activity remained constant. Plasma deoxycorticosterone and – to a lesser extent – plasma progesterone increased continuously during the whole period of medication, whereas a smaller rise in plasma corticosterone was not statistically significant. These findings were interpreted to be indicative of a partial blockade of aldosterone biosynthesis at two separate sites, i.e., at the levels of 11β-hydroxylation and the second methyl oxidation. No evidence of an inhibition of aldosterone biosynthesis was seen in patients with essential hypertension chronically treated with spironolactone (3 months to 14 years), who had a significantly higher urinary excretion rate of aldosterone and higher levels of plasma renin activity (supine and standing) than untreated hypertensives (Lewis et al. 1980). Spironolactone-treated patients also required less ACTH to induce a detectable rise in plasma cortisol and 18-OH-DOC and showed a greater response in plasma 18-OH-DOC to all doses and a greater response in plasma aldosterone to low doses of ACTH. Administration of canrenone (164 and 328 mg/day) for 10 days to diabetic patients without cardiovascular, liver, or kidney involvement led to a marked rise in plasma renin activity but to only slight increases in urinary and plasma aldosterone (Erbler et al. 1976). A clear-cut increase in urinary aldosterone excretion was seen only after the discontinuation of the drug. In these patients, canrenone did not alter urinary sodium and potassium excretion rates but nevertheless lowered the plasma sodium and raised the plasma potassium concentration, possibly through an action on the colon.

Long-term spironolactone medication led to a marked and persistent decrease in the plasma concentration and secretion rate of aldosterone in a woman with primary aldosteronism due to an adrenal adenoma (Sundsfjord et al. 1974). In other patients with aldosterone-producing adenomas, treatment with spironolactone induced an early marked decrease in urinary and plasma aldosterone levels while raising plasma renin activity and potassium concentration (Conn and Hinerman 1977). However, within several months of continued medication, the two parameters of aldosterone secretion rose to pretreatment or even higher values but remained in most instances inappropriately low in relation to plasma renin activity and potassium concentration. Abshagen et al. (1980) observed no consistent changes in plasma aldosterone or renin activity during 5–7 days of spironolactone administration to four patients with primary aldosteronism (two aldosterone-producing adenomas, two cases of idiopathic hyperaldosteronism). However, in a fifth patient (idiopathic hyperaldosteronism), spironolactone elicited a pronounced elevation in plasma renin activity and in the plasma concentrations of aldosterone, corticosterone, deoxycorticosterone, 18-OH-DOC, and 11-deoxycortisol. In the other four patients, corticosterone and deoxycorticosterone levels increased also, but to a markedly lesser extent. In two patients with aldosterone-producing adenomas studied by H. Vetter et al. (1977), treatment with canrenone for a week lowered the urinary aldosterone-18-glucuronide excretion rate to the normal range but had a much weaker suppressive effect on plasma aldosterone levels. Continued therapy with higher doses of canrenone did not significantly alter plasma aldosterone and was associated with a return of the urinary aldosterone-18-glucuronide excretion rate to pretreatment values. Cessation of the drug was followed by a marked increase in both parameters. After 6 months' therapy with spironolactone in another patient with an aldosterone-producing adenoma (when plasma renin activity had returned to normal), spontaneous diurnal variations in plasma aldosterone were significantly correlated with variations in plasma cortisol but not with variations in plasma renin activity (H. Vetter and Vetter 1975b). According to Kater et al. (1983), there were profound differences between patients with aldosterone-producing adenomas and patients with idiopathic hyperaldosteronism in their aldosterone secretory responses to spironolactone treatment. Urinary and plasma aldosterone rose markedly in patients with idiopathic aldosteronism but remained unchanged in patients with an aldosterone-producing adenoma, in spite of similar increases in plasma potassium and renin activity in the two groups. In one of the adenoma patients, urinary and plasma aldosterone increased significantly during prolonged treatment with amiloride. In spironolactone-treated patients with aldosterone-producing adenomas, plasma aldosterone and 18-hydroxycorticosterone concentrations did not increase in response to upright posture but rose in response to ACTH. The pattern of plasma deoxycorticosterone, corticosterone, and 18-hydroxycorticosterone levels (basal and after ACTH administration) was not affected by spironolactone treatment. Accordingly, the drug did not block aldosterone biosynthesis but rather caused (or a least did not reverse) a sustained impairment of the aldosterone-secretory response to angiotensin II and to elevated extracellular potassium concentrations. On the other hand, aldosterone output by the adenomas in vivo increased in response to ACTH to a similar extent before and after spironolactone medication. However, the in vitro aldosterone output by incubated adenoma tissue was not responsive to ACTH or to angiotensin II after prolonged (3 weeks to 1 year) preoperative spironolactone treatment of the patients (Wisgerhof 1984). Short-term treatment (8–10 days) blunted mainly the response to angiotensin II but not the response to ACTH.

Prolonged administration of spironolactone to man leads to a typical morphological alteration in the zona glomerulosa cells, i.e., the formation of so-called spironolactone bodies (Janigan 1963; Jenis and Hertzog 1969; D.A. Davis and Medline 1970). These are eosinophilic, laminated structures within the cytoplasm of the adrenal glomerulosa cells, consisting of whorls of concentrically arranged cytoplasmic reticulum containing a central core of lipid material and having a diameter of 2–20 μm. Spironolactone bodies were not induced by any other form of long-term diuretic therapy, occurred exclusively in intracapsular blastema cells, zona glomerulosa cells, and glomerulosa-like cells located along the adrenal venous drainage, and were not found in any other part of the adrenal gland or in any other organ of the body. In adrenal glands surgically removed from spironolactone-treated patients with aldosterone-producing adenomas, spironolactone bodies were found in the tumor cells but not in the cells of the adjacent zona glomerulosa (Conn and Hinerman 1977). Spironolactone bodies were most abundant after 40–60 days of treatment with the drug. Afterwards, their number decreased again, and they virtually disappeared after 170 days of spironolactone medication. They were not observed in the aldosterone-producing adenomas taken from two patients who had stopped taking the drug 18 and 24 days, respectively, before surgery. According to Sundsfjord et al. (1974), there were many spironolactone bodies in the adjacent glomerulosa cells but few in the adenoma cells of an adrenal gland removed from a patient with primary aldosteronism after 6 months of spironolactone therapy.

In acutely nephrectomized dogs, a bolus injection of canrenoate potassium significantly blunted the rise in plasma aldosterone in response to intravenous infusions of angiotensin II, potassium chloride, or ACTH and the increase in plasma cortisol elicited by ACTH (Lasaridis et al. 1984). In spironolactone-treated, but not in untreated dogs, all three agents significantly raised the plasma progesterone levels.

In moderately sodium-deficient sheep, infusions of spironolactone for 4–5 h, which completely inhibited the effect of endogenous aldosterone on the electrolyte composition of parotid saliva and significantly raised the plasma potassium concentration, did not consistently alter the adrenal secretion rates of aldosterone, corticosterone, or cortisol (Blair-West et al. 1965b).

The oral administration of spironolactone to rats for 6 weeks led to marked and persistent increases in urinary aldosterone, plasma aldosterone, and plasma renin activity (Erbler et al. 1979). Similar but less extensive and transient increases in these parameters wer elicited by treatment with canrenone. According to Chabert et al. (1984), the subcutaneous administration of spironolactone to rats on a normal sodium intake caused sustained five- to seven fold rises in the urinary excretion rate and in the plasma concentration of aldosterone and a much smaller rise (+70%) in plasma renin activity. It also moderately (+100%) raised the plasma concentrations of progesterone and deoxycorticosterone but did not affect plasma corticosterone, plasma 18-OH-DOC, or urinary corticosterone levels. In rats kept on a low-sodium diet, spironolactone administration caused significant additional increases in plasma and urinary aldosterone.

5.11.8 Heparin and Heparinoids

The natriuretic effect of heparin and heparinoids[4] is considered to be due mainly to suppression of aldosterone secretion. These drugs markedly lowered the urinary

[4] Although a number of different heparin derivatives exert a similar activity, the heparinoid most frequently used in these investigations has been Ro 1-8307 (*N*-formyl chitosan polysulfuric acid), which has a very low anticoagulant activity but is almost as active as heparin in promoting natriuresis.

excretion rate, secretion rate, and plasma concentration of aldosterone in normal human subjects on a liberal or restricted sodium intake and in patients with essential hypertension or edema or who were recovering from surgery (Cejka et al. 1960; Veyrat et al. 1963; Schlatmann et al. 1964; Abbott et al. 1966b; Bailey and Ford 1969; O'Kelly et al. 1983). Although heparin has been found to directly inhibit renin activity in vitro, probably by competition with renin substrate for the active site of the enzyme (Sealey et al. 1967), heparin treatment significantly raised plasma renin activity while lowering plasma aldosterone concentrations (O'Kelly et al. 1983). Moreover, heparin and heparinoids also decreased aldosterone secretion in patients with primary aldosteronism, i.e., in subjects with a very low plasma renin activity (Ford and Bailey 1966; Conn et al. 1966; Abbott et al. 1966b). They did not affect 17-hydroxycorticoid or 17-ketogenic steroid excretion (Abbott et al. 1966b; Conn et al. 1966), they did not lower plasma cortisol (O'Kelly et al. 1983), and they led to increases in the plasma potassium concentration (Abbott et al. 1966b; Phelps et al. 1980). These findings indicated that inhibition of aldosterone secretion by heparin or heparinoids was not secondary to inactivation of the renin-angiotensin system, suppression of ACTH secretion or hypokalemia. When added to the medium in which rat adrenal tissue was incubated, heparin or heparinoids did not affect the in vitro production of aldosterone (Gláz and Sugár 1964; J. Müller 1971; J. H. Levine et al. 1972). Significantly decreased amounts of aldosterone were produced in vitro, however, by adrenals taken from rats 6 h after the injection of a single dose of heparin or heparinoids in vivo (Gláz and Sugár 1964). Adrenals of rats treated with Ro 1-8307 (20 mg/day, s.c.) for 3 days produced normal amounts of aldosterone under basal conditions of incubation; when serotonin or KCl was added to the incubation medium, adrenals of heparinoid-treated rats put out approximately 30% less aldosterone but 30% more corticosterone and deoxycorticosterone than adrenals of control animals (J. Müller 1971). After 4 or 7 days of treatment with the same drug, rat adrenal aldosterone production in vitro was significantly decreased, whereas corticosterone output was unchanged (J. H. Levine et al. 1972). According to Uchida et al. (1986), plasma aldosterone was decreased, plasma renin activity and angiotensin II concentrations were increased, and plasma corticosterone was unchanged in rats treated with heparin for 6 weeks. Zona glomerulosa cells of heparin-treated rats produced normal amounts of aldosterone, 18-hydroxycorticosterone, and corticosterone under basal conditions of incubation. However, the maximal response in the output of these steroids to angiotensin II was substantially decreased. Heparin treatment also markedly lowered the response in aldosterone output to increasing potassium concentrations but hardly affected the response in aldosterone output to ACTH. The same investigators also found a direct suppressive effect of heparin on aldosterone production by rat adrenal zona glomerulosa cells in vitro under stimulation by angiotensin II but not under basal conditions of incubation. In man, heparinoid treatment decreased the secretion rate of 18-hydroxycorticosterone (Abbott et al. 1966b) and the urinary excretion of the major metabolite of 18-hydroxycorticosterone but did not alter the urinary excretion of tetrahydrocorticosterone (Conn et al. 1966). These observations indicated that aldosterone biosynthesis was inhibited at the stage of corticosterone methyl oxidation 1. Prolonged treatment of a patient with idiopathic hypercholesterolemia with heparin resulted in marked atrophy of the zona glomerulosa of the adrenal cortex and an increased juxtaglomerular cell index in the kidneys (I. D. Wilson and Goetz 1964). Long-term (2–4 weeks) administration of Ro 1-8307 to rats led to significant narrowing or even to complete atrophy of the zona glomerulosa (Abbott et al. 1966a) even in rats kept on a low-sodium diet or in adrenals regenerating after enucleation (Chremos et al. 1969). However, decreases in aldosterone output preceded measurable decreases in zona glomerulosa width (J. H. Levine et al. 1972). Ultrastructural changes observed in rat adrenals after 3 days of heparin administration consisted of a simplified structure of the mitochondrial cristae and an increased density of lipid droplets in the zona glomerulosa cells (Lustyick et al. 1976). Dark and light zona glomerulosa cells could no longer be distinguished.

6 Conclusions

6.1 Multiplicity of Aldosterone-Stimulating Substances

The number of substances reported to be capable of directly stimulating aldosterone production from endogenous precursors in autotransplanted or isolated perfused adrenal glands in vivo or in incubated or superfused adrenal zona glomerulosa cells or tissue in vitro is already amazingly large, but it appears to be constantly increasing. Only agents that could conceivably be of physiological importance as intercellular mediators of aldosterone secretion have been listed in Table 11. Structural analogues of hormones, drugs, toxins, and presumable intracellular mediators are not included. When tested in preparations of dispersed zona glomerulosa cells, most of these agents elicited dose-dependent significant increments of aldosterone production in physiological concentrations. In incubated intact zona glomerulosa tissue preparations, some stimulators, in particular peptide hormones such as angiotensin II or ACTH, displayed their steroidogenic activity only at relatively high concentrations, often far above the physiological range. Other substances, such as catecholamines, stimulated aldosterone production only when tested in a system of previously cultured zona glomerulosa cells or tissue (De Léan et al. 1984c; Pratt et al. 1985). There are also obvious species differences. Thus, serotonin, a potent stimulator of steroidogenesis in rat quartered adrenals (Müller and Ziegler 1968) and zona glomerulosa cells (Haning et al.

Table 11. Direct stimulators of aldosterone biosynthesis of potential physiological importance. (See Chap. 4, Sects. 1-6, for details and references)

1. Peptides and proteins	*3. Amines*
- Angiotensins II and III	- Serotonin
- ACTH	- Related indoleamines
- α-MSH	- Histamine
- Other POMC-derived peptides	- β-Adrenergic agonists
- ASF (pituitary glycoprotein)	*4. Prostaglandins*
- Prolactin	- E_2
- Vasopressin	- I_2
2. Monovalent cations	- Others
- K^+	
- NH_4^+	
- H^+	

Abbreviations: *ACTH,* adrenocorticotropic hormone; *MSH,* melanocyte-stimulating hormone; *POMC,* pro-opiomelanocortin (biological precursor protein of ACTH and β-lipotropin); *ASF,* "aldosterone-stimulating factor" (isolated from human urine)

1970), was completely inactive in sheep adrenocortical tissue (McDougall et al. 1976). However, conflicting results about the aldosterone-stimulating activity of some ACTH-related peptides and prostaglandins have been reported by different investigators from experiments performed in apparently very similar systems of zona glomerulosa cells from the same animal species (see Sects. 4.4.2 and 4.6.1).

Some agents, such as hydrogen ions or γ-MSH-related peptides, seem to have little or no inherent aldosterone-stimulating activity on their own but to potentiate the steroidogenic effect of other stimulators. However, in most instances, each of the agents listed in Table 11 acted independently of other stimulators, provided that the incubation medium contained glucose, calcium ions, and a potassium concentration in the normal range. Potassium ions stimulated aldosterone production maximally in a medium free of angiotensin II, ACTH, and serotonin. Serotonin, ACTH, and angiotensin II each induced a maximal increase in aldosterone output when added to an incubation medium with a potassium concentration of 3.6 mmol/l, which is at the lower limit of the normal range of plasma potassium levels. When two aldosterone-stimulating agents were added simultaneously to the incubation medium - each at a dose which by itself maximally stimulated aldosterone production - no cumulation of the steroidogenic effects was observed. The principle of multiple and independent aldosterone-stimulating agents that has been observed in vitro appears to pertain to short-term stimulation of aldosterone secretion in vivo, too. Thus, angiotensin II stimulated aldosterone production in hypophysectomized dogs (J. O. Davis 1962), and potassium was a potent aldosterone stimulator in nephrectomized and hypophysectomized animals (J. O. Davis et al. 1963). Also, in certain cases of renovascular hypertension, a high output of aldosterone is maintained presumably by the renin-angiotensin system despite marked hypokalemia (see Sect. 4.5.1). This apparent equality of aldosterone-stimulating substances sharply contrasts with the exclusive role of ACTH in the regulation of glucocorticoid secretion. In a hypophysectomized animal no known humoral agent, with the possible exception of cyclic AMP, can fully replace the cortisol- and corticosterone-stimulating effect of ACTH.

If we accept the principle of multiple and equal aldosterone stimulators, we must also accept the possibility that some potent aldosterone-stimulating substances are still undiscovered. Some such factors might be biologically unimportant, whereas others might be of physiological or clinical relevance. Reports by Carey et al. (1984) and by Griffing et al. (1985a) of elevated plasma levels of the "aldosterone-stimulating factor" (see Sect. 4.4.4) and of immunoreactive γ-MSH (see Sect. 4.4.2), respectively, in patients with nontumorous "primary" hyperaldosteronism are strongly suggestive of a pathogenetic role of these relatively recently discovered and only partially identified pituitary factors in a disease formerly thought to be "idiopathic" (Ross 1965; W. W. Davis et al. 1967; Katz 1967; Salti et al. 1969b; Baer et al. 1970; Weinberger 1984) or nonexistent as an entity separate from essential hypertension (J. J. Brown et al. 1979; Padfield et al. 1981). Other, yet unknown aldosterone-stimulating factors might conceivably account for the "normal" aldosterone plasma concentrations and secretion rates observed in patients with low-renin essential hypertension.

6.2 "Adrenoglomerulotropins"

The term "adrenoglomerulotropin" was introduced by Farrell (1958, 1960) to designate a hypothetical hormone of diencephalic or pineal origin which controlled specifically the secretion of aldosterone by the zona glomerulosa of the adrenal cortex. Later, evidence presented by Farrell and McIsaac (1961) indicated that "adrenoglomerulotropin" was identical with 6-methoxytetrahydroharman, a carboline derivative. However, in the hands of other investigators, this substance did not affect aldosterone secretion by sheep or dog adrenal glands in vivo (Blair-West et al. 1963; Mulrow et al. 1963) or by rat adrenal tissue in vitro (Jouan 1967; J. Müller and Ziegler 1968). Irrespective of its original use, the term "adrenoglomerulotropin" would very appropriately describe the biological activity of a group of substances that have been reproducibly found to stimulate aldosterone production. This group of "adrenoglomerulotropins" includes potassium ions, related monovalent cations, angiotensins II and III, serotonin and related indoleamines, and possibly other aldosterone-stimulating agents listed in Table 11. "Adrenoglomerulutropins" stimulate steroidogenesis exclusively or preferentially in the zona glomerulosa of the adrenal cortex, but they do not stimulate aldosterone production under all experimental conditions, and they also stimulate the biosynthesis of other steroids produced in the zona glomerulosa, such as 18-hydroxycorticosterone, corticosterone, 18-hydroxy-11-deoxycorticosterone, and deoxycorticosterone.

Angiotensin II, potassium ions, and serotonin stimulated steroidogenesis only in the capsular portion of rat adrenals (J. Müller 1970b) or only in rat zona glomerulosa cells (Haning et al. 1970) and had no effect on steroid production by tissue preparations or cells derived from the decapsulated portion of the glands. In contrast, ACTH stimulated steroidogenesis to a similar extent in both portions of the rat adrenal gland. In quartered adrenals of potassium-deficient rats, potassium ions and serotonin did not stimulate aldosterone production, but significantly augmented the production of deoxycorticosterone (J. Müller and Huber 1969). As shown in later experiments with separated zones, the response in deoxycorticosterone output to stimulation by these two agents and to angiotensin II was entirely due to production by the capsular portion. Thus, at least in short-term experiments with rat adrenals, serotonin, potassium ions, and angiotensin II are true "adrenoglomerulotropins", stimulating steroidogenesis exclusively in the zona glomerulosa even in adrenals in which they do not stimulate aldosterone production.

In man, as well as in animal species such as sheep, dogs, and cattle, angiotensin II appears to have at least a preferential action on the zona glomerulosa. In hypophysectomized dogs high doses of angiotensin II stimulated cortisol secretion. However, in such animals sodium deficiency also led to increased cortisol secretion, which may indicate that some cortisol is produced by the zona glomerulosa of the dog adrenal or that the zona fasciculata can be stimulated by high doses of angiotensin II, but only in the complete absence of ACTH. On the other hand, there is no convincing evidence that potassium ions can directly stimulate cortisol biosynthesis in vivo or in vitro in any animal species, with the possible exception of the dog (Burwell et al. 1969).

The zone specificity of hormonal "adrenoglomerulotropins" such as angiotensin and serotonin is dependent on the zonal distribution of specific high-affinity receptors closely associated with regulatory enzymes controlling the production of operational second messengers. On the other hand, the selective adrenoglomerulotropic activity of potassium ions appears to be based on the property of the zona glomerulosa cell plasma membrane to be depolarized by small increases in the extracellular potassium concentration within the physiological range. Thus, the zona glomerulosa cell is characterized both by its ability to respond to "adrenoglomerulotropins" and by its ability to produce aldosterone.

6.3 Physiological Inhibitors of Aldosterone Biosynthesis

In view of the abundance of stimulators in the mammalian organism, a balanced regulation of aldosterone secretion would seem to require the existence of at least one potent inhibitor of aldosterone biosynthesis. Indeed, there is a growing list of endogenous substances reported to directly inhibit aldosterone output by zona glomerulosa cells that could conceivably play a role as physiological antagonists of aldosterone stimulators (Table 12). These substances affected aldosterone production in different ways. Some steroid hormones blocked the conversion of added precursor steroids to aldosterone. Other steroid hormones or steroid-hormone metabolites inhibited the binding of angiotensin II to its receptors in the zona glomerulosa cell membrane. Agents such as somatostatin or dopamine selectively blocked the aldosterone-stimulating effect of angiotensin and did not impede the activity of other stimulators. However, these two inhibitors interacted with their own receptors and did not affect the receptor binding of angiotensin.

From the physiological point of view, the most interesting inhibitors of aldosterone production yet discovered are the atrial natriuretic peptides. The secretion of these substances is modified by the sodium status. High plasma levels of atrial natriuretic peptides are observed in situations in which aldosterone secretion is suppressed, e.g., during mineralocorticoid treatment (Sugimoto et al. 1986). More-

Table 12. Direct inhibitors of aldosterone biosynthesis of potential physiological importance. (See Chap. 4, Sects. 2.1, 4.4, 6.1, 7, 10, and 11 for details and references)

1. Peptides	*4. Steroids*
- Atrial natriuretic peptides	- Ouabain*
- Adrenal medullary inhibitor	- Androgens
- Somatostatin	- Gestagens
	- Glucocorticoids
2. Monovalent cations	- Steroid hormone metabolites
- Na^+	
	5. Prostaglandins
3. Amines	- E_2
- Dopamine	- $F_{1\alpha}$
- Dopamine-3-sulfate	- $F_{2\alpha}$
	- Others

* The plant glycoside ouabain has been included in this list because endogenous analogues of this agent exist in the mammalian organism (de Wardener and Clarkson 1982). Such a compound has been isolated from bovine hypothalamus (Haupert and Sancho 1979; Haupert et al. 1984).

over, atrial natriuretic peptides have been reported to inhibit the steroidogenic action of all stimulators tested, including angiotensin II, ACTH, potassium ions, and serotonin, in zona glomerulosa cell and tissue preparations of various animal species in vitro and to lower aldosterone plasma levels in man and experimental animals in vivo. Although details about the mode of action of atrial natriuretic peptides in inhibiting stimulated aldosterone production are as yet unknown, there is at least good evidence indicating (a) that these agents act through their own receptors and do not affect receptor binding of stimulators and (b) that they do not block aldosterone biosynthesis from pregnenolone.

Like atrial natriuretic peptides, the plant glycoside ouabain equally inhibited the steroidogenic effects of all stimulators tested, i.e., of angiotensin II, potassium ions, ACTH, and serotonin. The inhibitory activity of this drug may indicate that in the future the ouabain-like hypothalamic natriuretic hormone (de Wardener and Clarkson 1982; Haupert and Sancho 1979; Haupert et al. 1984) might turn out to also be a physiologically important regulator of aldosterone secretion.

Atrial natriuretic peptides and ouabain resembled "adrenoglomerulotropins" in their zone specificity. Both agents inhibited ACTH-stimulated steroidogenesis selectively or preferentially in the zona glomerulosa of the rat adrenal cortex.

6.4 The Shrinking Black Box

The development of new experimental models, in particular the introduction of incubated or superfused dispersed zona glomerulosa cell preparations in the early 1970s, also started an era of intensive study of the mechanisms of action of various aldosterone-stimulating substances. The efforts of many groups of outstanding investigators active in this field have considerably elucidated an area of cell physiology formerly characterized by the traditional symbol of the "black box". As illustrated by a simplified contemporary model of the modes of the short-term steroidogenic action of the most important regulators of aldosterone biosynthesis (Fig. 47), a definite region of ignorance still exists within this area, but it appears to be steadily shrinking.

In the zona glomerulosa, as well as in the zona fasciculata, of the adrenal cortex, the interaction of ACTH with a specific membrane receptor results in the activation of adenylate cyclase and thus in the formation of cyclic AMP, which acts as a second messenger and, in turn, activates a protein kinase. Serotonin acts in a similar manner, although through its own receptor. The interaction of angiotensin with its specific receptor leads to the activation of the membrane-bound enzyme phospholipase C, which catalyzes the breakdown of the membrane phospholipid phosphatidylinositol-4,5-bisphosphate into 1,2-diacylglycerol and myoinositol-1,4,5-trisphosphate. Diacylglycerol activates protein kinase C. Myoinositol-trisphosphate leads to the release of calcium ions from the endoplasmic reticulum and to an elevation of cytosolic calcium concentration. The liberated calcium ions are bound to calmodulin and lead to the activation of calmodulin-dependent protein kinases. The zona glomerulosa cell is particularly sensitive to changes in the extracellular potassium concentration and is depolarized by relatively small increases of this parameter. Potassium-induced decreases in membrane potential

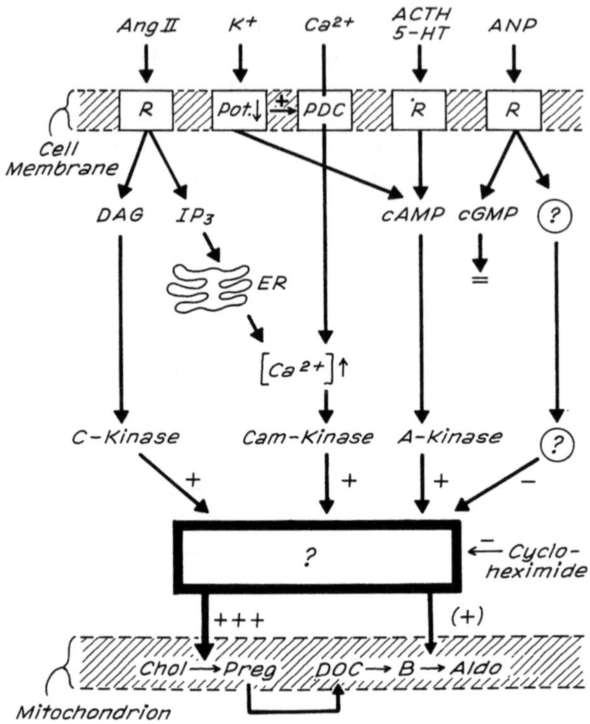

Fig. 47. Contemporary view (1986) of modes of action of angiotensin II *(Ang II)*, potassium ions, ACTH, serotonin *(5-HT)*, and atrial natriuretic peptides *(ANP)* on corticosteroid biosynthesis in mammalian adrenocortical zona glomerulosa cells: *R*, receptor; *Pot.*, membrane potential; *PDC*, potential-dependent calcium channel; *DAG*, 1,2-diacylglycerol; IP_3, *myo*inositol 1,4,5-trisphosphate; *ER*, endoplasmic reticulum; *C-Kinase*, protein kinase C; *Cam-Kinase*, calmodulin-dependent protein kinase; *A-Kinase*, cyclic AMP-dependent protein kinase; *Chol*, cholesterol; *Preg*, pregnenolone; *DOC*, deoxycorticosterone; *B*, corticosterone; *Aldo*, aldosterone. (From J. Müller 1987)

lead to the opening of potential-dependent calcium channels, and thus raise the cytosolic calcium concentration by facilitating the entry of this ion into the cell. At least according to some investigators (I. Kojima et al. 1985b; Hyatt et al. 1986), increases in the extracellular potassium concentration also stimulate the production of cyclic AMP by an unknown mechanism dependent on the influx of calcium ions.

Although different stimulators act on separate receptors, elicit the production of different messengers, and activate different protein kinases, their major biological effect appears to be identical, i.e., the facilitation of a step early in the biosynthetic pathway, preceding the formation of pregnenolone. This effect is dependent on protein synthesis and can be blocked by cycloheximide (but not by actinomycin D). Accordingly, a regulatory protein or peptide may be translated de novo in response to the stimulators. Alternatively, a regulatory protein with a short biological half-life may be activated, e.g., by phosphorylation. Such a regulatory protein might be a proteolytic enzyme regulating the formation of a peptide activating cytochrome $P\text{-}450_{SCC}$ (R. C. Pedersen and Brownie 1983b) or might be involved in the intramitochondrial transport of cholesterol (Vahouny et al. 1983; Lambert et al. 1986).

Most aldosterone-stimulating agents have also been shown to have a minor short-term stimulatory effect on the conversion of corticosterone to aldosterone. Generally, these observations resulted from experiments with elaborate protocols requiring pharmacological inhibitors of steroidogenic enzymes and exogenous precursor steroids. Very few details about this second effect of stimulators on aldosterone biosynthesis have been reported. Thus, it is not yet known which of the two corticosterone methyl oxidations is mainly affected, how soon the effect on late steps becomes apparent, and whether it also depends on protein synthesis.

The mechanism of the inhibitory action of atrial natriuretic peptides on stimulated aldosterone biosynthesis is still unknown. These substances have been shown to activate particulate guanylate cyclase in the zona glomerulosa as well as in their other target tissues. While there is some evidence indicating that the action of atrial natriuretic peptides on vascular smooth muscle cell relaxation (Hamet et al. 1984; Winquist et al. 1984) and on renin release from renal juxtaglomerular cells (Kurtz et al. 1986) is mediated by cyclic GMP, this does not seem to be the case in the rat adrenal zona glomerulosa. Thus, neither added dibutyryl cyclic GMP nor sodium nitroprusside imitated the inhibitory effect of atriopeptin II on the serotonin-stimulated steroidogenesis in rat capsular adrenal tissue (J. Müller 1986b; Fig. 21). Apparently, these agents block the steroidogenic action of stimulators in the black-box area distal to receptor binding, primary events, and known second messengers but proximal to aldosterone biosynthesis from pregnenolone (Fig. 47).

6.5 Long-Term Regulation of Late Steps in Aldosterone Biosynthesis

In man as well as in experimental animals, the response in aldosterone secretion to a constant dose of a given stimulator is variable and dependent on sodium and potassium intake (see Sects. 5.1.3 and 5.2.4). Some of this variability is due to changes in zona glomerulosa mass and the concentration of receptors for aldosterone-stimulating substances in the zona glomerulosa cell membrane. However, there is good evidence that specific alterations in aldosterone biosynthesis also play an important role. Thus, agents markedly stimulating aldosterone production in capsular adrenal tissue of normally fed rats did not stimulate aldosterone production when added to capsular adrenals of potassium-deficient rats but elicited strikingly greater increments in deoxycorticosterone output (see Figs. 8, 11, 14, and 16). On the other hand, the deoxycorticosterone-stimulating effect of the same agents was markedly reduced in capsular adrenals of sodium-deficient rats. This indicated that the activity of one or more late steps in aldosterone biosynthesis involved in the conversion of deoxycorticosterone to aldosterone was rate limiting for the aldosterone production elicited by the stimulators and was subject to considerable changes in response to alterations in sodium and potassium intake (Fig. 48).

According to conversion experiments, sodium and potassium intake affected mainly corticosterone methyl oxidations 1 and 2 but had only a minor influence on the capsular adrenal 11β-hydroxylation of deoxycorticosterone. Alterations in the rates of steroid 11β- and 18-hydroxylation induced by alterations in the sodium and potassium status were clearly restricted to the zona glomerulosa of the adrenal cortex.

Experimental manipulations which led to a significant increase or decrease in the conversions of substrate amounts of corticosterone to 18-hydroxycorticoster-

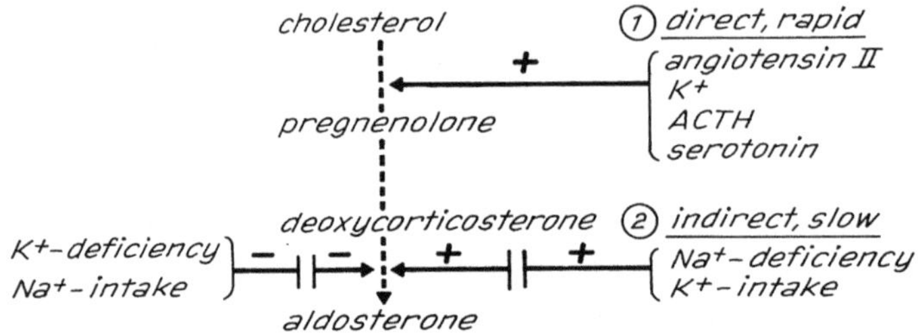

Fig. 48. Two sites of regulation of aldosterone biosynthesis in the rat adrenal zona glomerulosa

one and aldosterone by incubated capsular tissue or mitochondria in rats within a few days are listed in Table 13. According to the indirect evidence obtained in such experiments, the regulation of late steps in aldosterone biosynthesis is multifactorial. Most likely, alterations in response to changes in potassium intake are mediated mainly by the extracellular potassium concentrations. On the other hand, increases in the conversion of corticosterone to 18-hydroxycorticosterone and aldosterone induced by sodium restriction or sodium sequestration seem to result from a combination of elevated plasma concentrations of angiotensin II and potassium and yet unknown factors (J. Müller et al. 1982, 1984). Thus, some agents that have a short-term stimulatory effect on early steps of aldosterone biosynthesis also have a long-term stimulatory effect on late steps. These agents include potassium ions, angiotensin II, and perhaps serotonin. The exception is ACTH, which stimulates early steps in short-term experiments but suppresses late steps during prolonged in vivo administration (see Sect. 5.6.2). According to Aguilera and Catt (1984), dopamine may also be involved in the regulation of late steps in aldosterone biosynthesis during alterations of sodium intake. On the other hand, no evidence has yet been published which would indicate that atrial natriuretic peptides or ouabain have any long-term effect on corticosterone methyl oxidations.

The principle of independent and equal stimulators established in short-term in vitro experiments seems to also be valid for the long-term in vivo regulation of late

Table 13. Experimental manipulations leading to alterations in the capsular adrenal conversions of corticosterone to 18-hydroxycorticosterone and aldosterone in rats

Stimulation	Suppression
- Sodium restriction	- Potassium restriction
- Sodium sequestration[a]	- Sodium loading
- Furosemide administration	- Mineralocorticoid administration
- Potassium loading	- ACTH administration
- Uremia[b]	- Glucocorticoid administration
- Angiotensin II infusion	- Converting-enzyme inhibitor administration
- 5-Hydroxytryptophan administration[c]	- Chloride restriction[d]
- Water restriction[c]	
- Hypophysectomy[c]	

[a] Subcutaneous polyethylene glycol or formalin
[b] Bilateral nephrectomy or bladder resection
[c] Only in potassium-depleted rats
[d] Temporary effect except when combined with restriction of sodium and potassium (J. Müller 1986)

steps in aldosterone biosynthesis. Thus, hyperkalemia due to uremia considerably stimulated the conversion of corticosterone to aldosterone in bilaterally nephrectomized rats, i.e., in the (assumed) absence of circulating angiotensin II (J. Müller and Huber 1969). Also, fourfold increases in the activity of the final steps of aldosterone biosynthesis were induced in potassium-depleted rats by sodium sequestration or angiotensin II infusion in spite of marked and persistent hypokalemia, with plasma potassium concentrations as low as 2.2 mmol/l (J. Müller et al. 1982).

6.6 A Matter of Time

Although angiotensin II and potassium ions have been observed to enhance the conversion of corticosterone to aldosterone during short-term incubation experiments as well, the long-term in vivo stimulation of methyl oxidations 1 and 2 by these two agents is probably a different event, at least as indicated by its time course. In vivo alterations in the activity of late steps in aldosterone biosynthesis occur slowly over the course of days or at least several hours and are often characterized by a delayed onset. When previously potassium-depleted rats were repleted by an intravenous potassium chloride infusion, a statistically significant increase in the conversion of corticosterone to aldosterone was observed only after 6 h, although the presumable stimulator, the serum potassium concentration, had risen from 2.9 to 4.8 mmol/l within 2 h (Fig. 49; Komor and Müller 1979). On the other

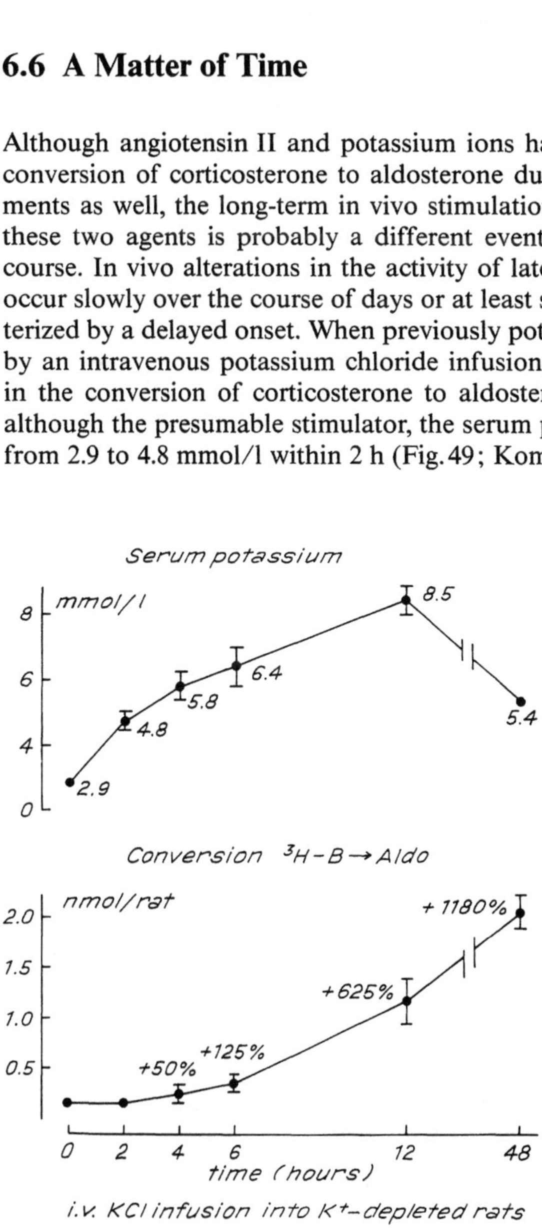

Fig. 49. Effects of an intravenous KCl infusion into potassium-deficient rats on the serum potassium concentration and on the conversion of tritiated corticosterone *(^3H-B)* to aldosterone *(ALDO)* by incubated quartered adrenals. (From J. Müller 1987; data from Komor and Müller 1979)

hand, the conversion continued to rise from hour 12 to hour 48, while the serum potassium concentration fell from 8.5 to 5.4 mmol/l. The capsular adrenal conversion of corticosterone to aldosterone remained unchanged within 24 h of sodium repletion of sodium- and potassium-restricted rats, whereas the plasma renin activity decreased to 13% of its original value (Fig. 50, left panel; J. Müller et al. 1984). During the following days of sodium repletion, the rate of conversion of corticosterone to aldosterone decreased with a biological half-life of 41 h. Decreases in the capsular adrenal conversions of corticosterone to 18-hydroxycorticosterone and aldosterone induced by prolonged ACTH treatment of rats started more promptly but were characterized by similarly long half-lives (Fig. 50, right panel; J. Müller 1978). Due to its long biological half-life, the conversion of corticosterone to aldosterone is a relatively stable parameter and is very reliable in the evaluation of experiments on long-term alterations in aldosterone production, since it is much less susceptible to short-term influences of unavoidable incidental stress than the plasma aldosterone concentration of rats, for example.

Time is also of the utmost importance for the interactions of the multiple stimulators and inhibitors in the physiological control of aldosterone secretion in the intact organism. This is illustrated by the transient effects of ACTH, atrial natriuretic peptides, and - to some extent - also of angiotensin II on aldosterone production during long-term infusion experiments. Different regulators appear to be alternately involved in the adaptation of aldosterone secretion to the sodium status. Thus, the 1-day delay in the decrease of the conversion of corticosterone to aldosterone observed during sodium repletion of sodium- and potassium-deficient rats (Fig. 50, left panel) makes it very unlikely that the suppression of aldosterone

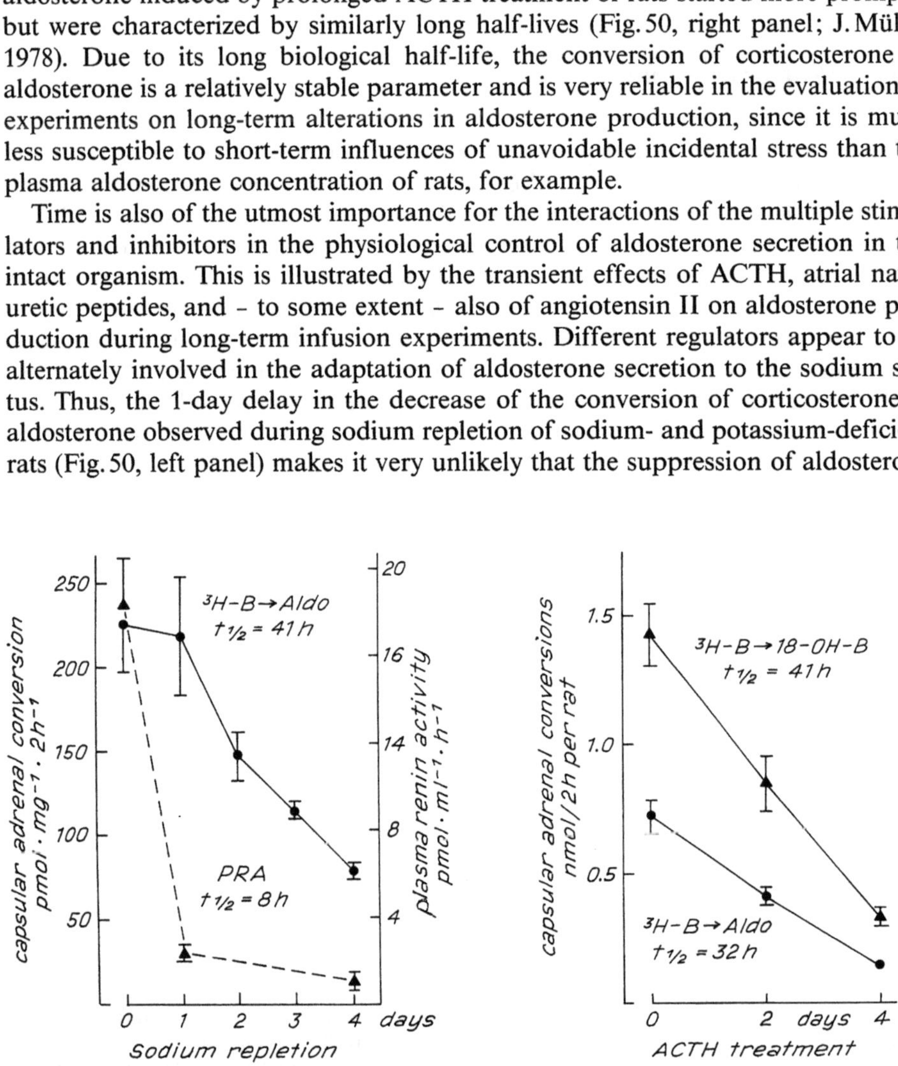

Fig. 50. Time course of the effects of sodium repletion of sodium- and potassium-deficient rats on the capsular adrenal conversion of tritiated corticosterone *(^{3}H-B)* to aldosterone *(ALDO)* and on plasma renin activity *(PRA; left panel)* and of the effects of ACTH treatment of rats (complete diet) on the capsular adrenal conversions of tritiated corticosterone *(^{3}H-B)* to 18-hydroxycorticosterone *(18-OH-B)* and aldosterone *(ALDO; right panel)*. (From J. Müller 1987; data from J. Müller et al. 1984 and from J. Müller 1978)

biosynthesis was triggered by the drop in plasma renin activity. This assumption was supported by the inability of captopril to decrease the conversion of corticosterone to aldosterone in rats with established sodium and potassium deficiency (Fig. 32). A relatively low level of plasma renin activity may, however, be a condition for, if not the cause of, low activity of the final steps in aldosterone biosynthesis in sodium-replete, potassium-deficient rats. This was indicated by other experiments which showed that increased activity of the renin-angiotensin system must have substantially contributed to the induction of a stimulated aldosterone biosynthesis during sodium depletion of potassium-deficient rats (J. Müller et al. 1984; Fig. 33).

The regulation of aldosterone secretion also changes in the course of life. This is documented by the steadily decreasing mean values of the plasma aldosterone concentration in normal man from birth to old age. This seems to reflect a changing requirement for aldosterone in the maintenance of sodium balance. Thus, mineralocorticoid deficiency leads to fatal salt loss in infants, whereas its main manifestation in middle-aged man is hyperkalemia, often associated with an increased body sodium content and high blood pressure, as seen in patients with hyporeninemic hypoaldosteronism.

6.7 Enzymology of the Final Steps in Aldosterone Biosynthesis

Although substantial evidence has been presented that the two final steps of aldosterone biosynthesis are catalyzed by mitochondrial cytochrome P-450 (Chap. 3), the identity of this enzyme is - at the time this monograph is written (1986) - still controversial. Two different groups of investigators (Wada et al. 1984, 1985; Yanagibashi et al. 1986) have reported that preparations of bovine and porcine cytochrome P-$450_{11\beta}$, removed from adrenocortical mitochondria and purified to electrophoretic homogeneity, catalyzed not only 11β-hydroxylation but also the two-step conversion of corticosterone to aldosterone, irrespective of the zonal origin of the reconstituted enzyme. According to these observations, one single enzyme, cytochrome P-$450_{11\beta}$, appears to be responsible for 11β-, 18-, and 19-hydroxylations of steroids in all the zones of the adrenal cortex as well as for the conversion of 18-hydroxycorticosterone to aldosterone in the zona glomerulosa, at least in cattle and pigs. However, such an assumption seems to conflict with many observations made mainly in rats but also in man and in other mammalian species. Thus, aldosterone is exclusively (Chap. 1) and 18-hydroxycorticosterone is preferentially (Sect. 2.3) produced by the zona glomerulosa. Marked alterations in the activity of the two final steps in aldosterone biosynthesis in response to alterations in sodium (Sect. 5.1.4) or potassium intake (Sect. 5.2.5) or to mineralocorticoid treatment (Sect. 5.11.1) were not accompanied by similar changes in the zona glomerulosa conversion of deoxycorticosterone to corticosterone and 18-OH-DOC (Fig. 51). Moreover, prolonged treatment of rats with ACTH resulted in a marked suppression of the conversions of corticosterone to 18-hydroxycorticosterone and aldosterone but in increased conversions of deoxycorticosterone to corticosterone and to 18-OH-DOC by capsular adrenal tissue (J. Müller 1978; Fig. 41). In reverse, hypophysectomy of rats induced a decreased

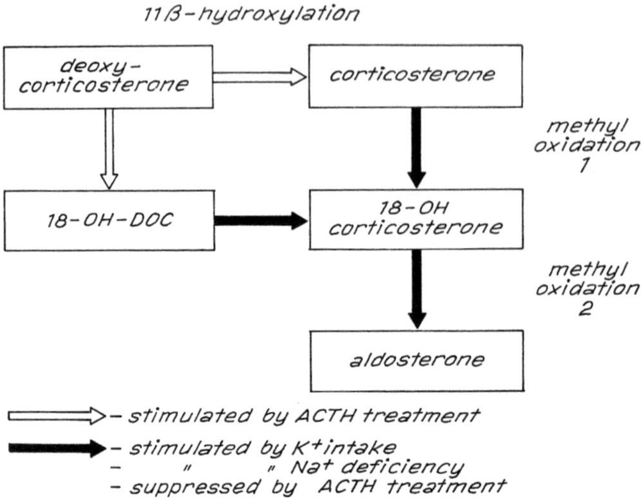

Fig. 51. Effects of ACTH treatment, potassium intake, and sodium deficiency on the steps involved in the two established biosynthetic pathways between deoxycorticosterone and aldosterone in the zona glomerulosa of the rat adrenal cortex. (From J. Müller 1987)

conversion of deoxycorticosterone to corticosterone by capsular adrenal mitochondria but an increased conversion of corticosterone to aldosterone (Aguilera et al. 1980b).

If all three steps involved in the conversion of deoxycorticosterone to aldosterone were catalyzed by the same enzyme ("one-enzyme" hypothesis, Fig. 52), the variable rates of corticosterone methyl oxidations 1 and 2 would have to be due to the action of specific mitochondrial regulators. Similar regulators would have to constantly suppress aldosterone biosynthesis in zona fasciculata mitochondria.

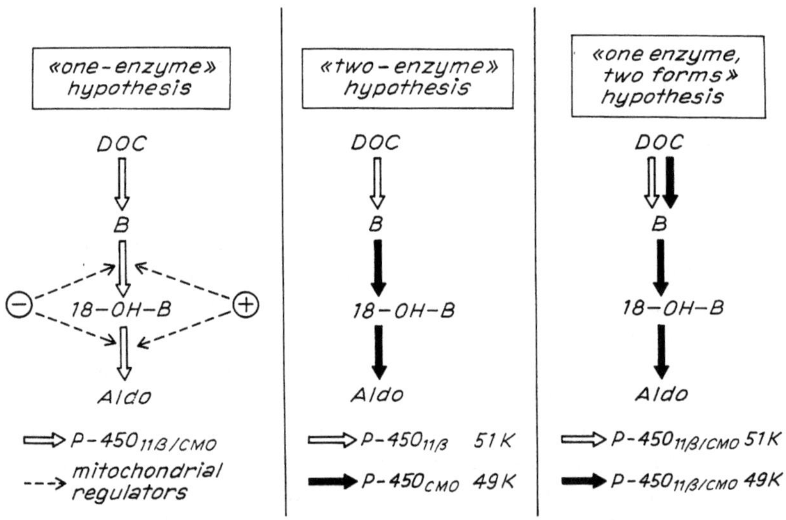

Fig. 52. Three different hypotheses about the number of enzymes involved in the three-step conversion of deoxycorticosterone *(DOC) to aldosterone (ALDO)* in the zona glomerulosa of the mammalian adrenal cortex. *B*, corticosterone; *18-OH-B*, 18-hydroxycorticosterone. (From Lauber et al. 1987)

There is as yet little direct evidence for the existence of these postulated mitochondrial regulators. Phospholipids have been observed to directly activate the conversion of 18-hydroxycorticosterone to aldosterone and, to a lesser extent, the conversion of corticosterone to 18-hydroxycorticosterone by a reconstituted system of purified bovine cytochrome P-$450_{11\beta}$ (Ohnishi et al. 1984). However, observations about a natural substance being capable of specifically suppressing the two final steps of aldosterone biosynthesis without affecting 11β-hydroxylation have, to my knowledge, never been published.

The rat adrenal cortex differs from the bovine adrenal cortex in a more extensive separation of zona glomerulosa and zona fasciculata functions. The higher zonal specificity of the rat adrenal cortex is expressed in the substrate affinity of steroidogenic enzymes, as well as in the distribution of operational receptors for stimulators and inhibitors. 18-Hydroxycorticosterone, a main (although not exclusive) product of the zona glomerulosa of the rat adrenal cortex (S. A. S. Tait et al. 1970), is abundantly produced by the bovine zona fasciculata (Karlmar 1982). Angiotensin II, which stimulated steroidogenesis exclusively in rat adrenal zona glomerulosa cells (Bing and Schulster 1977), also stimulated corticosterone production by bovine zona fasciculata cells (Hepp et al. 1977). Atrial natriuretic peptides inhibited ACTH-stimulated steroidogenesis in rat adrenal zona glomerulosa cells but not in zona fasciculata cells (W. B. Campbell et al. 1985). Such a zone specificity of these inhibitors was not observed in experiments with bovine adrenocortical cells (Racz et al. 1985). The enzyme responsible for the two final steps of aldosterone biosynthesis in the rat adrenal cortex has not yet been identified with cytochrome P-$450_{11\beta}$. It is therefore still conceivable that a specific enzyme catalyzing corticosterone methyl oxidations 1 and 2 exists in the rat zona glomerulosa cell ("two-enzyme" hypothesis, Fig. 52). Many experimental observations made in rats that are difficult to reconcile with the "one-enzyme" hypothesis would fit much better a "two-enzyme" hypothesis with a relatively stable "11β-hydroxylase" catalyzing the biosynthetic steps shown by open arrows in Fig. 51 and a more variable "corticosterone methyl oxidase" catalyzing the steps shown by full arrows. The second enzyme might be represented by the zona glomerulosa mitochondrial protein induced in rats by potassium repletion and sodium depletion (Meuli and Müller 1983, 1984).

Studies with a monoclonal antibody raised against purified bovine cytochrome P-$450_{11\beta}$ and cross-reacting with single adrenocortical mitochondrial proteins of similar molecular weight in other mammalian species (Sugano et al. 1985) suggested an alternative hypothesis (Lauber et al. 1987). In zona glomerulosa mitochondria from sodium-deficient, potassium-replete rats, this antibody cross-reacted exclusively with the "induced" 49-K protein. In contrast, it cross-reacted with a 51-K protein in zona fasciculata mitochondria and in zona glomerulosa mitochondria from sodium-replete, potassium-deficient rats. These findings indicated the existence of two different forms of cytochrome P-$450_{11\beta}$ in rats ("one-enzyme, two-forms" hypothesis, Fig. 52), with only one of them, the 49-K form, able to catalyze the final steps of aldosterone biosynthesis in situ. If the 49-K protein had the same amino acid sequence as the 51-K protein, it could be produced from the 51-K protein by a proteolytic process. The inherent aldosterone biosynthetic activity of the bovine cytochrome P-$450_{11\beta}$ was uncovered by removal of the enzyme from mitochondria of zona fasciculata cells in vitro (Yanagibashi et al.

1986). Similarly, the aldosterone biosynthetic activity of rat cytochrome P-$450_{11\beta}$ might be uncovered in zona glomerulosa mitochondria in vivo by its conversion from the 51-K to the 49-K form. Alternatively, the 49-K protein might have only a partial structural identity with the 51-K protein and might then be synthesized de novo in the zona glomerulosa cell in response to physiological stimuli.

6.8 Predominance of the Renin-Angiotensin System in the Physiological Control of Aldosterone Secretion?

The possibility that the renin-angiotensin system plays an important role in the regulation of aldosterone secretion was first suggested by the late Franz Gross (1958), who had observed an inverse relationship between sodium balance on the one side and aldosterone secretion and the renin content of the kidneys on the other. In a later review, Gross (1967) warned that the term "aldosterone-stimulating hormone" used in defining the physiological role of angiotensin II was "an oversimplification, as it might create the incorrect impression that angiotensin has an exclusive action on and elicits an exclusive response from the aldosterone-producing cells of the adrenal cortex." In view of many obvious discrepancies between aldosterone secretion and the activity of the renin-angiotensin system, I have similarly taken critical issue with the still widely used term "renin-angiotensin-aldosterone system", because it wrongly implies a stable functional connection between the juxtaglomerular apparatus and the zona glomerulosa of the adrenal cortex (J. Müller 1971, 1972). Coghlan et al. (1979) confirmed that there was an excellent correlation between renin-angiotensin and aldosterone under numerous circumstances, but they also listed 11 different experimental and clinical situations in which the nexus between angiotensin and aldosterone was broken. A similar list can be found in the first edition of this monograph. I leave it to the reader of the second edition to list newly discovered examples of obvious divergence between aldosterone and angiotensin II plasma concentrations and oppose them to recently found evidence further documenting a strong functional causal relationship between the renin-angiotensin system and aldosterone secretion. Altogether, the balance has not changed very much during the past 16 years.

Aldosterone is a potassium-excreting as well as a sodium-retaining hormone. This fact is duly considered in a model presented by G. H. Williams and Dluhy (1972), in which aldosterone secretion is controlled by the interaction of two independent and equally important feedback loops, one of them representing volume (sodium) homeostasis and acting through the renin-angiotensin system, the other one representing potassium homeostasis and acting through the extracellular potassium concentration. A modified version of this model is shown in Fig. 53. This model illustrates the well-documented priorities and equalities of sodium balance and potassium balance in affecting aldosterone secretion. During sodium deficiency with continued potassium intake, the two regulatory loops will potentiate each other, whereas they will neutralize each other during combined sodium and potassium deficiency. In this model, ACTH has a minor role in the overall control of aldosterone, since the secretion of this hormone is regulated mainly by mechanisms independent of sodium and potassium balance, i.e., by diurnal

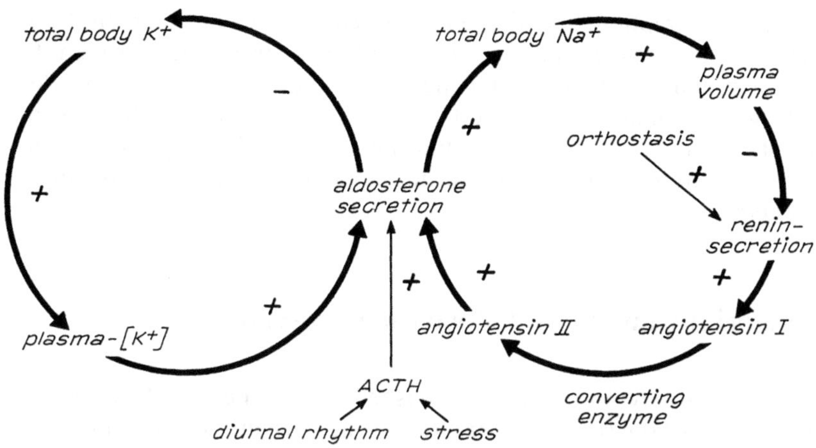

Fig. 53. Regulation of aldosterone secretion in the mammalian organism

rhythms, stress, and negative feedback effects of plasma glucocorticoids. Nevertheless, ACTH may become the major determinant of aldosterone secretion in normal supine human subjects with sufficient and balanced sodium and potassium intakes and particularly in patients with primary aldosteronism, in whom the two main regulators are suppressed.

The model shown in Fig. 53 contains many gross simplifications. The time element and the well-documented long-term effects of sodium and potassium intake on angiotensin receptors and on late steps in aldosterone biosynthesis are not included. Moreover, the model does not take into account the effects of potassium intake on the renin-angiotensin system (Sect. 5.2.7) or the effects of sodium intake on the urinary excretion and plasma levels of potassium (Sect. 5.1.6). Furthermore,

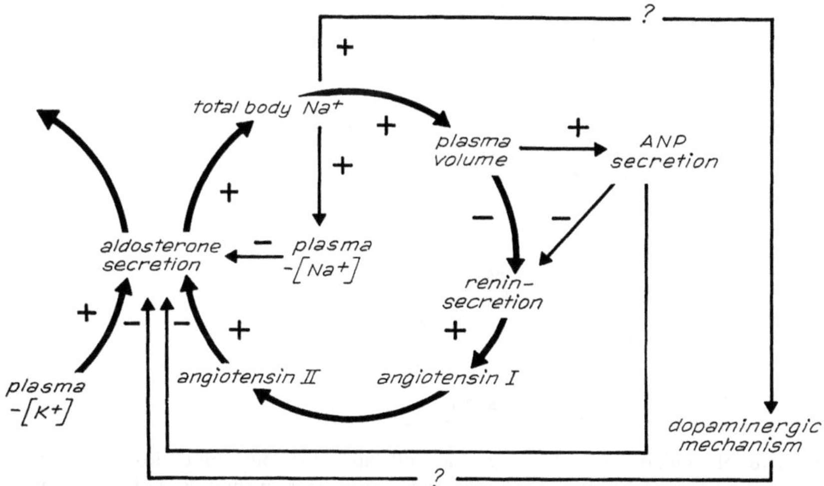

Fig. 54. Control mechanisms involved in the adaptation of aldosterone secretion to the total body sodium content in the mamalian organism. *ANP*, atrial natriuretic peptides

it is based on the assumption that the only efferent axis connecting the body sodium content with the zona glomerulosa of the adrenal cortex leads through the renin-angiotensin system. That this is not the case is illustrated by an expanded model (Fig. 54), which contains three additional regulatory loops involved in the adaptation of aldosterone secretion to sodium homeostasis, leading through the plasma sodium concentration, the atrial natriuretic peptides, and the dopaminergic nervous system.

6.9 Multifactorial Control of Aldosterone Biosynthesis

In spite of the mass of new information published during the past 16 years, the simplified 1986 model of the regulation of aldosterone biosynthesis in mammalian adrenocortical zona glomerulosa cells (Fig. 55) is not much different from the one presented in the first edition of this monograph. More than ever, "multifactorial" is the key word for describing the physiological regulation of aldosterone biosynthesis. In principle, multiple regulators influence aldosterone production in three different ways:

1. By rapidly facilitating or inhibiting an early step in the aldosterone biosynthetic pathway, preceding the production of pregnenolone
2. By slowly stimulating or suppressing the activity of the final two steps in aldosterone biosynthesis, i.e., the ones involved in the conversion of corticosterone to aldosterone
3. By inducing growth or atrophy of the zona glomerulosa cell and its subcellular structures

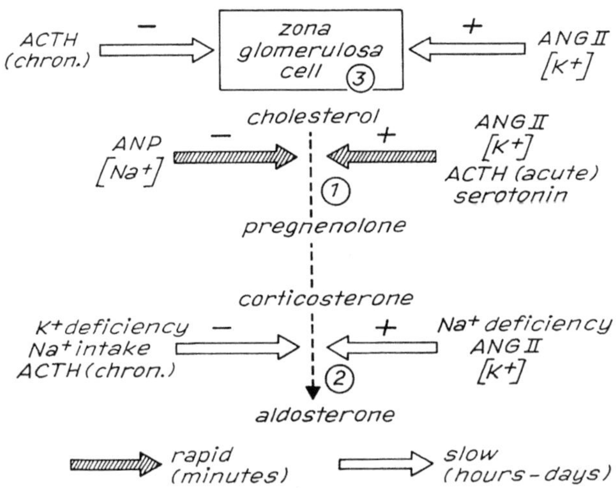

Fig. 55. Multifactorial regulation of aldosterone biosynthesis in the zona gomerulosa of the mammalian adrenal cortex at three different levels: (1) acute stimulation or inhibition of early biosynthetic steps, (2) long-term stimulation or suppression of late biosynthetic steps, and (3) growth or involution of zona glomerulosa cell. *ANG II*, angiotensin II; *ANP*, atrial natriuretic peptides

At the end of the first edition of this monograph, I predicted that the discovery of the nature and mechanism of action of the physiological factor or factors controlling the final steps of aldosterone biosynthesis could be an important forthcoming breakthrough in future research. Like many other premature statements made in the area of aldosterone physiology, this prediction was rather naive and did not take into account the immense complexity inherent in almost any aspect of the control of aldosterone production. Now more mature, I no longer believe in easy solutions. During the past 16 years it has been clearly established that the long-term regulation of the final steps in aldosterone biosynthesis is also multifactorial and at least as complex as the short-term regulation of the early steps. For years we have assumed the involvement of at least one specific aldosterone-biosynthetic enzyme, which, in the light of recent evidence, does not seem to exist. Thus, in many respects, research about the regulation of the final steps in aldosterone biosynthesis will have to start all over again. However, looking at the immense progress made in the fields of ultrastructural morphology of the zona glomerulosa cell (for reviews see Nussdorfer 1980, 1986) and of the short-term regulation of early steps in aldosterone biosynthesis, I am confident that a new generation of researchers will also solve the many puzzles involved in the control of the final steps, if not exactly by blood, sweat, and tears, at least by a mixture of inspiration, hard work, and new technology. Like any other human endeavor, the study of the regulation of aldosterone biosynthesis very often leads to disappointment or even frustration and is only rarely gratified by triumph. However, I feel safe in predicting that it will continue to be a fascinating challenge for many years.

References

Aarskog D, Støa KF, Thorsen T, Wefring KW (1967) Hypertension and hypokalemic alkalosis associated with underproduction of aldosterone. Pediatrics 39:884-890

Abbott EC, Monkhouse FC, Steiner JW, Laidlaw JC (1966a) Effect of a sulfated mucopolysaccharide (Ro 1-8307) on the zona glomerulosa of the rat adrenal gland. Endocrinology 78: 651-654

Abbott EC, Gornall AG, Sutherland DJA, Stiefel M, Laidlaw JC (1966b) The influence of a heparin-like compound on hypertension, electrolytes and aldosterone in man. Can Med Assoc J 94:1155-1164

Abboud HE, Luke RG, Galla JH, Kotchen TA (1979) Stimulation of renin by acute selective chloride depletion in the rat. Circ Res 44:815-821

Abraham SF, Blair-West JR, Coghlan JP, Denton DA, Mouw DR, Scoggins BA (1976) Aldosterone secretion during high sodium cerebrospinal fluid perfusion of the brain ventricles. Acta Endocrinol 81:120-132

Abshagen U, Spörl S, Schöneshöfer M, L'Age M, Oelkers W (1978) Interference of spironolactone therapy with adrenal steroid metabolism in secondary hyperaldosteronism. Klin Wochenschr 56:341-349

Abshagen U, Spörl S, Oelkers W (1979) Influence of spironolactone on serum corticosteroids in primary hyperaldosteronism. Klin Wochenschr 57:173-180

Ackerly JA, Tsai B-S, Peach MJ (1977) Role of converting enzyme in the responses of rabbit atria, aortas, and adrenal zona glomerulosa to [des-Asp[1]] angiotensin I. Circ Res 41:231-238

Acuña D, Soler A, Garcia-Torres L, Vargas F, Garcia del Rio C, Quesada T, Osorio C (1982) Inhibition of the renin-angiotensin system by pinealectomy in female rats. Rev Esp Fisiol 38: 251-255

Acuña D, Garcia del Rio C, Garcia-Torres L, Luna J, Osorio C (1984) Role of pineal gland in kidney-adrenal homeostasis. Horm Metabol Res 16:589-592

Agrimonti F, Boggio-Bertinet, Balzola F, Angeli A (1981) Circadian profile of plasma cortisol and aldosterone in post-traumatic comatose patients under high-dose dexamethasone treatment. J Endocrinol Invest 4:49-53

Aguilera G, Catt KJ (1978) Regulation of aldosterone secretion by the renin-angiotensin system during sodium restriction in rats. Proc Natl Acad Sci USA 75:4057-4061

Aguilera G, Catt KJ (1979) Loci of action of regulators of aldosterone biosynthesis in isolated glomerulosa cells. Endocrinology 104:1046-1052

Aguilera G, Catt KJ (1983) Regulation of aldosterone secretion during altered sodium intake. J Steroid Biochem 19:525-530

Aguilera G, Catt KJ (1984) Dopaminergic modulation of aldosterone secretion in the rat. Endocrinology 114:176-181

Aguilera G, Catt KJ (1986) Participation of voltage-dependent calcium channels in the regulation of adrenal glomerulosa function by angiotensin II and potassium. Endocrinology 118:112-118

Aguilera G, Marusic ET (1971) Role of the renin-angiotensin system in the biosynthesis of aldosterone. Endocrinology 89:1524-1529

Aguilera G, White A, Marusic ET (1975) The importance of sodium balance on the effect of angiotensin II on aldosterone production. Acta Endocrinol 80:104-113

Aguilera G, Hauger RL, Catt KJ (1978) Control of aldosterone secretion during sodium restriction: adrenal receptor regulation and increased adrenal sensitivity to angiotensin II. Proc Natl Acad Sci USA 75:975-979

Aguilera G, Capponi A, Baukal A, Fujita K, Hauger R, Catt KJ (1979) Metabolism and biological activities of angiotensin II and des-asp[1]-angiotensin II in isolated adrenal glomerulosa cells. Endocrinology 104:1279-1285

Aguilera G, Fujita K, Schirar A, Catt KJ (1980a) Role of angiotensin II in the regulation of aldosterone secretion. In: Cumming IA, Funder JW, Mendelsohn FAO (eds) Endocrinology 1980, proceedings of the 6th international congress of endocrinology, Melbourne 1980. Aust Acad Sci, Canberra, pp 389-392

Aguilera G, Menard RH, Catt KJ (1980b) Regulatory actions of angiotensin II on receptors and steroidogenic enzymes in adrenal glomerulosa cells. Endocrinology 107:55-60

Aguilera G, Harwood JP, Catt KJ (1981a) Somatostatin modulates effects of angiotensin II in adrenal glomerulosa zone. Nature 292:262-263

Aguilera G, Fujita K, Catt KJ (1981b) Mechanisms of inhibition of aldosterone secretion by adrenocorticotropin. Endocrinology 108:522-528

Aguilera G, Parker DS, Catt KJ (1982) Characterization of somatostatin receptors in the rat adrenal glomerulosa zone. Endocrinology 111:1376-1384

Aguilera G, Mendelsohn FAO, Catt KJ (1984) Dopaminergic regulation of aldosterone secretion. In: Martini L, Ganong WF (eds) Frontiers in neuroendocrinology, vol 8. Raven, New York, pp 265-293

Aikawa T, Hirose T, Matsumoto I, Suzuki T (1979) Secretion of aldosterone in respone to histamine in hypophysectomized-nephrectomized dogs. J Endocrinol 81:325-330

Aikawa T, Hirose T, Matsumoto I, Suzuki T (1981a) Direct stimulatory effect of histamine on aldosterone secretion of the perfused dog adrenal gland. Jpn J Physiol 31:457-463

Aikawa T, Hirose T, Matsumoto I, Suzuki T (1981b) Adrenal secretion of aldosterone in response to anaphylactic shock in hypophysectomized-nephrectomized dogs. Jpn J Physiol 31:145-151

Airaksinen MM, Sainio E-L, Leppäluoto J, Kari I (1984) 6-Methoxy-tetrahydro-β-carboline (pinoline): effects on plasma renin activity and aldosterone, TSH, LH and β-endorphin levels in rats. Acta Endocrinol 107:525-530

Alart I, Rosenthal J, Pal S, Raptis S (1978) Einfluß von Somatostatin auf Renin, Aldosteron und Blutdruck bei unilateraler renovaskulärer Hypertonie. Verh Dtsch Ges Inn Med 84:824-827

Albano JDM, Brown BL, Ekins RP, Tait SAS, Tait JF (1974) The effects of potassium, 5-hydroxytryptamine, adrenocorticotrophin and angiotensin II on the concentration of adenosine 3':5'-cyclic monophosphate in suspensions of dispersed rat adrenal zona glomerulosa and zona fasciculata cells. Biochem J 142:391-400

Al-Dujaili EAS, Boscaro M, Espiner EA, Edwards CRW (1980) *In vitro* and *in vivo* effects of indoleamines on aldosterone biosynthesis in the rat. 6th International congress of Endocrinology, Melbourne (abstract no 400)

Al-Dujaili EAS, Hope J, Estivariz FE, Lowry PJ, Edwards CRW (1981) Circulating human pituitary pro-γ-melanotropin enhances the adrenal response to ACTH. Nature 291:156-159

Al-Dujaili EAS, Williams BC, Edwards CRW, Salacinski P, Lowry PJ (1982a) Human γ-melanotropin precursor potentiates corticotropin-induced adrenal steroidogenesis by stimulating mRNA synthesis. Biochem J 204:301-305

Al-Dujaili EAS, Boscaro M, Edwards CRW (1982b) An *in vitro* stimulatory effect of indoleamines on aldosterone biosynthesis in the rat. J Steroid Biochem 17:351-355

Allsopp MNE, Levell MJ, Stitch SR, Hullin RP (1972) Aldosterone production rates in manic-depressive psychosis. Br J Psychiatry 120:399-404

Ames RP, Borkowski AJ, Sicinski AM, Laragh JH (1965) Prolonged infusions of angiotensin II and norepinephrine and blood pressure, electrolyte balance, and aldosterone and cortisol secretion in normal man and in cirrhosis with ascites. J Clin Invest 44:1171-1186

Anand-Srivastava MB, Franks DJ, Cantin M, Genest J (1984) Atrial natriuretic factor inhibits adenylate cyclase activity. Biochem Biophys Res Commun 121:855-862

Anand-Srivastava MB, Cantin M, Genest J (1985) Inhibition of pituitary adenylate cyclase by atrial natriuretic factor. Life Sci 36:1873-1879

Anderson GH Jr, Anderson T, Streeten DHP, Schroeder ET (1980) Acute effects of saralasin on plasma aldosterone in different pathophysiological conditions. J Clin Endocrinol Metab 50: 529-536

Anderson RC, Herbert PN, Mulrow PJ (1968) A comparison of properties of renin obtained from the kidney and uterus of the rabbit. Am J Physiol 215:774-778

Andoka G, Chauvin MA, Marie J, Saez JM, Morera AM (1984) Adrenocorticotropin regulates angiotensin II receptors in bovine adrenal cells in vitro. Biochem Biophys Res Commun 121: 441-447

Angles d'Auriac G, Boudoin M, Meyer P (1972) Mechanism of action of angiotensin in smooth muscle cell. Biochemical changes following interaction of the hormone with its membrane receptors. Circ Res 30/31 [Suppl II]: II-151-II-157

Antreassian J, Aupetit B, Ozon R, Legrand JC (1972) Métabolisme rénal de la 18-hydroxycorticostérone. CR Seances Acad Sci 275:1423-1425

Antreassian J, Lagoguey A, Cesselin F, Legrand JC (1979) Localisation submitochondriale des réactions de transformation de la corticostérone en 18 hydroxycorticostérone et en aldostérone. Biochimie 61:1081-1084

Aoi W, Wade MB, Rosner DR, Weinberger MH (1974) Renin release by rat kidney slices in vitro: effects of cations and catecholamines. Am J Physiol 227:630-634

Aragones A, Gros EG, Lantos CP, Locascio GA (1978) Less polar forms and derivatives of 18-hydroxycorticosterone. J Steroid Biochem 9:175-180

Armanini D, Karbowiak I, Krozowski Z, Funder JW, Adam WR (1982) The mechanism of mineralocorticoid action of carbenoxolone. Endocrinology 111:1683-1686

Armanini D, Kuhnle U, Strasser T, Dörr H, Butenandt I, Weber PC, Stockigt JR, Pearce P, Funder JW (1985) Aldosterone-receptor deficiency in pseudohypoaldosteronism. N Engl J Med 313:1178-1181

Armato U, Mantero F (1984) Primary tissue culture of human adrenocortical Conn's adenomata. Bromocriptine as a possible agonist-antagonist of angiotensin at the cellular level. Cell Tissue Res 236:67-72

Armbruster H, Vetter W, Beckerhoff R, Nussberger J, Vetter H, Siegenthaler W (1975) Diurnal variations of plasma aldosterone in supine man: relationship to plasma renin activity and plasma cortisol. Acta Endocrinol 80:95-103

Atarashi K, Mulrow PJ, Franco-Saenz R, Snajdar R, Rapp J (1984) Inhibition of aldosterone production by an atrial extract. Science 224:992-994

Atarashi K, Mulrow PJ, Franco-Saenz R (1985) Effect of atrial peptides on aldosterone production. J Clin Invest 76:1807-1811

Atlas SA, Case DB, Sealey JE, Laragh JH, McKinstry DN (1979) Interruption of the renin-angiotensin system in hypertensive patients by captopril induces sustained reduction in aldosterone secretion, potassium retention and natriuresis. Hypertension 1:274-280

Aupetit B (1972) Contribution à l'étude de la 18 ol déshydrogénase. Thèse de doctorat de 3e cycle, Université de Paris VI, Paris

Aupetit B, Aubry-Marais F, Legrand JC (1977a) Comportement comparé de la 18-hydroxycorticostérone de synthèse et de la 18-hydroxycorticostérone "endogène" dans la synthèse de l'aldostérone. Biochimie 59:311-321

Aupetit B, Antreassian J, Legrand JC (1977b) Caractère enzymatique et localisation subcellulaire de la réaction de transformation de la 18-hydroxycorticostérone en aldostérone. Biochimie 59: 705-712

Aupetit B, Bastien C, Legrand JC (1979) Cytochrome P_{450} et transformation de la 18-hydroxycorticostérone en aldostérone. Biochimie 61:1085-1089

Aupetit B, Toury R, Legrand JC (1980) Relation between energy metabolism in mitochondria and conversion of 18 hydroxycorticosterone to aldosterone in adrenals. Biochimie 62:823-827

Aupetit B, Accarie C, Emeric N, Vonarx V, Legrand J-C (1983) The final step of aldosterone biosynthesis requires reducing power. It is not a dehydrogenation. Biochim Biophys Acta 752: 73-78

Aupetit B, Emeric N, Accarie C, Vonarx V (1984) Action of different nucleotides on the last step of aldosterone synthesis. J Steroid Biochem 20:335-337

Aviv A, Kobayashi T, Higashino H, Bauman JW Jr, Yu SS (1982) Chronic sodium deficit in the immature rat: its effect on adaptation to sodium excess. Am J Physiol 242:E241-E247

Ayachi S, Hall CE, Holland OB, Gomez-Sanchez C (1976) Effect of deoxycorticosterone acetate and 16β-hydroxy-dehydroepiandrosterone on blood pressure and plasma renin activity of rats. Proc Soc Exp Biol Med 152:218-220

Ayres PJ, Gould AP, Simpson SA, Tait JF (1956) The in vitro demonstration of different corticosteroid production within the ox adrenal gland. Biochem J 63:19 *P* (abstract)

Ayres PJ, Hechter O, Saba N, Simpson SA, Tait JF (1957) Intermediates in the biosynthesis of aldosterone by capsule strippings of ox adrenal gland. Biochem J 65:22 *P* (abstract)

Ayres PJ, Eichhorn J, Hechter O, Saba N, Tait JF, Tait SAS (1960) Some studies on the biosynthesis of aldosterone and other adrenal steroids. Acta Endocrinol 33:27-58

Ayres PJ, Hurter RC, Williams ES, Rundo J (1961) Aldosterone excretion and potassium retention in subjects living at high altitude. Nature 191:78-80

Baas EU, Holtermüller KH, Sinterhauf K, Walter U (1976) Das Verhalten von Gastrin, Renin, Aldosteron und Elektrolyten nach deglycyrrhiziniertem Succus liquiritiae und Carbenoxolon bei Gesunden. Z Gastroenterol 14:273-276

Baba T, Aoyagi K, Murabayashi S, Sasaki K, Nigawara K, Takebe K (1983) Renin-angiotensin system and plasma aldosterone in Cushing's syndrome. Endocrinol Jpn 30:715-721

Bach I, Braun S, Gáti T, Sós J, Udvardy A (1960) Die Wirkungen der einwertigen Kationen auf die Nebennierenrinde. 1st International congress of endocrinology, Copenhagen (abstract no 76)

Bach I, Gáti T, Savely C, Sós J, Udvardy A (1967) Effect of rubidium on aldosterone and corticosterone production in rats. Endocrinology 81:913-914

Baer L, Sommers SC, Krakoff LR, Newton MA, Laragh JH (1970) Pseudo-primary aldosteronism: an entity distinct from the true primary aldosteronism. Circ Res 26/27 [Suppl 1]: I-203-I-216

Bailey RE, Ford HC (1969) The effect of heparin on sodium conservation and on the plasma concentration, the metabolic clearance and the secretion and excretion rates of aldosterone in normal subjects. Acta Endocrinol 60:249-264

Bailey RE, Bartos D, Bartos F, Castro A, Dobson RL, Grettie DP, Kramer R, Macfarlane D, Sato K (1972) Activation of aldosterone and renin secretion by thermal stress. Experientia 28: 159-160

Baird D, Horton R, Longcope C, Tait JF (1968) Steroid prehormones. Perspect Biol Med 11: 384-421

Bakiri F, Benmiloud M, Vallotton MB (1983) The renin-angiotensin system in panhypopituitarism: dynamic studies and therapeutic effects in Sheehan's syndrome. J Clin Endocrinol Metab 56:1042-1047

Bakiri F, Riondel AM, Benmiloud M, Vallotton MB (1986) Aldosterone in panhypopituitarism: dynamic studies and therapeutic effects in Sheehan's syndrome. Acta Endocrinol 112:329-335

Balikian HM, Brodie AH, Dale SL, Melby JC, Tait JF (1968) Effect of posture on the metabolic clearance rate, plasma concentration and blood production rate of aldosterone in man. J Clin Endocrinol Metab 28:1630-1640

Balla T, Spät A (1982) The effect of various calmodulin inhibitors on the response of adrenal glomerulosa cells to angiotensin II and cyclic AMP. Biochem Pharmacol 31:3705-3707

Balla T, Tarján É, Renczes G, Spät A (1981) Effect of reduced extracellular sodium concentration on the function of adrenal zona glomerulosa: studies in conscious rats. J Endocrinol 89: 411-416

Balla T, Hunyady L, Spät A (1982a) Possible role of calcium uptake and calmodulin in adrenal glomerulosa cells: effects of verapamil abd trifluoperazine. Biochem Pharmacol 31:1267-1271

Balla T, Enyedi P, Spät A (1982b) Role of calcium ions and calmodulin in the aldosterone stimulating action of prostaglandin E_2. J Steroid Biochem 16:493-494

Balla T, Enyedi P, Hunyady L, Spät A (1984) Effects of lithium on angiotensin-stimulated phosphatidylinositol turnover and aldosterone production in adrenal glomerulosa cells: a possible causal relationship. FEBS Lett 171:179-182

Balla T, Enyedi P, Spät A, Antoni FA (1985) Pressor-type vasopressin receptors in the adrenal cortex: properties of binding, effect on phosphoinositide metabolism and aldosterone secretion. Endocrinology 117:421-423

Baniukiewicz S, Brodie A, Flood C, Motta M, Okamoto M, Tait JF, Tait SAS, Blair-West JR, Coghlan JP, Denton DA, Goding JR, Scoggins BA, Wintour EM, Wright RD (1968) Adrenal biosynthesis of steroids in vitro and in vivo using continuous superfusion and infusion procedures. In: McKerns KW (ed) Functions of the adrenal cortex. North Holland, Amsterdam, pp 153-232

Barbeau A, Gillo-Joffroy L, Boucher R, Nowaczynski W, Genest J (1969) Renin-aldosterone system in Parkinson's disease. Science 165:291-292

Barbieri C, Caldara R, Ferrari C, Crossignani RM, Recchia M (1981) Inhibition of the renin-angiotensin-aldosterone system by L-dopa with and without inhibition of extra cerebral dopa decarboxylase in man. Clin Sci 61:187-190

Barbieri C, Parodi M, Bruno S, Bertassi F, Benaglia D, Moser P, Meroni R, Dubini A (1984) Effects of acute administration of zetidoline, a new antidopaminergic drug, on plasma prolactin and aldosterone levels in man. Eur J Clin Pharmacol 26:29-32

Barbour BH, Slater JDH, Casper AGT, Bartter FC (1965) On the role of the central nervous system in control of aldosterone secretion. Life Sci 4:1161-1170

Baron JH, Nabarro JDN, Slater JDH, Tuffley R (1969) Metabolic studies, aldosterone secretion rate, and plasma renin after carbenoxolone sodium. Br Med J 2:793-795

Barraclough MA (1966) Sodium and water depletion with acute malignant hypertension. Am J Med 40:265-272

Barraclough MA, Bacchus B, Brown JJ, Davies DL, Lever AF, Robertson JIS (1965) Plasma-renin and aldosterone secretion in hypertensive patients with renal or renal artery lesions. Lancet 2: 1310-1313

Barron JL, Klaff LJ, Levitt NS, Ling N, Millar RP (1984) Arginine hydrochloride stimulation of serum potassium and aldosterone is enhanced by somatostatin-28. Acta Endocrinol 106: 407-410

Bartter FC, Fourman P (1957) A non-renal effect of adrenal cortical steroids upon potassium metabolism. J Clin Invest 36:872-873 (abstract)

Bartter FC, Liddle GW, Duncan LE Jr, Barber JK, Delea C (1956) The regulation of aldosterone secretion in man: the role of fluid volume. J Clin Invest 35:1306-1315

Bartter FC, Delea CS, Halberg F (1962) A map of blood and urinary changes related to circadian variations in adrenal cortical function in normal subjects. Ann NY Acad Sci 98:969-983

Bartter FC, Barbour BH, Carr AA, Delea CS (1964) On the role of potassium and of the central nervous system in the regulation of aldosterone secretion. In: Baulieu EE, Robel P (eds) Aldosterone. Blackwell, Oxford, pp221-242

Bartter FC, Henkin RI, Bryan GT (1968) Aldosterone secretion in "non-salt-losing" congenital adrenal hyperplasia. J Clin Invest 47:1742-1752

Bauer JH, Gauntner WC (1979) Effect of potassium chloride on plasma renin activity and plasma aldosterone during sodium restriction in normal man. Kidney Int 15:286-293

Bauer JH, Gauntner WC, Nolph KD, Van Stone JC (1977) Effects of potassium chloride on plasma renin activity during sodium restriction in normal man. J Lab Clin Med 90:312-317

Baulieu EE, de Vigan M, Bricaire H, Jayle MF (1957) Lack of plasma cortisol and urinary aldosterone in a pregnant woman with Addison's disease. J Clin Endocrinol Metab 17:1478-1482

Baumann G, Loriaux DL (1976) The effect of endogenous prolactin on renal salt and water excretion and adrenal function in man. J Clin Endocrinol Metab 43:643-649

Baumann G, Marynick SP, Winters SJ, Loriaux DL (1977) The effect of osmotic stimuli on prolactin secretion and renal water excretion in normal man and in chronic hyperprolactinemia. J Clin Endocrinol Metab 44:199-202

Baumann K, Müller J (1972a) Effect of potassium intake on the final steps of aldosterone biosynthesis in the rat. I. 18-Hydroxylation and 18-hydroxydehydrogenation. Acta Endocrinol 69: 701-717

Baumann K, Müller J (1972b) Effect of potassium intake on the final steps of aldosterone biosynthesis in the rat. II. 11β-Hydroxylation. Acta Endocrinol 69:718-730

Baumann K, Müller J (1974) Effects of hypophysectomy with or without ACTH maintenance therapy on the final steps of aldosterone biosynthesis in the rat. Acta Endocrinol 76:102-116

Baumber JS, Davis JO, Johnson JA, Witty RS (1971) Increased adrenocortical potassium in association with increased biosynthesis of aldosterone. Am J Physiol 220:1094-1099

Bay WH, Ferris TF (1979) Factors controlling plasma renin and aldosterone during pregnancy. Hypertension 1:410-415

Bayard F, Ances IG, Tapper AJ, Weldon VV, Kowarski A, Migeon CJ (1970) Transplacental passage and fetal secretion of aldosterone. J Clin Invest 49:1389-1393

Bayard F, Cooke CR, Tiller DJ, Beitins IZ, Kowarski A, Walker WG, Migeon CJ (1971) The regulation of aldosterone secretion in anephric man. J Clin Invest 50:1585-1595

Bayard F, Biermacker A, Gassia J-P, Boulard C (1975) Rôle de l'ACTH dans la régulation de la sécretion de l'aldostérone. CR Soc Biol 169:222-228

Baylé JD, Boissin J, Daniel JY, Assenmacher I (1971) Hypothalamic-hypophysial control of adrenocortical function in birds. Neuroendocrinology 7:308-321

Beck JC, McGarry EE, Dyrenfurth I, Morgen RO, Bird ED, Venning EH (1960) Primate growth hormone studies in man. Metabolism 9:699-737

Becker RA, Hayashi RH, Franks RC, Speroff L (1978) Effects of positional change and sodium balance on the renin-angiotensin-aldosterone system, big renin and prostaglandins in normal pregnancy. J Clin Endocrinol Metab 46:467-472

Beckerhoff R, Vetter W, Armbruster H, Luetscher JA, Siegenthaler W (1973) Plasma-aldosterone during oral-contraceptive therapy. Lancet 1:1218-1220

Beckerhoff R, Uhlschmid G, Vetter W, Armbruster H, Siegenthaler W (1974) Plasma renin and aldosterone after renal transplantation. Kidney Int 5:39-46

Beckerhoff R, Uhlschmid G, Furrer J, Nussberger J, Schmied U, Vetter WJ, Siegenthaler W (1975a) In vivo effects of angiotensin antagonists on plasma aldosterone in the dog. Eur J Pharmacol 34:363-367

Beckerhoff R, Kappeler M, Vetter W, Armbruster H, Siegenthaler W (1975b) Effect of immunization against angiotensin II on blood pressure and on plasma aldosterone in the rabbit. Clin Sci Mol Med 48:413-420

Beckerhoff R, Uhlschmid G, Vetter W, Siegenthaler W (1976) Studies on the effect of angiotensin II and of des[1]-angiotensin II on blood pressure, plasma renin activity and plasma aldosterone in the dog. Clin Sci Mol Med 51:147s-150s

Beitins IZ, Bayard F, Kowarski A, Migeon CJ (1972a) The effect of ACTH administration on aldosterone production in non salt-losing congenital adrenal adrenal hyperplasia. J Clin Endocrinol Metab 35:595-603

Beitins IZ, Bayard F, Levitsky L, Ances IG, Kowarski A, Migeon CJ (1972b) Plasma aldosterone concentration at delivery and during the newborn period. J Clin Invest 51:386-394

Bell JBG, Tait JF, Tait SAS, Barnes GD, Brown BL (1981) Lack of effect of angiotensin II on levels of cyclic AMP in isolated adrenal zona glomerulosa cells from the rat. J Endocrinol 91:145-154

Bell JGB, Chu FW, Tait JF, Tait SAS, Khosla M (1984) The use of the superfusion approach with rat adrenal capsular cells to compare the steroidogenic potencies of angiotensin analogues, without the effects of peptide degradation. Proc R Soc Lond Biol 221:21-30

Belmega W, Oelkers W, Belkien L, Shirpai M, Fiedler U, Häring R (1983) Effects of angiotensin II and ACTH on normal and tumorous human adrenocortical cells. Acta Endocrinol 104: 103-109

Benelli C, Michel O, Michel R (1982a) Effect of thyroidectomy on pregnenolone and progesterone biosynthesis in rat adrenal cortex. J Steroid Biochem 16:749-754

Benelli C, Michel O, Michel R (1982b) Effect of thyroidectomy on the rat adrenal cortex enzyme activities involved in corticosterone and aldosterone biosynthesis. J Steroid Biochem 16: 755-761

Bennett WM, McDonald WJ, Lawson RK, Porter GA (1974) Posttransplant hypertension: studies of cortical blood flow and the renal pressor system. Kidney Int 6:99-108

Benraad TJ, Kloppenborg PWC (1970) Plasma renin activity and aldosterone secretory rate in man during chronic ACTH administration. J Clin Endocrinol Metab 31:581-583

Benraad TJ, Wellen JJ (1971) Selective sensitivity of zona glomerulosa slices of calf adrenals to the stimulatory effect of potassium on corticosterone biosynthesis (abstract). Acta Endocrinol [Suppl] 155:110

Beretta-Piccoli C, Weidmann P, Flammer J, Glück Z, Bachmann C (1980a) Effects of standard glucose loading on the renin-angiotensin-aldosterone system and its relation to circulating insulin. Klin Wochenschr 58:467-474

Beretta-Piccoli C, Weidmann P, Meier A, Grimm M, Keusch G, Glück Z (1980b) Effects of shortterm norepinephrine infusion on plasma catecholamines, renin, and aldosterone in normal and hypertensive man. Hypertension 2:623-630

Berman LB, Vertes V, Mitra S, Gould AB (1972) Renin-angiotensin system in anephric patients. N Engl J Med 286:58-61

Berridge MJ (1984) Inositol triphosphate and diacylglycerol as second messengers. Biochem J 220:345-360

Best JB, Bett JHN, Coghlan JP, Cran EJ, Scoggins BA (1971) Circulating angiotensin II and aldosterone levels during dietary sodium restriction. Lancet 2:1353-1354

Bevilacqua M, Vago T, Raggi U, Scorza D, Proverbio M, Malacco E, Norbiato G (1982) In vitro steroidogenic properties of FK 33824, a stable analog of methionin-enkephalin. Opiate-dopamine interaction in the control of aldosterone production. J Endocrinol Invest 5:277-280

Bevilacqua M, Scorza D, Meroni R, Vago T, Dagani R, Malacco E, Norbiato G (1984) Evidence for a role of a serum factor stimulated by metoclopramide in regulating aldosterone secretion. Biochem Biophys Res Commun 119:95-100

Bia MJ, Tyler K, DeFronzo RA (1982) The effect of dexamethasone on renal electrolyte excretion in the adrenalectomized rat. Endocrinology 111:882-888

Bianchetti MG, Beretta-Piccoli C, Weidmann P, Boehringer K, Link L, Morton JJ (1982) Studies on aldosterone responsiveness to angiotensin II during clinical variations in calcium metabolism in normal man. Clin Sci 63:325-328

Biglieri EG, Ganong WF (1961) Effect of hypophysectomy on adrenocortical response to bilateral carotid constriction. Proc Soc Exp Biol Med 106:806-809

Biglieri EG, Slaton PE Jr (1967) Pregnancy and primary aldosteronism. J Clin Endocrinol Metab 27:1628-1632
Biglieri EG, Watlington CO, Forsham PH (1961) Sodium retention with human growth hormone and its subfractions. J Clin Endocrinol Metab 21:361-370
Biglieri EG, Hane S, Slaton PE, Forsham P (1963) In vivo and in vitro studies of adrenal secretions in Cushing's syndrome and primary aldosteronism. J Clin Invest 42:516-524
Biglieri EG, Slaton PE Jr, Silen WS, Galante M, Forsham PH (1966a) Postoperative studies of adrenal function in primary aldosteronism. J Clin Endocrinol Metab 26:553-558
Biglieri EG, Herron MA, Brust N (1966b) 17-Hydroxylation deficiency in man. J Clin Invest 45: 1946-1954
Biglieri EG, Slaton PE Jr, Kronfield SJ, Schambelan M (1967) Diagnosis of an aldosterone-producing adenoma in primary aldosteronism. An evaluative maneuver. JAMA 201:510-514
Biglieri EG, Slaton PE, Schambelan M, Kronfield SJ (1968) Hypermineralocorticoidism. Am J Med 45:170-175
Biglieri EG, Schambelan M, Slaton PE Jr (1969) Effect of adrenocorticotropin on desoxycorticosterone, corticosterone and aldosterone excretion. J Clin Endocrinol Metab 29:1090-1101
Biglieri EG, Stockigt JR, Schambelan M (1970) A preliminary evaluation for primary aldosteronism. Arch Intern Med 126:1004-1007
Biglieri EG, Schambelan M, Stockigt JR (1972) Hypertension and hyperaldosteronism: discrimination of deoxycorticosterone acetate (DOCA) testing. Circ Res 20:163 (abstract)
Biglieri EG, Wajchenberg BL, Malerbi DA, Okada H, Leme CE, Kater CE (1981) The zonal origins of the mineralocorticoid hormones in the 21-hydroxylation deficiency of congenital adrenal hyperplasia. J Clin Endocrinol Metab 53:964-969
Bing RF, Schulster D (1977) Steroidogenesis in isolated rat adrenal zona glomerulosa cells: Response to physiological concentrations of angiotensin II and effects of potassium, serotonin and (sar^1, ala^8)-angiotensin II. J Endocrinol 74:261-272
Bing RF, Schulster D (1978) Adenosine 3':5'-cyclic monophosphate production and steroidogenesis by isolated rat adrenal glomerulosa cells. Effects of angiotensin II and sar^1, ala^5 angiotensin II. Biochem J 176:39-45
Binnion PF, Davis JO, Brown TC, Olichney MJ (1965) Mechanisms regulating aldosterone secretion during sodium depletion. Am J Physiol 208:655-661
Birabeau MA, Capponi AM, Vallotton MB (1984) Solubilized adrenal angiotensin II receptors: studies on the site of action of sodium and calcium ions, and on the role of disulfide bridges. Mol Cell Endocrinol 37:181-189
Birkhäuser M, Gaillard R, Riondel AM, Scholer D, Vallotton MB, Muller AF (1973) Effect of volume expansion by hyperosmolar and hyperoncotic solutions under constant infusion of angiotensin II on plasma aldosterone and its counterbalance by potassium administration. Eur J Clin Invest 3:307-316
Birkhäuser M, Godard C, Loirat C, Vallotton MB (1974) Plasma aldosterone concentration in anephric children. Acta Endocrinol 75:561-569
Birkhäuser M, Gaillard R, Riondel AM, Zahnd GR (1975) Influences of acute administration of human growth hormone and alpha-MSH on plasma concentrations of aldosterone, cortisol, corticosterone and growth hormone in man. Acta Endocrinol 79:16-24
Birkhäuser M, Riondel A, Vallotton MB (1979) Bromocriptine-induced modulation of plasma aldosterone response to acute stimulation. Acta Endocrinol 91:294-302
Birkhäuser M, Riondel AM, Gaillard R, Mantero F, Vallotton MB (1981) Plasma aldosterone response to acute stimulation in panhypopituitarism. Acta Endocrinol 97:514-521
Birmingham MK, Ward PJ (1961) The identification of the Porter-Silber chromogen secreted by the rat adrenal. J Biol Chem 236:1661-1667
Birmingham MK, Elliott FH, Valère PHL (1953) The need for the presence of calcium for the stimulation in vitro of rat adrenal glands by adrenocorticotrophic hormone. Endocrinology 53: 687-689
Biron P, Koiw E, Nowaczynski W, Brouillet J, Genest J (1961) The effects of intravenous infusions of valine-5-angiotensin II and other pressor agents on urinary electrolytes and corticosteroids, including aldosterone. J Clin Invest 40:338-347
Biron P, Chrétien M, Koiw E, Genest J (1962) Effects of angiotensin infusions on aldosterone and electrolyte excretion in normal subjects and patients with hypertension and adrenocortical disorders. Br Med J 1:1569-1575
Björkhem I, Karlmar K-E (1975) 18-Hydroxylation of deoxycorticosterone by reconstituted systems from rat and bovine glands. Eur J Biochem 51:145-154

Black DAK, Milne MD (1952) Experimental potassium depletion in man. Clin Sci 11:397-415
Blaine EH, Coghlan JP, Denton DA, Scoggins BA (1974a) In vivo effects of ouabain on aldosterone, corticosterone and cortisol secretion in conscious sheep. Endocrionlogy 94:1304-1310
Blaine EH, Coghlan JP, Denton DA, Hardy KJ, Scoggins BA (1974b) Effect of ouabain on adrenal potassium balance in sheep. Proc Soc Exp Biol Med 150:257-260
Blair-West J, Coghlan J, Denton DA, Goding JR, Munro JA, Wright RD (1960) The effect of neuraxial ablations upon the secretion of electrolyte-active steroid. J Physiol (Lond) 153:52-54
Blair-West JR, Coghlan JP, Denton DA, Goding JR, Peterson RE, Wintour M (1962a) Humoral stimulation of adrenal cortical secretion. J Clin Invest 41:1606-1627
Blair-West JR, Coghlan JP, Denton DA (1962b) Evidence against an aldosterone feed-back mechanism within the adrenal gland. Acta Endocrinol 41:61-66
Blair-West JR, Coghlan JP, Denton DA, Goding JR, Wintour M, Wright RD (1963) The control of aldosterone secretion. Recent Prog Horm Res 19:311-383
Blair-West JR, Coghlan JP, Denton DA, Munro JA, Wintour M, Wright RD (1964) The effect of bilateral nephrectomy and midcollicular decerebration with pinealectomy and hypophysectomy on the corticosteroid secretion of sodium-deficient sheep. J Clin Invest 43:1576-1595
Blair-West JR, Coghlan JP, Denton DA, Goding JR, Wintour M, Wright RD (1965a) Effect of variations of plasma sodium concentration on the adrenal response to angiotensin II. Circ Res 17:386-393
Blair-West JR, Coghlan JP, Denton DA, Goding JR, Wintour M, Wright RD (1965b) A study of feedback in aldosterone secretion. Endocrinology 77:501-507
Blair-West JR, Coghlan JP, Denton DA, Goding JR, Wintour M, Wright RD (1966) The direct effect of increased sodium concentration in adrenal arterial blood on corticosteroid secretion in sodium deficient sheep. Aust J Exp Biol Med Sci 44:455-474
Blair-West JR, Coghlan JP, Denton DA, Goding JR, Orchard E, Scoggins B, Wintour M, Wright RD (1967a) Mechanisms regulating aldosterone secretion during sodium deficiency. In: Proceedings of the 3rd international congress of Nephrology, Washington 1966, vol1. Karger, Basel, pp201-214
Blair-West JR, Boyd GW, Coghlan JP, Denton DA, Wintour M, Wright RD (1967b) The role of circulating fluid volume and related physiological parameters in the control of aldosterone secretion. Aust J Exp Biol Med Sci 45:1-40
Blair-West JR, Coghlan JP, Denton DA, Nelson JF, Orchard E, Scoggins BA, Wright RD, Myers K, Junqueira CL (1968a) Physiological, morphological and behavioural adaptation to a sodium deficient environment by wild native Australian and introduced species of animals. Nature 217:922-928
Blair-West JR, Coghlan JP, Denton DA, Goding JR, Wintour M, Wright RD (1968b) The effect of nephrectomy on aldosterone secretion in the conscious sodium-depleted hypophysectomized sheep. Aust J Exp Biol Med Sci 46:295-318
Blair-West JR, Coghlan JP, Denton DA, Orchard E, Scoggins BA, Wright RD (1968c) Renin-angiotensin-aldosterone system and sodium balance in experimental renal hypertension. Endocrinology 83:1199-1209
Blair-West JR, Coghlan JP, Denton DA, Goding JR, Wintour M, Wright RD (1968d) The local action of ammonium, calcium and magnesium on adrenocortical secretion. Aust J Exp Biol Med Sci 46:371-374
Blair-West JR, Coghlan JP, Denton DA, Scoggins BA, Wintour EM, Wright RD (1970a) The onset of effect of ACTH, angiotensin II and raised plasma potassium concentration on the adrenal cortex. Steroids 15:433-448
Blair-West JR, Brodie A, Coghlan JP, Denton DA, Flood C, Goding JR, Scoggins BA, Tait JF, Tait SAS, Wintour EM, Wright RD (1970b) Studies on the biosynthesis of aldosterone using the sheep adrenal transplant: Effect of sodium depletion on the conversion of corticosterone to aldosterone. J Endocrinol 46:453-476
Blair-West JR, Coghlan JP, Denton DA, Scoggins BA, Wintour EM, Wright RD (1970c) Effect of change of sodium balance on the corticosteroid response to angiotensin II. Aust J Exp Biol Med Sci 48:253-265
Blair-West JR, Coghlan JP, Denton DA, Funder JW, Scoggins BA, Wright RD (1971a) The effect of the heptapeptide (2-8) and hexapeptide (3-8) fragments of angiotensin II on aldosterone secretion. J Clin Endocrinol Metab 32:575-578
Blair-West JR, Coghlan JP, Denton DA, Funder JW, Scoggins BA, Wright RD (1971b) The effect of adrenal arterial infusion of hypertonic $NaHCO_3$ solution on aldosterone secretion in sodium deficient sheep. Acta Endocrinol 66:448-461

Blair-West JR, Coghlan JP, Denton DA, Funder JW, Scoggins BA, Wright RD (1971c) Effects of prostaglandin E_2 upon the steroid secretion of the adrenal of the sodium deficient sheep. Endocrinology 88:367-371
Blair-West JR, Cain MD, Catt KJ, Coghlan JP, Denton DA, Funder JW, Scoggins BA, Stockigt JR, Wright RD (1971d) Further facets of aldosterone regulation. In: James VHT, Martini L (eds) Hormonal Steroids. Proceedings of the 3rd international congress, Hamburg 1970. Excerpta Medica, Amsterdam (International Congress Series no 219), pp 572-580
Blair-West JR, Cain MD, Catt KJ, Coghlan JP, Denton DA, Funder JW, Scoggins BA, Wright RD (1971e) The dissociation of aldosterone secretion and systemic renin and angiotensin II levels during the correction of sodium deficiency. Acta Endocrinol 66:229-247
Blair-West JR, Coghlan JP, Denton DA, Funder JW, Scoggins BA, Wright RD (1971f) Inhibition of renin secretion by systemic and intrarenal angiotensin infusion. Am J Physiol 220:1309-1315
Blair-West JR, Coghlan JP, Cran E, Denton DA, Funder JW, Scoggins BA (1972) Contrived suppression of renin secretion during sodium depletion. In: Genest J, Koiw E (eds) Hypertension '72. Springer, Berlin Heidelberg New York, pp 14-25
Blair-West JR, Coghlan JP, Denton DA, Funder JW, Oddie CJ, Scoggins BA (1973a) Current concepts in aldosterone control. In: Scow RO (ed) Endocrinology. Proceedings of the 4th international congress of endocrinology, Washington DC 1972. Excerpta Medica, Amsterdam, pp 768-773
Blair-West JR, Coghlan JP, Cran E, Denton DA, Funder JW, Scoggins BA (1973b) Increased aldosterone secretion during sodium depletion with inhibition of renin release. Am J Physiol 224:1409-1414
Blair-West JR, Coghlan JP, Denton DA, Scoggins BA (1974) Aldosterone regulation in sodium deficiency: role of ionic factors and angiotensin II. In: Page IH, Bumpus FM (eds) Angiotensin. Springer, Berlin Heidelberg New York, pp 337-399 (Handbook of experimental pharmacology, vol 37)
Blair-West JR, Coghlan JP, Denton DA, Hardy KJ, Scoggins BA, Wright RD (1979) Effect of adrenal arterial infusion of P-113 on aldosterone secretion in Na-deficient sheep. Am J Physiol 236:F333-F341
Blair-West JR, Coghlan JP, Denton DA, Fei DTW, Hardy KJ, Scoggins BA, Wright RD (1980) A dose-response comparison of the actions of angiotensin II and angiotensin III in sheep. J Endocrinol 87:409-417
Bledsoe T, Island DP, Riondel AM, Liddle GW (1964) Modification of aldosterone secretion and electrolyte excretion in man by a chemical inhibitor of 18-oxidation. J Clin Endocrinol Metab 24:740-746
Bledsoe T, Island DP, Liddle GW (1966) Studies of the mechanism through which sodium depletion increases aldosterone biosynthesis in man. J Clin Invest 45:524-530
Bliddal J, Nielsen I (1970) Renin, aldosterone and electrolytes in idiopathic orthostatic hypotension. Dan Med Bull 17:153-157
Blizzard R, Liddle GW, Migeon CJ, Wilkins L (1959) Aldosterone excretion in virilizing adrenal hyperplasia. J Clin Invest 38:1442-1451
Bloom WL, Mitchell WR (1960) Salt excretion of fasting patients. Arch Intern Med 106:321-326
Bloom WL, Azar G, Clark JE (1966) Electrolyte and lipid metabolism of fasting men and women. Metabolism 15:401-408
Bojesen E (1966) Concentrations of aldosterone and corticosterone in peripheral plasma of rats. The effect of salt depletion, salt repletion, potassium loading and intravenous injections of renin and angiotensin II. Eur J Steroids 1:145-169
Bojesen E, Degn H (1961) A double isotope derivative method for the determination of aldosterone in peripheral plasma. The effect of adrenalectomy on aldosterone level in anaesthetized dogs. Acta Endocrinol 37:541-558
Bonelli J, Waldhäusl W, Magometschnigg D, Schwarzmeier J, Korn A, Hitzenberger G (1977) Effect of exercise and prolonged oral administration of propranolol on haemodynamic variables, plasma renin concentration, plasma aldosterone and c-AMP. Eur J Clin Invest 7: 337-343
Bonnin JM, Hodge RL, Lumbers ER (1972) A renin-secreting renal tumor associated with hypertension. Aust N Z J Med 75:178-181
Boonshaft B, O'Connell JMB, Hayes JM, Schreiner GE (1968) Serum renin activity during normal pregnancy: effect of alterations of posture and sodium intake. J Clin Endocrinol Metab 28:1641-1644

Boscaro M, Al Dujaili EAS, Edwards CRW (1979) *In vitro* studies on the mechanism of inhibition by bromocriptine, lisuride and metergoline of angiotensin II induced aldosterone release. Acta Endocrinol [Suppl] 225:342 (abstract)

Boscaro M, Al-Dujaili EAS, Edwards CRW (1981) Effects of two different angiotensin II antagonists on aldosterone secretion by isolated perfused rat zona glomerulosa cells. Endocrinology 108:2381-2384

Boscaro M, Scaroni C, Edwards CRW, Mantero F (1982) Inhibitory effect of somatostatin on the aldosterone response to angiotensin II: in vitro studies. J Endocrinol Invest 5:173-177

Bougas J, Flood C, Little B, Tait JF, Tait SAS, Underwood R (1964) Dynamic aspects of aldosterone metabolism. In: Baulieu EE, Robel P (eds) Aldosterone. Blackwell, Oxford, pp25-50

Bouhnik J, Galen F-X, Clauser E, Menard J, Corvol P (1981) The renin-angiotensin system in thyroidectomized rats. Endocrinology 108:647-650

Boulfekhar L, Brudieux R (1980) Peripheral concentrations of progesterone, cortisol, aldosterone, sodium and potassium in the plasma of the Tadmit ewe during pregnancy and parturition. J Endocrinol 84:25-33

Boulter PR, Spark RF, Arky RA (1974) Dissociation of the renin-aldosterone system and refractoriness to the sodium-retaining action of mineralocorticoid during starvation in man. J Clin Endocrinol Metab 38:248-254

Boyd GW, Adamson AR, Arnold M, James VHT, Peart WS (1972) The role of angiotensin II in the control of aldosterone in man. Clin Sci 42:91-104

Boyd JE, Mulrow PJ (1972) Further studies of the influence of potassium upon aldosterone production in the rat. Endocrinology 90:299-301

Boyd JE, Palmore WP, Mulrow PJ (1971) Role of potassium in the control of aldosterone secretion in the rat. Endocrinology 88:556-565

Boyd JE, Page RB, Mulrow PJ (1972) The effect of hypophysectomy on the conversion of corticosterone to aldosterone in the sodium depleted rat. Endocrinology 90:827-829

Boyd J, Mulrow PJ, Palmore WP, Silva P (1973) Importance of potassium in the regulation of aldosterone production. Circ Res 32/33 [Suppl 1]:I-39-I-45

Bradshaw SD, Grenot CJ (1976) Plasma aldosterone levels in two reptilian species, *Uromastix acanthinurus* and *Tiliqua rugosa*, and the effect of several experimental treatments. J Comp Physiol 111:71-76

Braley LM, Williams GH (1977) Rat adrenal cell sensitivity to angiotensin II, α^{1-24}-ACTH, and potassium: a comparative study. Am J Physiol 233:E402-E406

Braley LM, Williams GH (1978) The effects of ouabain on steroid production by rat adrenal cells stimulated by angiotensin II, α^{1-24} adrenocorticotropin and potassium. Endocrinolgy 103: 1997-2005

Braley LM, Williams GH (1979) The effect of angiotensin II and saralasin on 18-hydroxy-11-deoxycorticosterone production by isolated human adrenal glomerulosa cells. J Clin Endocrinol Metab 49:600-603

Braley LM, Williams GH (1980) The effect of unit gravity sedimentation on adrenal steroidogenesis by isolated rat glomerulosa and fasciculata cells. Endocrinology 106:50-55

Braley LM, Menachery A, Williams GH (1981) Angiotensin II's role in mediating angiotensin I- and tetradecapeptide-induced steroidogenesis by rat glomerulosa cells. Endocrinology 109: 960 965

Braley LM, Menachery AI, Underwood RH, Williams GH (1983a) Is the adrenal angiotensin receptor angiotensin II- or angiotensin III like? Acta Endocrinol 102:116-121

Braley LM, Menachery AI, Williams GH (1983b) Specificity of metoclopramide in assessing the role of dopamine in regulating aldosterone secretion. Endocrinology 112:1352-1357

Braley L, Menachery A, Brown E, Williams G (1984) The effects of extracellular K^+ and angiotensin II on cytosolic Ca^{++} and steroidogenesis in adrenal glomerulosa cells. Biochem Biophys Res Commun 123:810-815

Brandenberger G, Follenius M, Oyono S (1981) Effect of propranolol on aldosterone response to heat exposure in sodium-restricted men. J Endocrinol Invest 4:395-400

Braverman B, Davis JO (1973) Adrenal steroid secretion in the rabbit: sodium depletion angiotensin II, and ACTH. Am J Physiol 225:1306-1310

Bravo EL, Dustan HP, Tarazi RC (1975) Selective hypoaldosteronism despite prolonged pre- and postoperative hyperreninemia in primary aldosteronism. J Clin Endocrinol Metab 41:611-617

Bravo EL, Khosla MC, Bumpus FM (1976a) The role of angiotensins in aldosterone production. Circ Res 38 [Suppl 2]:II-104-II-107

Bravo EL, Khosla MC, Bumpus FM (1976b) Differential effects of asp[1]-angiotensin II and sar[1]-angiotensin II on vascular and adrenal receptors in the dog. Clin Sci Mol Med 51:41-45
Bravo EL, Saito I, Zanella T, Sen S, Bumpus FM (1980) In vitro steroidogenic properties of a new hypertension-producing compound isolated from normal human urine. J Clin Endocrinol Metab 51:176-178
Brecher P, Braga-Illa F, Chobanian AV (1973) The metabolism of cholesterol esters in rat adrenal cell suspensions. Endocrinology 93:1163-1172
Brecher PI, Pyun HY, Chobanian AV (1974) Studies on the angiotensin II receptor in the zona glomerulosa of the rat adrenal gland. Endocrinology 95:1026-1033
Breuer H, Kaulhausen H, Mühlbauer W, Fritzsche G, Vetter H (1974) Circadian rhythm of the renin-angiotensin-aldosterone system. In: Aschoff J, Ceresa F, Halberg F (eds) Chronobiological aspects of endocrinology. Schattauer, Stuttgart, pp 101-109
Brode E (1965) Über die Spezifität der Steroid-11β-hydroxylase. Arzneimittel Forsch 15:61-64
Brodie AH, Shimizu N, Tait SAS, Tait JF (1967) A method for the measurement of aldosterone in peripheral plasma using ^{3}H-acetic anhydride. J Clin Endocrinol Metab 27:997-1011
Brooks VL, Malvin RL (1981) Intracerebroventricular infusion of angiotensin II inhibits aldosterone secretion. Am J Physiol 239:E447-E453
Brown EH, Coghlan JP, Hardy KJ, Wintour EM (1978) Aldosterone, corticosterone, cortisol, 11-deoxycortisol and 11-deoxycorticosterone concentrations in the blood of chronically cannulated ovine foetuses: effect of ACTH. Acta Endocrinol 88:364-374
Brown G, Douglas J, Bravo E (1980) Angiotensin II receptors and in vitro aldosterone responses of aldosterone-producing adenomas, adjacent non-tumorous tissue, and normal human adrenal glomerulosa. J Clin Endocrinol Metab 51:718-723
Brown JJ, Davies DL, Lever AF, Robertson JIS (1964a) Influence of sodium deprivation and loading on the plasma-renin in man. J Physiol 173:408-419
Brown JJ, Davies DL, Lever AF, Robertson JIS (1964b) Variation in plasma renin during the menstrual cycle. Br Med J 2:1114-1115
Brown JJ, Davies DL, Doak PB, Lever AF, Robertson JIS, Tree M (1964c) The presence of renin in the amniotic fluid. Lancet 2:64-66
Brown JJ, Davies DL, Lever AF, Robertson JIS (1965a) Plasma renin concentration in human hypertension. 1: relationship between renin, sodium and potassium. Br Med J 2:144-148
Brown JJ, Davies DL, Doak PB, Lever AF, Robertson JIS, Trust P (1965b) Plasma renin concentration in hypertensive disease of pregnancy. Lancet 2:1219
Brown JJ, Davies DL, Lever AF, Robertson JIS (1966a) Renin and angiotensin. A survey of some aspects. Postgrad Med J 42:153-176
Brown JJ, Davies DL, Lever AF, McPherson D, Robertson JIS (1966b) Plasma renin concentration in relation to changes in posture. Clin Sci 30:279-284
Brown JJ, Davies DL, Doak PB, Lever AF, Robertson JIS (1966c) Plasma renin concentration in the hypertensive diseases of pregnancy. J Obstet Gynaecol Br Cwlth 73:410-417
Brown JJ, Fraser R, Lever AF, Robertson JIS, James VHT, McCusker J, Wynn V (1968) Renin, angiotensin, corticosteroids and electrolyte balance in Addison's disease. Q J Med 37:97-118
Brown JJ, Chinn RH, Davies DL, Düsterdieck G, Fraser R, Lever AF, Robertson JIS, Tree M, Wiseman A (1968b) Plasma electrolytes renin, and aldosterone in the diagnosis of primary aldosteronism. With a note on plasma corticosterone concentration. Lancet 2:55-59
Brown JJ, Düsterdieck G, Fraser R, Lever AF, Robertson JIS, Tree M, Weir RJ (1971) Hypertension and chronic renal failure. Br Med Bull 27:128-135
Brown JJ, Fraser R, Lever AF, Love DR, Morton JJ, Robertson JIS (1972) Raised plasma angiotensin II and aldosterone during dietary sodium restriction in man. Lancet 2:1106-1107
Brown JJ, Chinn RH, Fraser R, Lever AF, Morton JJ, Robertson JIS, Tree M, Waite MA, Park DM (1973a) Recurrent hyperkalaemia due to selective aldosterone deficiency: correction by angiotensin infusion. Br Med J 1:650-654
Brown JJ, Fraser R, Lever AF, Morton JJ, Robertson JIS, Tree M, Bell PRF, Davidson JK, Ruthven IS (1973b) Hypertension and secondary hyperaldosteronism associated with a renin-secreting renal juxtaglomerular-cell tumor. Lancet 2:1228-1232
Brown JJ, Lever AF, Robertson JIS, Beevers DG, Cumming AMM, Davies DL, Fraser R, Mason P, Morton JJ, Tree M (1979) Are idiopathic hyperaldosteronism and low-renin hypertension variants of essential hypertension? Ann Clin Biochem 16:380-388
Brown RD, Strott CA (1971) Plasma deoxycorticosterone in man. J Clin Endocrinol Metab 32: 744-750

Brown RD, Strott CA, Liddle GW (1972a) Site of stimulation of aldosterone biosynthesis by angiotensin and potassium. J Clin Invest 51:1413–1418
Brown RD, Strott CA, Liddle GW (1972b) Plasma deoxycorticosterone in normal and abnormal pregnancy. J Clin Endocrinol Metab 35:736–742
Brown RD, Tucker R, Tue K, Wisgerhof M, Salassa R (1978) Effect of saralasin on plasma aldosterone in hypertensive man. J Lab Clin Med 91:473–479
Brown RD, Wisgerhof M, Carpenter PC, Brown G, Jiang N-S, Kao P, Hegstad R (1979) Adrenal sensitivity to angiotensin II and undiscovered aldosterone stimulating factors in hypertension. J Steroid Biochem 11:1043–1050
Brown RD, Wisgerhof M, Jiang N-S, Kao P, Hegstad R (1981) Effect of metoclopramide on the secretion and metabolism of aldosterone in man. J Clin Endocrinol Metab 52:1014–1018
Brown RD, Billman GE, Kem DC, Stone HL, Jiang N-S, Kao P, Hegstad RL (1982) The effect of metoclopramide and dopamine on plasma aldosterone concentration in normal man and rhesus monkeys *(Macaca mulatta):* a new model to study dopamine control of aldosterone secretion. J Clin Endocrinol Metab 55:828–832
Brownell KA, Lee SL, Beck RR, Besch PK (1963) In vitro 17α-hydroxylation of steroids by the adrenals of hypertensive rats. Endocrinology 72:167–168
Brownie AC, Skelton FR, Gallant S, Nicholls P, Elliott WB (1968) Adrenal cytochrome levels, corticosteroidogenesis and respiratory activity in the rat following methylandrostenediol treatment. Life Sci 7:765–771
Brownie AC, Colby HD, Gallant S, Skelton FR (1970) Some studies on the effect of androgens on adrenal cortical function of rats. Endocrinology 86:1085–1092
Brubacher ES, Vander AJ (1968) Sodium deprivation and renin secretion in unanesthetized dogs. Am J Physiol 214:15–21
Brunner HR, Baer L, Sealey JE, Ledingham JGG, Laragh JH (1970) The influence of potassium administration and of potassium deprivation on plasma renin in normal and hypertensive subjects. J Clin Invest 49:2128–2138
Brunner HR, Kirshman JD, Sealey JE, Laragh JH (1971) Hypertension of renal origin: evidence for two different mechanisms. Science 174:1344–1346
Brunner HR, Gavras H, Waeber B, Kershaw GR, Turini GA, Vukovich RA, McKinstry DN, Gavras I (1979) Oral angiotensin-converting enzyme inhibitor in long-term treatment of hypertensive patients. Ann Intern Med 90:19–23
Bryan GT (1971) Aldosterone secretion rates in children with normal adrenal function. Pediatrics 47:587–593
Bryan GT, Kliman B, Gill JR Jr, Bartter FC (1964) Effect of human renin on aldosterone secretion rate in normal man and in patients with the syndrome of hyperaldosteronism, juxtaglomerular hyperplasia and normal blood pressure. J Clin Endocrinol Metab 24:729–732
Bull MB, Hillman RS, Cannon PJ, Laragh JH (1970) Renin and aldosterone secretion in man as influenced by changes in electrolyte balance and blood volume. Circ Res 27:953–960
Bunge G (1873) Über die Bedeutung des Kochsalzes und das Verhalten der Kalisalze im menschlichen Organismus. Z Biol 9:104–143
Burrow GN (1969) A steroid inhibitory effect on adrenal mitochondria. Endocrinology 84: 979–985
Burrow GN, Mulrow PJ, Bondy PK (1966) Protein synthesis and aldosterone production. Endocrinology 79:955–963
Burwell LR, Davis WW, Bartter FC (1969) Studies on the loci of action of stimuli to the biogenesis of aldosterone. Proc R Soc Med 62:1254–1257
Bush IE (1953) Species differences in adrenocortical secretion. J Endocrinol 9:95–100

Cade R, Perenich T (1965) Secretion of aldosterone by rats. Am J Physiol 208:1026–1030
Cain JP, Williams GH, Dluhy RG (1972) Plasma renin activity and aldosterone secretion in patients with acromegaly. J Clin Endocrinol Metab 34:73–81
Cain JP, Dluhy RG, Williams GH, Selenkow HA, Milech A, Richmond S (1973) Control of aldosterone secretion in hyperthyroidism. J Clin Endocrinol Metab 36:365–371
Campanile CP, Goodfriend TL (1981) Steroids as potential modulators of angiotensin receptors in bovine adrenal glomerulosa and kidney. Steroids 37:681–700
Campbell DJ (1982a) Effect of composition of incubation medium on aldosterone and corticosterone production by isolated rat zona glomerulosa cells. J Endocrinol 94:211–224
Campbell DJ (1982b) Increased steroidogenesis by rat zona glomerulosa cells with increased cell

concentration *in vitro:* evidence for a novel aldosterone-stimulating factor and implications regarding aldosterone biosynthesis. J Endocrinol 94:225-241
Campbell DJ (1982c) Stimulation by potassium of cholesterol ester depletion of rat zona glomerulosa cells *in vitro.* J Steroid Biochem 17:707-708
Campbell DJ (1982d) Effect of rat plasma lipoproteins on aldosterone production by rat zona glomerulosa cells *in vitro.* J Steroid Biochem 17:709-711
Campbell DJ, Mendelsohn FAO, Adam WR, Funder JW (1981a) Is aldosterone secretion under dopaminergic control? Circ Res 49:1217-1227
Campbell DJ, Mendelsohn FAO, Adam WR, Funder JW (1981b) Metoclopramide does not elevate aldosterone in the rat. Endocrinology 109:1484-1491
Campbell WB, Gomez-Sanchez CE (1985) Absence of prostacyclin involvement in angiotensin-induced aldosterone secretion in rat adrenal cells. Endocrinology 117:279-286
Campbell WB, Pettinger WA (1975) Sodium chloride suppression of renin release in the unanesthetized rat. Endocrinology 97:1394-1397
Campbell WB, Schmitz JM (1978) Effect of alterations in dietary potassium on pressor and steroidogenic effects of angiotensin II and angiotensin III. Endocrinology 103:2098-2104
Campbell WB, Brooks SN, Pettinger WA (1974) Angiotensin II- and angiotensin III-induced aldosterone release in vivo in the rat. Science 184:994-996
Campbell WB, Schmitz JM, Itskovitz HD (1977) (Des-asp^1) angiotensin I: a study of its pressor and steroidogenic activities in conscious rats. Endocrinology 100:46-51
Campbell WB, Gomez-Sanchez CE, Adams BV, Schmitz JM, Itskovitz HD (1979a) Attenuation of angiotensin II and III induced aldosterone release by prostaglandin synthesis inhibitors. J Clin Invest 64:1552-1557
Campbell WB, Schmitz JM, Itskovitz HD (1979b) Effect of sodium depletion on the steroidogenic and pressor actions of angiotensin in the rat. Clin Sci 56:325-333
Campbell WB, Gomez-Sanchez CE, Adams BV (1980) Role of prostaglandins in angiotensin-induced steroidogenesis. Absence of an effect by prostaglandin E_2. Hypertension 2:471-476
Campbell WB, Currie MG, Needleman P (1985) Inhibition of aldosterone biosynthesis by atriopeptins in rat adrenal cells. Circ Res 57:113-118
Campbell WB, Brady MT, Gomez-Sanchez CE (1986) Effects of angiotensin II, prostaglandin E_2 and indomethacin on the early and late steps of aldosterone biosynthesis in isolated adrenal cells. J Steroid Biochem 24:865-870
Campese VM, Romoff M, DeQuattro V, Massry SG (1980) Relationship between plasma catecholamines, plasma renin activity, aldosterone, and arterial pressure during postural stress in normal subjects. J Lab Clin Med 96:927-933
Cangiano JL, Gould AB, Vertes V (1971) Evaluation of select criteria for the diagnosis of primary aldosteronism in normokalemic hypertension. Am J Med Sci 262:261-267
Cannon PJ, Ames RP, Laragh JH (1966) Relation between potassium balance and aldosterone secretion in normal subjects and in patients with hypertensive or renal tubular disease. J Clin Invest 45:865-879
Cantin M (1970) Relationship of juxtaglomerular apparatus and adrenal cortex to biochemical and extracellular fluid volume changes in magnesium deficiency. Lab Invest 22:558-568
Cantin M, Genest J (1985) The heart and the atrial natriuretic factor. Endocrinol Rev 6:107-127
Cantin M, Veilleux R (1968) The juxtaglomerular apparatus and zona glomerulosa in magnesium-deficient rats. Rev Can Biol 27:179-182
Capelli JP, Wesson LG Jr, Aponte GE, Faraldo C, Jaffe E (1968) Characterization and source of a renin-like enzyme in anephric humans. J Clin Endocrinol Metab 28:221-230
Capponi AM, Catt KJ (1979) Angiotensin II receptors in adrenal cortex and uterus. Binding and activation properties of angiotensin analogues. J Biol Chem 254:5120-5127
Capponi AM, Catt KJ (1980) Solubilization and characterization of adrenal and uterine angiotensin II receptors after photoaffinity labeling. J Biol Chem 255:12081-12086
Capponi AM, Lew PD, Jornot L, Vallotton MB (1984) Correlation between cytosolic free Ca^{2+} and aldosterone production in bovine adrenal glomerulosa cells. Evidence for a difference in the mode of action of angiotensin II and potassium. J Biol Chem 259:8863-8869
Carballeira A, Fishman LM, Jacobi JD (1976) Dual sites of inhibition by metyrapone of human adrenal steroidogenesis: Correlation of in vivo and in vitro studies. J Clin Endocrinol Metab 42:687-695
Carey RM (1982) Acute dopaminergic inhibition of aldosterone secretion is independent of angiotensin II and adrenocorticotropin. J Clin Endocrinol Metab 54:463-469

Carey RM, van Loon GR (1982) Bromocriptine does not inhibit the aldosterone response so sodium depletion. J Clin Endocrinol Metab 55:162–165

Carey RM, Vaughan ED Jr, Peach MJ, Ayers CR (1978) Activity of (des-aspartyl[1])-angiotensin II and angiotensin II in man. Differences in blood pressure and adrenocortical response during normal and low sodium intake. J Clin Invest 61:20–31

Carey RM, Thorner MO, Ortt EM (1979) Effects of metoclopramide and bromocriptine on the renin-angiotensin-aldosterone system in man. Dopaminergic control of aldosterone. J Clin Invest 63:727–735

Carey RM, Thorner MO, Ortt EM (1980) Dopaminergic inhibition of metoclopramide-induced aldosterone secretion in man. Dissociation of responses to dopamine and bromocriptine. J Clin Invest 66:10–18

Carey RM, Sen S, Dolan LM, Malchoff CD, Bumpus FM (1984) Idiopathic hyperaldosteronism. A possible role for aldosterone-stimulating factor. N Engl J Med 311:94–100

Carpenter CCJ, Davis JO, Ayers CR (1961) Relation of renin, angiotensin II, and experimental renal hypertension to aldosterone secretion. J Clin Invest 40:2026–2042

Carr BR, Parker CR Jr, Madden JD, MacDonald PC, Porter JC (1979) Plasma levels of adrenocorticotropin and cortisol in women receiving oral contraceptive steroid treatment. J Clin Endocrinol Metab 49:346–349

Carretero O, Gross F (1967) Renin substrate in plasma under various experimental conditions in the rat. Am J Physiol 213:695–700

Carroll JE, Goodfriend TL (1984) Androgen modulation of adrenal angiotensin receptors. Science 224:1009–1011

Carroll JE, Campanile CP, Goodfriend TL (1982) The effect of prolactin on human aldosterone-producing adenomas in vitro. J Clin Endocrinol Metab 54:689–692

Carstensen H, Burgers ACJ, Li CH (1961) Demonstration of aldosterone and corticosterone as the principal steroids formed in incubates of adrenals of the American bullfrog *(Rana catesbeiana)* and stimulation of their production by mammalian adrenocorticotropin. Gen Comp Endocrinol 1:37–50

Casey ML, MacDonald PC (1982) Extraadrenal formation of a mineralocorticosteroid: deoxycorticosterone and deoxycorticosterone sulfate biosynthesis and metabolism. Endocrinol Rev 3: 396–403

Catalona WJ, Palmore WP, Levitin H (1972) Sodium wasting in potassium depletion: the role of aldosterone. Yale J Biol Med 45:33–41

Cejka V, de Vries LA, Smorenberg-Schoorl ME, van Daatselaar JJ, Borst JGG, Majoor CLH (1960) Effect of heparinoid and spirolactone on the renal excretion of sodium and aldosterone. Lancet 1:317–319

Chabert PR, Guelpa-Decorzant C, Riondel AM, Vallotton MB (1984) Effect of spironalectone on electrolytes, renin, ACTH and corticosteroids in the rat. J Steroid Biochem 20:1253–1259

Chagoya L, Nurko B, Santos E, Rivera A (1961) 6-Chloro,6-dehydro,17α-acetoxyprogesterone: its possible action as an aldosterone antagonist. J Clin Endocrinol Metab 21:1364–1375

Chan STH, Edwards BR (1970) Kinetic studies on the biosynthesis of corticosteroids in vitro from exogenous precursors by the interrenal glands of the normal, corticotrophin-treated and adenohypophysectomized *Xenopus laevis* Daudin. J Endocrinol 47:183–195

Chang RSL, Lotti VJ (1981) Solubilization of angiotensin II receptors in bovine adrenal cortex. Life Sci 29:613–618

Chartier L, Schiffrin E, Thibault G (1984a) Effect of atrial natriuretic factor (ANF)-related peptides on aldosterone secretion by adrenal glomerulosa cells: critical role of the intramolecular disulphide bond. Biochem Biophys Res Commun 122:171–174

Chartier L, Schiffrin E, Thibault G, Garcia R (1984b) Atrial natriuretic factor inhibits the stimulation of aldosterone secretion by angiotensin II, ACTH and potassium in vitro and angiotensin II-induced steroidogenesis in vivo. Endocrinology 115:2026–2028

Chavarri M, Luetscher JA, Dowdy AJ, Ganguly A (1977a) The effects of temperature and plasma cortisol on distribution of aldosterone between plasma and red blood cells: influence on metabolic clearance rate and on hepatic and renal extraction of aldosterone. J Clin Endocrinol Metab 44:752–759

Chavarri M, Ganguly A, Luetscher JA, Zager PG (1977b) Effect of bedrest on circadian rhythms of plasma renin, aldosterone, and cortisol. Aviat Space Environ Med 48:633–636

Chavin W, Honn KV, Clifford PJ (1978) Prostaglandin effects upon cyclic AMP levels, corticosterone output, and aldosterone output in the adrenal of the frog, *Rana berlandieri forreri*. Gen Comp Endocrinol 36:442–450

Cheng SC, Harding BW, Carballeira A (1974) Effects of metyrapone on pregnenolone biosynthesis and on cholesterol-cytochrome P-450 interaction in the adrenal. Endocrinology 94: 1451-1458
Cheng SC, Suzuki K, Sadee W, Harding BW (1976) Effects of spironolactone, cancrenone and canrenoate-K on cytochrome P-450, and 11β- and 18-hydroxylation in bovien and human adrenal cortical mitochondria. Endocrinology 99:1097-1106
Childers JW, Schneider EG (1981) Effect of hypertonic and hypotonic infusions on aldosterone in conscious sodium-depleted dogs. Clin Sci 61:191-199
Chinn RH, Brown JJ, Fraser R, Heron SM, Lever AF, Murchison L, Robertson JIS (1970) The natriuresis of fasting: relationship to changes in plasma renin and plasma aldosterone concentrations. Clin Sci 39:437-455
Chiu AT, Freer RJ (1979) Angiotensin-induced steroidgenesis in rabbit adrenal: effects of pH and calcium. Mol Cell Endocrinol 13:159-166
Chiu AT, Peach MJ (1973) Stimulation of aldosterone biosynthesis by angiotensin heptapeptide. Fed Proc 32:765 (abstract)
Chiu AT, Peach MJ (1974) Inhibition of induced aldosterone biosynthesis with a specific antagonist of angiotensin II. Proc Natl Acad Sci USA 71:341-344
Chouraqui J, Weniger J-P (1970) Identification des corticostéroides secrétés par les surrénales embryonaires de veau, lapin et souris cultivées in vitro. Acta Endocrinol 65:650-662
Chremos AN, Laidlaw JC, Ruse JL (1969) Effect of a sulfated mucopolysaccharide (Ro 1-8307) on the zona glomerulosa of regenerating rat adrenal glands and of intact glands in sodium deprived rats. Endocrinology 85:337-341
Christlieb AR, Amsterdam EA, Hickler RB (1972) Suppression of murine renal renin activity in the absence of sodium retention with desoxycorticosterone acetate but not with depot desoxycorticosterone. Am J Med Sci 263:457-464
Christlieb AR, Assal JP, Katsilambros N, Williams GH, Kozak GP, Suzuki T (1975) Plasma renin activity and blood volume in uncontrolled diabetes. Diabetes 24:190-193
Christlieb AR, Kaldany A, D'Elia JA, Williams GH (1978) Aldosterone responsiveness in patients with diabetes mellitus. Diabetes 27:732-737
Chu MD, Ulick S (1982) Isolation and identification of 18-hydroxycortisol from the urine of patients with primary aldosteronism. J Biol Chem 257:2218-2224
Chupin M, Bolo P, Charbonnel B, Lucas B (1981) La fonction minéralocorticoïde des hypopituitarismes avec insuffisance corticotrope. Presse Méd 10:2211
Churchill PC, Churchill MC (1979) Inhibition of in vitro renin release by low extracellular K. Life Sci 25:687-692
Churchill S, Bengele HH, Melby JC, Alexander EA (1981) Role of aldosterone in sodium retention of pregnancy in the rat. Am J Physiol 240:R175-R181
Clamage DM, Sanford CS, Vander AS, Mouw DR (1975) Effects of psychosocial stimuli on plasma renin activity in rats. Am J Physiol 231:1290-1294
Coghlan JP, Blair-West JR (1967) Aldosterone. In: Gray CH, Bacharach AL (eds) Hormones in blood, vol 2, 2nd ed. Academic, New York, pp 391-488
Coghlan JP, Scoggins BA (1967a) Measurement of aldosterone in peripheral blood of man and sheep. J Clin Endocrinol Metab 27:1470-1486
Coghlan JP, Scoggins BA (1967b) The measurement of aldosterone, cortisol and corticosterone in the blood of the wombat *(Vombatus hirsutus* Perry) and the kangaroo *(Macropus giganteus)*. J Endocrinol 39:445-448
Coghlan JP, Blair-West JR, Denton DA, Fei DT, Fernley RT, Hardy KJ, McDougall JG, Puy R, Robinson PM, Scoggins BA, Wright RD (1979a) Control of aldosterone secretion. J Endocrinol 81:55*P*-67*P*
Coghlan JP, Scoggins BA, Wintour EM (1979b) Aldosterone. In: Gray CH, James VHT (eds) Hormones in Blood, vol 3, 3rd ed. Academic, London, pp 493-609
Coghlan JP, Blair-West JR, Butkus A, Denton DA, Hardy KJ, Leksell L, McDougall JG, McKinley MJ, Scoggins BA, Tarjan E, Weisinger RS, Wright RD (1980) Factors regulating aldosterone secretion. In: Cumming IA, Funder JW, Mendelson FAO (eds) Endocrinology 1980. Proceedings of the 6th international congress of endocrinology, Melbourne 1980. Austr Acad Sci, Canberra, pp 385-388
Cohen EL, Conn JW, Rovner DR (1967) Postural augmentation of plasma renin activity and aldosterone excretion in normal people. J Clin Invest 46:418-428
Cohen JJ, Hulter HN, Smithline N, Melby JC, Schwartz WB (1976) The critical role of the

adrenal gland in the renal regulation of acid-base equilibrium. Evidence that chronic hyponatremia is a potent stimulus to aldosterone secretion. J Clin Invest 58:1201-1208
Cohen MP (1968) Aminoglutethimide inhibition of adrenal desmolase activity. Proc Soc Exp Biol Med 127:1086-1090
Cohen RB (1965) Effects of long-term sodium deprivation in the adrenal cortices of rats: a histochemical study. Endocrinology 77:1043-1047
Colby HD, Skelton FR, Brownie AC (1970) Testosterone-induced hypertension in the rat. Endocrinology 86:1093-1101
Coleman TG, McCaa RE, McCaa CS (1974) Effect of angiotensin II on aldosterone secretion in the conscious rat. J Endocrinol 60:421-427
Conaglen JV, Donald RA, Espiner EA, Livesey JH, Nicholls MG (1984) The effect of ovine corticotropin-releasing factor on catecholamine, vasopressin, and aldosterone secretion in normal man. J Clin Endocrinol Metab 58:463-466
Conn JW, Hinerman DL (1977) Spironolactone-induced inhibition of aldosterone biosynthesis of primary aldosteronism: morphological and functional studies. Metabolism 26:1293-1307
Conn JW, Cohen EL, Rovner DR (1964) Suppression of plasma renin activity in primary aldosteronism. Distinguishing primary from secondary aldosteronism in hypertensive disease. JAMA 190:213-221
Conn JW, Rovner DR, Cohen EL, Anderson JE Jr (1966) Inhibition by heparinoid of aldosterone biosynthesis in man. J Clin Endocrinol Metab 26:527-532
Conn JW, Rovner DR, Cohen EL (1968) Licorice-induced pseudoaldosteronism. Hypertension, hypokalemia, aldosteronopenia, and suppressed plasma renin activity. JAMA 205:492-496
Conn JW, Cohen EL, Lucas CP, McDonald WJ, Mayor GH, Blow WR Jr, Eveland WC, Bookstein JJ, Lapides J (1972) Primary reninism. Hypertension, hyperreninemia, and secondary aldosteronism due to renin-producing juxtaglomerular cell tumors. Arch Intern Med 130:682-696
Conneely OM, Headon DR, Olson CD, Ungar F, Dempsey ME (1984) Intramitochondrial movement of adrenal sterol carrier protein with cholesterol in response to corticotropin. Proc Natl Acad Sci USA 81:2970-2974
Connors M, Rosenkrantz H (1962) Serotonin uptake and action on the adrenal cortex. Endocrinology 71:407-413
Connors MH, Lee TC, Sheikholislam BM (1977) Plasma renin activity and aldosterone response to furosemide after chronic growth hormone administration to deficient subjects. Life Sci 20:535-540
Cooke CR, Horvath JS, Moore MA, Bledsoe T, Walker GW (1973a) Modulation of plasma aldosterone concentration by plasma potassium in anephric man in the absence of a change in potassium balance. J Clin Invest 52:3028-3032
Cooke CR, Ruiz-Maza F, Kowarski A, Migeon CJ, Walker WG (1973b) Regulation of plasma aldosterone concentration in anephric man and renal transplant recipients. Kidney Int 3:160-166
Cooke CR, Whelton PK, Moore MA, Caputo RA, Bledsoe T, Walker WG (1979) Dissociation of the diurnal variation of aldosterone and cortisol in anephric subjects. Kidney Int 15:669-675
Cooke CR, Gann DS, Hsu TH, Bledsoe T, Moore MA, Walker WG (1983) Hormonal responses to acute volume changes in anephric subjects. Kidney Int 23:71-78
Cope CL, Harwood M, Pearson J (1962) Aldosterone secretion in hypertensive diseases. Br Med J 1:659-665
Coppage WS Jr, Island DP, Smith M, Liddle GW (1959) Inhibition of aldosterone secretion and modification of electrolyte excretion in man by a chemical inhibitor of 11β-hydroxylation. J Clin Invest 38:2101-2110
Corrie JET, Edwards CRW, Jones DB, Padfield PL, Budd PS (1985) Factors affecting the secretion of 18-hydroxycortisol, a novel steroid of relevance to Conn's syndrome. Clin Endocrinol 23:579-586
Cortés JM, Péron FG, Dorfman RI (1963) Secretion of 18-hydroxydeoxycorticosterone by the rat adrenal gland. Endocrinology 73:713-720
Corvol P, Bertagna X, Bedrossian J (1974) Increased steroid metabolic clearance rate in anephric patients. Acta Endocrinol 75:756-762
Corvol P, Oblin ME, Degoulet P, Fressinaud Ph, Menard J (1977) Effect of acute potassium loading on plasma renin and urinary aldosterone in rats. Endocrinology 100:1008-1013

Costa G, Frisina N, de Pasquale R (1980) Increased aldosterone secretion induced by sulpiride. Clin Endocrinol 13:1-7

Cowley AW Jr, McCaa RE (1976) Acute and chronic dose-response relationships for angiotensin, aldosterone, and arterial pressure at varying levels of sodium intake. Circ Res 39:788-797

Cox JR, Leonard PJ, Singer B (1961) Effect of vasopressin on the volume of body fluid compartments and its relation to aldosterone excretion. Clin Sci 21:205-219

Cox JR, Leonhard PJ, Singer B (1962) Changes in sodium distribution in body fluid compartments and urinary aldosterone excretion. Clin Sci 23:13-26

Cozza EN, Burton G, Ceballos NR, Lantos CP, Harnik M, Scott AI (1985) Eighteen-deoxyaldosterone and other less polar forms of 18-hydroxycorticosterone as aldosterone precursors in rat adrenals. J Steroid Biochem 22:665-672

Crabbé J (1963) The sodium-retaining action of aldosterone. Arscia, Bruxelles

Crabbé J, Reddy WJ, Ross EJ, Thorn GW (1959) The stimulation of aldosterone secretion by adrenocorticotropic hormone. J Clin Endocrinol Metab 19:1185-1191

Crane MG, Harris JJ (1969a) Plasma renin activity and aldosterone excretion rate in normal subjects. I. Effect of ethinyl estradiol and medroxyprogesterone acetate. J Clin Endocrinol Metab 29:550-557

Crane MG, Harris JJ (1969b) Plasma renin activity and aldosterone excretion rate in normal subjects. II. Effect of oral contraceptive agents. J Clin Endocrinol Metab 29:558-562

Crane MG, Harris JJ (1970) Deoxycorticosterone secretion rate studies in edematous patients. Am J Med Sci 259:27-31

Crane MG, Harris JJ (1974) Suppression of plasma aldosterone by partial immersion. Metabolism 23:359-368

Crane MG, Harris JJ (1976) Effect of aging on renin activity and aldosterone excretion. J Lab Clin Med 87:947-959

Crivello JF, Gill GN (1983) Induction of cultured bovine adrenocortical zona glomerulosa cell 17-hydroxylase activity by ACTH. Mol Cell Endocrinol 30:97-107

Crivello JF, Jefcoate CR (1980) Intracellular movement of cholesterol in rat adrenal cells. Kinetics and effects of inhibitors. J Biol Chem 255:8144-8151

Crivello JF, Hornsby PJ, Gill GN (1982) Metyrapone and antioxidants are required to maintain aldosterone synthesis by cultured bovine adrenocortical zona glomerulosa cells. Endocrinology 111:469-479

Crivello JF, Hornsby PJ, Gill GN (1983) Suppression of cultured bovine adrenocortical zona glomerulosa cell aldosterone synthesis by steroids and its prevention by antioxidants. Endocrinology 113:235-242

Csánky MFD, van der Wal B, de Wied D (1968) The regulation of aldosterone production in normal and sodium-deficient rats. J Endocrinol 41:179-188

Cugini P, Manconi R, Serdoz R, Mancini A, Meucci T (1977) Influence of propranolol on circadian rhythms of plasma renin, aldosterone and cortisol in healthy supine man. Boll Soc Ital Biol Sper 53:263-269

Cugini P, Scavo D, Cornelissen G, Lee TY, Meucci T, Halberg F (1981) Circadian rhythms of plasma renin, aldosterone and cortisol on habitual and low dietary sodium intake. Horm Res 15:7-27

Cushman P Jr (1969) Inhibition of aldosterone secretion by ouabain in dog adrenal cortical tissue. Endocrinology 84:808-813

Cushman P Jr, Alter S, Hilton JG (1966) Effects of various growth hormone preparations on dog adrenal cortical secretion. Endocrinology 78:971-975

Dalakos TG, Streeten DHP (1975) The role of plasma volume in the increase of aldosterone secretion rate during sodium deprivation. Clin Sci Mol Med 48:161-165

Dalle M, Giry J, Gay M, Delost P (1978) Perinatal changes in plasma and adrenal corticosterone and aldosterone concentrations in the mouse. J Endocrinol 76:303-309

Damasco MC, Lantos CP (1975) The existence of two interconvertible forms of 18-hydroxycorticosterone: Is one of them the active precursor of aldosterone? J Steroid Biochem 6:69-74

Damasco MC, Diaz F, Cenal JP, Lantos CP (1979) Acute effects of three natural corticosteroids on the acid-base and electrolyte composition of urine in adrenalectomized rats. Acta Physiol Lat Am 29:305-314

Damasco MC, Vallverdú R, Ceñal JP, Otero de Bedners ME, Lantos CP (1983) Effects of 18-hydroxycorticosterone and of aldosterone on acid-base parameters in the arterial blood of adrenalectomized rats. Acta Physiol Lat Am 33:283-291

Davenport MW, Zipser RD (1983) Association of hypotension with hyperreninemic hypoaldosteronism in the critically ill patient. Arch Intern Med 143:735-737

David R, Golan S, Drucker W (1968) Familial aldosterone deficiency: enzyme defect, diagnosis, and clinical course. Pediatrics 41:403-412

David RR, Asnis M, Drucker WD (1972) Disturbance of cortisol production in congenital aldosterone deficiency. J Clin Endocrinol Metab 35:604-608

Davis DA, Medline NM (1970) Spironolactone (aldactone) bodies: concentric lamellar formations in the adrenal cortices of patients treated with spironolactone. Am J Clin Pathol 54:22-32

Davis JO (1961) Mechanisms regulating the secretion and metabolism of aldosterone in experimental secondary hyperaldosteronism. Recent Prog Horm Res 17:293-352

Davis JO (1962) The control of aldosterone secretion. Physiologist 5:65-86

Davis JO (1967) The regulation of aldosterone secretion. In: Eisenstein AB (ed) The adrenal cortex. Little Brown, Boston, pp 203-247

Davis JO (1975) Regulation of aldosterone secretion. In: Blaschko H, Sayers G, Smith AD (eds) Handbook of physiology. Adrenal gland, sect 7, vol 6. Am Physiol Soc, Washington DC, pp 77-106

Davis JO, Freeman RH (1977) The other angiotensins. Biochem Pharmacol 26:93-97

Davis JO, Bahn RC, Ball WC Jr (1959) Subacute and chronic effects of hypothalamic lesions on aldosterone and sodium excretion. Am J Physiol 197:387-390

Davis JO, Yankopoulos NA, Lieberman F, Holman J, Bahn RC (1960a) The role of the anterior pituitary in the control of aldosterone secretion in experimental secondary hyperaldosteronism. J Clin Invest 39:765-775

Davis JO, Carpenter CCJ, Ayers CR, Bahn RC (1960b) Relation of anterior pituitary function to aldosterone and corticosterone secretion in conscious dogs. Am J Physiol 199:212-216

Davis JO, Carpenter CCJ, Ayers CR, Holman JE, Bahn RC (1961a) Evidence for secretion of an aldosterone-stimulating hormone by the kidney. J Clin Invest 40:684-696

Davis JO, Ayers CR, Carpenter CCJ (1961b) Renal origin of an aldosterone-stimulating hormone in dogs with thoracic caval constriction and in sodium-depleted dogs. J Clin Invest 40: 1466-1474

Davis JO, Anderson E, Carpenter CCJ, Ayers CR, Haymaker W, Spence WT (1961c) Aldosterone and corticosterone secretion following midbrain transection. Am J Physiol 200:437-443

Davis JO, Urquhart J, Higgins JT Jr (1963) The effects of alterations of plasma sodium and potassium concentration on aldosterone secretion. J Clin Invest 42:587-609

Davis JO, Urquhart J, Higgins JT Jr, Ellis EC, Hartroft PM (1964) Hypersecretion of aldosterone in dogs with a chronic aortic-caval fistula and high output heart failure. Circ Res 14:471-485

Davis JO, Binnion PF, Brown TC, Johnston CI (1966) Mechanisms involved in the hypersecretion of aldosterone during sodium depletion. Circ Res [Suppl I] 28/29:I-143-I-157

Davis JO, Copeland DL, Taylor AA, Baumber JS (1970) Plasma electrolyte concentrations and steroid secretion in the bullfrog and opossum. Am J Physiol 219:555-559

Davis WW, Garren LD (1968) On the mechanism of action of adrenocorticotropic hormone. The inhibitory site of cycloheximide in the pathway of steroid biosynthesis. J Biol Chem 243: 5153-5157

Davis WW, Newsome HH Jr, Wright LD Jr, Hammond WG, Easton J, Bartter FC (1967) Bilateral adrenal hyperplasia as a cause of primary aldosteronism with hypertension, hypokaliemia and suppressed renin activity. Am J Med 42:642-647

Davis WW, Casper AGT, Bartter FC (1968) Sites of action of sodium depletion on aldosterone biosynthesis in the dog. J Clin Invest 47:1425-1433

Davis WW, Burwell LR, Bartter FC (1969) Inhibition of the effects of angiotensin II on adrenal steroid production by dietary sodium. Proc Natl Acad Sci USA 63:718-723

Dawson-Hughes BF, Moore TJ, Dluhy RG, Hollenberg NK, Williams GH (1981) Plasma angiotensin II concentration regulates vascular but not adrenal responsiveness to restriction in sodium intake in normal man. Clin Sci 61:527-534

Deane HW, Masson GMC (1951) Adrenal cortical changes in rats with various types of experimental hypertension. J Clin Endocrinol Metab 11:193-208

Deane HW, Shaw JH, Greep RO (1948) The effect of altered sodium or potassium intake on the width and cytochemistry of the zona zona glomerulosa of the rat's adrenal cortex. Endocrinology 43:133-153

Debreceni L, Csete B (1973a) Lack of a direct influence of sodium on aldosterone production in the rat adrenal in vitro. Acta Physiol Acad Sci Hung 43:247–251

Debreceni L, Csete B (1973b) The role of a feedback mechanism in the regulation of aldosterone secretion in the rat. Acta Endocrinol 73:282–288

Debreceni L, Csete B (1975) In vitro production of aldosterone by the rat adrenals after in vivo potassium- and sodium-loading and depletion. Endokrinologie 64:316–322

De Châtel R, Weidmann P, Flammer J, Ziegler WH, Beretta-Piccoli C, Vetter W, Reubi FC (1977) Sodium, renin, aldosterone, catecholamines and blood pressure in diabetes mellitus. Kidney Int 12:412–421

Decorzant C, Riondel AM, Philippe M-J, Bertrand J, Vallotton MB (1977) Detection of Na^+ and K^+ in the rat adrenal cortex with the electron microprobe. Clin Sci Mol Med 53:423–430

DeForrest JM, Davis JO, Freeman RH, Stephens GA, Watkins BE (1979) Circadian changes in plasma renin activity and plasma aldosterone concentration in two-kidney hypertensive rats. Hypertension 1:142–149

DeForrest JM, Davis JO, Freeman RH, Seymour AA, Rowe BP, Williams GM, Davis TP (1980) Effects of indomethacin and meclofenamate on renin release and renal hemodynamic function during chronic sodium depletion in conscious dogs. Circ Res 47:99–107

DeFronzo RA (1980) Hyperkalemia and hyporeninemic hypoaldosteronism. Kidney Int 17: 118–134

Degenhart HJ, Visser HKA, Wilmink R, Croughs W (1965) Aldosterone- and cortisol-secretion rates in infants and children with congenital adrenal hyperplasia suggesting different 21-hydroxylation defects in "salt-losers" and "non salt-losers". Acta Endocrinol 48:587–601

Degenhart HJ, Frankena L, Visser HKA, Cost WS, van Seters AP (1966) Further investigations of a new hereditary defect in the biosynthesis of aldosterone: evidence for a defect in 18-hydroxylation of corticosterone. Acta Physiol Pharmacol Neerl 14:88 (abstract)

Degli Uberti EC, Fabri BL, Margutti AR, Fersini CM, Pansini R (1979) Effect of bromocriptine on the control of plasma aldosterone diurnal variation in normal supine man. Horm Res 10: 64–78

Degli Uberti EC, Trasforini G, Margutti AR, Rotola CA, Pansini R (1982) Effect of ketanserin, a new inhibitor of 5-HT_2 receptors, on plasma renin activity and aldosterone levels in normal subjects. Horm Res 16:244–248

Degli Uberti EC, Trasforini G, Margutti AR, Rotola CA, Bianconi M, Pansini P (1983a) Effect of ketanserin, an inhibitor of 5-HT_2 receptors on the aldosterone-stimulating action of metoclopramide. Horm Res 17:160–167

Degli Uberti EC, Trasforini G, Salvadori S, Margutti A, Tomatis R, Bianconi M, Rotola C, Pansini R (1983b) Responses of plasma renin activity, aldosterone, adrenocorticotropin, and cortisol to dermorphin, a new synthetic potent opiate-like peptide, in man. J Clin Endocrinol Metab 57:1179–1185

Deheneffe J, Castermans R (1976) The control of aldosterone in anephric man. Acta Clin Belg 31:313–320

Deheneffe J, Cuesta V, Robertson I (1975) The control of aldosterone in anephric man. Clin Sci Mol Med 48:465–466

Deheneffe J, Cuesta V, Briggs JD, Brown JJ, Fraser R, Lever AF, Morton JJ, Robertson JIS, Tree M (1976) Response of aldosterone and blood pressure to angiotensin II infusion in anephric man. Effect of sodium deprivation. Clin Res 39:183–190

De Lange WE, Lappöhn RE, Sluiter WJ, Doorenbos H (1977) Primäre Amenorrhoe und Hypokaliämie infolge 17alpha-Hydroxylase-Mangels. Dtsch Med Wochenschr 102:1024–1028

Delarue C, Tonon MC, Leboulenger F, Jegou S, Leroux P, Vaudry H (1979a) *In vitro* study of frog *(Rana ridibunda* Pallas) interrenal function by use of a simplified perifusion system. II. Influence of adrenocorticotropin upon aldosterone production. Gen Comp Endocrinol 38: 399–409

Delarue C, Leboulenger F, Tonon M-C, Jegou S, Leroux P, Kusmierek M-C, Corvol P, Vaillant R, Vaudry H (1979b) Comparative effects of canrenoate-K and prorenoate-K upon aldosterone biosynthesis in perifused frog interrenal glands. Steroids 34:319–332

Delarue C, Perroteau I, Leboulenger F, Netchitailo P, Leroux P, Jégou S, Bélanger A, Tonon M-C, Vaudry H (1981) In vitro effect of prostaglandins on corticosterone and aldosterone production by frog interrenal gland. Biochem Biophys Res Commun 100:769–777

Delarue C, Netchitailo P, Leboulenger F, Perroteau I, Escher E, Vaudry H (1984) In vitro study of frog *(Rana ridibunda* Pallas) interrenal function by use of a simplified perifusion system. VII.

Lack of effect of somatostatin on angiotensin-induced corticosteroid production. Gen Comp Endocrinol 54:333-338

De Léan A, Racz K, Gutkowska J, Nguyen T-T, Cantin M, Genest J (1984a) Specific receptor-mediated inhibition by synthetic atrial natriuretic factor of hormone-stimulated steroidogenesis in cultured bovine adrenal cells. Endocrinology 115:1636-1638

De Léan G, Gutkowska J, McNicoll N, Schiller PW, Cantin M, Genest J (1984b) Characterizia-tion of specific receptors for atrial natriuretic factor in bovine adrenal zona glomerulosa. Life Sci 35:2311-2318

De Léan A, Racz K, McNicoll N, Desrosiers M-L (1984c) Direct β-adrenergic stimulation of aldosterone secretion in cultured bovine adrenal subcapsular cells. Endocrinology 115:485-492

De Léan A, Thibault G, Seidah NG, Lazure C, Gutkowska J, Chrétien M, Genest J, Cantin M (1985a) Structure-activity relationships of atrial natriuretic factor (ANF). III. Correlation of receptor affinity with relative potency on aldosterone production in zona glomerulosa cells. Biochem Biophys Res Commun 132:360-367

De Léan A, Ong H, McNicoll N, Racz R, Gutkowska J, Cantin M (1985b) Identification of aldosterone secretion inhibitory factor in bovine adrenal medulla. Life Sci 36:2375-2382

DeLeiva A, Christlieb AR, Melby JC, Graham CA, Day RP, Luetscher JA, Zager PG (1976) Big renin and biosynthetic defect of aldosterone in diabetes mellitus. N Engl J Med 295:639-643

Del Pozo E, Darragh A, Lancranjan I, Ebeling D, Burmeister P, Bühler F, Marbach P, Braun P (1977) Effect of bromocriptine on the endocrine system and fetal development. Clin Endocrinol [Suppl] 6:47s-55s

de Nicola AF, Birmingham MK (1968) Biosynthesis of 18-hydroxydeoxycorticosterone from deoxycorticosterone-4-^{14}C by the human adrenal gland. J Clin Endocrinol Metab 28: 1380-1383

de Nicola AF, Birmingham MK (1970) A comparative study of adrenal steroid biosynthesis in two species of monkeys. Can J Biochem 48:85-93

de Nicola AF, Dahl V (1971) Acute effects of SU-4885 and its reduced derivative (SU-5236) on the adrenocortical secretion of the rat. Endocrinology 89:1236-1241

de Nicola AF, Dahl V, Kaplan S (1973) Transformation of [^{14}C]-pregnenolone and production of corticosteroids by adrenal glands from rats bearing a transplantable mammotropic pituitary tumor. J Steroid Biochem 4:205-215

Denton D (1980) The hunger for salt. An anthropological, physiological and medical analysis. Springer, Berlin Heidelberg New York, p377

Denton DA, Goding JR, Wright RD (1959) Control of adrenal secretion of electrolyte-active steroids. Br Med J 2:447-456, 522-530

Denton DA, Blair-West JR, Coghlan JP, Scoggins BA, Wright RD (1977) Gustatory-alimentary reflex inhibition of aldosterone hypersecretion by the autotransplanted adrenal gland of sodium deficient sheep. Acta Endocrinol 84:119-132

Devynck M-A, Pernollet M-G, Mattews PG, Khosla MC, Bumpus FM, Meyer P (1977) Specific receptors for des-asp^1-angiotensin II ("angiotensin III") in rat adrenals. Proc Natl Acad Sci USA 74:4029-4032

Devynck M-A, Pernollet M-G, Matthews PG, Macdonald GJ, Raisman RS, Meyer P (1979) Sodium intake and plasma angiotensin level as modulators of adrenal and uterine angiotensin II receptors in the rat. J Cardiovasc Pharmacol 1:163-179

De Wardener HE, Clarkson EM (1982) The natriuretic hormone: recent developments. Clin Sci 63:415-420

Dexter RN, Fishman LM, Ney RL, Liddle GW (1967) Inhibition of adrenal corticosteroid synthesis by aminoglutethimide: studies of the mechanism of action. J Clin Endocrinol Metab 27: 473-480

Dietz R, Mast GJ, Möhring J, Vecsei P, Gless KH, Oster P, Gross F (1975) Aldosterone and corticosterone production in renal hypertensive rats. Acta Endocrinol 79:317-328

Dillon MJ, Ryness JM (1975) Plasma renin activity and aldosterone concentration in children. Br Med J 4:316-319

Dillon MJ, Gillin MEA, Ryness JM, de Swiet M (1976) Plasma renin activity and aldosterone concentration in the human newborn. Arch Dis Childhood 51:537-540

Dluhy RG, Williams GH (1969) Primary aldosteronism in a hypertensive acromegalic patient. J Clin Endocrinol Metab 29:1319-1324

Dluhy RG, Underwood RH, Williams GH (1970) Influence of dietary potassium on plasma renin activity in normal man. J Appl Physiol 28:299-302

Dluhy RG, Axelrod L, Underwood RH, Williams GH (1972) Studies of the control of plasma aldosterone concentration in normal man. II. Effect of dietary potassium and acute potassium infusion. J Clin Invest 51:1950-1957

Dluhy RG, Himathongkam T, Greenfield M (1974) Rapid ACTH test with plasma aldosterone levels. Improved diagnostic discrimination. Arch Intern Med 80:693-696

Dluhy RG, Greenfield M, Williams GH (1977) Effect of simultaneous potassium and saline loading on plasma aldosterone levels. J Clin Endocrinol Metab 45:141-146

Doi Y, Atarashi K, Franco-Saenz R, Mulrow P (1983) Adrenal renin: a possible regulator of aldosterone production. Clin Exp Hypertension (A) 5:1119-1126

Doi Y, Atarashi K, Franco-Saenz R, Mulrow PJ (1984) Effects of changes in sodium or potassium balance, and nephrectomy, on adrenal renin and aldosterone concentrations. Hypertension 6 [Suppl 1]: I-124-I-129

Dollery CT, Shackman R, Shillingford J (1959) Malignant hypertension and hypokalaemia: cured by nephrectomy. Br Med J 2:1367-1371

Dominguez OV (1965) The presence of two interconvertible forms of 18-hydroxy-11-deoxycorticosterone. Steroids 6 [Suppl 2]:29-49

Domoto DT, Boyd JE, Mulrow PJ, Kashgarian M (1973) The ultrastructure of the adrenal zona glomerulosa of rats on potassium-supplemented or sodium-deficient diets. Am J Pathol 72: 433-446

Donald RA, Redekopp C, Cameron V, Nicholls MG, Bolton J, Livesey J, Espiner EA, Vale W, Rivier J (1983) The hormonal actions of corticotropin releasing factor in sheep: effect of intravenous and intracerebroventricular injection. Endocrinology 113:866-870

Donaldson EM, Holmes WN (1965) Corticosteroidogenesis in the fresh water and saline-maintained duck *(Anas platyrhynchos)*. J Endocrinol 32:329-336

Donaldson EM, Holmes WN, Stachenko J (1965) In vitro corticosteroidogenesis by the duck (Anas platyrhynchos) adrenal. Gen Comp Endocrinol 5:542-551

Douglas JG (1979) Changes in potassium balance: inverse relationship between number and affinity of angiotensin II receptors of smooth muscle and adrenal targer tissues. Am J Physiol 237:E519-E523

Douglas JG (1980a) Effects of a high potassium diet on angiotensin II receptors and angiotensin-induced aldosterone production in rat adrenal glomerulosa cells. Endocrinology 106:983-990

Douglas JG (1980b) Potassium ion as a regulator of adrenal angiotensin II receptors. Am J Physiol 239:E317-E321

Douglas JG (1981) Mechanism of adrenal angiotensin II receptor changes after nephrectomy in rats. J Clin Invest 67:1171-1176

Douglas JG, Brown GP (1982) Effect of prolonged low dose infusion of angiotensin II and aldosterone on rat smooth muscle and adrenal angiotensin II receptors. Endocrinology 111:988-992

Douglas J, Catt KJ (1976) Regulation of angiotensin II receptors in the rat adrenal cortex by dietary electrolytes. J Clin Invest 58:834-843

Douglas J, Aguilera G, Kondo T, Catt K (1978a) Angiotensin II receptors and aldosterone production in rat adrenal glomerulosa cells. Endocrinology 102:685-696

Douglas J, Bartley P, Kondo T, Catt KJ (1978b) Formation of des-asp[1]-angiotensin II is not an obligatory step in the steroidogenic action of angiotensin II in the canine adrenal. Endocrinology 102:1921-1924

Douglas J, Hansen J, Catt KJ (1978c) Relationship between plasma renin activity and plasma aldosterone in the rat after dietary electrolyte changes. Endocrinology 103:60-65

Douglas J, Saltman S, Williams C, Bartley P, Kondo T, Catt K (1978d) An examination of possible mechanisms of angiotensin II-stimulated steroidogenesis. Endocrinol Res Commun 5: 173-188

Douglas JG, Michailov M, Khosla MC, Bumpus FM (1980) Comparative receptor-binding properties of heptapeptide and octapeptide antagonists of angiotensin II in the adrenal glomerulosa and uterine smooth muscle. Endocrinology 106:120-124

Douglas JG, Michailov M, Khosla MC, Bumpus FM (1981) Comparison of octa- and heptapeptide antagonists of angiotensin II as inhibitors of angiotensin II-induced steroidogenesis in the rat adrenal glomerulosa. Endocrinology 109:94-98

Douglas JG, Khosla MC, Bumpus FM (1985) Efficacy of octa- and heptapeptide antagonists of angiotensin II as inhibitors of angiotensin III binding in the rat adrenal glomerulosa. Endocrinology 116:1598-1602

Drake CR Jr, Carey RM (1984) Dopamine modulates sodium-dependent aldosterone responses to angiotensin II in humans. Hypertension 6 [Suppl 1]: I-119-I-123

Drake CR Jr, Ragsdale NV, Kaiser DL, Carey RM (1984) Dopaminergic suppression of angiotensin II-induced aldosterone secretion in man: differential responses during sodium loading and depletion. Metabolism 33:696-702

Drayer JIM, Weber MA, Longworth DL, Laragh JH (1978) The possible importance of aldosterone as well as renin in the long-term antihypertensive action of propranolol. Am J Med 64:187-192

Driessen WMM, Benraad TJ, Kloppenborg PWC (1969) Effect of potassium loading on aldosterone secretory rate (ASR) and plasma renin activity (PRA) in normals and hypertensives (abstract). Acta Endocrinol [Suppl] 138:126

Drop SLS, Frohn-Mulder IME, Visser HKA, Sippell WG, Dörr HG, Schöneshöfer M (1982) The effect of ACTH stimulation on plasma steroids in two patients with congenital hypoaldosteronism and in their relatives. Acta Endocrinol 99:245-250

Düsing R, Gill JR Jr, Bartter FC (1979) Prolactin in primary aldosteronism. Clin Sci 56:381-383

Dufau ML, Kliman B (1968a) Pharmacologic effects of angiotensin-II-amide on aldosterone and corticosterone secretion by the intact anesthetized rat. Endocrinology 82:29-36

Dufau ML, Kliman B (1968b) Acute effects of angiotensin-II-amide on aldosterone and corticosterone secretion by morphine-pentobarbital treated rats. Endocrinology 83:180-183

Dufau ML, Villee DB (1969) Aldosterone biosynthesis by human fetal adrenal *in vitro*. Biochim Biophys Acta 176:637-640

Dufau ML, Villee DB, Kliman B (1968) Pregnenolone metabolism in an aldosterone secreting tumor of the adrenal and its adjacent adrenal tissue. J Clin Endocrinol Metab 28:983-991

Dufau ML, Crawford JD, Kliman B (1969) Effect of high sodium intake on the response of the rat adrenal to angiotensin II. Endocrinology 84:462-463

Dunn JF, Nisula BC, Rodbard D (1981) Transport of steroid hormones: binding of 21 endogenous steroids to both testosterone-binding globulin and corticosteroid-binding globulin in human plasma. J Clin Endocrinol Metab 53:58-68

Dunn MG, Bosmann HB (1981) Peripheral dopamine receptor identification: properties of a specific dopamine receptor in the rat adrenal zona glomerulosa. Biochem Biophys Res Commun 99:1081-1087

Dunn MJ, Hood VL (1977) Prostaglandins and the kidney. Am J Physiol 233:F169-F184

Dyrenfurth I, Lucis OJ, Beck JC, Venning EH (1960) Studies in patients with adrenocortical hyperplasia function. III. In vitro secretion of steroids by human adrenal glands. J Clin Endocrinol Metab 20:765-779

Eberlein WR, Bongiovanni AM (1955) Congenital adrenal hyperplasia with hypertension: unusual steroid pattern in blood and urine. J Clin Endocrinol Metab 15:1531-1534

Eddy RL, Sanchez SA (1971) Renin-secreting renal neoplasm and hpyertension with hypokalemia. Ann Intern Med 75:725-729

Edwards CRW, Thorner MO, Miall P, Al-Dujaili EAS, Hanker JP, Besser GM (1975) Inhibition of the plasma-aldosterone response to furosemide by bromocriptine. Lancet 2:903-905

Edwards CRW, Boscaro M, Miall PA, Delitala G, Al-Dujaili EAS (1980a) Inhibition of aldosterone secretion by histamine-H_2 receptor antagonists. In: 6th International congress of endocrinology, Melbourne (abstract)

Edwards CRW, Al-Dujaili EAS, Boscaro M, Quyyumi S, Miall PA, Rees LH (1980b) In vivo and in vitro studies on the effect of metoclopramide on aldosterone secretion. Clin Endocrinol 13:45-50

Egdahl RH, Kahn PC, Melby JC (1968) Unilateral adrenalectomy for aldosteronomas localized preoperatively by differential adrenal vein catheterization. Surgery 64:117-125

Eguchi T, Bravo EL (1984) Humoral response to intracerebroventricularly administered of angiotensin II in dogs. Am J Physiol 247:E336-E342

Eguchi T, Saito I, Nakamura R, Yasui T, Saruta T (1980) Effects of angiotensin I converting enzyme inhibitor (SQ 14225) on control of aldosterone. Acta Endocrinol 94:213-220

Ehrlich EN, Dominguez OV, Samuels LT, Lynch D, Oberhelman H Jr, Warner NE (1963) Aldosteronism and precocious puberty due to an ovarian androblastoma (Sertoli cell tumor). J Clin Endocrinol Metab 23:358-367

Eilers EA, Peterson RE (1964) Aldosterone secretion in the rat. In: Baulieu EE, Robel P (eds) Aldosterone. Blackwell, Oxford, pp 251-264

Eisenstein AB, Hartroft PM (1957) Alterations in the rat adrenal cortex induced by sodium deficiency: steroid hormone secretion. Endocrinology 60:634-640

Eldridge JC, Lymangrover JR (1984) Prolactin stimulates and potentiates adrenal steroid secretion in vitro. Horm Res 20:252-260
Elema JD, Hardonk MJ, Koudstaal J, Arends A (1968a) Acute enzyme histochemical changes in the zona glomerulosa of the rat adrenal cortex. I. The effect of peritoneal dialysis with a glucose 5% solution. Acta Endocrinol 59:508-518
Elema JD, Hardonk MJ, Koudstaal J, Arends A (1968b) Acute enzyme histochemical changes in the zona glomerulosa of the rat adrenal cortex. II. The effect of bilateral nephrectomy either alone or followed by peritoneal dialysis with 5% glucose. Acta Endocrinol 59:519-528
Elliott ME, Goodfriend TL (1981) Angiotensin alters $^{45}Ca^{2+}$ fluxes in bovine adrenal glomerulosa cells. Proc Natl Acad Sci USA 78:3044-3048
Elliott ME, Goodfriend TL (1984) Identification of the cycloheximide-sensitive site in angiotensin-stimulated aldosterone synthesis. Biochem Pharmacol 33:1519-1524
Elliott ME, Alexander RC, Goodfriend TL (1982) Aspects of angiotensin action in the adrenal. Key roles for calcium and phosphatidyl inositol. Hypertension 4 [Suppl 2]: II-52-II-58
Elliott ME, Farese RV, Goodfriend TL (1983) Effects of angiotensin II and dibutyryl cyclic adenosine monophosphate on phosphatidylinositol metabolism, $^{45}Ca^{2+}$ fluxes, and aldosterone synthesis in bovine adrenal glomerulosa cells. Life Sci 33:1771-1778
Elliott ME, Siegel FL, Hadjokas NE, Goodfriend TL (1985) Angiotensin effects on calcium and steroidogenesis in adrenal glomerulosa cells. Endocrinology 116:1051-1059
El-Nouty FD, Elbanna IM, Davis TP, Johnson HD (1980) Aldosterone and ADH response to heat and dehydration in cattle. J Appl Physiol 48:249-255
Ensor DM, Ball JN (1972) Prolactin and osmoregulation in fishes. Fed Proc 31:1615-1623
Enyedi P, Spät A (1981) Effect of reduced extracellular sodium concentration on the function of adrenal zona glomerulosa: studies on isolated glomerulosa cells from the rat. J Endocrinol 89: 417-421
Enyedi P, Spät A, Antoni FA (1981) Role of prostaglandins in the control of the function of adrenal glomerulosa cells. J Endocrinol 91:427-437
Enyedi P, Szabó B, Spät A (1983) Failure of vasoactive intestinal peptide to stimulate aldosterone production. Acta Physiol Hung 61:77-79
Enyedi P, Büki B, Mucsi I, Spät A (1985) Polyphosphoinositide metabolism in adrenal glomerulosa cells. Mol Cell Endocrinol 41:105-112
Epstein M, Saruta T (1971) Effect of water immersion on renin-aldosterone and renal sodium handling in normal man. J Appl Physiol 31:368-374
Epstein M, Saruta T (1972) Effects of simulated high altitude on renin-aldosterone and Na homeostasis in normal man. J Appl Physiol 33:204-210
Epstein M, Saruta T (1973) Effects of hyperoxic hypobaric environment on renin-aldosterone in normal man. J Appl Physiol 34:49-52
Epstein M, Sancho J, Perez G, Haber E, Re R, Loutzenhiser R (1978) Volume as a determinant of plasma aldosterone in anephric man. J Clin Endocrinol Metab 46:309-316
Epstein M, Re R, Preston S, Haber E (1979) Comparison of the suppressive effects of water immersion and saline administration on renin-aldosterone in normal man. J Clin Endocrinol Metab 49:358-363
Epstein MT, Espiner EA, Donald RA, Hughes H (1977) Effect of eating liquorice on the renin-angiotensin aldosterone axis in normal subjects. Br Med J 1:488-490
Erbler HC (1972) Stimulation of aldosterone production in vitro and its inhibition by spironolactone. Naunyn-Schmiedeberg's Arch Pharmacol 273:366-375
Erbler HC (1973a) On the mechanism of the inhibitory action of the spirolactone SC 9376 (Aldadiene) on the production of corticosteroids in rat adrenals in vitro. Naunyn-Schmiedeberg's Arch Pharmacol 277:139-149
Erbler HC (1973b) Selective inhibition of aldosterone synthesis by 11-hydroxylated spirolactone in rat adrenals. Naunyn-Schmiedeberg's Arch Pharmacol 280:331-337
Erbler HC (1979) Effect of spironolactone and its main metabolite canrenone on the renin-angiotensin-aldosterone-system during long-term treatment in rats. Acta Endocrinol 90:147-156
Erbler HC, Wernze H, Hilfenhaus M (1976) Effect of the aldosterone antagonist canrenone on plasma aldosterone concentration and plasma renin activity, and on the excretion of aldosterone and electrolytes by man. Eur J Clin Pharmacol 9:253-257
Erdös EG (1979) Inhibitors of kininases. Fed Proc 38:2774-2777
Erkkola R, Lammintausta R, Punnonen R, Rauramo L (1978) The effect of estriol succinate therapy on plasma renin activity and urinary aldosterone in postmenopausal women. Maturitas 1: 9-14

Espiner EA, Lun S, Hart DS (1978a) Role of ACTH, angiotensin and potassium in stress-induced aldosterone secretion. J Steroid Biochem 9:109-113
Espiner EA, Lun S, Hart DS (1978b) Failure of angiotensin II to inhibit corticotropin-stimulated cortisol secretion. Endocrinology 102:1362-1365
Espiner EA, Al-Dujaili EAS, Martin VI, Edwards CRW (1980) Indoleamines stimulate steroidogenesis in isolated human adrenal cells. In: 6th International congress of endocrinology, Melbourne (abstract)
Exton JH (1985) Role of calcium and phosphoinositides in the actions of certain hormones and neurotransmitters. J Clin Invest 75:1753-1757

Fabre LF Jr, Banks RC, McIsaac WM, Farrell G (1965) Effects of ubiquinone and related substances on secretion of aldosterone and cortisol. Am J Physiol 208:1275-1280
Fagard R, Amery A, Lijnen P (1981) Angiotensin II and sodium as determinants of the agonistic-antagonistic balance of saralasin's actions. Clin Sci 60:377-385
Fagard R, Cattaert A, Lijnen P, Staessen J, Vanhees L, Moerman E, Amery A (1984) Responses of the systemic circulation and of the renin-angiotensin-aldosterone system to ketanserin at rest and exercise in normal man. Clin Sci 66:17-25
Faglia G, Gattinoni L, Travaglini P, Neri V, Acerbi L, Ambrosi B (1971) Evidence suggesting 11β-hydroxylase inhibition during aminoglutethimide administration. Metabolism 20:266-272
Fakundig JL, Catt KJ (1980) Dependence of aldosterone stimulation in adrenal glomerulosa cells on calcium uptake: effects of lanthanum and verapamil. Endocrinology 107:1345-1353
Fakundig JL, Catt KJ (1982) Calcium-dependent regulation of aldosterone production in isolated adrenal glomerulosa cells: effects of the ionophore A-23187. Endocrinology 110:2006-2010
Fakundig JL, Chow R, Catt KJ (1979) The role of calcium in the stimulation of aldosterone production by adrenocorticotropin, angiotensin II, and potassium in isolated glomerulosa cells. Endocrinology 105:327-333
Falbriard A, Muller AF, Neher R, Mach RS (1955) Etude des variations de l'aldostéronurie sous l'effet de surcharges en potassium et de déperditions rénales et extrarénales de sel et d'eau. Schweiz. Med Wochenschr 85:1218-1220
Fan JS, Coghlan JP, Denton DA, Oddie CJ, Scoggins BA, Shulkes AA (1975) Effect of intravenous infusion of corticosteroids on blood pressure, electrolytes, and water metabolism in sheep. Am J Physiol 228:1695-1701
Farese RV (1966) Effects of Actinomycin C on ACTH- induced corticosteroidogenesis. Endocrinology 78:929-936
Farese RV (1970) Inhibition of cholesterol side chain cleavage by cyanoketone. Steroids 15: 245-250
Farese RV (1983) The phosphatidate-phosphoinositide cycle: an intracellular messenger system in the action of hormones and neurotransmitters. Metabolism 32:628-641
Farese RV, Larson RE, Davis JS (1980) Potassium and angiotensin II increase the concentrations of phosphatidic acid, phosphatidylinositol, and polyphosphoinositides in rat adrenal capsules in vitro. J Clin Invest 66:1428-1431
Farese RV, Larson RE, Sabir MA, Gomez-Sanchez C (1981) Effects of angiotensin-II and potassium on phospholipid metabolism in the adrenal zona glomerulosa. J Biol Chem 256: 11093-11097
Farese RV, Larson RE, Sabir MA, Gomez-Sanchez CE (1983) Effects of angiotensin II, K^+, adrenocorticotropin, serotonin, adenosine 3'5'-monophosphate, guanosine 3',5'-monophosphate, A23187, and EGTA on aldosterone synthesis and phospholipid metabolism in the rat adrenal zona glomerulosa. Endocrinology 113:1377-1386
Farese RV, Larson RE, Davis JS (1984) Rapid effects of angiotensin-II on polyphosphoinositide metabolism in the rat adrenal glomerulosa. Endocrinology 114:302-304
Farinelli A, Squerzanti, de Paoli Vitali E, Bagni B, Cavallini AR, Gilli P, Bedani PL (1980) Response of plasma aldosterone to sequential ultrafiltration dialysis and conventional hemodialysis. Nephron 26:274-279
Farrell G (1958) Regulation of aldosterone secretion. Physiol Rev 38:709-728
Farrell G (1960) Adrenoglomerulotropin. Circulation 21:1009-1015
Farrell G, McIsaac WM (1961) Adrenoglomerulotropin. Arch Biochem Biophys 94:543-544
Farrell GL, Rauschkolb EW, Royce PC (1955) Secretion of aldosterone by the adrenal of the dog: effects of hypophysectomy and ACTH. Am J Physiol 182:269-272
Fattah DI, Whitehouse BJ, Vinson GP (1977) Biosynthesis of aldosterone from 18-hydroxylated precursors by rat adrenal tissue in vitro. J Endocrinol 75:187-195

Fazekas AG, Kokai K (1967) Biosynthesis of C-18-oxygenated corticosteroids by rabbit adrenals in vitro. Steroids 9:177-191

Fei DTW, Graham WF, McDougall JG, Scoggins BA, Coghlan JP (1980) [Des-asp[1]] angiotensin II in the sheep: blood levels and its effect on plasma renin concentration. Life Sci 27: 1495-1502

Fei DTW, Scoggins BA, Tregear GW, Coghlan JP (1981) Angiotensin I, II, and III in sheep. A model of angiotensin production and metabolism. Hypertension 3:730-737

Fekete G, Szeberényi Sz (1968) Comparison of the aldosterone production suppressing activity of synthetic glucocorticoids. Proceedings of the IV. Conference Hungarian Therapy of Invest Pharmacology, pp 373-376

Fenske M (1985) Effects of acute stress, (1-24)ACTH administration and changes in superfusion temperature and flow rate on the in vitro secretion of glucocorticosteroids and aldosterone from the Mongolian gerbil *(Meriones unguiculatus)* adrenal gland. Comp Biochem Physiol 82A:951-958

Ferguson JJ Jr, Morita Y (1964) RNA synthesis and adrenocorticotropin responsiveness. Biochim Biophys Acta 87:348-350

Ferrier CP, Bianchetti MG, Beretta-Piccoli C, Link L, Bachman C, Weidmann P (1983) Studies on the mechanism of acute glucose-induced aldosterone suppression. Role of corticotropin. Acta Endocrinol 103:391-399

Ferriss JB, Sullivan PA, Gonggrijp H, Cole M, O'Sullivan DJ (1982) Plasma angiotensin II and aldosterone in unselected diabetic patients. Clin Endocrinol 17:261-269

Fichman MP, Littenburg G, Brooker G, Horton R (1972) Effect of prostaglandin A_1 on renal and adrenal function in man. Circ Res 30/31 [Suppl 2]: II-19-II-35

Fichman MP, Michelakis AM, Horton R (1974) Regulation of aldosterone in the syndrome of inappropriate antidiuretic hormone secretion (SIADH). J Clin Endocrinol Metab 39:136-144

Finberg JPM, Berlyne GM (1976) Renin and aldosterone secretion following acute environmental heat exposure. Israel J Med Sci 12:844-847

Finberg JPM, Berlyne GM (1977) Modification of renin and aldosterone response to heat by acclimatization in man. J Appl Physiol 42:554-558

Finberg JPM, Yagil R, Berlyne GM (1978) Response of the renin-aldosterone system in the camel to acute dehydration. J Appl Physiol 44:926-930

Finkel MS, Aguilera G, Catt KJ, Keiser HR (1984) [^{3}H]Nitrendipine binding to adrenal capsular membranes. Life Sci 35:905-910

Finkelstein JW, Kowarski A, Spaulding JS, Migeon CJ (1965) Effect of various preparations of human growth hormone on aldosterone secretion rate of hypopituitary dwarfs. Am J Med 38: 517-521

Fiselier TJW, Lijnen P, Monnens L, van Munster P, Jansen M, Peer P (1983) Levels of renin, angiotensin I and II, angiotensin-converting enzyme and aldosterone in infancy and childhood. Eur J Pediatr 141:3-7

Fisher ER, Horvat B (1971) Ultrastructural features of aldosterone production. Arch Pathol 92: 172-179

Fishman LM, Liddle GW, Island DP, Fleischer N, Küchel O (1967) Effects of amino-glutethimide on adrenal function in man. J Clin Endocrinol Metab 27:481-490

Flack JD, Jessup R, Ramwell PW (1969) Prostaglandin stimulation of rat corticosteroidogenesis. Science 163:691-692

Flamenbaum W, Kotchen TA, Nagle R, McNeil JS (1970) Effect of potassium on the renin-angiotensin system and $HgCl_2$-induced acute renal failure. Am J Physiol 224:305-311

Flamenbaum W, Kleinman JG, McNeil JS, Hamburger RJ, Kotchen TA (1975) Effect of KCl infusion on renin secretory rates and aldosterone excretion in dogs. Am J Physiol 229:370-375

Fletcher MJ, MacKay N, Forbes CD (1967) Assessment of possible glucocorticoid activity of carbenoxolone sodium. Br Med J 1:412-413

Flood C, Gherondache C, Pincus G, Tait JF, Tait SAS, Willoughby S (1967) The metabolism and secretion of aldosterone in elderly subjects. J Clin Invest 46:960-966

Ford HC, Bailey RE (1966) The effect of heparin on aldosterone secretion and metabolism in primary aldosteronism. Steroids 7:30-40

Forget G, Heisler S (1979) Effect of various solubilizers on angiotensin II receptors in bovine adrenocortical plasma membranes. Experientia 35:125-126

Foster R, Lobo MV, Marusic ET (1979) Studies of relationship between angiotensin II and potassium ions on aldosterone release. Am J Physiol 237: E363-E366

Foster R, Lobo MV, Rasmussen H, Marusic ET (1981) Calcium: its role in the mechanism of action of angiotensin II and potassium in aldosterone production. Endocrinology 109: 2196-2201

Foster R, Lobo MV, Rasmussen H, Marusic ET (1982) The effect of calcium on potassium-induced depolarization of adrenal glomerulosa cells. FEBS Lett 149:253-256

Fragachan F, Nowaczynski W, Bertranau E, Kalina M, Genest J (1969) Evidence of *in vivo* inhibition of 11β-hydroxylation of steroids by dehydroepiandrosterone in the dog. Endocrinology 84:98-103

Francesconi RP, Hubbard RW (1985) Chronic low-sodium diet in rats: responses to severe heat exposure. J Appl Physiol 58:152-156

Francesconi RP, Sawka MN, Pandolf KB (1983) Hypohydration and heat acclimation; plasma renin and aldosterone during exercise. J Appl Physiol 55:1790-1794

Francis KT, MacGregor R III (1978) Effect of exercise in the heat on plasma renin and aldosterone with either water or a potassium-rich electrolyte solution. Aviat Space Environ Med 49: 461-465

Frankel RJ, Jenkins JS, Wright JJ, Khan MUA (1976) Effect of brain stimulation on aldosterone secretion in the rhesus monkey *(Macaca mulatta).* J Endocrinol 71:383-391

Franken FH, Irmscher K, Daweke H (1961) Aldosteronausscheidung und Elektrolythaushalt unter Pitressineinwirkung bei nicht hydrierten normalen Versuchspersonen. Acta Endocrinol 38:22-30

Frayser R, Rennie ID, Gray GW, Houston CS (1975) Hormonal and electrolyte response to exposure to 17,500 ft. J Appl Physiol 38:636-642

Fredlund P, Saltman S, Catt KJ (1975a) Aldosterone production by isolated adrenal glomerulosa cells: Stimulation by physiological concentrations of angiotensin II. Endocrinology 97: 1577-1586

Fredlund P, Saltman S, Catt KJ (1975b) Stimulation of aldosterone production by angiotensin II peptides *in vitro:* enhanced activity of the (1-sarcosine) analogue. J Clin Endocrinol Metab 40: 746-749

Fredlund P, Saltman S, Kondo T, Douglas J, Catt KJ (1977) Aldosterone production by isolated glomerulosa cells: Modulation of sensitivity to angiotensin II and ACTH by extracellular potassium concentration. Endocrinology 100:481-486

Freeman RH, Davis JO, Lohmeier TE, Spielman WS (1976) Evidence that des-asp^1-angiotensin II mediates the renin-angiotensin response. Circ Res 38 [Suppl 2]: II-99-II-103

Freeman RH, Davis JO, Vari RC (1985) Renal response to atrial natriuretic factor in conscious dogs with caval constriction. Am J Physiol 248:R495-R500

Freer RJ (1975) Calcium and angiotensin tachyphylaxis in rat uterine smooth muscle. Am J Physiol 228:1423-1430

Fregly MJ, Cade JR, Waters IW, Straw JA, Taylor RE Jr (1965) Secretion of aldosterone by adrenal glands of propylthiouracil-treated rats. Endocrinology 77:777-784

Frisina N, Costa G, Buemi M (1983) Evidence for dopaminergic control of aldosterone in man. Clin Endocrinol 19:741-745

Frölich JC, Hollifield JW, Dormois JC, Frölich BL, Seyberth H, Michelakis AM, Oates JA (1976) Suppression of plasma renin activity by indomethacin in man. Circ Res 39:447-452

Frölich M, Brand EG, van Hall EV (1976) Serum levels of unconjugated aetiocholanolone androstenedione, testosterone, dehydroepiandrosterone, aldosterone, progesterone and oestrogens during the normal menstrual cycle. Acta Endocrinol 81:548-562

Fromm M, Oelkers W, Hegel U (1983) Time course of aldosterone and corticosterone plasma levels in rats during general anaesthesia and abdominal surgery. Pflugers Arch 399:249-254

Fuchs-Hammoser R, Schweiger M, Oelkers W (1980) The effect of chronic low-dose infusion of ACTH (1-24) on renin, renin-substrate, aldosterone and other corticosteroids in sodium replete and deplete man. Acta Endocrinol 95:198-206

Fujita K, Aguilera G, Catt KJ (1979) The role of cyclic AMP in aldosterone production by isolated zona glomerulosa cells. J Biol Chem 254:8567-8574

Fuller PJ, Funder JW (1976) Mineralocorticoid and glucocorticoid receptors in human kidney. Kidney Int 10:154-157

Funder JW, Blair-West JR, Coghlan JP, Denton DA, Scoggins BA, Wright RD (1969) Effect of plasma [K^+] on the secretion of aldosterone. Endocrinology 85:381-384

Fyhrquist F, Holmberg C, Perheentupa J, Wallenius M (1977) Inappropriate secretion of antidiuretic hormone, hypertension, and hypoplastic corpus callosum. J Clin Endocrinol Metab 45:691-694

Gaillard RC, Merkelbach U, Riondel AM, Vallotton MB, Muller AF (1976) Effect on plasma aldosterone, renin activity and cortisol of acute volume depletion induced by ethacrynic acid under constant infusion of angiotensin II and dexamethasone in man. Eur J Clin Invest 6: 51-57

Gaillard RC, Riondel A, Merkelbach U, Vallotton MB (1979) Changes in plasma aldosterone following the administration of various combinations of stimuli. Acta Endocrinol 92:309-318

Gaillard RC, Riondel AM, Chabert P, Vallotton MB (1980) Effect of spironolactone on aldosterone regulation in man. Clin Sci 58:227-233

Gaillard RC, Riondel AM, Favrod-Coune CA, Vallotton MB, Muller AF (1983) Aldosterone escape to chronic ACTH administration in man. Acta Endocrinol 103:116-124

Ganguly A (1982) New insights and questions about glucocorticoid-suppressible hyperaldosteronism. Am J Med 72:851-854

Ganguly A (1984) Dopaminergic regulation of aldosterone secretion: how credible? Clin Sci 66: 631-637

Ganguly A, Hampton T (1985) Calcium dependence of serotonin-mediated aldosterone secretion and differential effects of calcium antagonists. Life Sci 36:1459-1464

Ganguly A, Melada GA, Luetscher JA, Dowdy AJ (1973) Control of plasma aldosterone in primary aldosteronism: distinction between adenoma and hyperplasia. J Clin Endocrinol Metab 37:765-775

Ganguly A, Chavarri M, Luetscher JA, Dowdy AJ (1977) Transient fall and subsequent return of high aldosterone secretion by adrenal adenoma during continued dexamethasone administration. J Clin Endocrinol Metab 44:775-779

Ganguly A, Grim CE, Bergstein J, Brown RD, Weinberger MH (1981a) Genetic and pathophysiologic studies of a new kindred with glucocorticoid-suppressible hyperaldosteronism manifest in three generations. J Clin Endocrinol Metab 53:1040-1046

Ganguly A, Grim CE, Weinberger MH (1981b) Anomalous postural aldosterone response in glucocorticoid-suppressible hyperaldosteronism. N Engl J Med 305:991-993

Ganguly A, Weinberger MH, Passmore JM, Caras JA, Khairi RM, Grim CE, Edmonson J, Johnston CC (1982) The renin-angiotensin-aldosterone system and hypertension in primary hyperparathyroidism. Metabolism 31:595-600

Ganguly A, Pratt JH, Weinberger MH, Grim CE, Fineberg NS (1983a) Differing effects of metoclopramide and adrenocorticotropin on plasma aldosterone levels in glucocorticoid-suppressible hyperaldosteronism and other forms of hyperaldosteronism. J Clin Endocrinol Metab 57: 388-392

Ganguly A, Luetscher JA, Weinberger MH (1983b) Primary aldosteronism: effects of inhibition of ACTH and potassium administration on plasma aldosterone concentration. Clin Exp Hypertension (A) 5:135-154

Ganguly A, Weinberger MH, Grim CE (1983c) The renin-angiotensin-aldosterone system in Cushing's disease and pheochromocytoma. Horm Res 17:1-10

Gann DS, Travis RH (1964) Mechanisms of hemodynamic control of secretion of aldosterone in the dog. Am J Physiol 207:1095-1101

Gann DS, Cruz JF, Casper AGT, Bartter FC (1962) Mechanism by which potassium increases aldosterone secretion in the dog. Am J Physiol 202:991-996

Gann DS, Delea CS, Gill JR Jr, Thomas JP, Bartter FC (1964) Control of aldosterone secretion by change of body potassium in normal man. Am J Physiol 207:104-108

Ganong WF (1972) Effect of sympathetic activity and ACTH on renin and aldosterone secretion. In: Genest J, Koiw E (eds) Hypertension '72. Springer, Berlin Heidelberg New York, pp 4-14

Ganong WF, Boryczka AT (1967) Effect of a low sodium diet on the aldosterone-stimulating activity of angiotensin II in dogs. Proc Soc Exp Biol Med 124:1230-1231

Ganong WF, Mulrow PJ (1961) Evidence of secretion of an aldosterone-stimulating substance by the kidney. Nature 190:1115-1116

Ganong WF, Mulrow PJ (1962) Role of the kidney in adrenocortical response to hemorrhage in hypophysectomized dogs. Endocrinology 70:182-188

Ganong WF, Lieberman AH, Dailey WJR, Yuen VS, Mulrow PJ, Luetscher JA Jr, Bailey RE (1959) Aldosterone secretion in dogs with hypothalamic lesions. Endocrinology 65:18-28

Ganong WF, Mulrow PJ, Boryczka A Cera G (1962) Evidence for a direct effect of angiotensin II on adrenal cortex of the dog. Proc Soc Exp Biol Med 109:381-384

Ganong WF, van Brunt EE, Lee TC, Mulrow PJ (1963) Inhibition of aldosterone-stimulating activity of hog and dog renin by the plasma of dogs immunized with hog renin. Proc Soc Exp Biol Med 112:1062-1064

Ganong WF, Boryczka AT, Shackleford R, Clark RM, Converse RP (1965a) Effect of dietary sodium restriction on adrenal cortical response to ACTH. Proc Soc Exp Biol Med 118:792-795
Ganong WF, Lee TC, van Brunt EE, Biglieri EG (1965b) Aldosterone secretion in dogs immunized with hog renin. Endocrinology 76:1141-1149
Ganong WF, Biglieri EG, Mulrow PJ (1966) Mechanisms regulating adrenocortical secretion of aldosterone and glucocorticoids. Recent Prog Horm Res 22:381-430
Ganong WF, Boryczka AT, Shackleford R (1967a) Effect of renin on adrenocortical sensitivity to ACTH and angiotensin II in dogs. Endocrinology 80:703-706
Ganong WF, Pemberton DL, Van Brunt EE (1967b) Adrenocortical responsiveness to ACTH and angiotensin II in hypophysectomized dogs and dogs treated with large doses of glucocorticoids. Endocrinology 81:1147-1150
Ganten D, Ganten U, Kubo S, Granger P, Nowaczynski W, Boucher R, Genest J (1974) Influence of sodium, potassium, and pituitary hormones on iso-renin in rat adrenal glands. Am J Physiol 227:224-229
Ganten D, Schelling P, Vecsei P, Ganten U (1976a) Iso-renin of extrarenal origin. "The tissue angiotensinogenase systems". Am J Med 60:760-772
Ganten D, Hutchinson JS, Schelling P, Ganten U, Fischer H (1976b) The iso-renin angiotensin systems in extrarenal tissue. Clin Exp Pharmacol Physiol 3:102-126
Garcia del Rio C, Smellie WSA, Morton JJ (1981) Des-asp[1]-angiotensin I: its identification in rat blood and confirmation as a substrate for converting enzyme. Endocrinology 108:406-412
Gardner JD, Lapey A, Simopoulos AP, Bravo EL (1971) Abnormal membrane sodium transport in Liddle's syndrome. J Clin Invest 50:2253-2258
Garren LD, Gill GN, Masui H, Walton GM (1971) On the mechanism of action of ACTH. Recent Prog Horm Res 27:433-478
Garrison JC, Borland MK, Florio VA, Twible DA (1979) The role of calcium ion as a mediator of the effects of angiotensin II, catecholamines, and vasopressin on the phosphorylation and activity of enzymes in isolated hepatocytes. J Biol Chem 254:7147-7156
Gaunt R, Chart JJ, Renzi AA (1965) Inhibitors of adrenal cortical function. Ergeb Physiol 56: 113-172
Gaunt R, Steinetz BG, Chart JJ (1968) Pharmacologic alteration of steroid hormone functions. Clin Pharmacol Ther 9:657-681
Gelhoud GW, Vander AJ (1967) The role of aldosterone in renin secretion. Life Sci 6:525-535
Genest J (1961) Angiotensin, aldosterone and human arterial hypertension. Can Med Assoc J 84: 403-419
Genest J, Boucher R, Nowaczynski W, Koiw E, de Champlain J, Biron P, Chrétien M, Marc-Aurèle J (1964) Studies on the relationship of aldosterone and angiotensin to human hypertensive disease. In: Baulieu EE, Robel P (eds) Aldosterone. Blackwell, Oxford, pp 393-416
George JM, Gillespie L, Bartter FC (1968) Aldosterone secretion in hypertension. Ann Intern Med 69:693-701
Giacomelli F, Wiener J, Spiro D (1965) Cytological alterations related to stimulation of the zona glomerulosa of the adrenal gland. J Cell Biol 26:299-521
Giebink GS, Gotlin RW, Biglieri EG, Katz FH (1973) A kindred with familial glucocorticoid-suppressible aldosteronism. J Clin Endocrinol Metab 36:715-723
Gilchrist M, Kem DC, Washburn D (1983) The effect of acidosis and alkalosis on in vitro aldosteronc production. Metabolism 32:697-700
Gilchrist NL, Espiner EA, Nicholls MG, Donald RA (1984) The effect of metoclopramide on aldosterone and regulatory factors in man. Clin Endocrinol 21:1-7
Gill JR Jr, Bartter FC (1981) Overproduction of sodium-retaining steroids by the zona glomerulosa is adrenocorticotropin-dependent and mediates hypertension in dexamethasone-suppressible aldosteronism. J Clin Endocrinol Metab 53:331-337
Ginn HE, Cade R, McCallum T, Fregley M (1967) Aldosterone secretion in magnesium-deficient rats. Endocrinology 80:969-971
Giroud CJP, Stachenko J, Venning EH (1956a) Secretion of aldosterone by the zona glomerulosa of rat adrenal glands incubated in vitro. Proc Soc Exp Biol Med 92:154-158
Giroud CJP, Saffran M, Schally AV, Stachenko J, Venning EH (1956b) Production of aldosterone by rat adrenal glands in vitro. Proc Soc Exp Biol Med 92:855-859
Giry J, Delost P (1977) Changes in the concentrations of aldosterone in the plasma and adrenal glands of the foetus, the newborn and the pregnant guinea-pig during the perinatal period. Acta Endocrinol 84:133-141

Giry J, Delost P (1980) Sécrétion, métabolisme et transport de l'aldostérone dans la période néonatale chez le Cobaye. Reprod Nutr Dev 20:271-176
Glasson P, Gaillard R, Riondel A, Vallotton MB (1979) Role of renal prostaglandins and relationship to renin, aldosterone, and antidiuretic hormone during salt depletion in man. J Clin Endocrinol Metab 49:176-181
Gláz E, Sugár K (1962) The effect of synthetic angiotensin II on synthesis of aldosterone by the adrenals. J Endocrinol 24:299-302
Gláz E, Sugár K (1964) Effect of heparin and heparinoids on the synthesis of aldosterone and corticosterone by the rat adrenal gland. Endocrinology 74:159-164
Glickman JA, Carson GD, Challis JRG (1979) Differential effects of synthetic adrenocorticotropin$^{1-24}$ and α-melanocyte-stimulating hormone on adrenal function in human and sheep fetuses. Endocrinology 104:34-39
Glossmann H, Baukal AJ, Catt KJ (1974) Properties of angiotensin II receptors in the bovine and rat adrenal cortex. J Biol Chem 249:825-834
Gniadek TC, Grekin J, Gross MD, Villareal JZ (1982) Hyper-responsiveness of aldosterone to metoclopramide in aldosteronism. Clin Endocrinol 16:475-481
Godard C, Riondel AM, Veyrat R, Mégevand A, Muller AF (1968) Plasma renin activity and aldosterone secretion in congenital adrenal hyperplasia. Pediatrics 41:883-896
Godard C, Gaillard R, Vallotton MB (1976) The renin-angiotensin-aldosterone system in mother and fetus at term. Nephron 17:353-360
Goldfarb D, Sack J, Iaina A, Eliahou H (1981) Sodium, potassium and age: possible determinants of plasma renin activity and aldosterone during childhood (age 4-16). Clin Endocrinol 15: 29-36
Goldman AS, Winter JSD (1967) Activities of various steroidogenic enzymes in two forms of experimental adrenal hyperplasia. J Clin Endocrinol Metab 27:1717-1722
Goldman AS, Yacovac WC, Bongiovanni AM (1965) Persistent effects of a synthetic androstene derivative on activities of 3β-hydroxysteroid dehydrogenase and glucose-6-phosphate dehydrogenase in rats. Endocrinology 77:1105-1118
Goldsmith O, Solomon DH, Horton R (1967) Hypogonadism and mineralocorticoid excess. The 17-hydroxylase deficiency syndrome. N Engl J Med 277:673-677
Golub MS, Speckart PF, Zia PK, Horton R (1976) The effect of prostaglandin A_1 on renin and aldosterone in man. Circ Res 39:574-579
Golub MS, Tuck ML, Fittingoff DB (1981) Effect of propranolol therapy on aldosterone responses to angiotensin II and adrenocorticotropic hormone in essential hypertension. Clin Sci 61:107-110
Golub MS, Sowers JR, Eggena P, Baron SH (1982) Effect of prostaglandin inhibition on the prolactin, renin, and aldosterone responses to dopamine antagonism. Metabolism 31:740-743
Gomez-Pan A, Snow MH, Piercy DA, Robson V, Wilkinson R, Hall R, Evered DC, Besser GM, Schally AV, Kastin AJ, Goy DH (1976) Actions of growth hormone-release inhibiting hormone on the renin aldosterone system. J Clin Endocrinol Metab 43:240-243
Gomez-Sanchez C, Holland OB, Higgins JR, Kem DC, Kaplan NM (1976) Circadian rhythms of serum renin activity and serum corticosterone, prolactin and aldosterone in the male rat on normal and low sodium diets. Endocrinology 99:567-572
Gomez-Sanchez C, Holland OB, Murry BA (1977) Effect of intraventricular infusions of artificial cerebrospinal fluid on serum renin activity, aldosterone, corticosterone, and prolactin. Proc Soc Exp Biol Med 156:31-34
Gomez-Sanchez CE, Montgomery M, Ganguly A, Holland OB, Gomez-Sanchez EP, Grim CE, Weinberger MH (1984) Elevated urinary excretion of 18-oxocortisol in glucocorticoid-suppressible aldosteronism. J Clin Endocrinol Metab 59:1022-1024
Gomoll AW, Schmid HE Jr (1965) Urinary excretion of aldosterone in dogs with elevated plasma titers of antirenin. Proc Soc Exp Biol Med 120:326-330
Gonzalez JJ, Werk EE Jr, Thrasher K, Behar R, Loadholt CB (1979) Renin aldosterone system and potassium levels in chronic lead intoxication. South Med J 72:433-437
Goodfriend TL, Peach MJ (1975) Angiotensin III: (des-aspartatic acid[1])-angiotensin II: Evidence and speculation for its role as an important agonist in the renin-angiotensin system. Circ Res 36, 37 [Suppl 1]:I-38-I-48
Goodfriend TL, Elliott ME, Atlas SA (1984) Actions of synthetic atrial natriuretic factor on bovine adrenal glomerulosa. Life Sci 35:1675-1682
Goodwin FJ, Knowlton AI, Laragh JH (1969) Absence of renin suppression by deoxycorticosterone acetate in rats. Am J Physiol 216:1476-1480

Goodwin FJ, Kirshman JD, Sealey JE, Laragh JH (1970a) Influence of the pituitary gland on sodium conservation, plasma renin and renin substrate concentrations in the rat. Endocrinology 86:824-834

Goodwin FJ, Ledingham JG, Laragh JH (1970b) The effects of prolonged administration of vasopressin and oxytocin on renin, aldosterone and sodium balance in normal man. Clin Sci 39: 641-651

Goodwin TJ, James VHT, Peart WS (1974) The control of aldosterone secretion in nephrectomized man. Clin Sci Mol Med 47:235-248

Gordon MB, Moore TJ, Dluhy RG, Williams GH (1983a) Dopaminergic blockade of the renin-angiotensin-aldosterone system: effect of high and low sodium intakes. Clin Endocrinol 19: 415-425

Gordon MB, Moore TJ, Dluhy RG, Williams GH (1983b) Dopaminergic modulation of aldosterone responsiveness to angiotensin II with changes in sodium intake. J Clin Endocrinol Metab 56:340-345

Gordon RD, Pawsey CGK (1971) Relative effects of serum sodium concentration and the state of body fluid balance on renin secretion. J Clin Endocrinol Metab 32:117-119

Gordon RD, Wolfe LK, Island DP, Liddle GW (1966) A diurnal rhythm in plasma renin activity in man. J Clin Invest 45:1587-1592

Gordon RD, Küchel O, Liddle GW, Island D (1967a) Role of the sympathetic nervous system in regulating renin and aldosterone production in man. J Clin Invest 46:599-605

Gordon RD, Fishman LM, Liddle GW (1967b) Plasma renin activity and aldosterone secretion in a pregnant woman with primary aldosteronism. J Clin Endocrinol Metab 27:385-388

Gordon RD, Thomas MJ, Thomas F, Doran F (1975) Effects of unilateral nephrectomy in man on renin, aldosterone and volume/pressure relationships. Clin Exp Pharmacol Physiol [Suppl] 2:109-114

Gordon RD, Nicholls MG, Tree M, Fraser R, Robertson JIS (1980) Influence of sodium balance on ACTH/adrenal corticosteroid dose-response curves in the dog. Am J Physiol 238: E543-E551

Gotshall RW, Davis JO, Shade RE, Spielman W, Johnson JA, Braverman B (1973) Effects of renal denervation on renin-release in sodium-depleted dogs. Am J Physiol 225:344-349

Gould AB, Skeggs LT, Kahn JR (1966) Measurements of renin and substrate concentrations in human serum. Lab Invest 15:1802-1813

Gowenlock AH, Wrong O (1962) Hyperaldosteronism secondary to renal ischaemia. Q J Med 31:323-343

Gowenlock AH, Mills JN, Thomas S (1959) Acute postural changes in aldosterone and electrolyte excretion in man. J Physiol (Lond) 146:133-141

Gower DB (1974) Modifiers of steroid-hormone metabolism: a review of their chemistry, biochemistry and clinical applications. J Steroid Biochem 5:501-523

Grahame-Smith DG, Butcher RW, Ney RL, Sutherland EW (1967) Cyclic AMP as the intracellular mediator of the action of ACTH on the adrenal cortex. J Biol Chem 242:5535-5541

Grant DB, Dillon MJ, Atherden SM, Levinsky RJ (1977) Congenital adrenal hyperplasia: renin and steroid values during treatment. Eur J Pediatr 126:89-96

Gray MJ, Strausfeld KS, Watanabe M, Sims EAH, Solomon S (1968) Aldosterone secretory rates in the normal menstrual cycle. J Clin Endocrinol Metab 28:1269-1275

Greco JM, Murphy GP (1971) The influence of aldosterone on renin release in the isolated canine kidney perfused with ascitic fluid. Res Commun Chem Pathol Pharmacol 2:315-322

Greene JA Jr, Vander AJ, Kowalczyk (1968) Plasma renin activity and aldosterone excretion after renal homotransplantation. J Lab Clin Med 71:586-595

Greengard P, Psychoyos S, Tallan HH, Cooper DY, Rosenthal O, Estabrook RW (1967) Aldosterone synthesis by adrenal mitochondria. III. Participation of cytochrome P-450. Arch Biochem Biophys 121:298-303

Greenway CV, Verney EB (1962) The effect of adrenocorticotrophic hormone on the secretion of corticosteroids by the isolated perfused adrenal gland of the dog. J Physiol 162:183-192

Greep RO, Deane HW (1947) Cytochemical evidence for the cessation of hormone production in the zona glomerulosa of the rat's adrenal cortex after prolonged treatment with deoxycorticosterone acetate. Endocrinology 40:417-425

Greiner JW, Rumbaugh RC, Kramer RE, Colby HD (1978) Relation of canrenone to the actions of spironolactone on adrenal cytochrome P-450 dependent enzymes. Endocrinology 103: 1313-1320

Grekin RJ, Dale SL, Melby JC (1973) The role of 18-hydroxy-11-deoxycorticosterone as a precursor in human adrenal tissue in vitro. J Clin Endocrinol Metab 37:261-264
Griffing GT, Berelowitz B, Hudson M, Salzman R, Manson JAE, Aurrechia S, Melby JC, Pedersen RC, Brownie AC (1985a) Plasma immunoreactive gamma melanotropin in patients with idiopathic hyperaldosteronism, aldosterone-producing adenomas, and essential hypertension. J Clin Invest 76:163-169
Griffing GT, McIntosh T, Berelowitz B, Hudson M, Salzman R, Manson JAE, Melby JC (1985b) Plasma β-endorphin levels in primary aldosteronism. J Clin Endocrinol Metab 60:315-319
Grim CE, Keitzer WF (1974) Circadian rhythm of renin and aldosterone in unilateral renovascular hypertension. Pre- and post-operative studies. Ann Intern Med 80:298-304
Grim CE, Weinberger MH (1980) Familial dexamethasone-suppressible, normokalemic hyperaldosteronism. Pediatrics 65:597-604
Gröger-Korber EM (1966) Beeinflussung der Nebennierenrindensekretion durch Polypeptide in vitro. Arch Int Pharmacodyn Ther 163:158-173
Gromova EA, Kraus M, Křeček J (1967) Effect of melatonin and 5-hydroxytryptamine on aldosterone and corticosterone production by adrenal glands of normal and hypophysectomized rats. J Endocrinol 39:345-350
Grospietsch G, Fenske M, König A (1980) Basal plasma aldosterone levels and tetracosactid-induced increase of aldosterone in conscious male rabbits: lack of correlation with glucocorticoid levels. Endokrinologie 75:324-334
Gross F (1958) Renin und Hypertensin, physiologische oder pathologische Wirkstoffe? Klin Wochenschr 36:693-706
Gross F (1967) The regulation of aldosterone secretion by the renin-angiotensin system under various conditions. Acta Endocrinol [Suppl] 124:41-64
Gross F, Schaechtelin G, Ziegler M, Berger M (1964) A renin-like substance in the placenta and uterus of the rabbit. Lancet 1:914-916
Gross F, Brunner H, Ziegler M (1965) Renin-angiotensin system, aldosterone and sodium balance. Recent Prog Horm Res 21:119-177
Gross F, Carretero O, Johnston CI (1968) Effect of ureteral ligation on blood pressure and renin content of the kidneys in renal hypertensive rats. Proc Soc Exp Biol Med 128:935-940
Güllner H-G, Gill JR Jr (1983) Beta endorphin selectively stimulates aldosterone secretion in hypophysectomized, nephrectomized dogs. J Clin Invest 71:124-128
Güllner H-G, Nicholson WE, Gill JR Jr, Orth DN (1983) Plasma immunoreactive proopiolipomelanocortin-derived peptides in patients with primary aldosteronism, idiopathic hyperaldosteronism with bilateral adrenal hyperplasia, and dexamethasone-suppressible hyperaldosteronism. J Clin Endocrinol Metab 56:853-855
Guillemette G, Escher E (1983) Analysis of the adrenal angiotensin II receptor with the photoaffinity labeling method. Biochemistry 22:5591-5596
Guthrie GP Jr, Cecil SG, Darden ED, Kotchen TA (1982) Dynamics of renin and aldosterone in the thoroughbred horse. Gen Comp Endocrinol 48:296-299
Guthrie GP Jr, McAllister RG Jr, Kotchen TA (1983) Effects of intravenous and oral verapamil upon pressor and adrenal steroidogenic responses in normal man. J Clin Endocrinol Metab 57:339-343
Gutman Y, Benzakein F (1971) Effect of an increase and of lack of antidiuretic hormone on plasma renin activity in the rat. Life Sci 10:1081-1085
Gutman Y, Shamir Y, Glushevitzky D, Hochman S (1972) Angiotensin increases microsomal (Na^+-K^+)-ATPase activity in several tissues. Biochim Biophys Acta 273:401-406
Gwynne JT, Mahaffee D, Brewer HB Jr, Ney RL (1976) Adrenal cholesterol uptake from plasma lipoproteins: regulation by corticotropin. Proc Natl Acad Sci USA 73:4329-4333

Haack D, Möhring J, Möhring B, Petri M, Hackenthal E (1977) Comparative study on development of corticosterone and DOCA hypertension in rats. Am J Physiol 233:F403-F411
Haack D, Engel R, Vecsei (1978) The effect of chronic ACTH treatment on blood pressure and urinary excretion of steroids in the rat. Klin Wochenschr 56 [Suppl I]:183-186
Hadjian AJ, Culty M, Chambaz EM (1984) Stimulation of phosphatidylinositol turnover by acetylcholine, angiotensin II and ACTH in bovine adrenal fasciculata cells. Bichim Biophys Acta 804:427-433
Hahn J, Zipser R, Zia P, Tuck M, Golub M, Horton R (1980) Induction of renin release by exogenous prostaglandins in hyporeninemic hypoaldosteronism. Prostaglandins 20:15-23

Haldy P, Müller J (1981a) Stimulation of aldosterone biosynthesis by experimental edema: role of the kidneys. Am J Physiol 240: E12-E17

Haldy P, Müller J (1981b) Stimulation of the conversion of corticosterone to aldosterone by sodium and water depletion of potassium-deficient rats. Comparisons of different experimental procedures. Acta Endocrinol 96: 519-526

Halkerston IDK, Feinstein M, Hechter OM (1965) Further observations on the inhibition of adrenal protein synthesis by ACTH in vitro. Endocrinology 76: 801-802

Hall JE, Guyton AC, Smith MJ Jr, Coleman TG (1979) Chronic blockade of angiotensin II formation during sodium deprivation. Am J Physiol 237: F424-F432

Hamet P, Tremblay J, Pang SC, Garcia R, Thibault G, Gutkowska J, Cantin M, Genest J (1984) Effect of native and synthetic atrial natruretic factor on cyclic GMP. Biochem Biophys Res Commun 123: 515-527

Hamilton BP, Zadik Z, Edwin CM, Hamilton JH, Kowarski AA (1979) Effect of adrenal suppression with dexamethasone in essential hypertension. J Clin Endocrinol Metab 48: 848-853

Haning R, Tait SAS, Tait JF (1970) In vitro effects of ACTH, angiotensins, serotonin and potassium on steroid output and conversion of corticosterone to aldosterone by isolated adrenal cells. Endocrinology 87: 1147-1167

Hannon RC, Deruyck RP, Joossens JV, Amery AK (1969) Disappearance rate of endogenous renin from the plasma after bilateral nephrectomy in humans. J Clin Endocrinol Metab 29: 1420-1424

Hansen EL, Hørlyck E, Grønbaek P, Iversen M (1967) Hyperaldosteronism following total fasting in obese subjects. Acta Med Scand 182: 65-71

Hansson B-G, Bergentz S-E, Dymling J-F, Hedeland H, Hökfelt B (1981) Pre- and postoperative studies in 72 hypertensive patients with renal artery stenosis, with special reference to renin activity and aldosterone. Acta Med Scand 210: 249-255

Hartroft PM, Eisenstein AB (1957) Alterations in the adrenal cortex of the rat induced by sodium deficiency: correlation of histologic changes with steroid hormone secretion. Endocrinology 60: 641-651

Hartroft PM, Hartroft WS (1955) Studies of renal juxtaglomerular cells. II. Correlation of the degree of granulation of juxtaglomerular cells with width of the zona glomerulosa of the adrenal cortex. J Exp Med 102: 205-213

Hashida Y, Yunis EJ (1972) Ultrastructure of the adrenal zona glomerulosa in children with renovascular hypertension. Human Pathol 3: 301-315

Hata T, Bravo EL, Sen S, Bumpus FM (1984) Steroidogenic properties of a new aldosterone-stimulating factor: interaction with angiotensin II and adrenocorticotropin during variations of dietary sodium. Endocrinology 115: 2065-2070

Hauger RL, Aguilera G, Catt KJ (1978) Angiotensin II regulates its receptor sites in the adrenal glomerulosa zone. Nature 271: 176-178

Hauger-Klevene JH (1970a) High plasma renin activity in an oat cell carcinoma: a renin-secreting carcinoma? Cancer 26: 1112-1114

Hauger-Klevene JH (1970b) ACTH, cyclic AMP, dexamethasone and actinomycin D effect on renin release. Acta Physiol Lat Am 20: 373-381

Hauger-Klevene JH, Cole VW (1970) Aldosterone secretion and plasma renin activity in chromophobe adenoma: effect of ACTH. Cancer 26: 1115-1118

Hauger-Klevene JH, Brown H, Fleischer N (1969) ACTH stimulation and glucocorticoid inhibition of renin release in the rat. Proc Soc Exp Biol Med 131: 539-542

Haupert GT Jr, Sancho JM (1979) Sodium transport inhibitor from bovine hypothalamus. Proc Natl Acad Sci USA 76: 4658-4660

Haupert GT Jr, Carilli CT, Cantley LC (1984) Hypothalamic sodium-transport inhibitor is a high-affinity reversible inhibitor of Na^+- K^+-ATPase. Am J Physiol 247: F919-F924

Hausdorff WP, Aguilera G, Catt KJ (1986) Selective enhancement of angiotensin II- and potassium-stimulated aldosterone secretion by the calcium channel agonist BAY K 8644. Endocrinology 118: 869-874

Havelka J, Vetter H, Studer A, Greminger P, Lüscher T, Wollnik S, Siegenthaler W, Vetter W (1982) Acute and chronic effects of the angiotensin-converting enzyme inhibitor captopril in severe hypertension. Am J Cardiol 49: 1467-1474

Hayashi K, Sala G, Catt K, Dufau ML (1979) Regulation of steroidogenesis by adrenocorticotropic hormone in isolated adrenal cells. The intermediate role of cyclic nucleotides. J Biol Chem 254: 6678-6683

Haynes RC Jr, Berthet L (1957) Studies on the mechanism of action of the adrenocorticotropic hormone. J Biol Chem 225:115-124
Haynes RC Jr, Koritz SB, Péron FG (1959) Influence of adenosine 3', 5' monophosphate on corticoid production by rat adrenal glands. J Biol Chem 234:1421-1423
Healey DP, Maciejewski AR, Printz MP (1985) Autoradiographic localization of [^{125}I]-angiotensin II binding sites in the rat adrenal gland. Endocrinology 116:1221-1223
Health and Public Policy Committee, American College of Physicians (1984) The dexamethasone suppression test for the detection, diagnosis, and management of depression. Ann Intern Med 100:307-308
Hegstad R, Brown RD, Jiang N-S, Kao P, Weinshilboum RM, Strong C, Wisgerhof M (1983) Aging and aldosterone. Am J Med 74:442-448
Helber A, Hayduk K, Nowaczynski W, Brecht HM, Küchel O, Genest J (1972) Zur Beeinflussung der Aldosteronsekretion des Hundes durch veränderte Plasma-Magnesium-Konzentration. Res Exp Med 157:336-346
Helbock HJ, Reynolds JW (1970) Pseudoaldosteronism (Liddle's syndrome): evidence for increased cell permeability to Na^+. Pediatr Res 4:455 (abstract)
Hellman L, Nakada F, Curti J, Weitzma ED, Kream J, Roffwarf H, Ellman S, Fukushima DK, Gallagher TF (1970) Cortisol is secreted episodically by normal man. J Clin Endocrinol Metab 30:411-422
Helmer OM (1958) Studies on renin antibodies. Circulation 17:648-652
Helmer OM, Griffith RS (1952) The effect of the administration of estrogens on the renin-substrate (hypertensinogen) content of rat plasma. Endocrinology 51:421-426
Helmer OM, Judson WE (1967) Influence of high renin substrate levels on renin-angiotensin system in pregnancy. Am J Obstet Gynecol 99:9-17
Henrich WL, Katz FH, Molinoff PB, Schrier RW (1977) Competitive effects of hypokalemia and volume depletion on plasma renin activity, aldosterone and catecholamine concentrations in hemodialysis patients. Kidney Int 12:279-284
Hepp R, Grillet C, Peytremann A, Vallotton MB (1977) Stimulation of corticosteroid biosynthesis by angiotensin I, (des-asp[1]) angiotensin I, angiotensin II and (des-asp[1]) angiotensin II in bovine adrenal fasciculata cells. Endocrinology 101:717-725
Hermus ARMM, Pieters GFFM, Smals AGH, Benraad TJ, Kloppenborg PWC (1984) Plasma adrenocorticotropin, cortisol, and aldosterone responses to corticotropin-releasing factor: modulatory effect of basal cortisol level. J Clin Endocrinol Metab 58:187-191
Hesse B, Nielsen I (1977) Suppression of plasma renin activity by intravenous infusion of antidiuretic hormone in man. Clin Sci Mol Med 52:357-360
Heyes MP, Farber MO, Manfredi F, Robertshaw D, Weinberger M, Fineberg N, Robertson G (1982) Acute effects of hypoxia on renal and endocrine function in normal humans. Am J Physiol 243:R265-R270
Hilfenhaus M (1974) in: Aschoff J, Ceresa F, Halberg F (eds) Chronobiological aspects of endocrinology. Schattauer, Stuttgart, pp 111-115
Hilfenhaus M (1976) Circadian rhythm of the renin angiotensin aldosterone system in the rat. Arch Toxicol 36:305-316
Hilfenhaus M (1977) Urinary aldosterone excretion rate and plasma aldosterone concentration in the rat: effect of ACTH, DOC, furosemide and of changes in sodium balance. Acta Endocrinol 85:134-142
Hilton JG, Scian LF, Westermann CD, Nakano J, Kruesi OR (1960) Vasopressin stimulation of the isolated adrenal glands: nature and mechanism of hydrocortisone secretion. Endocrinology 67:298-310
Hilton JG, Kruesi OR, Nedeljkovic RI, Scian LF (1961) Adrenocortical and medullary responses to adenosine 3', 5' monophosphate. Endocrinology 68:908-913
Hilton JG, Alter S, Cushman P (1965) Aldosterone secretion: differences in direct effects of rubidium and potassium perfusion into dog adrenals. Nature 208:1099-1100
Himathongkam T, Dluhy RG, Williams GH (1975) Potassium-aldosterone-renin interrelationships. J Clin Endocrinol Metab 41:153-159
Hinson JP, Vinson GP, Whitehouse BJ, Price G (1985) Control of zona glomerulosa function in the isolated perfused rat adrenal gland in situ. J Endocrinol 104:387-395
Hiramatsu K, Yamada T, Katakura M (1984) Acute effects of cold on blood pressure, renin-angiotensin-aldosterone system, catecholamines and adrenal steroids in man. Clin Exp Pharmacol Physiol 11:171-179

Hirata Y, Tomita M, Yoshimi H, Kuramochi M, Ito K, Ikeda M (1985) Effect of synthetic human atrial natriuretic peptide on aldosterone secretion by dispersed aldosterone-producing adenoma cells in vitro. J Clin Endocrinol Metab 61:677-680

Hirose S, Akiyama F, Shinjo M, Ohno H, Murakami K (1985) Solubilization and molecular weight estimation of atrial natriuretic factor receptor from bovine adrenal cortex. Biochem Biophys Res Commun 130:574-579

Hirose T, Matsumotso, Aikawa T (1978) Direct effect of histamine on cortisol and corticosterone production by isolated dog adrenal cells. J Endocrinol 76:371-372

Hirose T, Matsumoto I, Aikawa T (1982) Modification by Su-10'603 of the respone to histamine on the production of cortisol and corticosterone in isolated canine adrenocortical cells. Acta Endocrinol 101:586-591

Hogan RP III, Kotchen TA, Boyd AE III, Hartley LH (1973) Effect of altitude on renin-aldosterone system and metabolism of water and electrolytes. J Appl Physiol 35:385-390

Holcombe JH, Keeenan BS, Nichols BL, Kirkland RT, Clayton GW (1980) Neonatal salt loss in the hypertensive form of congenital adrenal hyperplasia. Pediatrics 65:777-781

Holland OB, Gomez-Sanchez CE, Kem DC, Weinberger MH, Kramer NJ, Higgins JR (1977) Evidence against prolactin stimulation of aldosterone in normal human subjects and patients with primary aldosteronism, including a patient with primary aldosteronism and a prolactin-producing pituitary microadenoma. J Clin Endocrinol Metab 45:1064-1076

Holland OB, Thomas C, Brown H, Schindewolf D, Hillier Y, Gomez-Sanchez C (1983) Aldosterone suppression with dopamine infusion in low-renin hypertension. J Clin Invest 72:754-766

Hollenberg NK, Chenitz WR, Adams DF, Williams GH (1974) Reciprocal influence of salt intake on adrenal glomerulosa and renal vascular responses to angiotensin II in normal man. J Clin Invest 54:34-42

Hollenberg NK, Williams G, Burger B, Hooshmand I (1975) The influence of potassium on the renal vasculature and the adrenal gland, and their responsiveness to angiotensin II in normal man. Clin Sci Mol Med 49:527-534

Hollenberg NK, Williams GH, Burger B, Ishikawa I, Adams DF (1976) Blockade and stimulation of renal, adrenal, and vascular angiotensin II receptors with 1-Sar, 8-Ala angiotensin II in normal man. J Clin Invest 57:39-46

Hollifield JW, McKenna TJ, Wolff JMcD, Chick WT, Liddle GW (1975) A new adrenal inhibitor effective in the treatment of primary aldosteronism. Clin Res 23:237A (abstract)

Holsboer F, Dörr HG, Sippell WG (1982) Blunted aldosterone response to dexamethasone in female patients with endogenous depression. Psychoneuroendocrinology 7:155-162

Honn KV, Chavin W (1976a) Role of prostaglandins in aldosterone production by the human adrenal. Biochem Biophys Res Commun 72:1319-1326

Honn KV, Chavin W (1976b) Prostaglandin modulation of the mechanism of ACTH action in the human adrenal. Biochem Biophys Res Commun 73:164-170

Honn KV, Chavin W (1977) Effects of A and B series of prostaglandins on cAMP, cortisol, and aldosterone production by the human adrenal. Biochem Biophys Res Commun 76:977-982

Horiba N, Nomura K, Saito Y, Demura H, Shizume K (1985) Rat atrial natriuretic polypeptide stimulation of adrenal zona glomerulosa cell growth. Biochem Biophys Res Commun 132: 261-263

Horner JM, Hintz RL, Luetscher JA (1979) The role of renin and angiotensin in salt-losing, 21-hydroxylase-deficient congenital adrenal hyperplasia. J Clin Endocrinol Metab 48:776-783

Hornsby PJ (1980) Regulation of cytochrome P-450-supported 11β-hydroxylation of deoxycortisol by steroids, oxygen, and antioxidants in adrenocortical cell cultures. J Biol Chem 255: 4020-4027

Hornsby PJ, Crivello JF (1983a) The role of lipid peroxidation and biological antioxidants in the function of the adrenal cortex. Part 1: a background review. Mol Cell Endocrinol 30:1-20

Hornsby PJ, Crivello JF (1983b) The role of lipid peroxidation and biological antioxidants in the function of the adrenal cortex, part 2. Mol Cell Endocrinol 30:123-147

Hornsby PJ, O'Hare MJ (1977) The roles of potassium and corticosteroids in determining the pattern of metabolism of [^{3}H] deoxycorticosterone by monolayer cultures of rat adrenal zona glomerulosa cells. Endocrinology 101:997-1005

Hornsby PJ, O'Hare MJ, Neville AM (1973) Effect of ACTH on biosynthesis of aldosterone and corticosterone by monolayer cultures of rat adrenal zona glomerulosa cells. Bichem Biophys Res Commun 54:1554-1559

Hornsby PJ, O'Hare MJ, Neville AM (1974) Functional and morphological observations on rat adrenal zona glomerulosa cells in monolayer culture. Endocrinology 95:1240-1251

Horton R (1969) Stimulation and suppression of aldosterone in plasma of normal man and in primary aldosteronism. J Clin Invest 48:1230-1236
Horton R (1970) Laboratory determinations of aldosterone and their clinical significance. Am J Clin Pathol 54:297-302
Horton R, Finck E (1972) Diagnosis and localization in primary aldosteronism. Ann Intern Med 76:885-890
Hsueh WA, Luetscher JA, Carlson EJ, Grislis G, Fraze E, McHargue A (1982) Changes in active and inactive renin throughout pregnancy. J Clin Endocrinol Metab 54:1010-1016
Hubl W, Lehmann GW, Büchner M, Dörner G (1978) Plasma aldosterone in newborns and children. Endokrinologie 72:367-370
Hubl W, Büchner M, Bellée H, Mühlbach F, Dörner G (1979) Study of plasma aldosterone in normal pregnancy, in pre-eclamptic women and in cord plasma of newborns. Endokrinologie 73:162-166
Huijmans JGM, Falke HE, Degenhart HJ (1984) Effects of some hydroxylated sterols on the steroid production in isolated rat adrenal glomerulosa and fasciculata/reticularis cells in the presence of potassium, angiotensin II or ACTH. Acta Endocrinol 105:411-416
Hullin RP, Jerram TC, Lee MR, Levell MJ, Tyrer SP (1977) Renin and aldosterone relationships in manic depressive psychosis. Br J Psychiatry 131:575-581
Hullin RP, Levell MJ, O'Brien MJ, Toumba KJ (1981) Inhibition of in vitro production of aldosterone by manic-depressive sera. Br J Psychiatry 138:373-380
Humpeler E, Skrabal F, Bartsch G (1980) Influence of exposure to moderate altitude on the plasma concentration of cortisol, aldosterone, renin, testosterone, and gonadotropins. Eur J Appl Physiol 45:167-176
Humphrey JH, Toh CC (1954) Absorption of serotonin (5-hydroxytryptamine) and histamine by dog platelets. J Physiol 124:300-304
Humphrey MJ, Lindup WE, Chakraborty J, Parke DV (1979) Effect of carbenoxolone on the concentrations of aldosterone in plasma and kidney. J Endocrinol 81:143-151
Humphrey TJ, Fan JSK, Coghlan JP, Denton DA, Scoggins BA, Stewart KW, Withworth JA (1983) Inter-relationships between sodium and potassium intake and the blood pressure effects of ACTH in sheep. J Hypertension 1:19-26
Hunyady L, Balla T, Nagy K, Spät A (1982) Control of phosphatidylinositol turnover in adrenal glomerulosa cells. Biochim Biophys Acta 713:352-357
Hunyady L, Balla T, Spät A (1983) Angiotensin II stimulates phosphatidylinositol turnover in adrenal glomerulosa cells by a calcium-independent mechanism. Biochim Biophys Acta 753: 133-135
Hunyady L, Balla T, Enyedi P, Spät A (1985) The effect of angiotensin II on arachidonate metabolism in adrenal glomerulosa cells. Biochem Pharmacol 34:3439-3444
Hunziker U, Müller J (1977) Effects of treatment with mineralocorticoids on late steps of aldosterone biosynthesis in the rat. J Steroid Biochem 8:1-7
Huq MS, Pfaff M, Jespersen D, Zucker IR, Kirschner MA (1976) Concurrence of aldosterone, androgen, and cortisol secretion in adrenal venous effluents. J Clin Endocrinol Metab 42: 230-238
Huston G, Martin VI, Al-Dujaili EAS, Edwards CRW (1981) Evaluation of the mineralocorticoid activity of 18-hydroxycorticosterone. Clin Sci 61:201-206
Hyatt PJ, Tait JF, Tait SAS (1986) The mechanism of the effect of K^+ on the steroidogenesis of rat zona glomerulosa cells of the adrenal cortex: role of cyclic AMP. Proc R Soc Lond [Biol] 227: 21-42
Hyde PM, Daigneault EA (1968) Adrenal plasma levels of corticosterone and deoxycorticosterone in methylandrostenediol-salt induced hypertension. Steroids 11:721-731
Hyman PE, Sha'afi RI, Tan SY, Hintz R (1979) Liddle syndrome. J Pediatr 95:77-78

Idelman S (1978) The structure of the mammalian adrenal cortex. In: Chester Jones I, Henderson IW (eds) General, comparative and clinical endocrinology of the adrenal cortex, vol 2. Academic, London, pp 1-199
Imai M, Igarashi Y, Sokabe H (1968) Plasma renin activity in congenital virilizing adrenal hyperplasia. Pediatrics 41:897-904
Imura H, Matsukura S, Matsuyama H, Setsuda T, Miyake T (1965) Adrenal steroidogenic effect of adenosine 3', 5' monophosphate and its derivatives in vivo. Endocrinology 76:933-937
Isaac R, Raymond J-P, Rainfray M, Ardaillou R (1984) Effects of an acute calcium load on plasma ACTH, cortisol, aldosterone and renin activity in man. Acta Endocrinol 105:251-257

Isler H (1973) Effect of posterior pituitary powder on the mitotic activity of the zona glomerulosa of the adrenal glands in intact and hypophysectomized rats: a preliminary observation. Anat Rec 177:321-324

Jackson B, Valentine R, Wagner G (1986) Primary aldosteronism due to a malignant ovarian tumour. Aust NZ J Med 16:69-71

Jacobs DR, van der Poll J, Gabrilove JL, Soffer LJ (1961) 17α-Hydroxyprogesterone - a salt-losing steroid: relation to congenital adrenal hyperplasia. J Clin Endocrinol Metab 21:909-922

Jadoul FAC, Broughton Pipkin F, Lamming GD (1982) Changes in the renin-angiotensin-aldosterone system in normotensive primigravidae in the four days after normal spontaneous delivery. Br J Obstet Gynaecol 89:633-639

James VHT, Tunbridge RDG, Wilson GA (1976) Studies on the control of aldosterone secretion in man. J Steroid Biochem 7:941-948

Janigan DT (1963) Cytoplasmic bodies in the adrenal cortex of patients treated with spironolactone. Lancet 1:850-852

Jean R, Legrand J-C, Meylan F, Rieu D, Astruc J (1969) Hypoaldostéronisme primaire par anomalie probable de la 18-hydroxylation. Arch Fr Pediatr 26:769-777

Jefcoate JR, Hume R, Boyd GS (1970) Separation of two forms of cytochrome *P*450 from adrenal cortex mitochondria. FEBS Lett 9:41-44

Jenis EH, Hertzog RW (1969) Effect of spironolactone on the zona glomerulosa of the adrenal gland. Light and electron microscopy. Arch Pathol 88:530-539

Jenkins JS, Frankel RJ, Wright JJ, Khan MUA (1976) Effect of adrenergic blockade and dexamethasone on the aldosterone response to brain stimulation in the rhesus monkey. J Endocrinol 71:393-397

Jenny M, Muller AF, Fabre J, Mach RS (1961) Hypokaliémie et alcalose par ingestion abusive d'extrait de réglisse (liquorice) et d'eau bicarbonatée. Pseudo-syndrome de Conn. Schweiz Med Wochenschr 91:869-875

Jessiman AG, Watson DD, Moore FD (1959) Hypophysectomy in the treatment of breast cancer. N Engl J Med 261:1199-1207

Johns DW, Baker KM, Ayers CR, Vaughan ED Jr, Carey RM, Peach MJ, Yancey MR, Ortt EM, Williams SC (1980) Acute and chronic effect of captopril in hypertensive patients. Hypertension 2:567-575

Johnson BB, Lieberman AH, Mulrow PJ (1957) Aldosterone excretion in normal subjects depleted of sodium and potassium. J Clin Invest 36:757-766

Johnson EIM, McDougall JG, Coghlan JP, Denton DA, Scoggins BA, Wright RD (1984) Potassium stimulation of aldosterone secretion in vivo is reversed by nisoldipine, a calcium transport antagonist. Endocrinology 114:1466-1468

Johnson EIM, McDougall JG, Coghlan JP, Denton DA, Hardy KJ, Scoggins BA, Wright RD (1985) Ca antagonists do not alter aldosterone secretion during established Na depletion. Am J Physiol 248:E676-E680

Johnson JA, Davis JO (1973) Effects of a specific competitive antagonist of angiotensin II on arterial pressure and adrenal steroid secretion in dogs. Circ Res 32/33 [Suppl 1]:I-159-I-168

Johnson JA, Davis JO, Taylor AA, Shade RE (1974) Secretion rates and plasma levels of deoxycorticosterone and removal rates of deoxycorticosterone and aldosterone in the pregnant dog. Endocrinology 94:580-583

Johnston CI, Davis JO, Wright FS, Howard SS (1967a) Effects of renin and ACTH on adrenal steroid secretion in the American bullfrog. Am J Physiol 213:393-399

Johnston CI, Davis JO, Hartroft PM (1967b) Renin-angiotensin system, adrenal steroids and sodium depletion in a primitive mammal, the American opossum. Endocrinology 81:633-642

Jones CR, Millar JA, Lawrie C, Sumner DJ, Reid JL (1984) Specific inhibition of aldosterone responses to endogenous and exogenous angiotensin II by somatostatin. Clin Endocrinol 21: 279-284

Jones KM, Lloyd-Jones R, Riondel A, Tait JF, Tait SAS, Bulbrook RD, Greenwood FC (1959) Aldosterone and secretion and metabolism in normal men and women and in pregnancy. Acta Endocrinol 30:321-342

Jornot LH, Capponi AM, Vallotton MB (1985) Responses of rat adrenal glomerulosa and inner zone cells to synthetic ACTH analogs and proopiomelanocortin-derived peptides. J Steroid Biochem 22:221-225

Jouan P (1963) Epiphyse, 5-hydroxytryptamine et corticoidogenèse in vitro. Ann Endocrinol 24: 365-370

Jouan P (1967) Propriétés adrénoglomérulotrophiques de la 5-hydroxytryptamine (sérotonine). Pathol Biol 15:1145-1153
Jouan P, Samperez S (1964) Recherches sur la spécificité d'action de la 5-hydroxytryptamine vis-à-vis de la sécrétion in vitro de l'aldostérone. Ann Endocrinol 25:70-75
Julian BA, Galla JH, Guthrie GP Jr, Kotchen TA (1982) Renin and aldosterone responses to short-term NaCl or $NaHCO_3$ loading in man. J Lab Clin Med 100:261-268
Jung RC, Dill DB, Horton R, Horvath SM (1971) Effects of age on plasma aldosterone levels and hemoconcentration at altitude. J Appl Physiol 31:593-597
Jungmann E, Wächtler M, Schwedes U, Usadel KH, Schöffling K (1983) The effect of metoclopramide and haloperidol on plasma renin activity and aldosterone levels in rats. Res Exp Med 183:133-138
Junnila RD, Travis RH, Brosnihan KB (1975) Episodic secretion of renin. Endocrinology 96: 129-134

Kahnt FW, Neher R (1962) On the specific inhibition of adrenal steroid biosynthesis. Experientia 18:499-501
Kahnt FW, Neher R (1965) Über die adrenale Steroid-Biosynthese in vitro: I. Umwandlung endogener und exogener Vorstufen im Nebennieren-Homogenat des Rindes. Helv Chim Acta 48:1457-1476
Kahnt FW, Neher R (1966a) Über die adrenale Steroidbiosynthese in vitro. II. Bedeutung von Steroiden als Hemmstoffe. Helv Chim Acta 40:123-133
Kahnt FW, Neher R (1966b) Über die adrenale Steroid-Biosynthese in vitro. III. Selektive Hemmung der Nebennieren-Funktion. Helv Chim Acta 49:725-732
Kahri AI, Pesonen S, Saure A (1970) Ultrastructural differentiation and progesterone-^{14}C metabolism in cultured cells of fetal rat adrenals under influence of ACTH. Steroidologia 1:25-64
Kaplan NM (1965) The biosynthesis of adrenal steroids: effects of angiotensin II, adrenocorticotropin, and potassium. J Clin Invest 44:2029-2039
Kaplan NM (1967) The steroid content of adrenal adenomas and measurements of aldosterone production in patients with essential hypertension and primary aldosteronism. J Clin Invest 46:728-734
Kaplan NM, Bartter FC (1962) The effect of ACTH, renin, angiotensin II, and various precursors on biosynthesis of aldosterone by adrenal slices. J Clin Invest 41:715-724
Karaboyas GC, Koritz SB (1965) Identity of the site of action of 3', 5'-adenosine monophosphate and adrenocorticotropic hormone in corticosteroidogenesis in rat adrenal and beef adrenal cortex slices. Biochemistry 4:462-468
Karlmar K-E (1982) In vitro conversion of cholesterol into aldosterone and cortisol in different zones of the bovine adrenal cortex. Acta Endocrinol 101:436-441
Karppanen H, Lahovaara S, Männistö P, Vapaatalo H (1975) Plasma renin activity and in vitro synthesis of aldosterone by the adrenal glands of rats with spontaneous, renal, or pinealectomy-induced hypertension. Acta Physiol Scand 94:184-188
Kasemsri S, Nickerson PA (1976) Quantitative ultrastructural study of the rat adrenal cortex in renal encapsulation-induced hypertension. Am J Pathol 82:143-156
Katayama S, Yamaji T (1982) A binding-protein for aldosterone in human plasma. J Steroid Biochem 16:185-192
Kater CE, Biglieri EG (1982) Zona fasciculata origin of 18-hydroxycorticosterone in the chronically suppressed zona glomerulosa. J Clin Endocrinol Metab 55:628-633
Kater CE, Biglieri EG (1983) Distinctive plasma aldosterone, 18-hydroxycorticosterone, and 18-hydroxydeoxycorticosterone profile in the 21-, 17α-, and 11β-hydroxylase deficiency types of congenital adrenal hyperplasia. Am J Med 75:43-48
Kater CE, Biglieri EG, Brust N, Chang B, Hirai J (1982) The unique patterns of plasma aldosterone and 18-hydroxycorticosterone concentrations in the 17α-hydroxylase deficiency syndrome. J Clin Endocrinol Metab 55:295-302
Kater CE, Biglieri EG, Schambelan M, Arteaga E (1983) Studies of impaired aldosterone response to spironolactone-induced renin and potassium elevations in adenomatous but not hyperplastic primary aldosteronism. Hypertension 5 [Suppl V]:V-115-V-121
Kater CE, Biglieri EG, Wajchenberg B (1985a) Effects of continued adrenocorticotropin stimulation on the mineralocorticoid hormones in classical and nonclassical simple virilizing types of 21-hydroxylase activity. J Clin Endocrinol Metab 60:1057-1062
Kater CE, Biglieri EG, Rost CR, Schambelan M, Hirai J, Chang BCF, Brust N (1985b) The con-

stant plasma 18-hydroxycorticosterone to aldosterone ratio: an expression of the efficacy of corticosterone methyloxidase type II activity in disorders with variable aldosterone production. J Clin Endocrinol Metab 60:225-228
Katz FH (1967) Primary aldosteronism with suppressed plasma renin activity due to bilateral adrenocortical hyperplasia. Ann Intern Med 67:1035-1042
Katz FH, Beck P (1974) Plasma renin activity, renin substrate and aldosterone during treatment with various oral contraceptives. J Clin Endocrinol Metab 39:1001-1004
Katz FH, Hoehn MH (1977) Effects of levodopa on the renin-aldosterone system. Clin Pharmacol Ther 21:388-391
Katz FH, Romfh (1972) Plasma aldosterone and renin activity during the menstrual cycle. J Clin Endocrinol Metab 34:819-821
Katz FH, Roper EF (1977) Testosterone effect on renin system in rats. Proc Soc Exp Biol Med 155:330-333
Katz FH, Romfh P, Smith JA (1972a) Episodic secretion of aldosterone in supine man: relationship to cortisol. J Clin Endocrinol Metab 35:178-181
Katz FH, Romfh P, Zimmering PE, Kelly WG (1972b) Radioimmunoassay of plasma aldosterone: effect of sexual differences and increasing duration of upright posture. Steroids Lipids Res 3:90-100
Katz FH, Romfh P, Smith JA (1972c) Effect of two consecutive acute stimuli on plasma aldosterone. Acta Endocrinol 71:153-159
Katz FH, Beck P, Makowski EL (1974) The renin-aldosterone system in mother and fetus at term. Am J Obstet Gynecol 118:51-55
Katz FH, Romfh P, Smith JA (1975a) Diurnal variation of plasma aldosterone, cortisol and renin activity in supine man. J Clin Endocrinol Metab 40:125-134
Katz FH, Romfh P, Smith JA, Roper EF, Barnes JS (1975b) Combination contraceptive effects on monthly cycle of plasma aldosterone, renin activity and renin substrate. Acta Endocrinol 79:295-300
Katznelson D, Sack J, Kraiem Z, Lunenfeld B (1979) Congenital hypoaldosteronism: thirteen-year follow-up in identical twins. Horm Res 11:22-28
Kaufmann W, Steiner B, Dürr F, Nieth H, Behn C (1967) Aldosteronstoffwechsel bei Nierenarterienstenose. Klin Wochenschr 45:966-973
Kaulhausen H, Weyand C (1982) Evidence for two independent effects of oestradiol benzoate on the renin-angiotensin-aldosterone system. Acta Endocrinol 100:77-84
Kaulhausen H, Venn HJ, Mühlbauer W, Beck KJ, Breuer H (1980) Effect of ethinyl estradiol on renin activity and on the concentrations of renin substrate and aldosterone in plasma of ovariectomized women. Arch Gynecol 229:29-36
Kaulhausen H, Öney T, Leydendecker G (1982) Inhibition of the renin-aldosterone axis and of prolactin secretion during pregnancy by L-dopa. Br J Obstet Gynaecol 89:483-488
Kawabe H, Kondo K, Hayashi M, Fujimaki M, Nagahama S, Saruta T (1983) Mechanism of the central effects of dopamine and metoclopramide on aldosterone regulation in the rat. Endocrinol Jpn 30:543-549
Kawasaki T, Uezono K, Ueno M, Omae T, Matsuoka M, Haus E, Halberg F (1983) Comparison of circadian rhythms of the renin-angiotensin-aldosterone system and electrolytes in clinically healthy young women in Fukuoka (Japan) and Minnesota (USA). Acta Endocrinol 102: 246-251
Keenan BS, Holcombe JH, Kirkland RT, Potts VE, Clayton GW (1979) Sodium homeostasis and aldosterone secretion in salt-losing congenital adrenal hyperplasia. J Clin Endocrinol Metab 48:430-436
Kelsch RC, Light GS, Luciano JR, Oliver WJ (1971) The effect of prednisone on plasma norepinephrine concentration and renin activity in salt-depleted man. J Lab Clin Med 77:267-277
Kem DC, Weinberger MH, Mayes DM, Nugent CA (1971) Saline suppression of plasma aldosterone in hypertension. Arch Intern Med 128:380-386
Kem DC, Weinberger MH, Gomez-Sanchez C, Kramer NJ, Lerman R, Furuyama S, Nugent CA (1973) Circadian rhythm of plasma aldosterone concentration in patients with primary aldosteronism. J Clin Invest 52:2272-2277
Kem DC, Gomez-Sanchez C, Kramer NJ, Holland OB, Higgins JR (1975) Plasma aldosterone and renin activity response to ACTH infusion in dexamethasone-suppressed normal and sodium-depleted man. J Clin Endocrinol Metab 40:116-124
Kem DC, Weinberger MH, Gomez-Sanchez C, Higgins JR, Kramer NJ (1976) The role of ACTH

in the episodic release of aldosterone in patients with idiopathic adrenal hyperplasia, hypertension and hyperaldosteronism. J Lab Clin Med 88:261-270
Kem DC, Weinberger MH, Higgins JR, Kramer NJ, Gomez-Sanchez C, Holland OB (1978) Plasma aldosterone response to ACTH in primary aldosteronism and in patients with low renin hypertension. J Clin Endocrinol Metab 46:552-560
Kem DC, Feldman M, Starkweather G, Li CH (1985) Effect of human β-endorphin on plasma aldosterone concentrations in normal human subjects. J Clin Endocrinol Metab 60:440-443
Kenyon CJ, Mosley W, Hargreaves G, Balment RJ, Henderson IW (1978a) The effects of dietary sodium restriction and potassium supplementation and hypophysectomy on adrenocortical function in the rat. J Steroid Biochem 9:337-344
Kenyon CJ, Hargreaves G, Henderson IW (1978b) Adrenocortical function in rats with inherited hypothalamic diabetes insipidus (Brattleboro strain). J Steroid Biochem 9:345-348
Keogh B, Kirdani R, Sandberg AA, Mittleman A, Murphy GP (1976) Renin-angiotensin-aldosterone system in long-term renal transplant patients. Urology 7:248-255
Keynes RJ, Smith GW, Slater JDH, Brown MM, Brown SE, Payne NN, Jowett TP, Monge CC (1982) Renin and aldosterone at high altitude in man. J Endocrinol 92:131-140
Khokhär AM, Slater JD, Forsling ML, Payne NN (1976) Effect of vasopressin on plasma volume and renin release in man. Clin Sci Mol Med 50:415-424
Khosla MC, Smeby RR, Bumpus FM (1974) Structure-activity relationship in angiotensin II analogs. In: Page IH, Bumpus FM (eds) Angiotensin. Springer, Berlin Heidelberg New York, pp 126-161
Kigoshi T, Morimoto S, Takeda R (1980) Role of the renin-angiotensin system in enhancing the aldosterone response to adrenocorticotropin during acute sodium depletion in normal subjects. J Clin Endocrinol Metab 51:1102-1106
Kihara I, Kitamura S, Hoshino T, Seida H, Watanabe T (1968) A hitherto unreported vascular tumor of the kidney: a proposal of "juxtaglomerular cell tumor". Acta Pathol Jpn 18:197-206
Kim CY, Sugiyama T, Okamoto M, Yamano T (1983) Regulation of 18-hydroxycorticosterone formation in bovine adrenocortical mitochondria. J Steroid Biochem 18:593-599
Kinson GA, Singer B (1967) The pineal gland and the adrenal response to sodium deficiency in the rat. Neuroendocrinology 2:283-293
Kinson GA, Singer B (1968) Sensitivity to angiotensin and adrenocorticotrophic hormone in the sodium-deficient rat. Endocrinology 83:1108-1116
Kinson GA, Singer B (1969) Influence of renal nerves on the secretion of aldosterone in the rat. Acta Endocrinol 61:239-247
Kinson G, Wahid AK, Singer B (1967) Effect of chronic pinealectomy on adrenocortical hormone secretion rates in normal and hypertensive rats. Gen Comp Endocrinol 8:445-454
Kinson GA, Singer B, Grant L (1968) Adrenocortical hormone secretion at various time intervals after pinealectomy in the rat. Gen Comp Endocrinol 10:447-449
Kirchner KA, Mueller R (1982) Effects of acute potassium infusions with salts other than chloride on plasma renin activity. Am J Physiol 242:F463-F469
Kisch ES, Dluhy RG, Williams GH (1976) Regulation of renin release by calcium and ammonium ions in normal man. J Clin Endocrinol Metab 43:1343-1350
Kittinger GW (1964) Puromycin inhibition of in vitro cortical hormone production by the rat adrenal gland. Steroids 4:539-545
Klepac R, Milkovic K, Milkovic S (1977) Development of steroidogenesis in the fetal rat adrenal gland: an in vitro study. J Steroid Biochem 8:841-845
Kliman B, Peterson RE (1960) Double isotope derivative assay of aldosterone in biological extracts. J Biol Chem 235:1639-1648
Klingbeil CK (1985a) Effects of chronic changes in dietary electrolytes and acute stress on plasma levels of corticosterone and aldosterone in the duck (Anas platyrhynchos). Gen Comp Endocrinol 58:10-19
Klingbeil CK (1985b) Corticosterone and aldosterone dose-dependent responses to ACTH and angiotensin II in the duck (Anas platyrhynchos). Gen Comp Endocrinol 59:382-390
Knochel JP, Osborn JR, Cooper EB (1964) Excretion of aldosterone in inappropriate secretion of antidiuretic hormone following head trauma. Metabolism 14:715-725
Kobayashi Y (1974a) Quantitative and electron microscopic studies on the pars intermedia of the hypophysis. I. Dietary effects of brown rice on the kidney, adrenal and pituitary of C57BL/6 mouse. J Electron Micros 23:107-115
Kobayashi Y (1974b) Quantitative and electron microscopic studies on the part intermedia of the

hypophysis. II. Alterations of the pars intermedia and the adrenal zona glomerulosa of albino mice following sodium restriction. Annot Zool Jpn 47:221-231

Kojima I, Ogata E (1982) Reversal by a dopamine antagonist of saline-induced attenuation of aldosterone response to angiotensin II infusion in man. Endocrinol Jpn 29:21-25

Kojima I, Inano H, Tamoaki B (1982) The final step of aldosterone biosynthesis is catalyzed by an NADPH-dependent and molecular oxygen-requiring enzyme. Biochem Biophys Res Commun 106:617-624

Kojima I, Lippes H, Kojima K, Rasmussen H (1983) Aldosterone secretion: effect of phorbol ester and A23187. Biochem Biophys Res Commun 116:555-562

Kojima I, Ogata E, Inano H, Tamaoki B (1984a) Oxygenation of 18-hydroxycorticosterone as the final reaction for aldosterone biosynthesis. Acta Endocrinol 107:395-400

Kojima I, Kojima K, Kreutter D, Rasmussen H (1984b) The temporal integration of the aldosterone secretory response to angiotensin occurs via two intracellular pathways. J Biol Chem 259: 14448-14457

Kojima I, Kojima K, Rasmussen H (1985a) Role of calcium and cAMP in the action of adrenocorticotropin on aldosterone secretion. J Biol Chem 260:4248-4256

Kojima I, Kojima K, Rasmussen H (1985b) Intracellular calcium and adenosine 3', 5'-cyclic monophosphate as mediators of potassium-induced aldosterone secretion. Biochem J 228:69-76

Kojima I, Kojima K, Rasmussen H (1985c) Effects of ANG II and K^+ on Ca efflux and aldosterone production in adrenal glomerulosa cells. Am J Physiol 248:E36-E43

Kojima I, Kojima K, Rasmussen H (1985d) Role of calcium fluxes in the sustained phase of angiotensin II-mediated aldosterone secretion from adrenal glomerulosa cells. J Biol Chem 260: 9177-9184

Kojima I, Kojima K, Rasmussen H (1985e) Characteristics of angiotensin II-, K^+- and ACTH-induced calcium influx in adrenal glomerulosa cells. Evidence that angiotensin II, K^+, and ACTH may open a common calcium channel. J Biol Chem 260:9171-9176

Kojima K, Kojima I, Rasmussen H (1984) Dihydropyridine calcium agonist and antagonist effects on aldosterone secretion. Am J Physiol 247:E645-E650

Kokko JP (1985) Primary acquired hypoaldosteronism. Kidney Int 27:690-702

Kolanowski J, Pizarro MA, de Gasparo M, Desmecht P, Harvengt C, Crabbé J (1970) Influence of fasting on adrenocortical and pancreatic islet response to glucose loads in the obese. Eur J Clin Invest 1:25-31

Kolena J (1968) Effect of ACTH on ^{14}C-progesterone hydroxylation in infantile and adult rat adrenals. Endocrinol Exp 2:199-205

Koletsky RJ, Brown EM, Williams GH (1983) Calmodulin-like activity and calcium-dependent phosphodiesterase in purified cells of the rat zona glomerulosa and zona fasciculata. Endocrinology 113:485-490

Komor J, Müller J (1979) Effects of prolonged infusions of potassium chloride, adrenocorticotrophin or angiotensin II upon serum aldosterone concentration and the conversion of corticosterone to aldosterone in rats. Acta Endocrinol 90:680-691

Kono T, Oseko F, Shimpo S, Nanno M, Endo J (1975) Biological activity of des-asp^1-angiotensin II (angiotensin III) in man. J Clin Endocrinol Metab 41:1174-1177

Kono T, Oseko F, Ikeda F, Nanno M, Endo J (1977) Biological activity of high dose of des-asp^1-ileu^8-angiotensin II in man. Acta Endocrinol 86:156-161

Kono T, Ikeda F, Oseko F, Nanno M, Imura H, Endo J (1978) Inhibition of angiotensin III action by des-asp^1-ileu^8-angiotensin II in man. Acta Endocrinol 87:359-366

Kono T, Ikeda F, Oseko F, Imura H, Endo J (1980) Biological activity of des-asp^1-angiotensin I in man. J Clin Endocrinol Metab 50:40-45

Kono T, Ikeda F, Oseko F, Ohmori Y, Nakano R, Muranaka H, Taniguchi A, Imura H, Khosla MC, Bumpus FM (1982) Biological activity of des-asp^1-, des-arg^2-angiotensin II in man. Acta Endocrinol 99:577-584

Kono T, Taniguchi A, Imura H, Oseko F, Khosla MC (1985a) Relative biological activities of asn^1, val -angiotensin II, ile^5-angiotensin II and sar^1-angiotensin II in man. Life Sci 37:365-369

Kono T, Ikeda F, Taniguchi A, Imura H, Oseko F, Yoshioka M, Khosla MC (1985b) Responses of patients with Bartter's syndrome to angiotensin III and angiotensin II-(3-8)-hexapeptide. Acta Endocrinol 109:249-253

Kornerup HJ, Pedersen EB (1977) Plasma renin, plasma aldosterone and exchangeable sodium in normotensives and hypertensive kidney transplant recipients with and without transplant renal artery stenosis. Acta Med Scand 202:509-516

Kosunen KJ (1977) Plasma renin activity, angiotensin II, and aldosterone after mental arithmetic. Scand J Lab Clin Invest 37:425-429

Kosunen KJ, Pakarinen AJ (1976) Plasma renin, angiotensin II, and plasma and urinary aldosterone in running exercise. J Appl Physiol 41:26-29

Kosunen KJ, Pakarinen A (1978) Correlations between plasma renin activity, angiotensin II, and aldosterone. J Clin Endocrinol Metab 47:665-666

Kosunen KJ, Pakarinen AJ, Kuoppasalmi K, Adlercreutz H (1976) Plasma renin activity, angiotensin II, and aldosterone during intense heat stress. J Appl Physiol 41:323-327

Kosunen KJ, Kuoppasalmi K, Näveri H, Rehunen S, Närvänen S, Adlercreutz H (1977) Plasma renin activity, angiotensin II, and aldosterone during the hypnotic suggestion of running. Scand J Clin Lab Invest 37:99-103

Kotchen TA, Lytton B, Morrow LB, Mulrow PJ, Shutkin PM, Stansel HC (1970) Angiotensin and aldosterone in renovascular hypertension. Arch Intern Med 125:265-272

Kotchen TA, Galla JH, Luke RG (1976) Failure of $NaHCO_3$ and $KHCO_3$ to inhibit renin in the rat. Am J Physiol 231:1950-1956

Kotchen TA, Galla JH, Luke RG (1977) Effects of calcium on renin and aldosterone in the rat. Am J Physiol 232:E388-E393

Kotchen TA, Krzyzaniak KE, Anderson JE, Ernst CB, Galla JH, Luke RG (1980) Inhibition of renin secretion by HCl is related to chloride in both dog and rat. Am J Physiol 239:F44-F49

Kowarski A, Finkelstein J, Loras B, Migeon CJ (1964) The in vitro stability of the tritium label in 1,2-^{3}H-D-aldosterone when used for the measurement of aldosterone secretion rate by the double isotope dilution technique. Steroids 3:95-107

Kowarski A, Finkelstein JW, Spaulding JS, Holman GH, Migeon CJ (1965) Aldosterone secretion rate in congenital adrenal hyperplasia. A discussion of the theories on the pathogenesis of the salt-losing form of the syndrome. J Clin Invest 44:1505-1513

Kowarski A, Russell A, Migeon CJ (1968) Aldosterone secretion rate in the hypertensive form of congenital adrenal hyperplasia. J Clin Endocrinol Metab 28:1445-1449

Kowarski A, Katz H, Migeon CJ (1974) Plasma aldosterone concentration in normal subjects from infancy to adulthood. J Clin Endocrinol Metab 38:489-491

Kowarski A, de Lacerda L, Migeon CJ (1975) Integrated concentrations of plasma aldosterone in normal subjects. J Clin Endocrinol Metab 40:205-210

Kraiem Z, Rosenthal T, Rotzak, Lunenfeld B (1979) Angiotensin II and K challenge following prolonged ACTH administration to normal subjects. Acta Endocrinol 91:657-665

Krakoff LR (1973) Measurement of plasma renin substrate by radioimmunoassay of angiotensin I: Concentration in syndromes associated with steroid excess. J Clin Endocrinol Metab 37: 110-117

Krakoff LR, Selvadurai R, Sutter E (1975) Effect of methylprednisolone upon arterial pressure and the renin angiotensin system in the rat. Am J Physiol 228:613-617

Kramer HJ, Goldman R, Gonick HC (1972) Salt-losing syndromes following renal transplantation. Am J Med 53:368-376

Kramer RE, Gallant S, Brownie AC (1979) The role of cytochrome P-450 in the action of sodium depletion on aldosterone biosynthesis in rats. J Biol Chem 254:3953-3958

Kramer RE, Gallant S, Brownie AC (1980) Actions of angiotensin II on aldosterone biosynthesis in the rat adrenal cortex. Effects on cytochrome P-450 enzymes of the early and late pathway. J Biol Chem 255:3442-3447

Krause JZ, Matarese RA, Zabetakis PM, Michelis MF (1980) Reversibility of hyporeninemia and hypoaldosteronemia in chronic hemodialysis patients by correction of fluid excess. J Lab Clin Med 96:734-742

Kraut JA, Wish JB, Zoller KM, Weinstein SS, Melby JC, Jordan JJ (1982) Aldosterone and the renal response to sulfuric acid feeding in the dog. Am J Physiol 243:F494-F502

Krieger DT, Krieger HP (1964) Aldosterone excretion in pretectal disease. J Clin Endocrinol Metab 24:1055-1066

Kubo S, Ganten D, Ganten U, Nowaczynski W, Genest J (1974) Plasma steroids in rats with experimental ectopic pituitary tumor and arterial hypertension. Endocrinology 94:459-463

Kuchel O, Cuche JL, Buu NT, Guthrie GP Jr, Unger T, Nowaczynski W, Boucher R, Genest J (1977) Catecholamine excretion in "idiopathic" edema: decreased dopamine excretion, a pathogenetic factor? J Clin Endocrinol Metab 44:639-646

Kuchel O, Buu NT, Hamet P, Nowaczynski W, Genest J (1979) Free and conjugated dopamine in pheochromocytoma, primary aldosteronism and essential hypertension. Hypertension 1: 267-273

Kuchel O, Buu NT, Vecsei P, Bourque M, Hamet P, Genest J (1980) Are plasma aldosterone surges in primary aldosteronism due to a loss of an inhibitory dopaminergic control? J Clin Endocrinol Metab 51:337-344
Kudo T, Baird A (1984) Inhibition of aldosterone production in the adrenal glomerulosa by atrial natriuretic factor. Nature 312:756-757
Kuhlmann U, Vetter W, Fischer E, Siegenthaler W (1978) Control of plasma aldosterone in diabetic patients with hyporeninemic hypoaldosteronism. Klin Wochenschr 56:229-234
Kuhn C III, Kissane JM (1964) Quantitative histochemistry of glucose-6-phosphate dehydrogenase and 6-phosphogluconate dehydrogenase in rat adrenal cortex: effect of ACTH, cortisone and sodium deprivation. Endocrinology 75:741-746
Kuhnle U, Chow D, Rapaport R, Pang S, Levine LS, New MI (1981) The 21-hydroxylase activity in the glomerulosa and fasciculata of the adrenal cortex in congenital adrenal hyperplasia. J Clin Endocrinol Metab 52:534-544
Kumar D, Feltham LW, Gornall AG (1959) Aldosterone excretion and tissue electrolytes in normal pregnancy and pre-eclampsia. Lancet 1:541
Kunita H, Obara T, Komatsu T, Hata S, Okamoto M (1976) The effects of dietary sodium on the diurnal activity of the renin-angiotensin-aldosterone system and the excretion of urinary electrolytes. J Clin Endocrinol Metab 43:756-759
Kurtz A, della Bruna R, Pfeilschifter J, Taugner R, Bauer C (1986) Atrial natriuretic peptide inhibits renin release from juxtaglomerular cells by a cGMP-mediated process. Proc Natl Acad Sci USA 83:4769-4773

Laidlaw JC, Cohen M, Gornall (1958) Studies on the origin of aldosterone during human pregnancy. J Clin Endocrinol Metab 18:222-225
Laidlaw JC, Yendt ER, Gornall AG (1960) Hypertension caused by renal artery occlusion simulating primary aldosteronism. Metabolism 9:612-623
Laidlaw JC, Ruse JL, Gornall AG (1962) The influence of estrogen and progesterone on aldosterone excretion. J Clin Endocrinol Metab 22:161-171
Lamberg B-A, Petterson T, Karlsson R (1964) The effect of corticotropin and synthetic angiotensin on the glucose-6-phosphate and the succinic acid dehydrogenase in the adrenal cortex of the rat. Acta Med Scand [Suppl] 412:215-220
Lambert F, Lammerant J, Kolanowski J (1986) The mechanism of the prolonged stimulatory effect of corticotrophin on pregnenolone production by guinea-pig adrenocortical mitochondria. J Endocrinol 108:377-384
Landau RL, Lugibihl K (1958) Inhibition of the sodium-retaining influence of aldosterone by progesterone. J Clin Endocrinol Metab 18:1237-1245
Landau RL, Bergenstal DM, Lugibihl K, Kascht ME (1955) The metabolic effects of progesterone in man. J Clin Endocrinol Metab 15:1194-1215
Landau R, Lugibihl K, Dimick DF (1958) Metabolic effects in man of steroids with progestational activity. Ann NY Acad Sci 71:588-598
Langford HG (1963) An explanation of hypertension produced by narrowing the renal artery. Perspect Biol Med 6:372-379
Laplante C, Giroud CJP, Stachenko J (1964) Lack of appreciable 17α-hydroxylase activity in the normal and regenerated rat adrenal cortex. Endocrinology 75:825-827
Laragh JH, Kelly WG (1964) Aldosterone: its biochemistry and physiology. In: Levine R, Luft R (eds) Advances in metabolic disorders, vol 1. Academic, New York, pp 217-262
Laragh JH, Stoerk HC (1957) A study of the mechanism of secretion of the sodium-retaining hormone (aldosterone). J Clin Invest 36:383-392
Laragh JH, Angers M, Kelly WG, Lieberman S (1960) Hypotensive agents and pressor substances. The effect of epinephrine, norepinephrine, angiotensin II and others on the secretory rate of aldosterone in man. JAMA 174:234-240
Laragh JH, Sealey JE, Sommers SC (1966) Patterns of adrenal secretion and urinary excretion of aldosterone and plasma renin activity in normal and hypertensive subjects. Circ Res 18/19 [Suppl 1]: I-158-I-174
Laragh JH, Sealey JE, Ledingham JGG, Newton MA (1967) Oral contraceptives. Renin, aldosterone, and high blood pressure. JAMA 201:918-922
Lasaridis AN, Tourkantonis A, Spanos P, Apostolopoulou K, Pharmakiotis A (1984) The effects of canrenoate K on corticosteroid biosynthesis in nephrectomised dogs. J Steroid Biochem 20: 923-929

Lauber M, Sugano S, Ohnishi T, Okamoto M, Müller J (1987) Aldosterone biosynthesis and cytochrome P-$450_{11\beta}$: evidence for two different forms of the enzyme in rats. J Steroid Biochem 26 (in press)
Lauer CG, Braley LM, Menachery AI, Williams GH (1982) Metoclopramide inhibits aldosterone biosynthesis in vitro. Endocrinology 111:238-243
Laulin JP, Ezzarani EA, Brudieux R (1984) Production of aldosterone in male rats with inherited diabetes insipidus (Brattleboro strain). Horm Metab Res 16:254-257
Layne DS, Meyer CJ, Vaishwanar PS, Pincus G (1962) The secretion and metabolism of cortisol and aldosterone in normal and in steroid-treated women. J Clin Endocrinol Metab 22:107-118
Lebel M, Grose JH (1976) Selective hypoaldosteronism: a study of steroid biosynthetic pathways under adrenocorticotrophin and angiotensin II infusion. Clin Sci Mol Med 51 [Suppl 3]: 335s-337s
Leboulenger F, Delarue C, Jégou S, Netchitailo P, Leroux P, Tonon MC, Vaudry H (1980) In vitro study of frog (*Rana ridibunda* Pallas) interrenal function by use of a simple perifusion system. III. Influence of previous hypophysectomy or dexamethasone treatment of donors upon corticosterone and aldosterone production. Gen Comp Endocrinol 42:389-400
Leboulenger F, Delarue C, Belanger A, Perroteau I, Netchitailo P, Leroux P, Jegou S, Tonon MC, Vaudry H (1982) Direct radioimmunoassays for plasma corticosterone and aldosterone in frog. I. Validation of the methods and evidence of daily rhythms in a natural environment. Gen Comp Endocrinol 46:521-532
Leboulenger F, Leroux P, Delarue C, Tonon MC, Charnay Y, Dubois PM, Coy DH, Vaudry H (1983) Co-localization of vasoactive intestinal peptide (VIP) and enkephalins in chromaffin cells of the adrenal gland of amphibia. Stimulation of corticosteroid production by VIP. Life Sci 32:375-383
Ledoux F, Genest J, Nowaczynski W, Kuchel O, Lebel M (1975) Plasma progesterone and aldosterone in pregnancy. Can Med Assoc J 112:943-947
Lee JB, Patak RV, Mookerjee BK (1976) Renal prostaglandins and the regulation of blood pressure and sodium and water homeostasis. Am J Med 60:798-816
Lee TC, de Wied D (1968) Somatotropin as the non-ACTH factor of anterior pituitary origin for the maintenance of enhanced aldosterone scretory responsiveness of dietary sodium restriction in chronically hypophysectomized rats. Life Sci 7:35-45
Lee TC, Biglieri EG, van Brunt EE, Ganong WF (1965) Inhibition of aldosterone secretion by passive transfer of antirenin antibodies to dogs on a low sodium diet. Proc Soc Exp Biol Med 119:315-318
Lee TC, van der Wal B, de Wied D (1968) Influence of the anterior pituitary on the aldosterone secretory response to dietary sodium restriction in the rat. J Endocrinol 42:465-475
Leenen FHH, Scheeren JW, Omylanowski D, Elema JD, van der Wal B, de Jong W (1975) Changes in the renin-angiotensin-aldosterone system an in sodium and potassium balance during development of renal hypertension in rats. Clin Sci Mol Med 48:17-26
Lefebvre J, Dewailly D, Racadot A, Liquette M (1979) Aldostérone et activité rénine plasmatiques dans l'hypopituitarisme antérieur. Ann Endocrinol 40:433-434
Lehoux J-G (1974) Aldosterone biosynthesis and presence of cytochrome P-450 in the adrenocortical tissue of the chick embryo. Mol Cell Endocrinol 2:43-58
Lever JD (1956) Cytological studies on the hypophysectomized rat adrenal cortex: the alterations of its fine structure following ACTH administration and on lowering the Na/K ratio. Endocrinology 58:163-180
Levine JH, Laidlaw JC, Ruse JL (1972) Effects of a sulfated mucopolysaccharide (Ro 1-8307) on zona glomerulosa width and steroid production by rat adrenal. Can J Physiol Pharmacol 50:-270-275
Levine LS, Rauh W, Gottesdiener K, Chow D, Gunczler P, Rapaport R, Pang S, Schneider B, New MI (1980) New studies of the 11β-hydroxylase and 18-hydroxylase enzymes in the hypertensive form of congenital adrenal hyperplasia. J Clin Endocrinol Metab 50:258-263
Lewis PS, Gorchein A, James VHT, May CN, Horth CE (1980) Long term spironolactone and the adrenal cortex in essential hypertension. Clin Endocrinol 13:273-283
Li CH, Ng TB, Cheng CHK (1982) Melanotropins: aldosterone- and corticosterone-stimulating activity in isolated rat adrenal cells. Int J Peptide Protein Res 19:361-365
Liard J-F, Cowley AW Jr, McCaa RE, McCaa CS, Guyton AC (1974) Renin, aldosterone, body fluid volumes, and the baroreceptor reflex in the development and reversal of Goldblatt hypertension in conscious dogs. Circ Res 34:549-560

Lichtenstein LS, Colwell JA, Levine JH (1976) Prolactin stimulates aldosterone biosynthesis. In: 5th International congress of endocrinology, Hamburg. p 86 (abstract)
Liddle GW (1959) Effects of anti-inflammatory steroids on electrolyte metabolism. Ann N Y Acad Sci 82:854-867
Liddle GW, Duncan LE Jr, Bartter FC (1956) Dual mechanism regulating adrenocortical function in man. Am J Med 21:380-386
Liddle GW, Bledsoe T, Coppage WS Jr (1964) A familial renal disorder simulating primary aldosteronism but with negligible aldosterone secretion. In: Baulieu EE, Robel P (eds) Aldosterone. Blackwell, Oxford, pp 353-368
Lieberman AH, Luetscher JA (1960) Some effects of abnormalities of pituitary, adrenal or thyroid function on excretion of aldosterone and the response to corticosterone or sodium deprivation. J Clin Endocrinol Metab 20:1004-1016
Lightman SL (1981) Studies on the responses of plasma renin activity and aldosterone and cortisol levels to dopaminergic and opiate stimuli in man. Clin Endocrinol 15:45-52
Lightman SL, James VHT, Linsell C, Mullen PE, Peart WS, Sever PS (1981) Studies of diurnal changes in plasma renin activity and plasma noradrenaline, aldosterone and cortisol concentrations in man. Clin Endocrinol 14:213-223
Lihrmann I, Netchitailo P, Leboulenger F, Delarue C, Vaudry H (1985) Effect of calcium on corticosteroid secretion by isolated frog interrenal gland. J Steroid Biochem 23:169-175
Lijnen PJ, Amery AK, Fagard RH, Reybrouck TM, Moerman EJ, de Schaepdryver AF (1979) The effects of β-adrenoceptor blockade on renin, angiotensin, aldosterone and catecholamines at rest and during exercise. Br J Clin Pharmacol 7:175-181
Lijnen P, Verschueren LJ, Fagard R, Staessen J, Amery A (1980) Acute changes in the renin-angiotensin-aldosterone system, catecholamines and prostaglandins during captopril in man. Meth Find Exp Clin Pharmacol 2:293-301
Limal J-M, Rappaport R, Bayard F (1977) Plasma aldosterone, renin activity, and 17α-hydroxyprogesterone in salt-losing congenital adrenal hyperplasia. I. Response to ACTH in hydrocortisone treated patients and effect of 9α-fluorocortisol. J Clin Endocrinol Metab 45:551-559
Lin S-Y, Goodfriend TL (1970) Angiotensin receptors. Am J Physiol 218:1319-1328
Lis M, Hamet P, Gutkowska J, Maurice G, Seidah NG, Larivière N, Chrétien M, Genest J (1981) Effect of N-terminal portion of pro-opiomelanocortin on aldosterone release by human adrenal adenoma in vitro. J Clin Endocrinol Metab 52:1053-1056
Llano M, Kolanowski J, Ortega N, Crabbé J (1982) Changes in corticosteroid secretory pattern induced by prolonged corticotropin treatment in the rabbit. J Steroid Biochem 17:631-638
Llanos AJ, Ramachandran J, Creasy RK, Rudolph AM, Séron-Ferré M (1979) α-Melanocyte-stimulating hormone and adrenocorticotropin in the regulation of glucocorticoid secretion during the perinatal period in sheep. Endocrinology 105:613-717
Lobo MV, Marusic ET (1982) Dissociation of aldosterone response to angiotensin II and potassium in the presence of strontium. J Steroid Biochem 16:629-632
Lobo MV, Marusic ET, Aguilera G (1978) Further studies on the relationship between potassium and sodium levels and adrenocortical activity. Endocrinology 102:1061-1068
Lohmeier TE, Davis JO (1976) Renin-angiotensin-aldosterone system in experimental renal hypertension in the rabbit. Am J Physiol 230:311-318
Lohmeier TE, Kastner PR, Smith MJ Jr, Guyton AC (1980) Is aldosteronism important in the maintenance of arterial blood pressure and electrolyte balance during sodium depletion? Hypertension 2:497-505
Lommer D (1966) Hemmung der Corticosteroid-11β-hydroxylierung durch einen Extrakt aus corpus pineale. Experientia 22:122
Lommer D, Wolff HP (1966) Stimulation of the in vitro biosynthesis of corticosteroids by angiotensin II. Experientia 22:699-700
Lommer D, Düsterdieck G, Jahnecke J, Vecsei P, Wolff HP (1968) Sekretion, Plasmakonzentration, Verteilung, Stoffwechsel und Ausscheidung von Aldosteron bei Gesunden und Kranken. Klin Wochenschr 46:741-751
Lommer D, Distler A, Müller B, Philipp T, Wolff HP (1972) Sekretion, Verteilung und Abbau von Aldosteron bei verschiedenen Hochdruckformen. Klin Wochenschr 50:1037-1042
Loras B, Haour F, Bertrand J (1967) Mesure du sodium échangeable et du taux de sécrétion d'aldosterone chez des enfants. CR Soc Biol 161:2490-2496
López JM, Rodriguez JA, Marusic ET (1980) Plasma aldosterone response to angiotensin II and potassium chloride in hypopituitary patients. Clin Endocrinol 13:331-337

Louis LH, Conn JW (1961) Primary aldosteronism: content of adrenocortical steroids in adrenal tissue. Recent Prog Horm Res 17:415-422
Lowder SC, Brown RD (1975) Hypertension corrected by discontinuing chronic sodium bicarbonate ingestion. Subsequent transient hypoaldosteronism. Am J Med 58:272-279
Lowenstein J, Steele JM Jr (1977) Aldosterone production during dietary sodium restriction and beta-adrenergic blockade. Kidney Int 11:128-135
Lowenstein J, Boyd GW, Rippon AE, James VHT, Peart WS (1972) Increased aldosterone in response to sodium deficiency in the angiotensin II-immunized rabbit. In: Genest J, Koiw E (eds) Hypertension '72. Springer, Berlin Heidelberg New York, pp 481-489
Lucis OJ, Venning EH (1960) In vitro and in vivo effect of growth hormone on aldosterone secretion. Can J Biochem Physiol 38:1069-1075
Lucis OJ, Dyrenfurth I, Venning EH (1961) Effect of various preparations of pituitary and diencephalon on the in vitro secretion of aldosterone and corticosterone by the rat adrenal gland. Can J Biochem Physiol 39:901-913
Lucis R, Carballeira A, Venning EH (1965) Biotransformation of progesterone-4-^{14}C and 11-deoxycorticosterone-4-^{14}C by rat adrenal glands in vitro. Steroids 6:737-756
Lucke C, Erbler H, Herting T, Döhler KD (1980) Secretion of arginin-vasopressin, aldosterone and corticosterone and plasma-renin activity in water-deprived rats. Contrib Nephrol 19:63-70
Luetscher JA Jr, Axelrad BJ (1954) Increased aldosterone output during sodium deprivation in normal men. Proc Soc Exp Biol Med 87:650-653
Luetscher JA Jr, Lieberman AH (1958) Aldosterone. Arch Intern Med 102:314-330
Luetscher JA, Cohn AP, Camargo CA, Dowdy AJ, Callaghan AM (1963) Aldosterone secretion and metabolism in hyperthyroidism and myxedema. J Clin Endocrinol Metab 23:873-880
Luetscher JA, Weinberger MH, Dowdy AJ, Nokes GW, Balikian HM, Brodie AH, Willoughby S (1969) Effects of sodium loading, sodium depletion and posture on plasma aldosterone concentration and renin activity in hypertensive patients. J Clin Endocrinol Metab 29:1310-1318
Lun S, Espiner EA, Hart DS (1979) Effect of ACTH on the aldosterone response to potassium in sheep with adrenal transplants. Am J Physiol 237:E500-E503
Lun S, Espiner EA, Nicholls MG, Hart DS (1983) Lack of direct effect of dopamine on aldosterone secretion in vivo. Endocrinology 112:60-63
Lund JO, Nielsen MD (1980) Fludrocortisone suppression test in normal subjects, in patients with essential hypertension and in patients with various forms of aldosteronism. Acta Endocrinol 93:100-107
Lustyik Gy, Szabo J, Gláz, Kiss R (1977) Electron microscopic investigation on the zona glomerulosa of rats after treatment with heparin and potassium. Exp Pathol 13:189-197
Lymangrover JR, Keku E, Eldridge JC (1983) Naloxone potentiates ACTH and angiotensin II but not potassium stimulated aldosterone secretion in vitro. Life Sci 33:1605-1612

Maack T, Marion DN, Camargo MJF, Kleinert HD, Laragh JH, Vaughan ED Jr, Atlas SA (1984) Effects of auriculin (atrial natriuretic factor) on blood pressure, renal function, and the renin-aldosterone system in dogs. Am J Med 77:1069-1075
Maack T, Camargo MJF, Kleinert H, Laragh JH, Atlas SA (1985) Atrial natriuretic factor: Structure and functional properties. Kidney Int 27:607-615
Macchi IA, Phillips JG (1966) In vitro effect of adrenocorticotropin on corticoid secretion in the turtle, snake and bullfrog. Gen Comp Endocrinol 6:170-182
Macdonald TL, Burka LT, Wright ST, Guengerich FP (1982) Mechanism of hydroxylation by cytochrome P-450: exchange of iron-oxygen intermediates with water. Biochem Biophys Res Commun 104:620-625
MacGregor GA, Markandu ND, Roulston JE, Jones JC, Morton JJ (1981) Maintenance of blood pressure by the renin-angiotensin system in normal man. Nature 291:329-331
Mackie CM, Simpson ER, Mee MSR, Tait SAS, Tait JF (1977) Intracellular potassium and steroidogenesis of isolated rat adrenal cells: effect of potassium ions and angiotensin II on purified zona glomerulosa cells. Clin Sci Mol Med 53:289-296
Mackie C, Warren RL, Simpson ER (1978) Investigations into the role of calcium ions in the control of steroid production by isolated adrenal zona glomerulosa cells of the rat. J Endocrinol 77:119-127
Maher JT, Jones LG, Hartley LH, Williams GH, Rose LI (1975) Aldosterone dynamics during graded exercise at sea level and high altitude. J Appl Physiol 39:18-22
Major P, Kuchel O, Boucher R, Nowaczynski W, Genest J (1978) Selective hypopituitarism with severe hyponatremia and secondary hyporeninism. J Clin Endocrinol Metab 46:15-19

Malinowska KW, Nathanielz PW (1974) Plasma aldosterone, cortisol and corticosterone concentrations in the new-born guinea-pig. J Physiol 236:83-93
Mantero F, Gion M, Armanini D, Opocher G (1977) Effect of bromocriptine on plasma aldosterone in primary aldosteronism. Acta Endocrinol [Suppl] 212:94 (abstract)
Mantero F, Armanini D, Boscaro M (1978) Plasma renin activity and urinary aldosterone in Cushing's syndrome. Horm Metab Res 10:65-71
Mantero F, Opocher G, Armanini D, Paviotti G, Boscaro M, Muggeo M (1979) Plasma renin activity and urinary aldosterone in acromegaly. J Endocrinol Invest 2:13-18
Mantero F, Boscaro M, Scaroni C, Armanini D, Sonino N, Masarotto P, Ridolfi P (1980) Effect of a synthetic substituted α1-18 ACTH on mineralocorticoid secretion. Horm Metab Res 12: 464-470
Mantero F, Opocher G, Boscaro M, Armanini D (1982) Effect of serotonin on plasma aldosterone in man. J Endocrinol Invest 5:97-99
Manuelidis L, Mulrow P (1973) ACTH effects on aldosterone production and mitochondrial ultrastructure in adrenal gland cultures. Endocrinology 93:1104-1108
Manuelidis L, Mulrow P (1974) Aldosterone production by adrenal cultures via dbcAMP, and the reversibility of ACTH and dbcAMP-induced changes. Endocrinology 95:728-740
Maran JW, Yates FE (1977) Cortisol secretion during intrapituitary infusion of angiotensin II in conscious dogs. Am J Physiol 233:E273-E285
Maresh CM, Noble BJ, Robertson KL, Seip RL (1984) Adrenocortical responses to maximal exercise in moderate-altitude natives at 447 Torr. J Appl Physiol 56:482-488
Marie J, Jard S (1983) Angiotensin II inhibits adenylate cyclase from adrenal cortex glomerulosa zone. FEBS Lett 159:97-101
Marieb NJ, Mulrow PJ (1965) Role of the renin-angiotensin system in the regulation of aldosterone secretion in the rat. Endocrinology 76:657-664
Marks P, Anderson J, Vincent R (1978) Aldosterone in myxoedema. Lancet 2:1277-1278
Marone C, Luisoli S, Bomio F, Beretta-Piccoli C, Bianchetti MG, Weidmann P (1985) Body sodium-blood volume state, aldosterone, and cardiovascular responsiveness after calcium entry blockade with nifedipine. Kidney Int 28:658-665
Martin IH, Baulan D, Basso N, Taquini AC (1982) The renin-angiotensin-aldosterone system in rats of both sexes subjected to chronic hypobaric hypoxia. Arch Int Physiol Biochim 90: 129-133
Martin VI, Edwards CRW, Biglieri EG, Vinson GP, Bartter FC (1975) The development and application of a radioimmunoassay for 18-hydroxy-corticosterone. Steroids 26:591-604
Marusic ET, Mulrow PJ (1967a) In vitro conversion of corticosterone-4-^{14}C to 18-hydroxycorticosterone by zona fasciculata-reticularis of beef adrenal. Endocrinology 80:214-218
Marusic ET, Mulrow PJ (1967b) Stimulation of aldosterone biosynthesis in adrenal mitochondria by sodium depletion. J Clin Invest 46:2101-2108
Marusic ET, Mulrow PJ (1969) 18-Hydroxycorticosterone biosynthesis in an aldosterone-secreting tumor and in the surrounding non-tumorous adrenal gland. Proc Soc Exp Biol Med 131: 778-780
Marusic ET, Mulrow PJ (1972) The effect of puromycin on the biosynthesis of aldosterone in isolated adrenal mitochondria. Endocrinology 91:1125-1129
Marusic ET, Galli SM, Aguilera G (1973a) Control of aldosterone biosynthesis. In: Scow RO (ed) Endocrinology. Proceedings of the 4th international congress of endocrinology, Washington DC 1972. Excerpta Medica, Amsterdam, pp 774-779
Marusic E, White A, Aedo AR (1973b) Oxidative reactions in the formation of an aldehyde group in the biosynthesis of aldosterone. Arch Biochem Biophys 157:320-321
Marx AJ, Deane HW (1963) Histophysiologic changes in the kidney and adrenal cortex in rats on a low-sodium diet. Endocrinology 73:317-328
Marx AJ, Deane HW, Mowles TF, Sheppard H (1963) Chronic administration of angiotensin in rats: changes in blood pressure, renal and adrenal histophysiology and aldosterone production. Endocrinology 73:329-337
Maser-Gluth C, Toygar A, Vecsei P (1984) Time course of plasma corticosterone, 18-hydroxycorticosterone and aldosterone concentrations following CRF administration in the rat. A phase of corticosterone inhibition. Life Sci 35:879-884
Mason PA, Fraser R, Morton JJ, Semple PF, Wilson A (1976) The effect of angiotensin II infusion on plasma corticosteroid concentrations in normal man. J Steroid Biochem 7:859-861
Mason PA, Fraser R, Morton JJ, Semple PF, Wilson A (1977) The effect of sodium deprivation

and of angiotensin II infusion on the peripheral plasma concentrations of 18-hydroxycorticosterone, aldosterone and other corticosteroids in man. J Steroid Biochem 8:799-804
Mason PA, Fraser R, Semple PF, Morton JJ (1979) The interaction of ACTH and angiotensin II in the control of corticosteroid plasma concentration in man. J Steroid Biochem 10:235-239
Masson GMC, Travis RH (1968) Effects of renin on aldosterone secretion in the rat. Can J Physiol Pharmacol 46:11-14
Mathias CJ, Christensen NJ, Corbett JL, Frankel HL, Goodwin TJ, Peart WS (1975) Plasma catecholamines, plasma renin activity and plasma aldosterone in tetraplegic man, horizontal and tilted. Clin Sci Mol Med 49:291-299
Matsuda S (1984) Interrelationships between the renin-angiotensin-aldosterone system and sympathetic and dopaminergic activities in elderly subjects. Clin Exp Hypertension (A) 6: 1431-1440
Matsuoka H, Tan SY, Mulrow PJ (1980a) Effects of prostaglandins on adrenal steroidogenesis in the rat. Prostaglandins 19:291-298
Matsuoka H, Mulrow PJ, Li CH (1980b) β-Lipotropin: a new aldosterone-stimulating factor. Science 209:307-308
Matsuoka H, Mulrow PJ, Franco-Saenz R, Li CH (1981a) Stimulation of aldosterone production by β-melanotropin. Nature 291:155-156
Matsuoka H, Mulrow PJ, Franco-Saenz R, Li CH (1981b) Effects of β-lipotropin and β-lipotropin-derived peptides on aldosterone production in the rat adrenal gland. J Clin Invest 68: 752-759
Matsuoka H, Ishii M, Yamakado M, Uehara Y (1984) Effect of natrium ionophore on aldosterone production in the rat adrenal gland. Endocrinology 115:2039-2044
Matsuoka H, Ishii M, Goto A, Sugimoto T (1985a) Role of serotonin type 2 receptors in regulation of aldosterone production. Am J Physiol 249:E234-E238
Matsuoka H, Ishii M, Sugimoto T, Hirata Y, Sugimoto T, Kangawa K, Matsuo H (1985b) Inhibition of aldosterone production by α-human atrial natriuretic polypeptide is associated with an increase in cGMP production. Biochem Biophys Res Commun 127:1052-1056
Mayberry HE, van den Brande JL, Van Wyk JJ, Waddell WJ (1971) Early localization of ^{125}I-labeled human growth hormone in adrenals and other organs of immature hypophysectomized rats. Endocrinology 88:1309-1317
Mayes D, Furuyama D, Kem DC, Nugent CA (1970) A radioimmunoassay for plasma aldosterone. J Clin Endocrinol Metab 30:682-685
Mazzocchi G, Nussdorfer GG, Rebuffat P, Belloni AS, Meneghelli V (1974) Further investigations on the trophic effect of 3':5'-cyclic adenosine monophosphate and 3':5'-cyclic guanosine monophosphate on the zona glomerulosa of hypophysectomized rat adrenal cortex. Experientia 30:87-88
Mazzocchi G, Rebuffat P, Belloni AS, Robba C, Nussdorfer GG (1980) An ultrastructural stereologic study of the effects of angiotensin II on the zona glomerulosa of rat adrenal cortex. Acta Endocrinol 95:523-527
Mazzocchi G, Robba C, Rebuffat, Nussdorfer GG (1982) Effects of sodium repletion and timolol maleate administration on the zona glomerulosa of the rat adrenal cortex: an electron microscopic morphometric study. Endokrinologie 79:81-88
Mazzocchi G, Robba C, Rebuffat P, Gottardo G, Nussdorfer GG (1985) Effect of somatostatin on the zona glomerulosa of rats treated with angiotensin II or captopril: stereology and plasma hormone concentrations. J Steroid Biochem 23:353-356
Mazzocchi G, Robba C, Rebuffat P, Nussdorfer GG (1986) Effects of prolactin administration on the zona glomerulosa of the rat adrenal cortex: stereology and plasma hormone concentrations. Acta Endocrinol 111:101-105
McAreavey D, Cumming AMM, Boddy K, Brown JJ, Fraser R, Leckie BJ, Lever AF, Morton JJ, Robertson JIS, Williams ED (1983) The renin-angiotensin system and total body sodium and potassium in hypertensive women taking oestrogen-progestagen oral contraceptives. Clin Endocrinol 18:111-118
McCaa CS, Richardson TQ, McCaa RE, Sulya LL, Guyton AC (1965) Aldosterone secretion by dogs during the developmental phase of Goldblatt hypertension. J Endocrinol 33:97-102
McCaa RE (1977) Role of the renin-angiotensin system in the regulation of aldosterone biosynthesis and arterial pressure during sodium deficiency. Circ Res 40 [Suppl 1]:I-157-I-162
McCaa RE (1978) Aldosterone response to long term infusion of angiotensin II and angiotensin III in conscious dogs before and after dietary sodium restriction. Endocrinology 103: 458-464

McCaa RE, McCaa CS, Read DG, Bower JD, Guyton AC (1972) Increased plasma aldosterone concentration in response to hemodialysis in nephrectomized man. Circ Res 31:473-480
McCaa RE, McCaa CS, Cowley AW Jr, Ott CE, Guyton AC (1973a) Stimulation of aldosterone secretion by hemorrhage in dogs after nephrectomy and decapitation. Circ Res 32:356-362
McCaa RE, Bower JD, McCaa CS (1973b) Relative influence of acute sodium and volume depletion on aldosterone secretion in nephrectomized man. Circ Res 33:555-562
McCaa RE, Read VH, Cowley AW Jr, Bower JD, Smith GV, McCaa CS (1973c) Influence of acute stimuli on plasma aldosterone concentration in anephric man and kidney allograft recipients. Circ Res 33:313-322
McCaa RE, Young DB, Guyton AC, McCaa CS (1974) Evidence for a role of an unidentified pituitary factor in regulating aldosterone secretion during altered sodium balance. Circ Res 34/35 [Suppl 1]: I-15-I-25
McCaa RE, McCaa CS, Guyton AC (1975) Role of angiotensin II and potassium in the long-term regulation of aldosterone secretion in intact conscious dogs. Circ Res 36/37 [Suppl I]: I-57-I-67
McCaa RE, Hall JE, McCaa CS (1978a) The effects of angiotensin I-converting enzyme inhibitors on arterial blood pressure and urinary sodium excretion. Role of the renal renin-angiotensin and kallikrein-kinin systems. Circ Res 43 [Suppl 1]: I-32-I-39
McCaa RE, Montalvo JM, McCaa CS (1978b) Role of growth hormone in the regulation of aldosterone biosynthesis. J Clin Endocrinol Metab 46:247-253
McCaa RE, McCaa CS, Bengis RG, Guyton AC (1979) Role of aldosterone in experimental hypertension. J Endocrinol 81:69*P*-78*P*
McCaa RE, Langford HG, Montalvo JM, Andy OJ, Read VH, McCaa CS (1981) Regulation of aldosterone biosynthesis during sodium deficiency. Evidence for an essential role of the pituitary gland. Hypertension 3 [Suppl 1]: I-74-I-80
McCarty R, Kirby RF, Carey RM (1984) Dopamine may be a neurohormone in rat adrenal cortex. Am J Physiol 247: E709-E713
McDougall JG, Coghlan JP, McGarry EE, Scoggins BA (1976) Aldosterone production by sheep adrenal glands in vitro. J Steroid Biochem 7:421-428
McDougall JG, Butkus A, Coghlan JP, Denton DA, Müller J, Oddie CJ, Robinson PM, Scoggins BA (1980) Biosynthetic and morphological evidence for inhibition of aldosterone production following administration of ACTH to sheep. Acta Endocrinol 94:559-570
McDougall JG, Scoggins BA, Butkus A, Coghlan JP, Denton DA, Fei DT, Hardy KJ, Wright RD (1981) Dopaminergic modulation of aldosterone secretion? J Endocrinol 91:271-280
McKeever A, Oliver JA, Henderson IW, Mosley W (1983) Influence of captopril on fluid and electrolyte balances and adrenocortical responses during sodium deprivation in the rat. J Endocrinol 98:211-220
McKenna TJ, Island DP, Nicholson WE, Liddle GW (1978a) Angiotensin stimulates cortisol biosynthesis in human adrenal cells in vitro. Steroids 32:127-136
McKenna TJ, Island DP, Nicholson WE, Liddle GW (1978b) Angiotensin stimulates both early and late steps in aldosterone biosynthesis in isolated bovine glomerulosa cells. J Steroid Biochem 9:967-972
McKenna TJ, Island DP, Nicholson WE, Liddle GW (1978c) The effects of potassium on early and late steps in aldosterone biosynthesis in cells of the zona glomerulosa. Endocrinology 103: 1411-1416
McKenna TJ, Island DP, Nicholson WE, Miller RB, Lacroix A, Liddle GW (1979a) ACTH stimulates late steps in cortisol biosynthesis. Acta Endocrinol 90:122-132
McKenna TJ, Island DP, Nicholson WE, Liddle GW (1979b) Dopamine inhibits angiotensin-stimulated aldosterone biosynthesis in bovine adrenal cells. J Clin Invest 84:287-291
McKenna TJ, Island DP, Nicholson WE, Liddle GW (1980) Stimulation of aldosterone production by catecholamines in vitro. Abstract No 395, 6th Int Congr Endocrinol, Melbourne
McKenzie JK, Montgomerie JZ (1969) Renin-like activity in the plasma of anephric men. Nature 223:1156
Mehdi AZ, Sandor T (1977) The effect of melatonin on the biosynthesis of corticosteroids in beef adrenal preparations in vitro. J Steroid Biochem 8:821-823
Melada GA, Goldman RH, Luetscher JA, Zager PG (1975) Hemodynamics, renal function, plasma renin, and aldosterone in man after 5 to 14 days of bedrest. Aviat Space Environ Med 46:1049-1055
Melby JC, Wilson TE, Dale SL, Hickler RV (1965) Extrarenal regulation of aldosterone secretion in primary aldosteronism. J Clin Invest 44:1074-1075 (abstract)

Melby JC, Spark RF, Dale SL, Egdahl RH, Kahn PC (1967) Diagnosis and localization of aldosterone-producing adenomas by adrenal vein catheterization. N Engl J Med 277:1050-1056
Melby JC, Dale SL, Grekin RJ, Gaunt R, Wilson TE (1972) 18-Hydroxy-11-deoxycorticosterone (18-OH-DOC) secretion in experimental and human hypertension. Recent Prog Horm Res 28: 287-339
Menard J, Meyer P, Milliez P (1967) L'hypertension artérielle d'origine rénale: Sécrétion inappropriée de rénine. Presse Med 75:2083-2085
Mendelsohn FAO (1980) Different potencies of [asp[1], ile[5]]- and [asn[1], val[5]]-angiotensin II in stimulating aldosterone production from rat adrenal zona glomerulosa cells in vitro. Clin Exp Pharmacol Physiol 7:199-203
Mendelsohn FAO, Kachel CD (1980) Action of angiotensins I, II, and III on aldosterone production by isolated rat adrenal zona glomerulosa cells: Importance of metabolism and conversion of peptides in vitro. Endocrinology 106:1760-1768
Mendelsohn FAO, Kachel CD (1981) Stimulation by serum of aldosterone production from rat adrenal glomerulosa cells in vitro: relationships to K^+, serotonin and angiotensin II. Acta Endocrinol 97:231-242
Mendelsohn FA, Mackie C (1975) Relation of intracellular K^+ and steroidogenesis in isolated adrenal zona glomerulosa and fasciculata cells. Clin Sci Mol Med 49:13-26
Mendelsohn FAO, Warren RL (1975) Micromethods for the simultaneous measurement of intracellular potassium and corticosterone output of isolated adrenal cortical cells. Clin Sci Mol Med 49:1-12
Mendelsohn FAO, Johnston CI, Doyle AE, Scoggins BA, Denton DA, Coghlan JP (1972) Renin, angiotensin II, and adrenal corticosteroid relationships during sodium deprivation and angiotensin infusion in normotensive and hypertensive man. Circ Res 31:728-739
Merriam GR, Baer L (1980) Adrenocorticotropin deficiency: correction of hyponatremia and hypoaldosteronism with chronic glucocorticoid therapy. J Clin Endocrinol Metab 50:10-14
Mersey JH, Williams GH, Hollenberg NK, Dluhy RG (1977) Relationship between aldosterone and bradykinin. Circ Res 40 [Suppl 1]: I-84-I-88
Mesbah S, Brudieux R (1982) Diurnal variation of plasma concentrations of cortisol, aldosterone and electrolytes in the ram. Horm Metab Res 14:320-323
Messerli FH, Nowaczynski W, Honda M, Genest J, Kuchel O (1976) Effects of ACTH on steroid metabolism. J Clin Endocrinol Metab 42:1074-1080
Messerli FH, Nowaczynski W, Honda M, Genest J, Boucher R, Kuchel O, Rojo-Ortega JM (1977) Effects of angiotensin II on steroid metabolism and hepatic blood flow in man. Circ Res 40:204-207
Messerli FH, Weidmann P, DeChâtel R, Maxwell MH (1978) Responsiveness of plasma aldosterone: dependency upon basal secretory activity. Klin Wochenschr 56:719-726
Meuli C, Müller J (1982) Characterization of rat capsular adrenal (zona glomerulosa) (Na,K)-ATPase activity. J Steroid Biochem 16:129-131
Meuli C, Müller J (1983 a) Potassium intake and aldosterone biosynthesis: the role of cytochrome P-450. J Steroid Biochem 18:167-171
Meuli C, Müller J (1983 b) A potassium-induced mitochondrial protein related to aldosterone biosynthesis. Am J Physiol 245:E449-E456
Meuli C, Müller J (1984) Sodium depletion of rats induces a mitochondrial protein in the zona glomerulosa of the adrenal cortex. J Steroid Biochem 21:773-775
Meyer CJ, Layne DS, Tait JF, Pincus G (1961) The binding of aldosterone to plasma proteins in normal, pregnant, and steroid-treated women. J Clin Invest 40:1663-1671
Michelakis AM, Horton R (1970) The relationship between plasma renin and aldosterone in normal man. Circ Res 26/27 [Suppl 1]: I-185-I-194
Michelakis AM, Robertson D (1970) Plasma renin activity and levodopa in Parkinson's disease. JAMA 213:83-85
Michelakis AM, Yoshida H, Dormois JC (1975) Plasma renin activity and plasma aldosterone during the normal menstrual cycle. Am J Obstet Gynecol 123:724-726
Milla PJ, Trompeter R, Dillon MJ, Robins D, Shackleton C (1977) Salt-losing syndrome in two infants with defective 18-dehydrogenation in aldosterone biosynthesis. Arch Dis Child 52: 580-586
Millar JA, Jones RI, Reid JL (1983) Somatostatin attenuates aldosterone responses to angiotensin II in normal subjects. Proc Univ Otago Med School (Dunedin NZ) 61:79-81
Milledge JS, Catley DM (1982) Renin, aldosterone, and converting enzyme during exercise and acute hypoxia in humans. J Appl Physiol 52:320-323

Milledge JS, Bryson EI, Catley DM, Hesp R, Luff N, Minty BD, Older MWJ, Payne NN, Ward MP, Withey WR (1982) Sodium balance, fluid homeostasis and the renin-aldosterone system during the prolonged exercise of hill walking. Clin Sci 62:595-604

Milledge JS, Catley DM, Ward MP, Williams ES, Clarke CRA (1983a) Renin, aldosterone and angiotensin converting enzyme during prolonged altitude exposure. J Appl Physiol 55:699-702

Milledge JS, Catley DM, Blume FD, West JB (1983b) Renin, angiotensin-converting enzyme, and aldosterone in humans on Mount Everest. J Appl Physiol 55:1109-1112

Miller PD, Waterhouse C, Owens R, Cohen E (1975) The effect of potassium loading on sodium excretion and plasma renin activity in Addisonian man. J Clin Invest 56:346-353

Miller PD, Dubovsky KL, McDonald KM, Katz FH, Robertson GL, Schrier RW (1979) Central, renal and adrenal effects of lithium in man. Am J Med 66:797-803

Miller RA (1965) Effect of hypophysectomy and a low sodium diet on the volumes of the glomerular zones, fascicular and reticular zones and medulla of rats. Acta Endocrinol 50:486-492

Miller RT, Douglas JG, Dunn MJ (1980) Dissociation of aldosterone and prostaglandin biosynthesis in the rat adrenal glomerulosa. Prostaglandins 20:449-462

Missale C, Liberini, Memo M, Caaruba MO, Spano P (1985) Identification of D-2 dopaminergic receptors in bovine adrenal cortex. Life Sci 37:2539-2548

Mitchell JD, Baxter TJ, Blair-West JR, McCredie DA (1970) Renin levels in nephroblastoma (Wilms' tumour). Arch Dis Child 45:376-384

Mitra S, Genuth SM, Berman LB, Vertes V (1972) Aldosterone secretion in anephric patients. N Engl J Med 286:61-64

Miura K, Yoshinaga K, Goto K, Katsushima I, Maebashi M, Demura H, Iino M, Demura R, Torikai T (1968) A case of glucocorticoid-responsive hyperaldosteronism. J Clin Endocrinol Metab 28:1807-1815

Miyamori I, FitzGerald GA, Brown MJ, Lewis PJ (1979) Prostacyclin stimulates the renin angiotensin aldosterone system in man. J Clin Endocrinol Metab 49:943-944

Mizuno K, Yamazaki M, Ikeda K, Yaginuma K, Hashimoto S, Fukuchi S (1983) Effects of metoclopramide, a dopamine antagonist, on secretion of aldosterone and renin release in patients with primary aldosteronism. Jpn Heart J 24:917-924

Modigliani E, Delzant G, Sebaoun J (1972) Étude de la fonction minéralotrope cortico-surrénalien dans l'acromégalie. A propos de 15 observations. Ann Endocrinol 33:301-303

Möhring B, Möhring J, Dauda G, Haack D (1974) Potassium deficiency in rats with hereditary diabetes insipidus. Am J Physiol 227:916-920

Möhring J, Möhring B (1972) Reevaluation of DOCA escape phenomenon. Am J Physiol 223: 1237-1245

Möhring J, Siegenthaler W, Weidmann P (1968) Verkürzte biologische Halbwertszeit von 1,2-^{3}H-Aldosteron bei drei nephrektomierten Patienten. Klin Wochenschr 46:47-49

Möhring J, Möhring B, Petri M, Saur W, Haack D, Hackenthal E (1976) Sodium and water balance, ECFV and BV, and hormones in the pathogenesis of renal hypertension in rats. In: Giovannetti S, Bonomini V, d'Amico G (eds) Proceedings of the 6th international congress of nephrology, Florence 1975. Karger, Basel, pp 255-265

Mogil RA, Itskovitz HD, Russell JH, Murphy JJ (1969) Renal innervation and renin activity in salt metabolism and hypertension. Am J Physiol 216:693-697

Molhuysen JA, Gerbrandy J, de Vries LA, de Jong JC, Lenstra JB, Turner KP, Borst JGG (1950) A liquorice extract with deoxycortone-like action. Lancet 2:381-386

Moncaup M, Giry J, Barlet J-P, Delost P (1980) Aldosterone metabolism in pregnant ewes and fetal and newborn lambs. Reprod Nutr Dev 20:277-286

Montiel M, Jiménez E, Narvaez JA, Morell M (1983) Renin-angiotensin-aldosterone system in hyper- and hypothyroid rats during sodium depletion. Endocr Res Commun 9:249-260

Moore TJ, Braley LM, Williams GH (1978) Stimulation of 18-hydroxy-11-deoxycorticosterone secretion by angiotensin II and potassium. Endocrinology 103:152-155

Moore TJ, Thein-Wai W, Dluhy RG, Dawson-Hughes BF, Hollenberg NK, Williams GH (1980) Abnormal adrenal and vascular responses to angiotensin II and an angiotensin antagonist in acromegaly. J Clin Endocrinol Metab 51:215-222

Morgan ET, Koop DR, Coon MJ (1982) Catalytic activity of cytochrome P-450 isozyme 3a isolated from liver microsomes of ethanol treated rabbits. J Biol Chem 257:13951-13957

Morganti A, Sealey JS, Lopez-Ovejero JA, Pickering TG, Laragh JH (1979) The substitutive role of ACTH in supporting aldosterone response to head-up tilt during acute renin suppression in patients with essential hypertension. Hypertension 1:130-135

Mori M, Kobayashi I, Ohshima K, Maruta S, Shimomura Y, Fukuda H (1980) Potentiation of sulpiride-induced prolactin secretion by sodium deprivation in man. Acta Endocrinol 94: 25-29

Morimoto S, Takeda R, Murakami M (1970) Does prolonged pretreatment with large doses of spironolactone hasten a recovery from juxtaglomerular-adrenal suppression in primary aldosteronism? J Clin Endocrinol Metab 31:659-664

Morimoto S, Kim KS, Yamamoto I, Uchida K, Takeda R, Kornel L (1979) Selective hypoaldosteronism with hyperreninemia in a diabetic patient. J Clin Endocrinol Metab 49:742-747

Moriwaki K, Iida S, Gomi M, Tsugawa M, Itoh Y (1985) Gamma-MSH and its related peptides do not augment ACTH-induced steroidogenesis in isolated rat adrenal cells. Endocrinol Jpn 32:563-568

Morris RC Jr, Sebastian A (1983) Renal tubular acidosis and Fanconi syndrome. In: Stanbury JB, Wyngaarden JB, Fredrickson DS, Goldstein JL, Brown MS (eds) The metabolic basis of inherited disease, 5th edn. McGraw-Hill, New York, pp 1808-1843

Moss S, Oster JR, Perez GO, Katz FH, Vaamonde CA (1978) Renin-aldosterone responsiveness in uncomplicated juvenile-type diabetes mellitus. Horm Res 9:130-136

Mouw DR, Vander AJ (1970) Evidence for brain Na receptors controlling renal Na excretion and plasma renin activity. Am J Physiol 219:822-832

Mouw DR, Vander AJ (1971) Evidence for hormonal mediation of the renal response to low-sodium stimulation of the brain. Proc Soc Exp Biol Med 137:179-182

Müller J (1965a) Aldosterone stimulation in vitro. I. Evaluation of assay procedure and determination of aldosterone-stimulating activity in a human urine extract. Acta Endocrinol 48: 283-296

Müller J (1965b) Stimulation of aldosterone production in vitro by ammonium ions. Nature 206: 92

Müller J (1965c) Aldosterone stimulation in vitro. II. Stimulation of aldosterone production by monovalent cations. Acta Endocrinol 50:301-309

Müller J (1966) Aldosterone stimulation in vitro. III. Site of action of different aldosterone-stimulating substances on steroid biosynthesis. Acta Endocrinol 52:515-526

Müller J (1968) Alterations of aldosterone biosynthesis by rat adrenal tissue due to increased intake of sodium and potassium. Acta Endocrinol 58:27-37

Müller J (1970a) Decreased aldosterone production by rat adrenal tissue in vitro due to treatment with 9α-fluorocortisol, dexamethasone and adrenocorticotrophin in vivo. Acta Endocrinol 63: 1-10

Müller J (1970b) Steroidogenic effect of stimulators of aldosterone biosynthesis upon separate zones of the rat adrenal cortex. Influence of sodium and potassium deficiency. Eur J Clin Invest 1:180-187

Müller J (1971) Regulation of aldosterone biosynthesis. In: Gross F, Labhart A, Mann T, Samuels LT, Zander J (eds) Monographs on endocrinology, vol 5. Springer, Berlin Heidelberg New York

Müller J (1972) Gibt es ein Renin-Angiotensin-Aldosteron-System? Schweiz Med Wochenschr 102:1105-1112

Müller J (1978) Suppression of aldosterone biosynthesis by treatment of rats with adrenocorticotropin: comparison with glucocorticoid effects. Endocrinology 103:2061-2068

Müller J (1980a) The conversion of 18-hydroxycorticosterone and 18-hydroxy-11-deoxycorticosterone to aldosterone by rat adrenal tissue: Evidence for an alternative biosynthetic pathway. J Steroid Biochem 13:245-251

Müller J (1980b) 11β-Hydroxylation of 18-hydroxy-11-deoxycorticosterone: Zone specificity and effect of sodium and potassium restriction. J Steroid Biochem 13:253-257

Müller J (1986) Divergent effects of dietary chloride restriction on aldosterone biosynthesis and the renin-angiotensin system in rats. Acta Endocrinol 112:105-113

Müller J (1987) Steroidogenesis in the zona glomerulosa cell: the complexity of its regulation. In: Corticosteroids and Peptide Hormones in Hypertension. Serono symposium. (to be published)

Müller J, Baumann K (1973) Effect of potassium intake on the final steps of aldosterone biosynthesis in the rat. III. Combined sodium and potassium restriction. Acta Endocrinol 73:80-90

Müller J, Baumann K (1974) Multifactorial regulation of the final steps of aldosterone biosynthesis in the rat. J Steroid Biochem 5:795-800

Müller J, Gross F (1969) Effects of experimental renal hypertension on aldosterone biosynthesis by rat adrenal tissue. Acta Endocrinol 60:669-680

Müller J, Huber R (1969) Effects of sodium deficiency, potassium deficiency and uremia upon the steroidogenic response of rat adrenal tissue to serotonin, potassium ions and adrenocorticotropin. Endocrinology 85:43-49

Müller J, Weick WJ (1967) Aldosterone stimulation in vitro. IV. Corticosteroid-stimulating activity in rat serum. Acta Endocrinol 54:63-72

Müller J, Ziegler WH (1968) Stimulation of aldosterone biosynthesis in vitro by serotonin. Acta Endocrinol 59:23-35

Müller J, Lund E-G, Hofstetter L, Brunner DB, Haldy P (1982) Stimulation of aldosterone biosynthesis by sodium sequestration: role of angiotensin II. Am J Physiol 243:E450-E457

Müller J, Hofstetter L, Schwendener-Canlas P, Brunner DB, Lund E-G (1984) Role of the renin-angiotensin system in the regulation of late steps in aldosterone biosynthesis by sodium intake of potassium-deficient rats. Endocrinology 115:350-356

Müller OA, Dörr HG, Hagen B, Stalla GK, von Werder K (1982) Corticotropin realeasing factor (CRF)-stimulation test in normal controls and patients with disturbances of the hypothalamo-pituitary-adrenal axis. Klin Wochenschr 60:1485-1491

Muller AF (1963) Regulation der Aldosteronsekretion. Verh Dtsch Ges Inn Med 68:599-615

Muller AF, Riondel AM, Mach RS (1956a) Control of aldosterone excretion by changes in volume of body fluids. Lancet 1:831-832

Muller AF, Riondel AM, Manning EL (1956b) Effect of corticotrophin on secretion of aldosterone. Lancet 2:1021-1024

Muller AF, Manning EL, Riondel AM (1958a) L'excrétion de l'aldostérone chez le sujet normal pendant la déplétion et la réplétion en potassium. Helv Med Acta 25:547-557

Muller AF, Manning EL, Riondel AM (1958b) Diurnal variation of aldosterone related to position and activity in normal subjects and patients with pituitary insufficiency. In: Muller AF, O'Connor CM (eds) An international symposium on aldosterone. Little, Brown, Boston, pp 111-127

Muller AF, Manning EL, Moret P, Mégevand R (1964) Renal blood supply and aldosterone secretion. In: Baulieu EE, Robel P (eds) Aldosterone. Blackwell, Oxford, pp 187-202

Mullins MM, Kleinman LI, Russell PT, Srivastava LS (1982) Plasma aldosterone concentrations in neonatal spontaneously hypertensive rats. Life Sci 31:2751-2755

Mulrow PJ (1966) Neural and other mechanisms regulating aldosterone secretion. In: Martini L, Ganong WF (eds) Neuroendocrinology, vol 1. Academic, New York, pp 407-444

Mulrow PJ (1981) Glucocorticoid-suppressible hyperaldosteronism: a clue to the missing hormone? N Engl J Med 305:1012-1014

Mulrow PJ, Cohn GL (1959) Conversion of corticosterone-4-^{14}C to aldosterone by human adrenal slices. Proc Soc Exp Biol Med 101:731-734

Mulrow PJ, Ganong WF (1961) The effect of hemorrhage upon aldosterone secretion in normal and hypophysectomized dogs. J Clin Invest 40:579-585

Mulrow PJ, Ganong WF, Cera G, Kuljian A (1962) The nature of the aldosterone-stimulating factor in dog kidneys. J Clin Invest 41:505-518

Mulrow PJ, Ganong WF, Boryczka A (1963) Further evidence for a role of the renin-angiotensin system in control of aldosterone secretion. Proc Soc Exp Biol Med 112:7-10

Murakami M, Takeda R, Morimoto S, Hirasawa K, Uchida K, Nakabayashi H, Nishizawa C, Noto M, Hasatani K, Usukura N (1972) Plasma renin activity in panhypopituitarism. Acta Endocrinol 69:531-541

Murphy DL, Goodwin FK, Bunney WE Jr (1969) Aldosterone and sodium response to lithium administration in man. Lancet 2:458-461

Nagahama S, Fujimaki M, Kawabe H, Nakamura R, Saito I, Saruta T (1983a) Effect of metoclopramide on the secretion of aldosterone and other adrenocortical steroids. Clin Endocrinol 18:287-293

Nagahama S, Fujimaki M, Suzuki H, Kawabe H, Saito I, Saruta T (1983b) In vivo and in vitro effects of metoclopramide on aldosterone secretion in rats. Acta Endocrinol 102:589-593

Nagy K, Koroknai L, Spät A (1984) Effect of lipoproteins on aldosterone production by isolated glomerulosa cells. J Steroid Biochem 20:789-791

Nakada T, Momose G, Katayama T, Sotoma T, Ito H, Yoshida T (1970) Histochemical study of the human adrenal cortex. Acta Histochem Cytochem 3:163-172

Nakahara M, Shibasaki T, Shizume K, Kiyosawa Y, Odagiri E, Suda T, Yamaguchi H, Tsushima T, Demura H, Maeda T, Wakabayashi I, Ling N (1983) Corticotropin-releasing factor test

in normal subjects and patients with hypothalamic-pituitary-adrenal disorders. J Clin Endocrinol Metab 57:963-968
Nakamura M, Misono KS, Naruse M, Workman RJ, Inagami T (1985) A role for the adrenal renin-angiotensin system in the regulation of potassium-stimulated aldosterone production. Endocrinology 117:1772-1778
Naruse M, Sussman CR, Naruse K, Jackson TV, Inagami T (1983) Renin exists in human adrenal tissue. J Clin Endocrinol Metab 57:482-487
Nasjletti A, Masson GMC (1969) Effects of corticosteroids on plasma angiotensinogen and renin activity. Am J Physiol 217:1396-1400
Navar LG, Langford HG (1974) Effect of angiotensin on the renal circulation. In: Page IH, Bumpus FM (eds) Angiotensin. Springer, Berlin Heidelberg New York, pp 455-474 (Handbook of experimental pharmacology, vol 37)
Neher R (1958) Aldosterone and other adrenocortical hormones in human adrenals and adrenal tumours. In: Muller AF, O'Connor CO (eds) An international symposium on aldosterone. Little, Brown, Boston, pp 11-23
Neher R (1979) Aldosterone: chemical aspects and related enzymology. J Endocrinol 81: 25*P*-35*P*
Netchitailo P, Delarue C, Perroteau I, Jegou S, Tonon M-C, Leroux P, Leboulenger F, Kusmierek M-C, Capron M-H, Vaudry H (1982) Effect of aldosterone antagonists on mineralocorticoid synthesis in vitro. Inhibition of aldosterone production by prorenoate-K. Eur J Pharmacol 77:243-249
Netchitailo P, Perroteau I, Delarue C, Leboulenger F, Capron MH, Vaudry H (1983) Action directe des antagonistes des minéralocorticoides sur la biosynthèse de l'aldostérone: activités comparées de plusieurs nouveaux composés. Can J Physiol Pharmacol 61:23-28
Netchitailo P, Delarue C, Perroteau I, Leboulenger F, Capron M-H, Vaudry H (1985) Relative inhibitory potency of five mineralocorticoid antagonists on aldosterone biosynthesis in vitro. Biochem Pharmacol 34:189-194
Neville AM, Engel LL (1968) Inhibition of steroid Δ-isomerase of the bovine adrenal gland by substrate analogues. Endocrinology 83:873-876
New MI (1970) Male pseudohermaphroditism due to 17α-hydroxylase deficiency. J Clin Invest 49:1930-1941
New MI, Levine LS (1984) Congenital adrenal hyperplasia. Springer, Berlin Heidelberg New York (Monographs on Endocrinology, vol 26)
New MI, Peterson RE (1967) A new form of congenital adrenal hyperplasia. J Clin Endocrinol Metab 27:300-305
New MI, Seaman MP (1970) Secretion rates of cortisol and aldosterone precursors in various forms of congenital adrenal hyperplasia. J Clin Endocrinol Metab 30:361-371
New MI, Miller B, Peterson RE (1966) Aldosterone excretion in normal children and in children with adrenal hyperplasia. J Clin Invest 45:412-428
New MI, Peterson RE, Saenger P, Levine LS (1976) Evidence for an unidentified ACTH-induced steroid hormone causing hypertension. J Clin Endocrinol Metab 43:1283-1293
New MI, Oberfield SE, Levine LS, Dupont P, Pollack M, Gill JR Jr, Bartter FC (1980) Autosomal dominant transmission and absencence of HLA linkage in dexamethasone-suppressible hyperaldosteronism. Lancet 1:550-551
Newmark SR, Himathongkam T, Martin RP, Cooper KH, Rose LI (1976) Adrenocortical response to marathon running. J Clin Endocrinol Metab 42:393-394
Newton MA, Laragh JH (1968a) Effect of corticotropin on aldosterone excretion and plasma renin in normal subjects, in essential hypertension and in primary aldosteronism. J Clin Endocrinol Metab 28:1006-1013
Newton MA, Laragh JH (1968b) Effects of glucocorticoid administration on aldosterone excretion and plasma renin in normal subjects, in essential hypertension and in primary aldosteronism. J Clin Endocrinol Metab 28:1014-1022
Ney RL (1969) Effects of dibutyryl cyclic AMP on adrenal growth and steroidogenic capacity. Endocrinology 84:168-170
Nicholls MG, Espiner EA, Donald RA (1975) Plasma aldosterone response to low dose ACTH stimulation. J Clin Endocrinol Metab 41:186-188
Nicholls MG, Tree M, Brown JJ, Douglas BH, Fraser R, Hay GD, Lever AF, Morton JJ, Robertson JIS (1978) Angiotensin II/aldosterone dose-response curves in the dog: effects of changes in sodium balance. Endocrinology 102:485-493

Nicholls MG, Grekin RJ, Padfield PL (1980) Angiotensin II-induced aldosterone stimulation in man is not dependent upon adrenocorticotropin. J Clin Endocrinol Metab 51:1171-1174
Nicholls MG, Espiner EA, Ikram H, Maslowski AH, Lun S, Scandrett MS (1981) Angiotensin II is more potent than potassium in regulating aldosterone in cardiac failure: evidence during captopril therapy. J Clin Endocrinol Metab 52:1253-1256
Nicholls MG, Tree M, Livesey JH, Fraser R, Morton JJ, Robertson JIS (1982) Effect of changes in sodium balance on potassium/aldosterone dose-response curves in the dog. Clin Sci 62: 373-380
Nicholls MG, Malvin RL, Espiner EA, Lun S, Myles KD (1983) Adrenal hormone responses to cerebroventricular angiotensin II in the sheep. Proc Soc Exp Biol Med 172:330-333
Nickerson PA (1977) A quantitative ultrastructural study of the adrenal gland in rats resistant and sensitive to the hypertensive effects of sodium chloride. Lab Invest 37:120-125
Nickerson PA, Brownie AC (1975) Effect of hypophysectomy on the volume and ultrastructure of zona glomerulosa in rat adrenal. Endocrinol Exp 9:187-196
Nickerson PA, Brownie AC, Molteni A (1970) Adrenocortical structure and function in rats bearing an adrenocorticotrophic hormone, growth hormone, and prolactin-secreting tumor. Lab Invest 23:368-377
Nicolis GL, Ulick S (1965) Role of 18-hydroxylation in the biosynthesis of aldosterone. Endocrinology 76:514-521
Nicolis GL, Mitty HA, Modlinger RS, Gabrilove JL (1972) Percutaneous adrenal venography. A clinical study of 50 patients. Ann Intern Med 76:899-909
Nishida S, Matsuki M, Nagase Y, Horino M, Endoh M, Kakita K, Tenku A, Oyama H (1983a) Stress-mediated effect of metoclopramide on cortisol secretion in man. J Clin Endocrinol Metab 56:839-843
Nishida S, Matsuki M, Nagase Y, Horino M, Endoh M, Kakita K, Tenku A, Oyama H (1983b) Adrenocorticotropin-mediated effect of metoclopramide on plasma aldosterone in man. J Clin Endocrinol Metab 57:981-985
Nishizuka Y, Takai Y, Kishimoto A, Kikkawa U, Kaibuchi K (1984) Phospholipid turnover in hormone action. Recent Prog Horm Res 40:301-345
Nolten WE, Lindheimer MD, Oparil S, Ehrlich EN (1978) Desoxycorticosterone in normal pregnancy. I. Sequential studies of the secretory patterns of desoxycorticosterone, aldosterone and cortisol. Am J Obstet Gynecol 132:414-420
Norbiato G, Bevilacqua M, Raggi U, Micossi P, Moroni C (1977) Metoclopramide increases plasma aldosterone concentration in man. J Clin Endocrinol Metab 45:1313-1316
Norbiato G, Bevilacqua M, Raggi U, Micossi P, Moroni C, Fasoli A (1978) Effect of prostaglandin synthetase inhibitors on renin and aldosterone in man on a normal or low sodium diet. Acta Endocrinol 87:577-588
Norbiato G, Bevilacqua M, Raggi U, Micossi P, Nitti F, Lanfredini M, Barbieri S (1979) Effect of metoclopramide, a dopaminergic inhibitor, on renin and aldosterone in idiopathic edema: possible therapeutic approach with levodopa and carbidopa. J Clin Endocrinol Metab 48:37-42
Norman N (1961) The influence of the concentration of sodium and potassium in the cerebrospinal fluid on the secretion of steroids, including aldosterone, from the adrenal cortex. Acta Endocrinol 37:559-564
Noth RH, Tan SY, Mulrow PJ (1977a) Effects of angiotensin II blockade by saralasin in normal man. J Clin Endocrinol Metab 45:10-15
Noth RH, Lassman MN, Tan SY, Fernandez-Cruz A, Mulrow PJ (1977b) Age and the renin-aldosterone system. Arch Intern Med 137:1414-1417
Noth RH, McCallum RW, Contino C, Havelick J (1980) Tonic dopaminergic suppression of plasma aldosterone. J Clin Endocrinol Metab 51:64-69
Nothstine SA, Davis JO, DeRoos RM (1971) Kidney extracts and ACTH on adrenal steroid secretion in a turtle and a crocodilian. Am J Physiol 221:726-732
Nugent CA, Nichols T, Tyler FH (1965) Diagnosis of Cushing's syndrome: Single dose dexamethasone suppression test. Arch Intern Med 116:172-176
Nussdorfer GG (1980) Cytophysiology of the adrenal zona glomerulosa. Int Rev Cytol 64: 307-368
Nussdorfer GG (1986) Cytophysiology of the adrenal cortex. Int Rev Cytol 98:1-405
Nussdorfer GG, Mazzocchi G (1982) Stereologic and biochemical investigations on a possible aldosterone-containing secretory organelle in the zona glomerulosa of rat adrenal gland. Anat Anz 151:74-81

Nussdorfer GG, Mazzocchi G, Rebuffat P (1973) An ultrastructural stereologic study of the effects of ACTH and adenosine 3',5'-cyclic monophosphate on the zona glomerulosa of rat adrenal cortex. Endocrinology 92:141-151

Nussdorfer GG, Mazzocchi G, Robba C, Belloni AS, Rebuffat P (1977) Effects of ACTH and dexamethasone on the zona glomerulosa of the rat adrenal cortex: an ultrastructural stereologic study. Acta Endocrinol 85:608-614

Nussdorfer GG, Robba C, Mazzocchi G, Rebuffat P (1981) Effects of angiotensin II on the zona fasciculata of the rat adrenal cortex: an ultrastructural stereologic study. J Anat 132:235-242

Nussdorfer GG, Neri G, Belloni AS, Mazzocchi G, Rebuffat P, Robba C (1982) Effects of ACTH on the zona glomerulosa of sodium-loaded timolol maleate-treated rats: stereology and plasma hormone concentrations. Acta Endocrinol 99:256-262

Oates JA, Whorton AR, Gerkens JF, Branch RA, Hollifield JW, Frölich JC (1979) The participation of prostaglandins in the control of renin release. Fed Proc 38:72-74

Oberfield SE, Levine LS, Carey RM, Bejar R, New MI (1979) Pseudohypoaldosteronism: multiple target organ unresponsiveness to mineralocorticoid hormones. J Clin Endocrinol Metab 48:228-234

Oberfield SE, Levine LS, Stoner E, Chow D, Rauh W, Greig F, Lee SM, Lightner E, Witte M, New MI (1981) Adrenal glomerulosa function in patients with dexamethasone-suppressible hyperaldosteronism. J Clin Endocrinol Metab 53:158-164

O'Brien MJ, Levell MJ, Hullin RP (1979) Inhibition of aldosterone production in adrenal cell suspensions by serum from patients with manic-depressive psychosis. J Endocrinol 80:41-50

Oddie CJ, Coghlan JP, Scoggins BA (1972) Plasma deoxycorticosterone levels in man with simultaneous measurement of aldosterone, corticosterone, cortisol and 11-deoxycortisol. J Clin Endocrinol Metab 34:1039-1054

Oelkers W, Brown JJ, Fraser R, Lever AF, Morton JJ, Robertson JIS (1974a) Sensitization of the adrenal cortex to angiotensin II in sodium-deplete man. Circ Res 34:69-77

Oelkers W, Schöneshöfer M, Blümel A (1974b) Effects of progesterone and four synthetic progestagens on sodium balance and the renin-aldosterone system in man. J Clin Endocrinol Metab 39:882-890

Oelkers W, Schöneshöfer M, Schultze G, Brown JJ, Fraser R, Morton JJ, Lever AF, Robertson JIS (1975) Effect of prolonged low-dose angiotensin II infusion on the sensitivity of adrenal cortex in man. Circ Res 36/37 [Suppl I]: I-49-I-56

Oelkers W, Blümel A, Schöneshöfer M, Schwartz U, Hammerstein J (1976) Effects of ethinylestradiol on the renin-angiotensin-aldosterone-system and on plasma transcortin in women and men. J Clin Endocrinol Metab 43:1036-1040

Oelkers W, Schöneshöfer M, Schultze G, Bauer B (1978a) Effect of prolonged low-dose infusions of ile^5-angiotensin II on blood pressure, aldosterone and electrolyte excretion in sodium replete man. Klin Wochenschr 56:37-41

Oelkers W, Schöneshöfer M, Schultze G, Wenzler M, Bauer B, L'Age M, Fehm HL (1978b) Prolonged infusions of ile^5-angiotensin II in sodium replete and deplete man: effects on aldosterone, ACTH, cortisol, blood pressure and electrolyte balance. J Clin Endocrinol Metab 46: 402-413

Oelkers W, Belkien L, Baumann J, Meyland M (1982) The effect of captopril on renin, angiotensin II, cortisol and aldosterone during ACTH-infusion in man. Clin Exp Hypertens (A) 4: 1505-1517

Öney T, Beer A, Kaulhausen H (1981) Effect of postural change on plasma renin and aldosterone concentrations in third-trimester pregnancy. Obstet Gynecol 58:31-34

Ogihara T, Yamamoto T, Miyai K, Kumahara Y (1973) Plasma renin activity and aldosterone concentration in patients with hyperthyroidism and hypothyroidism. Endocrinol Jpn 20: 433-438

Ogihara T, Matsumara S, Ohnishi T, Miyai K, Uozumi T, Kumahara Y (1977) The effect of metoclopramide-induced prolactin on aldosterone secretion in normal subjects. Life Sci 20: 523-526

Ogihara T, Hata T, Maruyama A, Mikami H, Nakamura M, Okada Y, Kumahara Y (1979) Blood pressure response to an angiotensin II antagonist in patients. J Clin Endocrinol Metab 48: 159-162

Ogle TF, Kitay JI (1978) In vitro effects of melatonin and serotonin on adrenal steroidogenesis. Proc Soc Exp Biol Med 157:103-105

Oh MS, Carroll HJ, Clemmons JE, Vagnucci AH, Levison SP, Whang ESM (1974) A mechanism for hyporeninemic hypoaldosteronism in chronic renal disease. Metabolism 23:1157-1166

Ohnishi T, Wada A, Nonaka Y, Okamoto M, Yamano T (1984) Effect of phospholipid on aldosterone biosynthesis by a cytochrome P-$450_{11\beta}$-reconstituted system. Biochem Int 9:715-723

O'Kelly R, Magee F, McKenna TJ (1983) Routine heparin therapy inhibits adrenal aldosterone production. J Clin Endocrinol Metab 56:108-112

Ølgaard K (1975) Plasma aldosterone in anephric and non-nephrectomized dialysis patients in relation to changes in plasma potassium without change in total potassium balance. Acta Med Scand 198:213-218

Ølgaard K, Madsen S (1977a) Aldosterone response to ACTH stimulation in anephric and non-nephrectomized patients on regular hemodialysis. Acta Med Scand 201:77-82

Ølgaard K, Madsen S (1977b) Regulation of plasma aldosterone in anephric and non-nephrectomized patients during hemodialysis treatment. Acta Med Scand 201:457-462

Ølgaard K, Madsen S, Roosen J, Hammer M (1977a) Circadian rhythm of plasma aldosterone and plasma renin activity in steroid and non-steroid treated kidney transplanted patients. Scand J Clin Lab Invest 37:431-437

Ølgaard K, Madsen S, Hammer M (1977b) The adrenocortical response to angiotensin II infusion in anephric and non-nephrectomized patients on regular hemodialysis. Acta Med Scand 202:213-219

Ølgaard K, Hagen C, Madsen S, Hummer L (1977c) Lack of effect of prolactin inhibition by alpha-bromoergocriptine (CB 154) on plasma aldosterone in anephric and non-nephrectomized patients on regular haemodialysis. Acta Endocrinol 85:587-594

Ølgaard K, Madsen S, Hammer M, Ladefoged J (1978) Calcium-dependent aldosterone secretion in anephric and non-nephrectomized patients on regular hemodialysis. J Clin Endocrinol Metab 46:740-746

Omura T, Sato R, Cooper DY, Rosenthal O, Estabrook RW (1965) Function of cytochrome P-450 of microsomes. Fed Proc 24:1181-1189

Oparil S, Ehrlich EN, Lindheimer MD (1975) Effect of progesterone on renal sodium handling in man: relationship to aldosterone excretion and plasma renin activity. Clin Sci Mol Med 49: 139-147

Opocher G, Murgia A, Boscaro M, D'Agostino D, Fallo F, Mantero F (1983) Effect of metergoline on the aldosterone-stimulating properties of metoclopramide. J Steroid Biochem 19: 531-536

Oster JR, Perez GO, Rosen MS (1976) Hyporeninemic hypoaldosteronism after chronic sodium bicarbonate abuse. Arch Intern Med 136:1179-1180

Osterwalder H (1971) Die juxtaglomerulären Zellen beim virilisierenden adrenogenitalen Syndrom mit und ohne Salzverlustsyndrom. Schweiz Med Wochenschr 101:1298-1305

Padfield PL, Brown JJ, Davies D, Fraser R, Lever AF, Morton JJ, Robertson JIS (1981) The myth of idiopathic hyperaldosteronism. Lancet 2:83-84

Page RB, Boyd JE, Mulrow PJ (1974) The effect of alpha-melanocyte stimulating hormone on aldosterone production in the rat. Endocr Res Commun 1:53-62

Paglin S, Jamieson JD (1982) Covalent crosslinking of angiotensin II to its binding sites in rat adrenal membranes. Proc Natl Acad Sci USA 79:3739-3743

Palkovits M, Földvári PI (1963) Effect of the subcomissural organ and the pineal body on the adrenal cortex. Endocrinology 72:28-32

Palkovits M, Monos E, Fachet J (1965) The effect of subcommissural-organ lesions on aldosterone production in the rat. Acta Endocrinol 48:169-176

Palkovits M, de Jong W, van der Wal B, de Wied D (1970) Effect of adrenocorticotrophic and growth hormones on aldosterone production and plasma renin activity in chronically hypophysectomized sodium-deficient rats. J Endocrinol 47:243-250

Palkovits M, de Jong W, van der Wal B, de Wied D (1971a) The aldosterone secretory response to sodium restriction in chronically hypophysectomized corticotrophin-maintained rats as a function of duration and amount of growth hormone treatment. J Endocrinol 50:407-411

Palkovits M, Strik JJTWA, de Jong W, de Wied D (1971b) Effect of growth hormone on the aldosterone secretory response to sodium restriction in corticotrophin-maintained hypophysectomized nephrectomized rats. J Endocrinol 51:369-374

Palkovits M, de Jong W, de Wied D (1974) Hypothalamic control of aldosterone production in sodium-deficient rats. Neuroendocrinology 14:297-309

Palluk R, Gaida W, Hoefke W (1985) Atrial natriuretic factor. Life Sci 36: 1415-1425
Palmore WP, Mulrow PJ (1967) Control of aldosterone secretion by the pituitary gland. Science 158: 1482-1484
Palmore WP, Marieb NJ, Mulrow PJ (1969) Stimulation of aldosterone secretion by sodium depletion in nephrectomized rats. Endocrinology 84: 1342-1351
Palmore WP, Anderson R, Mulrow PJ (1970) Role of the pituitary in controlling aldosterone production in sodium-depleted rats. Endocrinology 86: 728-734
Parkinson CA, Belton SJ, Pratt JH (1984) The effect of captopril treatment on potassium-induced stimulation of aldosterone production in vitro. Endocrinology 114: 1567-1570
Pasqualini JR (1964) Conversion of tritiated 18-hydroxycorticosterone to aldosterone by slices of human cortico-adrenal gland and adrenal tumour. Nature 201: 501
Pasqualini JR, Wiqvist N, Diczfalusy E (1966) Biosynthesis of aldosterone by human foetuses perfused with corticosterone at mid-term. Biochem Biophys Acta 121: 430-431
Patwardhan VV, Lanthier A (1971) Effect of 17-hydroxyprogesterone and medroxyprogesterone acetate on the route of in vitro biosynthesis of corticosterone and aldosterone from labelled pregnenolone by bovine adrenals. Steroids 18: 433-440
Pavlatos FC, Smilo RP, Forsham PH (1965) A rapid screening test for Cushing's syndrome. JAMA 193: 720-723
Payet N, Isler H (1976) Adrenal glomerulosa mitotic stimulation by posterior pituitary hormones. Cell Tissue Res 172: 93-101
Payet N, Lehoux J-G (1980) A comparative study of the role of vasopressin and ACTH in the regulation of growth and function of rat adrenal glands. J Steroid Biochem 12: 461-467
Payet N, Lehoux J-G (1982a) Effect of ACTH or zinc treatment on plasma aldosterone and corticosterone levels and on the in vitro steroid output from adrenocortical cells. Can J Biochem 60: 1058-1064
Payet N, Lehoux J-G (1982b) Aldosterone and corticosterone stimulation by ACTH in isolated rat adrenal glomerulosa cells: interaction with vasopressin. J Physiol (Paris) 78: 317-321
Payet N, Lehoux J-G, Isler H (1980) Effect of ACTH on the proliferative and secretory activities of the adrenal glomerulosa. Acta Endocrinol 93: 365-374
Payet N, Déziel Y, Lehoux J-G (1984) Vasopressin: a potent growth factor in adrenal glomerulosa cells in culture. J Steroid Biochem 20: 449-454
Peach MJ, Chiu AT (1974) Stimulation and inhibition of aldosterone biosynthesis in vitro by angiotensin II and analogs. Circ Res 34-35 [Suppl 1]: I-7-I-13
Peach MJ, Sarstedt CA, Vaughan ED Jr (1976) Changes in cardiovascular and adrenocortical responses to angiotensin III induced by sodium deprivation in the rat. Circ Res 38 [Suppl 2]: II-117-II-121
Peart WS (1965) The renin-angiotensin system. Pharmacol Rev 17: 143-182
Pedernera EA, Lantos CP (1973) The biosynthesis of adrenal steroids by the 15-day-old chick embryo. Gen Comp Endocrinol 20: 331-341
Pedersen EB, Kornerup HJ (1976a) Relationship between plasma aldosterone concentration and plasma potassium in patients with essential hypertension during alprenolol treatment. Acta Med Scand 200: 263-267
Pedersen EB, Kornerup HJ (1976b) The renin-aldosterone system and renal hemodynamics in patients with posttransplant hypertension. Acta Med Scand 200: 501-508
Pedersen EB, Darling S, Kierkegaard-Hansen A, Amdisen A (1977a) Plasma aldosterone during lithium treatment. Neuropsychobiology 3: 153-159
Pedersen EB, Amdisen A, Darling S (1977b) Plasma aldosterone and magnesium in serum and urine during lithium treatment. Int Pharmacopsychiat 12: 80-85
Pedersen EB, Rasmussen AB, Johannesen P, Kornerup HJ, Kristensen S, Lauritsen JG, Wohlert M (1982) The renin-aldosterone system in pre-eclampsia, essential and transient hypertension during pregnancy, and normotensive pregnant and non-pregnant control subjects. Acta Endocrinol 101: 273-280
Pedersen RC, Brownie AC (1983a) Lys-γ_3-melanotropin binds with high affinity to the rat adrenal cortex. Endocrinology 112: 1279-1287
Pedersen RC, Brownie AC (1983b) Cholesterol side-chain cleavage in the rat adrenal cortex: Isolation of a cycloheximide-sensitive activator peptide. Proc Natl Acad Sci USA 80: 1882-1886
Pedersen RC, Brownie AC, Ling N (1980) Pro-adrenocorticotropin/endorphin-derived peptides: coordinate action on adrenal steroidogenesis. Science 208: 1044-1046
Pedrinelli R, Magagna A, Salvetti A (1982) The effect of oxprenolol and indomethacin on renin and aldosterone of normal subjects during low sodium diet. Eur J Clin Invest 12: 107-111

Pelleya R, Oster JR, Perez GO (1983) Hyporeninemic hypoaldosteronism, sodium wasting and mineralocorticoid-resistant hyperkalemia in two patients with obstructive uropathy. Am J Nephrol 3:223-227
Perchellet J-P, Shanker G, Sharma RK (1978) Regulatory role of guanosine 3′,5′-monophosphate in adrenocorticotropic hormone-induced steroidogenesis. Science 199:311-312
Perez GO, Oster JR, Vaamonde CA (1974) Renal acidosis and renal potassium handling in selective hypoaldosteronism. Am J Med 57:809-816
Perez GO, Oster JB, Vaamonde CA, Katz FH (1977a) Effect of NH_4Cl on plasma aldosterone, cortisol and renin activity in supine man. J Clin Endocrinol Metab 45:762-767
Perez GO, Lespier L, Jacobi J, Oster JR, Katz FH, Vaamonde CA, Fishman LM (1977b) Hyporeninemia and hypoaldosteronism in diabetes mellitus. Arch Intern Med 137:852-855
Perez GO, Lespier LE, Oster JR, Vaamonde CA (1977c) Effect of alterations of sodium intake in patients with hyporeninemic hypoaldosteronism. Nephron 18:259-265
Perez GO, Kem DC, Oster JR, Vaamonde CA (1980) Effect of acute metabolic acidosis on the renin-aldosterone system. Mechanism of increases in plasma aldosterone in dogs infused with lactic acid. J Lab Clin Med 96:371-378
Péron FG (1960) The isolation and identification of some adrenocorticosteroids released by rat adrenal tissue incubated in vitro. Endocrinology 66:458-469
Péron FG, Koritz SB (1958) On the exogenous requirements for the action of ACTH in vitro on rat adrenal glands. J Biol Chem 233:256-259
Perroteau I, Netchitailo P, Homo-Delarche F, Delarue C, Lihrmann I, Leboulenger F, Vaudry H (1984a) Role of exogenous and endogenous prostaglandins in steroidogenesis by isolated frog interrenal gland: evidence for dissociation in adrenocorticotropin and angiotensin action. Endocrinology 115:1765-1773
Perroteau I, Netchitailo P, Delarue C, Leboulenger F, Philibert D, Deraedt R, Vaudry H (1984b) The effect of the antimineralocorticoid RU 28318 on aldosterone biosynthesis in vitro. J Steroid Biochem 20:853-856
Peschel E, Race GG (1954) Studies on the adrenal zona glomerulosa of hypertensive patients and rats with special reference to the effect of dietary salt restriction. Am J Med 17:355-364
Peterson RE (1957) Plasma corticosterone and hydrocortisone levels in man. J Clin Endocrinol Metab 17:1150-1157
Peterson RE (1958) The influence of the thyroid on adrenal cortical function. J Clin Invest 37: 736-743
Peterson RE (1959) The miscible pool and turnover rate of adrenocortical steroids in man. Recent Prog Horn Res 15:231-274
Peterson RE, Nokes G, Chen PS Jr, Black RL (1960) Estrogens and adrenocortical function in man. J Clin Endocrinol Metab 20:495-514
Pettinger WA, Marchelle M, Augusto L (1971) Renin suppression by DOC and NaCl in the rat. Am J Physiol 221:1071-1074
Peytremann A, Nicholson WE, Brown RD, Liddle GW, Hardman JG (1973) Comparative effects of angiotensin and ACTH on cyclic AMP and steroidogenesis in isolated bovine adrenal cells. J Clin Invest 52:835-842
Peytremann A, Brown RD, Nicholson WE, Island DP, Liddle GW, Hardman JG (1974) Regulation of aldosterone synthesis. Steroids 24:451-462
Pham-Huu-Trung MT, Bogyo A, Girard F (1979) Isolated guinea-pig adrenal cells: effects of physiological concentrations of ACTH, angiotensin II and potassium chloride. J Steroid Biochem 11:1543-1546
Pham-Huu-Trung M-T, Bogyo A, de Smitter N, Girard F (1985) Effects of proopiomelanocortin peptides and angiotensin II on steroidogenesis in isolated aldosteronoma cells. J Clin Endocrinol Metab 61:467-471
Phelps KR, Oh MS, Carroll HJ (1980) Heparin-induced hyperkalemia: report of a case. Nephron 25:254-258
Pines A, Slater JDH, Jowett TP (1977) The kidney and aldosterone in acclimatization at altitude. Br J Dis Chest 71:203-307
Plouin P-F, Corvol P, Menard J (1977) Influence of beta-blockade on aldosterone metabolism under normal and low sodium diet in man. Biomedicine 27:233-235
Prader A, Spahr A, Neher R (1955) Erhöhte Aldosteronausscheidung beim kongenitalen adrenogenitalen Syndrom. Ein Beitrag zur Pathogenese des Syndroms. Schweiz Med Wochenschr 85:1085-1088

Pratt JH (1982) Role of angiotensin II in potassium-mediated stimulation of aldosterone secretion in the dog. J Clin Invest 70:667-672
Pratt JH, Dale SL, Melby JC (1976) The effect of administered ACTH on aldosterone metabolism and secretion. J Clin Endocrinol Metab 42:355-360
Pratt JH, Parkinson CA, Luft FC (1978) A new approach to the difficult assessment of aldosterone secretion in anephric patients. Steroids 32:487-498
Pratt JH, Grim CE, Parkinson CA (1980) Effects of propranolol on aldosterone plasma concentration and aldosterone metabolic clearance in hypertensive patients. J Lab Clin Med 95: 693-697
Pratt JH, Ganguly A, Parkinson CA, Weinberger MH (1981) Stimulation of aldosterone secretion by metoclopramide in humans: apparent independence of renal and pituitary mediation. Metabolism 30:129-134
Pratt JH, Turner DA, McAteer JA, Henry DP (1985) β-Adrenergic stimulation of aldosterone production by rat adrenal capsular explants. Endocrinology 117:1189-1194
Preedy JRK, Aitken EH (1956a) The effect of estrogen on water and electrolyte metabolism. I. The normal. J Clin Invest 35:423-429
Preedy JRK, Aitken EH (1956b) The effect of estrogen on water and electrolyte metabolism. II. Hepatic disease. J Clin Invest 35: 430-442
Psychoyos S, Tallan HH, Greengard P (1966) Aldosterone biosynthesis by adrenal mitochondria. J Biol Chem 241:2949-2956

Rabinowe SL, Taylor T, Dluhy RG, Williams GH (1985) β-Endorphin stimulates plasma renin and aldosterone release in normal human subjects. J Clin Endocrinol Metab 60:485-489
Rácz K, Varga I, Gláz E, Kiss R, Vida S, Lada G, di Gleria K, Medzihradszky K, Lichtwald K, Vecsei P (1982) Met-enkephalin inhibits mineralocorticoid production in isolated human aldosteronoma cells. J Clin Endocrinol Metab 54:656-660
Rácz K, Buu NT, Kuchel O, De Léan A (1984) Dopamine 3-sulfate inhibits aldosterone secretion in cultured bovine adrenal cells. Am J Physiol 247:E431-E435
Rácz K, Kuchel O, Cantin M, de Léan A (1985) Atrial natriuretic factor inhibits the early pathway of steroid biosynthesis in bovine adrenal cortex. FEBS Lett 192:19-22
Radke KJ, Schneider EG, Taylor RE Jr, Kramer RE (1986) Effect of hydrogen ion concentration on corticosteroid secretion. Am J Physiol 250:E259-E264
Rado JP, Boer P, Dorhout Mees EJ, Simatupang T (1979) "Outpatient hyperkalemia" syndrome in renal and hypertensive patients with suppressed aldosterone production. J Med 10:145-157
Raiti S, Kowarski A, Weldon VV, Migeon CJ (1968) Secretion of cortisol, corticosterone and aldosterone in children with hypopituitarism. Johns Hopkins Med J 122:229-231
Raman PB, Sharma DC, Dorfman RI (1966) Studies on aldosterone biosynthesis in vitro. Biochemistry 5:1795-1804
Ramsay DJ, Keil LC, Sharpe MC, Shinsako J (1978) Angiotensin II infusion increases vasopressin, ACTH, and 11-hydroxycorticosteroid secretion. Am J Physiol 234:R66-R71
Rand M, Reid G (1951) Source of serotonin in serum. Nature 168:385
Rapp JP, Dahl LK (1976) Mutant forms of cytochrome P-450 controlling both 18- and 11β-steroid hydroxylation in the rat. Biochemistry 15:1235-1242
Rappaport R, Dray F, Legrand JC, Royer P (1968) Hypoaldostéronisme congénital familial par défaut de la 18-OH-déhydrogénase. Etude hormonale avant et après guérison du syndrome de perte de sel. Pediatr Res 2:456-463
Rasmussen H, Barrett PQ (1984) Calcium messenger system: an integrated view. Physiol Rev 64: 938-984
Rasmussen H, Goodman DBP (1977) Relationships between calcium and cyclic nucleotides in cell activation. Physiol Rev 57:421-509
Rauh W, Levine LS, Gottesdiener K, New MI (1978) Mineralocorticoids, salt balance and blood pressure after prolonged ACTH administration in juvenile hypertension. Klin Wochenschr 56 [Suppl I]:161-167
Rauh W, Levine LS, Gottesdiener K, Chow D, Oberfield SE, Gunczler P, Pareira J, New MI (1979) Adrenocortical function, electrolyte metabolism, and blood pressure during prolonged adrenocorticotropin infusion in juvenile hypertension. J Clin Endocrinol Metab 49:52-57
Rayfield EJ, Rose LI, Dluhy RG, Williams GH (1973) Aldosterone secretory and glucocorticoid excretory responses to alpha 1-24 ACTH (cortrosyn) in sodium-depleted normal man. J Clin Endocrinol Metab 36:30-35

Rayyis SS, Horton R (1971) Effect of angiotensin II on adrenal and pituitary function in man. J Clin Endocrinol Metab 32:539–546

Re RN, Kourides IA, Weihl AC, Maloof F (1979) The relationship between endogenous hyperprolactinaemia and plasma aldosterone. Clin Endocrinol 10:187–193

Read VH, McCaa CS, Bower JD, McCaa RE (1973) Effect of hemodialysis on the metabolic clearance rate, plasma concentration and blood production rate of aldosterone in anephric man. J Clin Endocrinol Metab 36:773–778

Rebuffat P, Belloni AS, Mazzocchi G, Vasanelli P, Nussdorfer GG (1979) A stereological study of the trophic effects of the renin-angiotensin system on the rat adrenal zona glomerulosa. J Anat 129:561–570

Rebuffat P, Robba C, Mazzocchi G, Nussdorfer GG (1984) Inhibitory effect of somatostatin on the growth and steroidogenic capacity of rat adrenal zona glomerulosa. J Steroid Biochem 21: 387–390

Reck G, Beckerhoff R, Vetter W, Armbruster H, Siegenthaler W (1975) Control of plasma aldosterone in normal man during upright posture. Klin Wochenschr 53:955–959

Regöly-Mérey J, Sólyom J (1975) Prolonged oral administration of potassium upon aldosterone biosynthesis by rat glomerulosa tissue in vitro. Endocrinol Exp 9:43–50

Reid IA, Tu WH, Assaykeen TA, Ganong WF (1973) Studies concerning the regulation and importance of plasma angiotensinogen concentration in the dog. Endocrinology 93:107–114

Reinking LN (1983) Aldosterone response to renin, angiotensin, ACTH, hemorrhage and sodium depletion in a freshwater teleost, *Catostomus macrocheilus*. Comp Biochem Physiol 74A: 873–880

Reiter EO, Teree TM (1979) Adrenal 18-dehydrogenase defect: clinical remission but chemical persistence. Pediatr Res 13:384 (abstract)

Relkin R, Adachi M (1973) Effects of sodium deprivation on pituitary and plasma prolactin, growth hormone, and thyrotropin levels in the rat. Neuroendocrinology 11:240–247

Rembiesa R, Young PCM, Saffran M (1968) Effect of 17α-methyltestosterone on steroid formation by rat adrenal tissue. Can J Biochem 46:433–440

Richards AM, Nicholls MG, Espiner EA, Ikram H, Yandle TG, Joyce SL, Cullens MM (1985) Effects of alpha-human atrial natriuretic peptide in essential hypertension. Hypertension 7: 812–817

Richardson KSC, Nowaczynski W, Genest J (1977) Specific aldosterone-binding proteins in human plasma: partial characterization. J Steroid Biochem 8:951–957

Richer CL, Veilleux R, Bois P (1968) Effect of magnesium deficiency on corticosterone in rats. Endocrinology 82:954–958

Robb CA, Davis JO, Johnston CI, Hartroft PM (1969) Effects of deoxycorticosterone on plasma renin activity in conscious dogs. Am J Physiol 216:884–889

Robb CA, Davis JO, Johnson JA, Blaine EH, Schneider EG, Baumber JS (1970) Mechanisms regulating the renal excretion of sodium during pregnancy. J Clin Invest 49:871–880

Robba C, Rebuffat P, Mazzocchi G, Nussdorfer GG (1986) Long-term trophic action of α-melanocyte-stimulating hormone on the zona glomerulosa of the rat adrenal cortex. Acta Endocrinol 112:404–408

Robertson D, Michelakis AM (1975) The effect of chlorpromazine on plasma renin activity and aldosterone in man. J Clin Endocrinol Metab 41:1166–1168

Robertson PW, Klidjian A, Harding LK, Walters G, Lee MR, Robb-Smith AHT (1967) Hypertension due to a renin-secreting renal tumour. Am J Med 43:963–976

Robillard JE, Ramberg E, Sessions C, Consamus B, VanOrden D, Weismann D, Smith FG Jr (1980) Role of aldosterone on renal sodium and potassium excretion during fetal life and newborn period. Dev Pharmacol Ther 1:201–216

Robillard JE, Weismann DN, Gomez RA, Ayres NA, Lawton WJ, VanOrden DE (1983) Renal and adrenal responses to converting-enzyme inhibition in fetal and newborn life. Am J Physiol 244:R249–R256

Robillard JE, Ayres NA, Gomez RA, Nakamura KT, Smith FG Jr (1984) Factors controlling aldosterone secretion during hypoxemia in fetal lambs. Pediatr Res 18:607–611

Roll D, Licht A, Rösler A, Durst AL, Kleeman CR, Czaczkes JW (1979) Transient hypoaldosteronism after renal allotransplantation. Israel J Med Sci 15:29–34

Romero JC, Dunlap CL, Strong CG (1976) The effect of indomethacin and other anti-inflammatory drugs on the renin-angiotensin system. J Clin Invest 58:282–288

Rosenberg S, Franks RC, Ulick S (1980) Mineralocorticoid unresponsiveness with severe neonatal hyponatremia and hyperkalemia. J Clin Endocrinol Metab 50:401–404

Rosenfeld G, Rosemberg E, Ungar F, Dorfman RI (1956) Regulation of the secretion of aldosterone-like material. Endocrinology 58:255-261
Rosenkrantz H (1959) A direct influence of 5-hydroxytryptamine on the adrenal cortex. Endocrinology 64:355-362
Rosenkrantz H, Laferte RC (1960) Further observations on the relationship between serotonin and the adrenal. Endocrinology 66:832-841
Rosenthal J, Raptis S, Escobar-Jimenez F, Pfeiffer EF (1976) Inhibition of frusemide-induced hyperreninaemia by growth-hormone release-inhibiting hormone in man. Lancet 1:772-774
Rosenthal J, Escobar-Jimenez F, Raptis S (1977) Prevention by somatostatin of rise in blood pressure and plasma renin mediated by beta-receptor stimulation. Clin Endocrinol 6:455-462
Rösler A (1984) The natural history of salt-wasting disorders of adrenal and renal origin. J Clin Endocrinol Metab 59:689-700
Rösler A, Rabinowitz D, Theodor R, Ramirez LC, Ulick S (1977a) The nature of the defect in a salt-wasting disorder in Jews of Iran. J Clin Endocrinol Metab 44:279-291
Rösler A, Theodor R, Boichis H, Gerty R, Ulick S, Alagem M, Tabachnik E, Cohen B, Rabinowitz D (1977b) Metabolic responses to the administration of angiotensin II, K and ACTH in two salt-wasting syndromes. J Clin Endocrinol Metab 44:292-301
Rösler A, Levine LS, Schneider B, Novogroder M, New MI (1977c) The interrelationship of sodium balance, plasma renin activity and ACTH in congenital adrenal hyperplasia. J Clin Endocrinol Metab 45:500-512
Rosnagle RS, Farrell GL (1956) Alterations in electrolyte intake and adrenal steroid secretion. Am J Physiol 187:7-10
Ross EJ (1965) Conn's syndrome due to adrenal hyperplasia with hypertrophy of zona glomerulosa relieved by unilateral adrenalectomy. Am J Med 39: 994-1002
Ross EJ, Van't Hoff W, Crabbé J, Thorn GW (1960) Aldosterone excretion in hypopituitarism and after hypophysectomy in man. Am J Med 28:229-238
Rosset E, Veyrat R (1971) Release of renin by human kidney slices, in vitro effect of angiotensin II, norepinephrine and aldosterone. Rev Eur Etud Clin Biol 16:792-794
Rovner DR, Conn JW, Cohen EL, Berlinger FG, Kem DC, Gordon DL (1979) 17α-Hydroxylase deficiency. A combination of hydroxylation defect and reversible blockade in aldosterone biosynthesis. Acta Endocrinol 90:490-504
Rowe BP, Davis JO, Freeman RH, Stephens GA, DeForrest JM, Seymour AA, Williams GM (1980) Aldosterone response to angiotensin blockade in the conscious sodium-depleted rabbit. Proc Soc Exp Biol Med 163:120-125
Roy AK, Ramirez LC, Ulick S (1976) Structure and mechanism of formation of the two forms of 18-hydroxy-11-deoxycorticosterone. J Steroid Biochem 7:81-87
Roy MW, Guthrie GP Jr, Holladay FP, Kotchen TA (1983) Effects of verapamil on renin and aldosterone in the dog and rat. Am J Physiol E410-E416
Rubin RT, Poland RE, Gouin PR, Tower BB (1978) Secretion of hormones influencing water and electrolyte balance (antidiuretic hormone, aldosterone, prolactin) during sleep in normal adult men. Psychosom Med 40:44-59
Ruilope LM, García-Robles R, Mancheño E, Rodríguez FJ, Barrientos A, Torres J, Rodicio JL, Oriol-Bosch A (1977) Response of plasma aldosterone to orthostatism during follicular and luteal phases of the normal menstrual cycle. Rev Esp Fisiol 33:341-346
Ryan JW (1967) Renin-like enzyme in the adrenal gland. Science 158:1589-1590
Ryan JW (1972) Distribution of a renin-like enzyme in the bovine adrenal gland. Experientia 29: 407-408

Saito I, Saruta T (1982) Regulation of aldosterone secretion by a new aldosterone stimulating factor. Jpn Circ J 46:523-530
Saito I, Bravo EL, Zanella T, Sen S, Bumpus FM (1981) Steroidogenic characteristics of a new aldosterone-stimulating factor (ASF) isolated from normal human urine. Hypertension 3: 300-305
Sala GB, Hayashi K, Catt KJ, Dufau ML (1979) Adrenocorticotropin action in isolated adrenal cells. The intermediate role of cyclic AMP in stimulation of corticosterone synthesis. J Biol Chem 254:3861-3865
Salassa RM, Mattox VR, Rosevear JW (1962) Inhibition of the "mineralocorticoid" activity of licorice by spironolactone. J Clin Endocrinol Metab 22:1156-1159
Salmenperä M, Kahri AI (1976) Corticosterone, 18-OH-deoxycorticosterone, deoxycorticosterone

and aldosterone secretion in tissue culture of foetal rat adrenals in the presence and the absence of ACTH. Acta Endocrinol 83:781-793

Salti IS, Stiefel M, Ruse JL, Laidlaw JC (1969a) Non-tumorous "primary" aldosteronism: I. Type relieved by glucocorticoid (glucocorticoid-remediable aldosteronism). Can Med Assoc J 101: 1-10

Salti IS, Ruse JL, Stiefel M, Delarue NC, Laidlaw JC (1969b) Non-tumorous "primary" aldosteronism: II. Type not relieved by glucocorticoid. Can Med Assoc J 101:11-16

Saltman S, Baukal A, Waters S, Bumpus FM, Catt KJ (1975) Competitive binding activity of angiotensin II analogues in an adrenal cortex radioligand-receptor assay. Endocrinology 97: 275-282

Saltman S, Fredlund P, Catt KJ (1976) Actions of angiotensin II antagonists upon aldosterone production by isolated adrenal glomerulosa cells. Endocrinology 98:894-903

Salvetti A, Pedrinelli R, Poli L, Gianessi D, dalle Luche A, Duranti P (1977) The relationship of plasma and urinary aldosterone with sodium balance, plasma renin activity and age in normal subjects. J Nucl Med Allied Sci 21:7-17

Sambhi MP, Levitan BA, Beck JC, Venning EH (1963) The rate of aldosterone secretion in hypertensive patients with demonstrable renal artery stenosis. Metabolism 12:498-506

Sancho JM, Garcia Robles R, Mancheño E, Paya C, Rodicio JL, Ruilope L (1984) Interference of ranitidine with aldosterone secretion in vivo. Eur J Clin Pharmacol 27:495-497

Sandberg AA, Slaunwhite WR Jr (1959) Transcortin: a corticosteroid-binding protein of plasma. II. Levels in various conditions and the effect of estrogens. J Clin Invest 38:1290-1297

Sandor T, Lanthier A (1963) The in vitro biosynthesis of 18-hydroxycorticosterone-4-C^{14} by slices of zona glomerulosa of beef adrenals and by human adrenals. Acta Endocrinol 42:355-363

Sandor T, Fazekas AG, Lehoux J-G, Leblanc H, Lanthier A (1972) Studies on the biosynthesis of 18-oxygenated steroids from exogenous corticosterone by domestic duck *(Anas platyrhynchos)* adrenal mitochondria. J Steroid Biochem 3:661-682

Sarstedt CA, Vaughan ED Jr, Peach MJ (1975) Selective inhibition by des-1-asp-8-ile-angiotensin II of the steroidogenic response to restricted sodium intake in the rat. Circ Res 37:350-358

Saruta T, Kaplan NM (1972) Adrenocortical steroidogenesis: the effects of prostaglandins. J Clin Invest 51:2246-2251

Saruta T, Saade GA, Kaplan NM (1970) A possible mechanism for hypertension induced by oral contraceptives. Diminshed feedback suppression of renin release. Arch Intern Med 127: 621-626

Saruta T, Cook R, Kaplan NM (1972) Adrenocortical steroidogenesis: studies on the mechanism of action of angiotensin and electrolytes. J Clin Invest 51:2239-2245

Saruta T, Kitajima W, Hayashi M, Kato E, Matsuki S (1980a) Renin and aldosterone in hypothyroidism: relation to excretion of sodium and potassium. Clin Endocrinol 12:483-489

Saruta T, Kondo K, Saito I, Nagahama S, Suzuki H, Konishi K, Matsuki S (1980b) Control of aldosterone in 17-hydroxylase deficiency. Horm Res 13:98-108

Saruta T, Nagahama S, Eguchi T, Oka M, Kanbegawa A (1981) Renin, aldosterone and other mineralocorticoids in hyperkalemic patients with chronic renal failure showing mild azotemia. Nephron 29:128-132

Sassard J, Vincent M, Annat G, Bizollon CA (1976) A kinetic study of plasma renin and aldosterone during changes of posture in man. J Clin Endocrinol Metab 42:20-27

Sato H, Ashida N, Suhara K, Itagaki E, Takemori S, Katagiri M (1978) Properties of an adrenal cytochrome *P*-450 (P-$450_{11\beta}$) for the hydroxylations of corticosteroids. Arch Biochem Biophys 190:307-314

Schaison G, Sebaoun J, Delzant G, Gilbert-Dreyfus V (1970) Etude de la fonction minéralo-corticoide dans la maladie d'Addison à propos de 10 observations. Ann Endocrinol 31:609-623

Schaison G, Couzinet B, Gourmelen M, Elkik F, Bougneres P (1980) Angiotensin and adrenal steroidogenesis: study of 21-hydroxylase-deficient congenital adrenal hyperplasia. J Clin Endocrinol Metab 51:1390-1394

Schalekamp MADH, Bever Donker SC, Jansen-Goemans A, Fawzi TD, Muller A (1976) Dissociation of renin and aldosterone during dehydration: studies in a case of diabetes insipidus and adipsia. J Clin Endocrinol Metab 43:287-294

Schambelan M, Biglieri EG (1972) Deoxycorticosterone production and regulation in man. J Clin Endocrinol Metab 34:695-703

Schambelan M, Stockigt JR, Biglieri EG (1972) Isolated hypoaldosteronism in adults. A renin-deficiency syndrome. N Engl J Med 287:573-578

Schambelan M, Howes EL Jr, Stockigt JR, Noakes CA, Biglieri EG (1973) Role of renin and aldosterone in hypertension due to a renin-secreting tumor. Am J Med 55:86-92
Schambelan M, Sebastian A, Biglieri EG (1980) Prevalence, pathogenesis and functional significance of aldosterone deficiency in hyperkalemic patients with chronic renal insufficiency. Kidney Int 17:89-101
Schambelan M, Sebastian A, Rector FC Jr (1981) Mineralocorticoid-resistant renal hyperkalemia without salt wasting (type II pseudohypoaldosteronism): Role of increased renal chloride reabsorption. Kidney Int 19:716-727
Schane HP, Potts GO (1978) Oral progestational activity of spironolactone. J Clin Endocrinol Metab 47:691-694
Schane HP, Harding HR, Creange JE, Botton I, Castracane VD, Snyder BW (1984) Nivazol: a glucocorticoid in rats with only hypothalamic-pituitary-adrenal-inhibiting activity in primates. Endocrinology 114:1983-1989
Schiffrin EL, Lis M, Gutkowska J, Genest J (1981) Role of Ca^{2+} in response of adrenal glomerulosa cells to angiotensin II, ACTH, K^+, and ouabain. Am J Physiol 241:E42-E46
Schiffrin EL, Gutkowska J, Lis M, Genest J (1982) Relative roles of sodium and calcium ions in the steroidogenic response of isolated rat adrenal glomerulosa cells. Hypertension 4 [Suppl II]: II-36-II-42
Schiffrin EL, Chrétien M, Seidah NG, Lis M, Gutkowska J, Cantin M, Genest J (1983) Response of human aldosteronoma cells in culture to the N-terminal glycopeptide of pro-opiomelanocortin and γ^3-MSH. Horm Metab Res 15:181-184
Schiller PW, Maziak L, Nguyen TM, Godin J, Garcia R, de Léan A, Cantin M (1985) Synthesis and biological activity of a linear fragment of the atrial natriuretic factor (ANF). Biochem Biophys Res Commun 131:1056-1062
Schlatmann RJAFM, Jansen AP, Prenen H, van der Korst JK, Majoor CLH (1964) The natriuretic and aldosterone-suppressive action of heparin and some related polysulfated polysaccharides. J Clin Endocrinol Metab 24:35-47
Schlegel W, Schwyzer R (1977) Purification of bovine adrenal-cortex plasma-membrane vesicles containing a highly corticotropin-sensitive adenylate-cyclase system and angiotensin II-binding sites. Eur J Biochem 72:415-424
Schmidlin J, Wettstein A (1961) Synthese Aldosteron-ähnlicher Corticosteroide *d,l*-18-Hydroxycorticosteron und *d,l*-18-Desoxyaldosteron. Helv Chim Acta 44:1596-1607
Schneider EG, Lynch RE, Willis LR, Knox FG (1972) The effect of potassium infusion on proximal sodium reabsorption and renin release in the dog. Kidney Int 2:197-202
Schneider EG, Taylor RE Jr, Radke KJ, Davis PG (1984) Effect of sodium concentration on aldosterone secretion by isolated perfused canine adrenal glands. Endocrinology 115: 2195-2204
Schneider EG, Radke KJ, Ulderich DA, Taylor RE Jr (1985) Effect of osmolality on aldosterone secretion. Endocrinology 116:1621-1626
Schneider G, Mulrow PJ (1973) Regulation of aldosterone production during pregnancy in the rat. Endocrinology 92:1208-1215
Schnurr E, Küppers H, Wiesen K, Grabensee B (1976) Verhalten des Plasma-Aldosterons während der Hämodialyse bei terminal niereninsuffizienten Patienten. Dtsch Med Wochenschr 101:1120-1128
Schönbeck M, Siegenthaler W, Weidmann P, Endres P, Baumann K, Gysling E, Werning C (1968) Correlation of plasma renin activity and aldosterone secretion rate in hypertensive patients. Helv Med Acta 34:386-400
Schöneshöfer M, Claus M (1985) Multiple sites of inhibition by intravenous metyrapone of human adrenal steroidogenesis. Acta Endocrinol 109:378-385
Schöneshöfer M, Oelkers W (1980) Effect of prolonged infusions of angiotensin II on eight serum corticosteroids in man on different sodium diets. Horm Metab Res 12:614-620
Schöneshöfer M, Wagner GG (1977) Sex differences in corticosteroids in man. J Clin Endocrinol Metab 45:814-817
Scholer D, Riondel AM, Manning EL (1972) Plasma aldosterone determination with ^{35}S-p-toluene-sulphonic anhydride. Acta Endocrinol 70:552-566
Scholer D, Birkhäuser M, Peytremann A, Riondel A, Vallotton MB, Muller AF (1973) Response of plasma aldosterone to angiotensin II, ACTH and potassium in man. Acta Endocrinol 72: 293-307
Schriefer H, Bayer JM, Pittel M (1963) Vergleichende Untersuchungen zur Biogenese von Steroid-

hormonen bei der Durchströmung überlebender Nebennieren einer Patientin mit Cushing- und einer Patientin mit Conn-Syndrom. Acta Endocrinol 43:419-429
Schulster D, Burstein S, Cooke BA (1976) Molecular endocrinology of the steroid hormones. Wiley, London, pp 167-207
Schwartz D, Geller DM, Manning PT, Siegel NR, Fok KF, Smith CE, Needleman P (1985) Ser-leu-arg-arg-atriopeptin III: the major circulating form of atrial peptide. Science 229:397-400
Sciarra F, Sparano F, Giaquinto G, Odoardi A, Mercuri MA, Concolino G (1977) Prolactinemia during hemodialysis: lack of correlation with sodium, potassium and renin-aldosterone system. Horm Res 8:129-138
Scoggins BA, Blair-West JR, Coghlan JP, Denton DA, Myers K, Nelson JF, Orchard E, Wright RD (1970) The physiological and morphological response of mammals to changes in their sodium status. Mem Soc Endocrinol 18:577-602
Scoggins BA, Coghlan JP, Denton DA, Fan JSK, McDougall JG, Oddie CJ, Shulkes AA (1974) Metabolic effects of ACTH in the sheep. Am J Physiol 226:198-205
Scoggins BA, McDougall JG, Butkus A, Coghlan JP, Denton DA, Hardy KJ, Wright RD (1981) [Asp^1, Val^5]Angiotensin-(1-8)octapeptide does not stimulate aldosterone secretion in sodium depleted sheep. Clin Sci 61:111-113
Sealey JE, Gerten JN, Ledingham JGG, Laragh JH (1967) Inhibition of renin by heparin. J Clin Endocrinol Metab 27:699-705
Sealey JE, Clark I, Bull MB, Laragh JH (1970) Potassium balance and the control of renin secretion. J Clin Invest 49:2119-2127
Sealey JE, White RP, Laragh JH, Case DB, Rubin AL (1978) Studies of plasma aldosterone in anephric people: evidence for the fundamental role of the renin system in maintaining aldosterone secretion. J Clin Endocrinol Metab 47:52-60
Sebastian A, Schambelan M, Lindenfeld S, Morris RC Jr (1977) Amelioration of metabolic acidosis with fludrocortisone therapy in hyporeninemic hypoaldosteronism. N Engl J Med 297: 576-583
Seidah NG, Rochemont J, Hamelin J, Lis M, Chrétien M (1981) Primary structure of the major human pituitary pro-opiomelanocortin NH_2-terminal glycopeptide. Evidence for an aldosterone-stimulating activity. J Biol Chem 256:7977-7984
Seifert C, Oelkers W (1981) Aldosterone response to sodium deprivation and angiotensin II in patients with hypopituitarism. Acta Endocrinol 96:361-369
Semple PF, Mason PA (1978) Bromocriptine: lack of effect on the angiotensin II and aldosterone responses to sodium deprivation. Clin Endocrinol 9:155-161
Semple PF, Morton JJ (1976) Angiotensin II and angiotensin III in rat blood. Circ Res 38 [Suppl 2]: II-122-II-126
Semple PF, Nicholls MG, Tree M, Fraser R (1978) (Des-asp^1)-angiotensin II in the dog: blood levels and effect on aldosterone. Endocrinology 103:1476-1482
Semple PF, Buckingham JC, Mason PA, Fraser R (1979) Suppression of plasma ACTH concentration by angiotensin II infusion in normal humans and in a subject with a steroid 17α-hydroxylase defect. Clin Endocrinol 10:137-144
Sen S, Bravo EL, Bumpus FM (1977) Isolation of a hypertension producing compound from normal human urine. Circ Res 40 [Suppl 1]: I-5-I-10
Sen S, Shainoff JR, Bravo EL, Bumpus FM (1981a) Isolation of aldosterone-stimulating factor (ASF) and its effect on rat adrenal glomerulosa cells in vitro. Hypertension 3:4-10
Sen S, Valenzuela R, Smeby R, Bravo EL, Bumpus FM (1981b) Localization, purification, and biological activity of a new aldosterone-stimulating factor. Hypertension 3 [Suppl I]: I-81-I-86
Sequeira SJ, McKenna TJ (1985) Examination of the effects of epinephrine, norepinephrine, and dopamine on aldosterone production in bovine glomerulosa cells in vitro. Endocrinology 117: 1947-1952
Sernia C, McDonald IR (1981) Secretion of aldosterone in the monotreme mammal, *Tachyglossus aculeatus*. J Endocrinol 90:267-273
Shade RE, Grim CE (1975) Suppression of renin and aldosterone by small amounts of DOCA in normal man. J Clin Endocrinol Metab 40:652-658
Shade RE, Davis JO, Johnson JA, Gotshall RW, Spielman WS (1973) Mechanism of action of angiotensin II and antidiuretic hormone on renin secretion. Am J Physiol 224:926-929
Sharma DC (1970) Studies on aldosterone biosynthesis in vitro. Acta Endocrinol 63:299-312
Sharma DC, Forchielli E, Dorfman RI (1962) Inhibition of enzymatic steroid 11β-hydroxylation by androgens. J Biol Chem 238:572-575

Sharma DC, Nerenberg CA, Dorfman RI (1967) Studies on aldosterone biosynthesis in vitro. II. Biochemistry 6:3472-3479
Sharp GW, Leaf A (1966) Mechanism of action of aldosterone. Physiol Rev 46:593-633
Shelton JH, Jones AL (1971) The fine structure of the mouse adrenal cortex and the ultrastructural changes in the zona glomerulosa with low and high sodium diets. Anat Rec 170:147-182
Shenker Y, Villareal JZ, Sider RS, Grekin RJ (1985a) α-Melanocyte-stimulating hormone stimulation of aldosterone secretion in hypophysectomized rats. Endocrinology 116:138-141
Shenker Y, Gross D, Grekin RJ (1985b) Central serotonergic stimulation of aldosterone secretion. J Clin Invest 76:1485-1490
Shenker Y, Gross MD, Grekin RJ (1985c) Peripheral serotonin$_2$ receptor blockade does not inhibit 5-hydroxytryptophan-induced aldosterone stimulation. J Clin Endocrinol Metab 61: 1201-1204
Sheppard H, Swenson R, Mowles TF (1963) Steroid biosynthesis by rat adrenal: functional zonation. Endocrinology 73:819-824
Shima S, Kawashima Y, Hirai M (1978) Studies on cyclic nucleotides in the adrenal gland. VIII. Effects of angiotensin on adenosine 3′,5′-monophosphate and steroidogenesis in the adrenal cortex. Endocrinology 103:1361-1367
Shima S, Kawashima Y, Hirai M (1979) Effects of ACTH and calcium on cyclic AMP production and steroid output by the zona glomerulosa of the adrenal cortex. Endocrinol Jpn 26:219-225
Shima S, Komoriyama K, Hirai M, Kouyama H (1984) Studies on cyclic nucleotides in the adrenal gland. XI. Adrenergic regulation of adenylate cyclase activity in the adrenal cortex. Endocrinology 114:325-329
Shire JGM, Stewart J (1972) The zona glomerulosa and corticotrophin: a genetic study in mice. J Endocrinol 55:185-193
Siegel SR (1979) Potassium chloride stimulation of aldosterone in the newborn lamb. J Pediatr 95:1051-1054
Siegel SR, Fisher DA, Oh W (1974) Serum aldosterone concentrations related to sodium balance in the newborn infant. Pediatrics 53:410-413
Siegler RL, Crouch RH, Hackett TN Jr, Walker M, Jubiz W (1977) Potassium-renin-aldosterone relationships during the first year of life. J Pediatr 91:52-55
Simpson E (1979) Cholesterol side-chain cleavage, cytochrome P450, and the control of steroidogenesis. Mol Cell Endocrinol 13:213-227
Singer B, Stack-Dunne MP (1955) The secretion of aldosterone and corticosterone by the rat adrenal. J Endocrinol 12:130-145
Singer B, Losito C, Salmon S (1963a) Aldosterone and corticosterone secretion rates in rats with experimental renal hypertension. Acta Endocrinol 44:505-518
Singer B, Losito C, Salmon S (1963b) Effect of progesterone on adrenocortical hormone secretion in normal and hypophysectomized rats. J Endocrinol 28:65-72
Sippell WG, Becker H, Versmold HT, Bidlingmaier F, Knorr D (1978) Longitudinal studies of plasma aldosterone, corticosterone, deoxycorticosterone, progesterone, 17-hydroxyprogesterone, cortisol, and cortisone determined simultaneously in mother and child at birth and during the early neonatal period. I. Spontaneous delivery. J Clin Endocrinol Metab 46:971-985
Sippell WG, Dörr HG, Bidlingmaier F, Knorr D (1980) Plasma levels of aldosterone, corticosterone, 11-deoxycorticosterone, progesterone, 17-hydroxyprogesterone, cortisol, and cortisone during infancy and childhood. Pediatr Res 14:39-46
Sippell WG, Müller-Holve W, Dörr HG, Bidlingmaier F, Knorr D (1981) Concentrations of aldosterone, corticosterone, 11-deoxycorticosterone, progesterone, 17-hydroxyprogesterone, 11-deoxycortisol, cortisol, and cortisone determined simultaneously in human amniotic fluid throughout gestation. J Clin Endocrinol Metab 52:385-392
Sizonenko P-C, Riondel A-M, Kohlberg IJ, Paunier L (1972) 11β-Hydroxylase deficiency: steroid response to sodium restriction and ACTH stimulation. J Clin Endocrinol Metab 35:281-287
Skadhauge E, Thomas DH, Chadwick A, Jallageas M (1983) Time course of adaptation to low and high NaCl diets in the domestic fowl. Effects on electrolyte excretion and on plasma hormone levels (aldosterone, corticosterone and prolactin). Pflugers Arch 396:301-307
Skinner SL, Lumbers ER, Symonds EM (1972) Analysis of changes in the renin-angiotensin system during pregnancy. Clin Sci 42:479-488
Slaga TJ, Krum AA (1973) Aldosterone biosynthesis from 4-^{14}C-pregnenolone and 4-^{14}C-progesterone in the rat and the rabbit. Proc Soc Exp Biol Med 143:1201-1206
Slater JDH, Barbour BH, Henderson HH, Casper AGT, Bartter FC (1965) Physiological influence

of the kidney on the secretion of aldosterone, corticosterone and cortisol by the adrenal cortex. Clin Sci 28:219-236
Slater JDH, Williams ES, Edwards RHT, Ekins RP, Sönksen PH, Beresford CH, McLaughlin M (1969a) Potassium retention during the respiratory alkalosis of mild hypoxia in man: its relationship to aldosterone secretion and other metabolic changes. Clin Sci 37:311-326
Slater JDH, Tuffley RE, Williams ES, Beresford CH, Sönksen PH, Edwards RHT, Ekins RP, McLaughlin M (1969b) Control of aldosterone secretion during acclimatization to hypoxia in man. Clin Sci 37:327-341
Slaton PE Jr, Biglieri EG (1964) Hypertension and hyperaldosteronism of renal and adrenal origin. Am J Med 38:324-336
Slaton PE Jr, Biglieri EG (1967) Reduced aldosterone excretion in patients with autonomic insufficiency. J Clin Endocrinol Metab 27:37-45
Smeaton TC, Andersen GJ, Fulton IS (1977) Study of aldosterone levels in plasma during pregnancy. J Clin Endocrinol Metab 44:1-7
Smith R, Ross EJ, Marshall-Jones P (1969) Aldosterone and sodium excretion in obese subjects on water diet. Metabolism 18:700-705
Snow MH, Percy DA, Robson V, Wilkinson R (1977) An investigation into the pathogenesis of hypertension in acromegaly. Clin Sci Mol Med 53:87-91
Solounias BM, Schwartz R (1975) The effect of magnesium deficiency on serum aldosterone in rats fed two levels of sodium. Life Sci 17:1211-1217
Sonkodi S, Nicholls MG, Cumming AMM, Robertson JIS (1981) Effect of change in body posture on plasma and serum electrolytes in normal subjects and in primary aldosteronism. Clin Endocrinol 14:613-620
Sowers JR (1982) Modulation of corticosteroid secretion by dopaminergic mechanisms in rhesus monkeys. Am J Physiol 243:E375-E379
Sowers JR, Barrett JD (1983) The renin-angiotensin system in association with hyperreninemic hypoaldosteronism in neoplasia induced hypercalcaemia. Acta Endocrinol 102:439-446
Sowers JR, Beck FWJ (1983a) Dopaminergic regulation of 18-hydroxycorticosterone and aldosterone secretion in man. Acta Endocrinol 102:258-264
Sowers JR, Beck FWJ (1983b) Dopaminergic modulation of corticosteroid responses to angiotensin II in man. Clin Exp Hypertens A5:651-664
Sowers JR, Stern N (1982) Influence of sofium homeostasis and circadian rhythm on dopaminergic modulation of 18-hydroxycorticosterone secretion in man. J Clin Endocrinol Metab 55: 1046-1051
Sowers JR, Sollars E, Barrett JD, Sambhi MP (1980a) Effect of L-dopa and bilateral nephrectomy on the aldosterone response to metoclopramide. Life Sci 27:497-501
Sowers JR, Tuck ML, Golub MS, Sollars EG (1980b) Dopaminergic modulation of aldosterone secretion is independent of alterations in renin secretion. Endocrinology 107:937-941
Sowers JR, Brickman AS, Asp N, Tuck ML, Jasberg K, Magnone S (1981a) Altered dopaminergic modulation of prolactin and aldosterone secretion in pseudohypoparathyroidism. J Clin Endocrinol Metab 52:914-918
Sowers JR, Brickman AS, Sowers DK, Berg G (1981b) Dopaminergic modulation of aldosterone secretion in man is unaffected by glucocorticoids and angiotensin blockade. J Clin Endocrinol Metab 52:1078-1083
Sowers JR, Sharp B, Levin ER, Golub MS, Eggena P (1981c) Metoclopramide, a dopamine antagonist, stimulates aldosterone secretion in rhesus monkeys but not in dogs and rabbits. Life Sci 29:2171-2175
Sowers JR, Sollars E, Nyby M, Jasberg K, Tuck ML (1981d) Hyperreninemic hypoaldosteronism in association with neoplasia induced hypercalcemia in the Fischer rat. Metabolism 30: 987-991
Sowers JR, Golub MS, Eggena PH, Catania RA (1982a) Influence of sodium homeostasis on dopaminergic modulation of aldosterone, renin, and prolactin secretion in man. J Clin Endocrinol Metab 54:121-126
Sowers JR, Berg G, Tuck ML, Martin VI, Chandler DW, Mayes DM (1982b) Dopaminergic modulation of 18-hydroxycorticosterone secretion in man. J Clin Endocrinol Metab 54:523-527
Sowers JR, Catania R, Paris J, Tuck M (1982c) Effects of bromocriptine on renin, aldosterone and prolactin responses to posture and metoclopramide in idiopathic edema: possible therapeutic approach. J Clin Endocrinol Metab 54:510-516
Sowers JR, Beck FWJ, Berg G (1982d) Altered dopaminergic modulation of 18-hydroxycorticos-

terone secretion in idiopathic edema: therapeutic effects of bromocriptine. J Clin Endocrinol Metab 55:749–756
Sowers JR, Berg G, Martin VS, Mayes DM (1982e) Dopaminergic modulation of aldosterone secretion in the rhesus monkey: evidence that dopamine affects the late pathway of aldosterone biosynthesis by inhibiting the conversion of corticosterone to 18-hydroxycorticosterone. Endocrinology 110:1173–1177
Sowers JR, Sharp B, McCallum RW (1982f) Effect of domperidone, an extracerebral inhibitor of dopamine receptors, on thyrotropin, prolactin, renin, aldosterone, and 18-hydroxycorticosterone secretion in man. J Clin Endocrinol Metab 54:869–871
Sowers JR, Eggena P, Phillips D (1982g) Effect of metoclopramide and domperidone on aldosterone, 18-hydroxycorticosterone and prolactin secretion in the rhesus monkey. Metabolism 31: 1219–1223
Sowers JR, Martin VI, Stern N, Berg G (1982h) Dopaminergic control of 18-hydroxycorticosterone responses to posture, isometric exercise, and diuretic administration in normal man. J Clin Endocrinol Metab 55:475–480
Sowers JR, Beck FWJ, Stern N (1983a) Glucocorticoid suppression enhances the 18-hydroxycorticosterone and aldosterone response to metoclopramide in man. Life Sci 33:2489–2495
Sowers JR, Beck FWJ, Stern N, Asp N (1983b) Effects of metoclopramide on plasma corticosteroid levels in sheep. Endocrinology 113:903–906
Sowers JR, Beck FWJ, Eggena P (1984) Evidence for direct inhibitory effects of dopamine on zona glomerulosa secretion of 18-hydroxycorticosterone in rhesus monkeys. Life Sci 34: 2339–2346
Spät A, Fejes Tóth G (1978) Effect of indomethacin on aldosterone secretion in chronic sodium depletion. Acta Physiol Acad Sci Hung 52:87–91
Spät A, Józan S (1975) Effect of prostaglandins E_2 and A_2 on steroid synthesis by the rat adrenal gland. J Endocrinol 65:55–63
Spät A, Sturcz J, Szigeti R (1965) New observations on the function of the angiotensin-aldosterone system. Acta Physiol Acad Sci Hung 27:199–203
Spät A, Sturcz J, Szigeti R, Bánki L (1966a) Study on the role of the kidney in the regulation of aldosterone secretion in the rat. Endokrinologie 50:276–279
Spät A, Sturcz J, Bánki L, Szigeti R (1966b) Effect on adrenal steroid secretion of changes in renal haemodynamics in the rat. Acta Physiol Acad Sci Hung 30:209–214
Spät A, Sólyom S, Sturcz J, Mészáros I, Ludwig E (1969) Effect of angiotensin superfusion on the rate of aldosterone production by incubated rat adrenals. Acta Physiol Acad Sci Hung 35: 149–151
Spät A, Sarkadi B, Intödy Zs, Körner A, Szántó J (1971) Effect of renal papillary lipids and prostaglandin E_1 on corticosteroid production in the rat. Acta Physiol Acad Sci Hung 40:187–199
Spät A, Polgár Z, Koltai Gy, Révai K (1976) Acute effect of reduced renal perfusion pressure on aldosterone secretion rate in the rat. J Endocrinol 70:307–308
Spät A, Antoni FA, Balla T, Bonta E, Siklós P (1977a) A study on the interaction of ACTH and prostaglandin $F_{2\alpha}$. J Steroid Biochem 8:247–248
Spät A, Siklos P, Antoni FA, Nagy K, Szirnyi K (1977b) Effect of prostaglandin synthetase inhibitors on basal and ACTH-stimulated steroid synthesis by separated adrenocortical zones. J Steroid Biochem 8:293–299
Spät A, Józan S, Gaál K, Mózes T (1977c) Effect of indomethacin on the adrenal response to frusemide in the rat. J Endocrinol 73: 401–402
Spät A, Balla T, Lantos A, Kanyár B (1979a) Hyperaldosteronism in the sodium depleted rat: effect of a low sodium diet or diuretic treatment on steroid output by the superfused adrenal zona glomerulosa. Acta Physiol Acad Sci Hung 54:141–149
Spät A, Nagy K, Tarján É (1979b) Hyperaldosteronism in the sodium-depleted rat: mechanism of aldosterone stimulation by peritoneal dialysis with glucose solution. J Endocrinol 82:17–25
Spät A, Tarján E, Tóth G (1979c) Hyperaldosteronism in the sodium-depleted rat: mode of aldosterone-stimulating action of frusemide. J Endocrinol 82:7–15
Sparano F, Sciarra F, Natoli G, Mercuri MA, Odoardi A, Colangelo E (1978) Behaviour of plasma renin activity and aldosterone during the first 72 h of life. Clin Endocrinol 8:207–211
Spark RF, Gordon SJ, Dale SL, Melby JC (1968) Aldosterone production after suppression of corticotropic secretory activity. Arch Intern Med 122:394–398
Speckart P, Zia P, Croxson C, Mayeda S, Horton R (1978) Effect of prostaglandin inhibition on the renin and aldosterone response to posture in normal and hypertensive man. Miner Electrolyte Metab 1:208–215

Spielman WS, Davis JO (1974) The renin-angiotensin system and aldosterone secretion during sodium depletion in the rat. Circ Res 35:615-624
Spielman WS, Davis JO, Freeman RH (1976) Des-asp-1-angiotensin II: possible role in mediating the renin-angiotensin response in the rat. Proc Soc Exp Biol Med 151:177-182
Srikant CB, Patel YC (1985) Somatostatin receptors in the rat adrenal cortex: characterization and comparison with brain and pituitary receptors. Endocrinology 116:1717-1723
Stachenko J, Giroud CJP (1964) Further observations on the functional zonation of the adrenal cortex. Can J Biochem 42:1777-1786
Stachenko J, Laplante C, Giroud CJP (1964) Double isotope derivative assay of aldosterone, corticosterone, and cortisol. Can J Biochem 42:1275-1291
Staessen J, Lijnen P, Fagard R, Verschueren LJ, Amery A (1981) Rise in plasma concentration of aldosterone during long-term angiotensin II suppression. J Endocrinol 91:457-465
Stark G (1966) Zum Aldosteron- und Elektrolyt-Stoffwechsel in der Schwangerschaft. Arch Gynäkol 203:192-204
Stark G, Kossmann H (1963) Die Ausscheidung von Aldosteron, Natrium und Kalium nach fortlaufender hoher Progesterongabe. Acta Endocrinol 42:537-551
Stark G, Lehmann-Achilles J (1964) Der Einfluß hoher Kochsalz-Gaben auf die Aldosteron- und Elektrolytausscheidung bei gesunden Schwangeren. Acta Endocrinol 47:469-484
Stark P, Beckerhoff R, Leumann EP, Vetter W, Siegenthaler W (1975) Control of plasma aldosterone in infancy and childhood. A study of plasma renin activity, plasma cortisol and plasma aldosterone. Helv Paediatr Acta 30:349-356
Steele JM Jr, Lowenstein J (1974) Differential effects of an angiotensin II analogue on pressor and adrenal receptors in the rabbit. Circ Res 35:592-600
Steele JM Jr, Neusy A-J, Lowenstein J (1976) The effects of des-asp^1-angiotensin II on blood pressure, plasma aldosterone concentration and plasma renin activity in the rabbit. Circ Res 38 [Suppl 2]: II-113-II-116
Steinacker HG, Vecsei P, Lommer D, Wolff HP (1968) In vitro corticosteroid biosynthesis in adrenals of rats uraemic after bilateral nephrectomy. Acta Endocrinol 58:630-636
Stephens GA, Davis JO, Freeman RH, Watkins BE, Khosla MC (1977) The effects of angiotensin II blockade in conscious sodium-replete dogs. Endocrinology 101:378-388
Stephens GA, Davis JO, Freeman RH, Watkins BE (1978) Effects of sodium and potassium salts with anions other than chloride on renin secretion in the dog. Am J Physiol 234:F10-F15
Stern N, Beck FWJ, Sowers JR, Tuck M, Hsueh WA, Zipser RD (1983) Plasma corticosteroids in hyperreninemic hypoaldosteronism: evidence for diffuse impairment of the zona glomerulosa. J Clin Endocrinol Metab 57:217-220
Stockigt JR, Hewett MJ, Topliss DJ, Higgs EJ, Taft P (1979) Renin and renin substrate in primary adrenal insufficiency. Contrasting effects of glucocorticoid and mineralocorticoid deficiency. Am J Med 66:915-922
Stone D, Hechter O (1954) Studies on ACTH action in perfused bovine adrenals: The site of action of ACTH in corticosteroidogenesis. Arch Biochem Biophys 51:457-469
Strauch G, Vallotton MB, Touitou Y, Bricaire H (1972) The renin-angiotensin-aldosterone system in acromegaly. N Engl J Med 287:795-799
Streeten DHP, Schletter FE, Clift GV, Stevenson CT, Dalakos TG (1969) Studies of the renin-angiotensin-aldosterone system in patients with hypertension and in normal subjects. Am J Med 46:844-860
Stricker EM, Vagnucci AH, McDonald RH Jr, Leenen FH (1979) Renin and aldosterone secretions during hypovolemia in rats: relation to NaCl intake. Am J Physiol 237:R45-R51
Strickland AL, Kotchen TA (1972) A study of the renin-aldosterone system in congenital adrenal hyperplasia. J Pediatr 81:962-969
Sturcz J, Spät A, Szigeti R (1966) Effect of local aldosterone concentration on aldosterone production in incubated adrenals. Acta Physiol Acad Sci Hung 30:125-128
Sturcz J, Spät A, Szöcs G, Kövér G, Farkas A (1967) Effect of potassium on adrenal aldosterone production in rats. Acta Endocrinol 55:193-201
Sufrin G, Kirdani R, Sandberg AA, Murphy GP (1978) Studies of renin-aldosterone axis in stable normotensive and hypertensive renal allograft recipients. Urology 11:46-52
Sugano S, Ohnishi T, Hatae N, Ishimura K, Fujita H, Yamano T, Okamoto M (1985) Monoclonal antibodies against bovine adrenal cytochrome P-$450_{11\beta}$ and cytochrome P-450_{SCC}. Their isolation, characterization and application to immunohistochemical analysis of adrenal cortex. J Steroid Biochem 23:1013-1021

Sugimoto T, Ishii M, Hirata Y, Matsuoka H, Sugimoto T, Myata A, Toshimori T, Masuda H, Kangawa K, Matsuo H (1986) Increased release of atrial natriuretic polypeptides in rats with DOCA-salt hypertension. Life Sci 38:1351-1358

Suhara K, Gomi T, Sato H, Itagaki E, Takemori S, Katagiri M (1978) Purification and immunochemical characterization of the two adrenal cortex mitochondrial cytochrome *P*-450 proteins. Arch Biochem Biophys 190:290-299

Sullivan JM, Nakano KK, Tyler HR (1973) Plasma renin activity during levodopa therapy. Significance of long- and short-term treatment. JAMA 224:1726-1729

Sulon J, Sparano F, Sciarra F, Giaquinto G, Genard P (1978) 24-Hour profile of 18-hydroxy-11-deoxycorticosterone in normal supine man: relationship with cortisol and aldosterone. Clin Endocrinol 8:367-372

Sundsfjord JA (1971) Plasma renin activity and aldosterone excretion during prolonged progesterone administration. Acta Endocrinol 67:483-490

Sundsfjord JA, Aakvaag A (1972) Plasma renin activity, plasma renin substrate and urinary aldosterone excretion in the menstrual cycle in relation to the concentration of progesterone and oestrogens in the plasma. Acta Endocrinol 71:519-529

Sundsfjord JA, Aakvaag A (1973) Variations in plasma aldosterone and plasma renin activity throughout the menstrual cycle, with special reference to the pre-ovulatory period. Acta Endocrinol 73:499-508

Sundsfjord JA, Strømme SB, Refsum HE, Aakvaag A (1973) Plasma aldosterone, plasma renin activity and cortisol in exercise. Acta Endocrinol [Suppl] 177:184 (abstract)

Sundsfjord JA, Marton P, Jørgensen H, Aakvaag A (1974) Reduced aldosterone secretion during spironolactone treatment in primary aldosteronism: report of a case. J Clin Endocrinol Metab 39:734-739

Sutherland DJA, Ruse JL, Laidlaw JC (1966) Hypertension, increased aldosterone secretion and low plasma renin activity relieved by dexamethasone. Can Med Assoc J 95:1109-1119

Sutton JR, Viol GW, Gray GW, McFadden M, Keane PM (1977) Renin, aldosterone, electrolyte, and cortisol responses to hypoxic decompression. J Appl Physiol 43:421-424

Suzuki S, Franco-Saenz R, Tan SY, Mulrow PJ (1981a) Effects of indomethacin on plasma renin activity in the conscious rat. Am J Physiol 240:E286-E289

Suzuki S, Franco-Saenz R, Mulrow PJ (1981b) The role of renal prostaglandins in the renin response to isoproterenol in the rat in vitro. Endocrinology 108:1654-1657

Swales JD, Thurston H, Queiroz FP, Medina A, Holland J (1971) Dual mechanism for experimental hypertension. Lancet 2:1181-1184

Swann HG (1940) Pituitary-adrenocortical relationship. Physiol Rev 20:493-521

Swartz SL, Williams GH (1983) Role of prostaglandins in adrenal steroidogenesis. Endocrinology 113:992-996

Swartz SL, Williams GH, Hollenberg NK, Dluhy RG, Moore TJ (1980) Primacy of the renin-angiotensin system in mediating the aldosterone response to sodium restriction. J Clin Endocrinol Metab 50:1071-1074

Szalay KSz (1969) Inhibiting effect of angiotensin on potassium accumulation of adrenal cortex. Biochem Pharmacol 18:962-964

Szalay KSz (1971) The effect ouabain on aldosterone production in the rat. Acta Endocrinol 68:477-484

Szalay KSz, Stark E (1981) Effect of beta-endorphin on the steroid production of isolated zona glomerulosa and zona fasciculata cells. Life Sci 29:1355-1361

Szalay KSz, Stark E (1982) Effect of alpha-MSH on the corticosteroid production of isolated zona glomerulosa and zona fasciculata cells. Life Sci 30:2101-2108

Szalay KSz, Bacsy E, Stark E (1975) Adrenal potassium and sodium in experimental hyper- and hypoaldosteronism in the rat. Determination by electron probe X-ray microanalysis. Acta Endocrinol 80:114-125

Szalay KSz, De Wied D, Stark E, Folly G (1985) Structure-activity studies with ACTH/-MSH fragments on corticosteroid secretion of isolated zona glomerulosa and fasciculata cells. Regul Pept 11:187-192

Szylman P, Better OS, Chaimowitz C, Rosler A (1976) Role of hyperkalemia in the metabolic acidosis of isolated hypoaldosteronism. N Engl J Med 294:361-365

Tait JF, Little B (1968) The metabolism of orally and intravenously administered labeled aldosterone in pregnant subjects. J Clin Invest 47:2423-2429

Tait JF, Tait SAS (1976) The effect of changes in potassium concentration on the maximal steroidogenic response of purified zona glomerulosa cells to angiotensin II. J Steroid Biochem 7: 687-690

Tait JF, Tait SAS (1979) Recent perspectives in the history of the adrenal cortex. The Sir Henry Dale lecture for 1979. J Endocrinol 83:1*P*-24*P*

Tait JF, Little B, Tait SAS, Flood C (1962) The metabolic clearance rate of aldosterone in pregnant and non pregnant subjects estimated by both single-injection and constant-infusion methods. J Clin Invest 41:2093-2100

Tait JF, Tait SAS, Gould RP (1972) Purification of rat adrenal zona glomerulosa cells by unit gravity sedimentation. J Endocrinol 55:xxxi (abstract)

Tait JF, Tait SAS, Bell JBG, Hyatt PJ, Williams BC (1980a) Further studies on the stimulation of rat adrenal capsular cells: four types of response. J Endocrinol 87:11-27

Tait JF, Tait SAS, Bell JBG (1980b) Steroid hormone production by mammalian adrenocortical dispersed cells. Essays Biochem 16:99-175

Tait SAS, Schulster D, Okamoto M, Flood C, Tait JF (1970) Production of steroids by in vitro superfusion of endocrine tissue. II. Steroid output from bisected whole, capsular and decapsulated adrenals of normal, hypophysectomized and hypophysectomized-nephrectomized rats as a function of time of superfusion. Endocrinology 86:360-382

Tait SAS, Tait JF, Bradley JES (1972) The effect of serotonin and potassium on corticosterone and aldosterone production by isolated zona glomerulosa cells of the rat adrenal cortex. Aust J Exp Biol Med 50:833-846

Tait SAS, Tait JF, Gould RP, Brown BL, Albano JDM (1974) The preparation and use of purified and unpurified dispersed adrenal cells and a study of the relationship of their cAMP and steroid output. J Steroid Biochem 5:775-787

Takeda R, Morimoto S, Uchida K, Miyamori I, Hashiba T (1980) Effect of age on plasma aldosterone response to exogenous angiotensin II in normotensive subjects. Acta Endocrinol 94: 553-558

Takeda R, Miyamori I, Ikeda M, Koshida H, Takeda Y, Yasuhara S, Morise T, Takimoto H (1984) Circadian rhythm of plasma aldosterone and time dependent alterations of aldosterone regulators. J Steroid Biochem 20:321-323

Tallan HH, Psychoyos S, Greengard P (1967) Aldosterone synthesis by adrenal mitochondria. II. The effect of citric acid cycle intermediates; identification of the soluble stimulatory factor as fumarase. J Biol Chem 242:1912-1914

Tan SY, Burton M (1981) Hyporeninemic hypoaldosteronism - an overlooked cause of hyperkalemia. Arch Intern Med 141:30-33

Tan SY, Mulrow PJ (1975) The contribution of the zona fasciculata and glomerulosa to plasma 11-deoxycorticosterone levels in man. J Clin Endocrinol Metab 41:126-130

Tan SY, Mulrow PJ (1977) Inhibition of the renin-aldosterone response to furosemide by indomethacin. J Clin Endocrinol Metab 45:174-176

Tan SY, Mulrow PJ (1978) Regulation of 18-hydroxydeoxycorticosterone in rat. Endocrinology 102:1113-1117

Tan SY, Noth R, Mulrow PJ (1978) Direct non-chromatographic radioimmunoassay of aldosterone: validation of a commercially available kit and observations on age-related changes in concentrations in plasma. Clin Chem 24:1531-1533

Tan SY, Shapiro R, Franco R, Stockard H, Mulrow PJ (1979) Indomethacin-induced prostaglandin inhibition with hyperkalemia. A reversible cause of hyporeninemic hypoaldosteronism. Ann Intern Med 90:783-785

Tan SY, Antonipillai I, Mulrow PJ (1980) Inactive renin an prostaglandin E_2 production in hyporeninemic hypoaldosteronism. J Clin Endocrinol Metab 51:849-853

Targum SD, White ME, Collins-Rothe H, Sullivan P (1984) Serum aldosterone following dexamethasone in depressed patients. Psychoneuroendocrinology 9:305-309

Tarjan E, Spät A, Balla T, Székely A (1980) Role of the renin-angiotensin system in the adaptation of aldosterone biosynthesis to sodium restriction in the rat. Acta Endocrinol 94:381-388

Taylor AA, Davis JO (1971) Effects of carp kidney extracts and angiotensin II on adrenal steroid secretion. Am J Physiol 221:652-657

Taylor AA, Davis JO, Breitenbach RB, Hartroft PM (1970) Adrenal steroid secretion and a renal-pressor system in the chicken *(Gallus domesticus)*. Gen Comp Endocrinol 14:321-333

Taylor AA, Davis JO, Johnson JA (1972a) Control of deoxycorticosterone in the dog. Am J Physiol 223:466-472

Taylor AA, Davis JO, Braverman B (1972b) Deoxycorticosterone secretion in the bullfrog: effects of ACTH, hypophysectomy, and renin. Am J Physiol 223:858-863

Taylor AN, Farrell G (1962) Effects of brain stem lesions on aldosterone and cortisol secretions. Endocrinology 70:556-566

Telegdy G, Lissák K (1965) The effect of progesterone on adrenal corticosterone and aldosterone secretion in the female rat. Acta Physiol Acad Sci Hung 26:313-318

Thomas JP (1984) Aldosterone deficiency in a patient with idiopathic haemochromatosis. Clin Endocrinol 21:271-277

Thorn GW, Ross EJ, Crabbé J, van't Hoff W (1957) Studies on aldosterone secretion in man. Br Med J 2:955-966

Thurau K (1974) Intrarenal action of angiotensin. In: Page IH, Bumpus FM (eds) Angiotensin. Springer, Berlin Heidelberg New York, pp 475-489 (Handbook of experimental pharmacology, vol 37)

Tobian L, Coffee K, McCrea P (1969) Contrasting exchangeable sodium in rats with different types of Goldblatt hypertension. Am J Physiol 217:458-460

Todesco S, Terribile V, Borsatti A, Mantero F (1975) Primary aldosteronism due to a malignant ovarian tumor. J Clin Endocrinol Metab 41:809-819

Tonutti E, Bayer JM (1964) The structure of the adrenal cortex in aldosteronism (Conn Syndrome). Eur Rev Endocrinol 1:201-209

Touitou Y, Legrand J-C (1971) Étude d'un inhibiteur de la corticostéroidogenèse, l'aminoglutéthimide, sur la 18-hydroxylation de la corticostérone, par incubation de surrénales de moutons in vitro. CR Acad Sci 272:992-995

Touitou Y, Malmejac A, Aupetit B, Biancani J, Legrand J-C (1970) Taux de sécrétion de 18-hydroxycorticostérone. Application de la méthode d'Ulick modifiée à différents cas cliniques. Ann Biol Clin 28:49-54

Touitou Y, Bogdan A, Legrand J-C, Desgrez P (1975) Aminoglutethimide and glutethimide: effects on 18-hydroxycorticosterone biosynthesis by human and sheep adrenals in vitro. Acta Endocrinol 80:517-526

Traikov H, de Nicola AF, Birmingham MK (1969) Effects of 2-methyl-1,2 bis (3'pyridyl)-l-propanol (reduced metopirone) on the formation of 18-hydroxylated steroids by quartered rat adrenals. Steroids 13:457-463

Tree M, Morton JJ (1980) Evidence that the acute hypotensive effect of captopril in dogs is not wholly explained by a reduction of plasma angiotensin II and its direct vasoconstrictor effect. Clin Sci 59:451-456

Tremblay J, Gerzer R, Pang SC, Cantin M, Genest J, Hamet P (1985) ANF stimulation of detergent-dispersed particulate guanylate cyclase from bovine adrenal cortex. FEBS Lett 194: 210-214

Trono D, Cereda JM, Favre L (1983) Pseudo-syndrome de Conn par intoxication au pastis sans alcool. Schweiz Med Wochenschr 113:1092-1095

Trost BN, Müller J (1976) Uptake and metabolism of serotonin by rat adrenal tissue in vitro. Acta Endocrinol 82:353-363

Tsai R, Morris DJ (1978) The effect of spironolactone on the hepatic metabolism of aldosterone in male rats. Endocrinology 103:1239-1244

Tsai R, Davis RP, Morris DJ (1979) The effect of the antimineralocorticoid spironolactone on the hepatic synthesis of polar metabolites of aldosterone in male rats. J Steroid Biochem 13: 481-487

Tsunoda S, Demura R, Demura H (1984) 6-Keto-prostaglandin E_1 and prostaglandin E_2 stimulate aldosterone production in bovine adrenal glands. Clin Exp Pharmacol Physiol 11:533-536

Tucci JR, Espiner EA, Jagger PI, Pauk GL, Lauler DP (1967a) ACTH stimulation of aldosterone secretion in normal subjects and in patients with adrenocortical insufficiency. J Clin Endocrinol Metab 27:568-575

Tucci JR, Espiner EA, Jagger PI, Lauler DP (1967b) The effect of metyrapone on aldosterone secretion in man. Acta Endocrinol 56:376-384

Tuck ML, Mayes DM (1980) Mineralocorticoid biosynthesis in patients with hyporeninemic hypoaldosteronism. J Clin Endocrinol Metab 50:341-347

Tuck ML, Williams GH, Cain JP, Sullivan JM, Dluhy RG (1973) Relation of age, diastolic pressure and known duration of hypertension to presence of low renin essential hypertension. Am J Cardiol 32:637-642

Tuck ML, Dluhy RG, Williams GH (1974) A specific role for saline or the sodium ion in the regulation of renin and aldosterone secretion. J Clin Invest 53:988-995

Tuck ML, Dluhy RG, Williams GH (1975) Sequential responses of the renin-angiotensin-aldosterone axis to acute postural change: effect of dietary sodium. J Lab Clin Med 86:754-763

Tuck ML, Chandler DW, Mayes DM (1977) The influences of ACTH, dietary sodium, upright posture, and angiotensin II on plasma 18-hydroxy-11-deoxycorticosterone levels in normal subjects. J Clin Endocrinol Metab 45:893-902

Tuck ML, Sambhi MP, Levin L (1979) Hyporeninemic hypoaldosteronism in diabetes mellitus. Studies of the autonomic nervous system's control of renin release. Diabetes 28:237-241

Tuck ML, Sowers JR, Fittingoff DB, Fisher JS, Berg GJ, Asp ND, Mayes DM (1981) Plasma corticosteroid concentrations during spironolactone administration: evidence for adrenal biosynthetic blockade in man. J Clin Endocrinol Metab 52:1057-1061

Tuma J, Záruba K, Studer A, Lüscher T, Siegenthaler W, Vetter H, Vetter W (1981) Regulation der Plasmaaldosteronkonzentration bei nierenlosen Patienten. Klin Wochenschr 59:27-34

Turpie AGG, Thomson TJ (1964) Carbenoxolone sodium in the treatment of gastric ulcer with special reference to side-effects. Gut 6:591-594

Tvedegaard E, Frederiksen V, Ølgaard K, Damkjaer Nielsen M, Starup J (1981) Two cases of 17α-hydroxylase deficiency - one combined with complete gonadal agenesis. Acta Endocrinol 98:267-273

Uchida K, Azukizawa S, Imaizumi N, Kigoshi T, Yamamoto I, Hosojima H, Morimoto S (1986) Effect of angiotensin II on aldosterone and its precursor steroid production in adrenal zona glomerulosa cells from heparin-treated rats. Acta Endocrinol 111:222-227

Ulick S (1973) Normal and alternate pathways in aldosterone biosynthesis. In: Scow RO (ed) Endocrinology. Proceedings of the 4th international congress of endocrinology, Washington DC 1972. Excerpta Medica, Amsterdam, pp 761-767

Ulick S (1976) Diagnosis and nomenclature of the disorders of the terminal portion of the aldosterone biosynthetic pathway. J Clin Endocrinol Metab 43:92-96

Ulick S, Chu MD (1982) Hypersecretion of a new corticosteroid, 18-hydroxycortisol in two types of adrenocortical hypertension. Clin Exp Hypertension (A) 4:1771-1777

Ulick S, Feinholtz E (1968) Metabolism and rate of secretion of aldosterone in the bullfrog. J Clin Invest 47:2523-2529

Ulick S, Vetter KK (1965) Simultaneous measurement of secretory rates of aldosterone and 18-hydroxycorticosterone. J Clin Endocrinol Metab 25:1015-1026

Ulick S, Nicolis GL, Vetter KK (1964a) Relationship of 18-hydroxycorticosterone to aldosterone. In: Baulieu EE, Robel P (eds) Aldosterone. Blackwell, Oxford, pp 3-17

Ulick S, Gautier EE, Vetter KK, Markello JR, Yaffe S, Lowe CU (1964b) An aldosterone biosynthetic defect in a salt-losing disorder. J Clin Endocrinol Metab 24:669-672

Ulick S, Eberlein WR, Bliffeld AR, Chu MD, Bongiovanni AM (1980) Evidence for an aldosterone biosynthetic defect in congenital adrenal hyperplasia. J Clin Endocrinol Metab 51:1346-1353

Ulick S, Chu MD, Land M (1983a) Biosynthesis of 18-oxocortisol by aldosterone-producing adrenal tissue. J Biol Chem 258:5498-5502

Ulick S, Land M, Chu MD (1983b) 18-oxocortisol, a naturally occurring mineralocorticoid agonist. Endocrinology 113:2320-2322

Ulmann A, Menard J, Breminer J, Corvol P (1973) Dissociation between plasma renin activity and plasma aldosterone in man after oral administration of D-L propranolol. Biomedicine 19:547-549

Urquhart J, Davis JO, Higgins JT Jr (1963) Effects of prolonged infusion of angiotensin II in normal dogs. Am J Physiol 205:1241-1246

Urquhart J, Davis JO, Higgins JT Jr (1964) Simulation of spontaneous secondary hyperaldosteronism by intravenous infusion of angiotensin II in dogs with an arteriovenous fistula. J Clin Invest 43:1355-1366

Vaamonde CA, Perez GO, Oster JR (1981) Syndromes of aldosterone deficiency. Miner Electrolyte Metab 5:121-134

Vagnucci AH, Shapiro AP, McDonald RH Jr (1969) Effects of upright posture on renal electrolyte cycles. J Appl Physiol 26:720-731

Vagnucci AH, McDonald RH Jr, Drash AL, Wong AKC (1974) Intradiem changes of plasma aldosterone, cortisol, corticosterone and growth hormone in sodium restriction. J Clin Endocrinol Metab 38:761-776

Vahouny GV, Chanderbhan R, Hinds R, Hodges VA, Treadwell CR (1978) ACTH-induced hydrolysis of cholesteryl esters in rat adrenal cells. J Lipid Res 19:570-577

Vahouny GV, Chanderbhan R, Noland BJ, Irwin D, Dennis P, Lambeth D, Scallen TJ (1983) Sterol carrier protein$_2$. Identification of adrenal sterol carrier protein$_2$ and site of action for mitochondrial cholesterol utilization. J Biol Chem 258:11731-11737

Van Acker KJ, Scharpe SL, Deprettere AJR, Neels HM (1979) Renin-angiotensin-aldosterone system in the healthy infant and child. Kidney Int 16:196-203

Vander AJ (1967) Control of renin release. Physiol Rev 47:359-382

Vander AJ (1970) Direct effects of potassium ion on renin secretion and renal function. Am J Physiol 219:455-459

Van der Wal B, De Wied D (1963) Effects of various peptides on aldosterone production of sodium deficient rats. Acta Endocrinol 59:186-192

Vandongen R (1975) Inhibition of renin secretion in the isolated rat kidney by antidiuretic hormone. Clin Sci Mol Med 49:73-76

Van Stone JC, Bauer J, Carey J (1982) The effect of dialysate sodium concentration on body fluid compartment volume, plasma renin activity and plasma aldosterone concentration in chronic hemodialysis patients. Am J Kidney Dis 2:58-64

Varma MM (1977) Ultrastructural evidence for aldosterone- and corticosterone-secreting cells in the adrenocortical tissue of the American Bullfrog *(rana catesbeiana)*. Gen Comp Endocrinol 33:61-75

Vaughan ED Jr, Peach MJ, Ackerly JA, Tsai B-S, Larner A (1977) Pressor and steroidogenic actions of [des-Asp^1] angiotensin I. Dependency on conversion to angiotensin III. Circ Res 40 [Suppl 1]:I-94-I-97

Vazir H, Whitehouse BJ, Vinson GP, McCredie E (1981) Effects of prolonged ACTH treatment on adrenal steroidogenesis and blood pressure in rats. Acta Endocrinol 97:533-542

Vecsei P, Kessler H (1970) In vivo conversion of corticosterone into aldosterone in rats treated with ACTH or submitted to stress. Experientia 26:1015-1016

Vecsei P, Kessler H (1971) In vivo conversion of radioactive progesterone and corticosterone to adrenal cortical hormones in normal and ACTH-treated rats. Acta Endocrinol 68:759-770

Vecsei P, Lommer D, Steinacker HG, Wolff HP (1966a) Changes of 18-hydroxycorticosterone and aldosterone synthesis in rat adrenals in vitro after renal hypertension, nephrectomy, and variation of sodium intake in vivo. Eur J Steroids 1:91-101

Vecsei P, Lommer D, Steinacker HG, Vecsei-Görgenyi A, Wolff HP (1966b) In vitro Corticosteroidbiosynthese in proliferierenden Rattennebennieren. Acta Endocrinol 53:24-36

Vecsei P, Lommer D, Wolff HP (1968) The intermediate role of 18-hydroxycorticosteroids in aldosterone biosynthesis. Experientia 24:1199-1201

Vecsei P, Düsterdieck G, Jahnecke J, Lommer D, Wolff HP (1969) Secretion and turnover of aldosterone in various pathological states. Clin Sci 36:241-256

Veldhuis JD, Melby JC (1981) Isolated aldosterone deficiency in man: acquired and inborn errors in the biosynthesis or action of aldosterone. Endocr Rev 2:495-517

Veldhuis JD, Kulin HE, Santen RJ, Wilson TE, Melby JC (1980) Inborn error in the terminal step of aldosterone biosynthesis. Corticosterone methyl oxidase type II deficiency in a North American pedigree. N Engl J Med 303:117-121

Venning EH, Lucis OJ (1962) Effect of growth hormone on the biosynthesis of aldosterone in the rat. Endocrinology 70:486-491

Venning EH, Dyrenfurth I, Beck JC (1956a) Effect of corticotropin and prednisone on the excretion of aldosterone in man. J Clin Endocrinol Metab 16:1541-1553

Venning EH, Dyrenfurth I, Giroud CJP (1956b) Aldosterone excretion in healthy persons. J Clin Endocrinol Metab 16:1326-1332

Venning EH, Dyrenfurth I, Lewenstein L, Beck J (1959) Metabolic studies in pregnancy and the puerperium. J Clin Endocrinol Metab 19:403-424

Venning E, Dyrenfurth I, Dossetor JB, Beck JC (1962) Influence of alterations in sodium intake on urinary aldosterone response to corticotropin in normal individuals and patients with essential hypertension. Metabolism 11:254-264

Verdesca AS, Westermann CD, Crampton RS, Black WC, Nedeljkovic RI, Hilton JG (1961) Direct adrenocortical stimulatory effect of serotonin. Am J Physiol 201:1065-1067

Vetter H, Vetter W (1975a) Regulation of aldosterone secretion in primary aldosteronism. Horm Metab Res 7:418-424

Vetter H, Vetter W (1975b) Control of plasma aldosterone in a patient with Conn's syndrome before and during spironolactone medication. Klin Wochenschr 53:391-393

Vetter H, Berger M, Armbruster H, Siegenthaler W, Werning C, Vetter W (1974) Episodic secretion of aldosterone in primary aldosteronism: relationship to cortisol. Clin Endocrinol 3:41-48

Vetter H, Appenheimer M, Lucas R, Weiand H, Herschbach ML, Glänzer K, Wittasek F, Krück F (1977) Effect of a spirolactone on plasma and urinary aldosterone in primary aldosteronism. Horm Res 8:23-28

Vetter W, Záruba K, Armbruster H, Beckerhoff R, Reck G, Siegenthaler W (1974) Control of plasma aldosterone in supine anephric man. Clin Endocrinol 3:411-419

Vetter W, Záruba K, Armbruster H, Beckerhoff R, Siegenthaler W (1975) Einfluß der Hämodialyse auf die Plasmaaldosteronkonzentration bei nierenlosen Patienten. Klin Wochenschr 53: 17-20

Vetter W, Záruba K, Armbruster H, Beckerhoff R, Nussberger J, Furrer J, Fontana A, Siegenthaler W (1977) Control of plasma aldosterone during hemodialysis in patients with terminal renal failure. Nephron 18:114-123

Veverbrants E, Arky RA (1969) Effects of fasting and refeeding. I. Studies on sodium, potassium and water excretion on a constant electrolyte and fluid intake. J Clin Endocrinol Metab 29: 55-62

Veyrat R, Manning EL, Fabre JF, Muller AF (1963) Messure de la sécrétion de l'aldostérone sous administration d'un adrénostatique semi-synthétique l'héparinoïde Ro 1-8307. Rev Franç Étud Clin Biol 8:667-673

Veyrat R, Brunner HR, Grandchamp A, Muller AF (1967a) Inhibition of renin release by potassium in man. Acta Endocrinol [Suppl] 119:86 (abstract)

Veyrat R, Brunner HR, Manning EL, Muller AF (1967b) Inhibition de l'activité de la rénine plasmatique par le potassium. J Urol Nephrol 73:271-275

Veyrat R, Muller AF, Mach RS (1968) Les hyperaldostéronismes secondaires. Schweiz Med Wochenschr 98:65-75

Vincent M, Dessart Y, Annat G, Sassard J, François R, Ciér JF (1980) Plasma renin activity, aldosterone, and dopamine-beta-hydroxylase activity as a function of age in normal children. Pediatr Res 14:894-895

Vinson GP, Whitehouse BJ (1969) The relationship between steroid production from endogenous precursors and from added radioactive precursors by rat adrenal tissue in vitro. The effect of corticotrophin. Acta Endocrinol 61:695-708

Vinson GP, Whitehouse BJ (1982) Effects of stimulation on the steroid profile formed by rat adrenal capsule tissue incubated in vitro. J Steroid Biochem 17:159-164

Vinson GP, Whitehouse BJ, Goddard C (1978a) Steroid 17-hydroxylation and androgen production by incubated rat adrenal tissue. J Steroid Biochem 9:677-683

Vinson GP, Goddard C, Whitehouse BJ (1978b) Corticosteroid production in vitro by adrenal tissue from rats with inherited hypothalamic diabetes insipidus (Brattleboro strain). J Steroid Biochem 9:657-665

Vinson GP, Whitehouse BJ, Goddard C, Sibley CP (1979) Comparative and evolutionary aspects of aldosterone secretion and zona glomerulosa function. J Endocrinol 81:5*P*-24*P*

Vinson GP, Whitehouse BJ, Dell A, Etienne T, Morris HR (1980) Characterization of an adrenal zona glomerulosa-stimulating component of posterior pituitary extracts as α-MSH. Nature 284:464-467

Vinson GP, Whitehouse BJ, Dell A, Etienne AT, Morris HR (1981) Specific stimulation of steroidogenesis in rat adrenal zona glomerulosa cells by pituitary peptides. Biochem Biophys Res Commun 99:65-72

Vinson GP, Whitehouse BJ, Dell A, Bateman A, McAuley ME (1983) α-MSH and zona glomerulosa function in the rat. J Steroid Biochem 19:537-544

Visser HKA, Cost WS (1964) A new hereditary defect in the biosynthesis of aldosterone: urinary C_{21}-corticoid pattern in three related patients with a salt-losing syndrome suggesting a 18-oxidation defect. Acta Endocrinol 47:589-612

Voigt J, Köster H (1980) Induction of plasma proangiotensin by steroid hormones in nephrectomized rats. Eur J Biochem 110:57-65

Volpe M, Odell G, Kleinert HD, Muller F, Camargo MJ, Laragh JH, Maack T, Vaughan ED Jr, Atlas SA (1985) Effect of atrial natriuretic factor on blood pressure, renin, and aldosterone in Goldblatt hypertension. Hypertension 7:143-148

Wada A, Okamoto M, Nonaka Y, Yamano T (1984) Aldosterone biosynthesis by a reconstituted cytochrome P-$450_{11\beta}$ system. Biochem Biophys Res Commun 119:365-371

Wada A, Ohnishi T, Nonaka Y, Okamoto M, Yamano T (1985) Synthesis of aldosterone by a reconstituted system of cytochrome P-$450_{11\beta}$ from bovine adrenocortical mitochondria. J Biochem 98:245-256

Wakerlin GE (1958) Antibodies to renin as proof of the pathogenesis of sustained renal hypertension. Circulation 17:653-657

Waldhäusl W, Herkner K, Nowotny P, Bratusch-Marrain P (1978) Combined 17α- and 18-hydroxylase deficiency associated with complete male pseudohermaphrotitism and hypoaldosteronism. J Clin Endocrinol Metab 46:236-246

Waldman SA, Rapoport RM, Murad F (1984) Atrial natriuretic factor selectively activates particulate guanylate cyclase and elevates cyclic GMP in rat tissues. J Biol Chem 259:1432-1434

Wambach G, Higgins JR (1978) Antimineralocorticoid action of progesterone in the rat: correlation of the effect of electrolyte excretion and interaction with renal mineralocorticoid receptors. Endocrinology 102:1686-1693

Wambach G, Higgins JR (1979) Antihypertensive effect of progesterone in rats with mineralocorticoid-induced hypertension. Am J Physiol 5:E366-E370

Wang C, Chan TK, Yeung RTT, Coghlan JP, Scoggins BA, Stockigt JR (1981) The effect of triamterene and sodium intake on renin, aldosterone, and erythrocyte sodium transport in Liddle's syndrome. J Clin Endocrinol Metab 52:1027-1032

Washburn DD, Kem DC, Orth DN, Nicholson WE, Chrétien M, Mount CD (1982) Effect of β-lipotropin on aldosterone production in the isolated rat adrenal cell preparation. J Clin Endocrinol Metab 54:613-618

Watanabe M, Meeker CI, Gray MJ, Sims EAH, Solomon S (1963) Secretion rate of aldosterone in normal pregnancy. J Clin Invest 42:1619-1631

Watanabe M, Meeker CI, Gray MJ, Sims EAH, Solomon S (1965) Aldosterone secretion rates in abnormal pregnancy. J Clin Endocrinol Metab 25:1665-1670

Watanuki M, Tilley BE, Hall PF (1978) Cytochrome P-450 for 11β- and 18-hydroxylase activities of bovine adrenocortical mitochondria: One enzyme or two? Biochemistry 17:127-130

Watkins L Jr, Burton JA, Haber E, Cant JR, Smith FW, Barger AC (1976) The renin-angiotensin-aldosterone system in congestive failure in conscious dogs. J Clin Invest 57:1606-1617

Weaver J, Nickerson PA, Molteni A, Solliday N, Albertson D (1981) Failure of chronic administration of the angiotensin I-converting enzyme inhibitor, teprotide (SQ 20881), to affect the ultrastructure of the adrenal cortex and the aldosterone secretion in the rat. Lab Invest 45: 527-533

Weber PC, Larsson C, Änggård E, Hamberg M, Corey EJ, Nicolaou KC, Samuelsson B (1976) Stimulation of renin release from rabbit renal cortex by arachidonic acid and prostaglandin endoperoxides. Circ Res 39:868-874

Weidmann P, Siegenthaler W (1967) Das Renin-Angiotensin-Aldosteron-System bei hypertensiven Zuständen. Dtsch Med Wochenschr 92:1953-1961

Weidmann P, Maxwell MH, Lupu AN (1973a) Plasma aldosterone in terminal renal failure. Ann Intern Med 78:13-18

Weidmann P, Horton R, Maxwell MH, Franklin SS, Fichman M (1973b) Dynamic studies of aldosterone in anephric man. Kidney Int 4:289-298

Weidmann P, Reinhart R, Maxwell MH, Rowe P, Coburn JW, Massry SG (1973c) Syndrome of hyporeninemic hypoaldosteronism and hyperkalemia in renal disease. J Clin Endocrinol Metab 36:965-977

Weidmann P, Maxwell MH, Rowe P, Riner R, Massry SG (1975a) Role of the renin-angiotensin-aldosterone system in the regulation of plasma potassium in chronic renal disease. Nephron 15:35-49

Weidmann P, Maxwell MH, de Lima J, Hirsch D, Franklin SS (1975b) Control of aldosterone responsiveness in terminal renal failure. Kidney Int 7:351-359

Weidmann P, De Myttenaere-Bursztein S, Maxwell MH, De Lima J (1975c) Effect of aging on plasma renin and aldosterone levels in normal man. Kidney Int 79:411-424

Weidmann P, de Châtel R, Schiffmann A, Bachmann E, Beretta-Piccoli C, Reubi FC, Ziegler WH, Vetter W (1977) Interrelations between age and plasma renin, aldosterone and cortisol, urinary catecholamines, and the body sodium/volume state in normal man. Klin Wochenschr 55:725-733

Weidmann P, de Châtel R, Ziegler WH, Flammer J, Reubi F (1978a) Alpha and beta adrenergic blockade with orally administered labetalol in hypertension. Studies on blood volume, plasma renin and aldosterone and catecholamine excretion. Am J Cardiol 41:570-576

Weidmann P, Beretta-Piccoli C, Ziegler WH, Keusch G, Glück Z, Reubi FC (1978b) Age versus urinary sodium for judging renin, aldosterone and catecholamine levels: Studies in normal subjects and patients with essential hypertension. Kidney Int 14:619-628

Weidmann P, Beretta-Piccoli C, Glück Z, Keusch G, Reubi FC, de Châtel R, Cottier C (1980) Hypoaldosteronism without hyperkalemia. Klin Wochenschr 58:185-194

Weinberger MH (1984) Primary aldosteronism: diagnosis and differentiation of subtypes. Ann Intern Med 100:300-302

Weinberger MH, Collins RD, Dowdy AJ, Nokes GW, Luetscher JA (1969) Hypertension induced by oral contraceptives containing estrogen and gestagen. Effects on plasma renin activity and aldosterone excretion. Ann Intern Med 71:891-902

Weinberger MH, Kem DC, Gomez-Sanchez C, Kramer NJ, Martin BT, Nugent CA (1975) The effect of dexamethasone on the control of plasma aldosterone concentration in normal recumbent man. J Lab Clin Med 85:957-967

Weinberger MH, Kramer NJ, Grim CE, Petersen LP (1977) The effect of posture and saline loading on plasma renin activity and aldosterone concentration in pregnant, non-pregnant and estrogen-treated women. J Clin Endocrinol Metab 44:69-77

Weinstein RL, Kliman B (1972) In vitro inhibition of aldosterone biosynthesis by 17-hydroxylated steroids in rat adrenal. Metabolism 21:99-106

Weir RJ, Brown JJ, Fraser R, Kraszewski A, Lever AF, McIlwaine GM, Morton JJ, Robertson JIS, Tree M (1973) Plasma renin, renin substrate, angiotensin II and aldosterone in hypertensive disease of pregnancy. Lancet 1:291-294

Weir RJ, Brown JJ, Fraser R, Lever AF, Logan RW, McIlwaine GM, Morton JJ, Robertson JIS, Tree M (1975) Relationship between plasma renin, renin-substrate, angiotensin II, aldosterone and electrolytes in normal pregnancy. J Clin Endocrinol Metab 40:108-115

Welch WJ, Ott CE, Guthrie GP Jr, Kotchen TA (1983) Effect of adrenalectomy on renin release in the rat. Endocrinology 113:2086-2091

Weldon VV, Kowarski A, Migeon CJ (1967) Aldosterone secretion rates in normal subjects from infancy to adulthood. Pediatrics 39:713-723

Wellen JJ, Benraad TJ (1969a) Effect of ouabain on corticosterone biosynthesis and on potassium and sodium concentration in calf adrenal tissue in vitro. Biochim Biophys Acta 183:110-117

Wellen JJ, Benraad TJ (1969b) Effect of ouabain on corticosterone biosynthesis and in intracellular potassium and sodium concentration in adrenal tissue in vitro. Acta Endocrinol [Suppl] 138:50 (abstract)

Wenting GJ, Man in't Veld AJ, Derkx FH, Brummelen PV, Schalekamp MADH (1978) ACTH-dependent aldosterone excess due to adrenocortical adenoma: a variant of primary aldosteronism. J Clin Endocrinol Metab 46:326-335

Werning C, Baumann K, Weidmann P, Gysling E, Siegenthaler W (1969) Die Plasmareninaktivität bei dekompensiertem und nicht-dekompensiertem Diabetes insipidus. Zugleich ein Beitrag zur Regulation der Reninsekretion. Schweiz Med Wochenschr 99:661-668

Werning C, Schweikert HU, Stiel D, Vetter W, Siegenthaler W (1970) Aldosteron und Renin bei Akromegalie. Klin Wochenschr 48:1365-1367

Werning C, Stadelmann O, Miederer SE, Vetter W, Schweikert HU, Stiel D, Siegenthaler W (1971) Der Einfluß einer Carbenoxolon-Therapie auf das Renin-Angiotensin-Aldosteron-System. Klin Wochenschr 49:1285-1289

Werning C, Bayer JM, Fischer N, Schweikert HU, Siegenthaler W (1972) Die Wirkung von Carbenoxolon-Natrium auf den Blutdruck, die Plasmareninaktivität und die Serumelektrolyte bei adrenalektomierten Patienten. Dtsch Med Wochenschr 97:91-92

Wettstein A (1961) Biosynthèse des hormones steroides. Experientia 17:329-344

Whipp GT, Wintour EM, Coghlan JP, Scoggins BA (1976) Regulation of aldosterone in the guinea pig - effect of oestrus cycle, pregnancy and sodium status. Aust J Exp Biol Med Sci 54: 71-78

Whipp GT, Coghlan JP, Shulkes AA, Skinner SL, Wintour EM (1978) Regulation of aldosterone in the rat. Effect of oestrous cycle, pregnancy, and sodium status. Aust J Exp Biol Med Sci 56: 545-551

White FN, Gold EM, Vaughn DL (1967) Renin-aldosterone system in endotoxin shock in the dog. Am J Physiol 212:1195-1198

White RE, Coon MJ (1980) Oxygen activation by cytochrome P-450. Annu Rev Biochem 49: 315-356

Whitehouse BJ, Vinson GP, Thody AJ (1982) Dopaminergic control of aldosterone: modulation

of the response of rat adrenal zona glomerulosa cells to α-MSH by pretreatment with bromocriptine or metoclopramide. Steroids 39:155–164
Whitfield L, Sowers JR, Tuck ML, Golub MS (1980) Dopaminergic control of plasma catecholamine and aldosterone responses to acute stimuli in normal man. J Clin Endocrinol Metab 51: 724–729
Whitley GStJ, Bell JG, Chu FW, Tait JF, Tait SAS (1984) The effects of ACTH, serotonin, K^+ and angiotensin analogues on ^{32}P incorporation into phospholipids of the rat adrenal cortex: basis for an assay method using zona glomerulosa cells. Proc R Soc Lond Biol 222: 273–294
Wiederholt M, Hampel J, Belkien L, Oelkers W (1984) Aldosterone secretion, measurements of membrane potential and intracellular potassium activity in the isolated zona glomerulosa. FEBS Lett 174:211–214
Williams BC, McDougall JG, Hyatt PJ, Bell JBG, Tait SAS, Tait JF (1979) Purification of dispersed rat adrenal cells by column filtration. J Endocrinol 81:109 *P* (abstract)
Williams BC, McDougall JG, Tait JF, Tait SAS (1981) Calcium efflux and steroid output from superfused rat adrenal cells: effects of potassium, adrenocorticotropic hormone, 5-hydroxytryptamine, adenosine 3′:5′-cyclic monophosphate and angiotensins II and III. Clin Sci 61: 541–551
Williams FA Jr, Schambelan M, Biglieri EG, Carey RM (1983) Acquired primary hypoaldosteronism due to an isolated zona glomerulosa defect. N Engl J Med 309:1623–1627
Williams GH, Braley LM (1977) Effects of dietary sodium and potassium intake and acute stimulation on aldosterone output by isolated human adrenal cells. J Clin Endocrinol Metab 45: 55–64
Williams GH, Dluhy RG (1972) Aldosterone biosynthesis. Interrelationship of regulatory factors. Am J Med 53:595–605
Williams GH, Dluhy RG, Underwood RH (1970) The relationship of dietary potassium intake to the aldosterone stimulating properties of ACTH. Clin Sci 39:489–496
Williams GH, Rose LI, Dluhy RG, Dingman JF, Lauler DP (1971) Aldosterone response to sodium restriction and ACTH stimulation in panhypopituitarism. J Clin Endcrinol Metab 32: 27–35
Williams GH, McDonnell LM, Tait SAS, Tait JF (1972a) The effect of medium composition and in vitro stimuli on the conversion of corticosterone to aldosterone in rat glomerulosa tissue. Endocrinology 91:948–960
Williams GH, Cain JP, Dluhy RG, Underwood RH (1972b) Studies of the control of plasma aldosterone concentration in normal man. I. Response to posture, acute and chronic volume depletion, and sodium loading. J Clin Invest 51:1731–1742
Williams GH, Tuck ML, Rose LI, Dluhy RG, Underwood RH (1972c) Studies of the control of plasma aldosterone concentration in normal man. III. Response to sodium chloride infusion. J Clin Invest 51:2645–2652
Williams GH, Bailey GL, Hampers CL, Lauler DP, Merrill JP, Underwood RH, Blair-West JR, Coghlan JP, Denton DA, Scoggins BA, Wright RD (1973) Studies on the metabolism of aldosterone in chronic renal failure and anephric man. Kidney Int 4:280–288
Williams GH, McDonnell LM, Raux MC, Hollenberg NK (1974) Evidence for different angiotensin II receptors in rat adrenal glomerulosa and rabbit vascular smooth muscle cells. Studies with competitive antagonists. Circ Res 34:384–390
Williams GH, Braley LM, Underwood RH (1976a) The regulation of plasma 18-hydroxy-11-deoxycorticosterone in man. J Clin Invest 58:221–229
Williams GH, Hollenberg NK, Braley LM (1976b) Influence of sodium intake on vascular and adrenal angiotensin II receptors. Endocrinology 98:1343–1350
Williams GH, Dluhy RG, Moore TJ (1977) Aldosterone regulation in essential hypertension. Altered adrenal responsiveness to angiotensin II. Mayo Clin Proc 52:312–316
Williams GH, Hollenberg NK, Brown C, Mersey JH (1978) Adrenal responses to pharmacological interruption of the renin-angiotensin system in sodium-restricted normal man. J Clin Endocrinol Metab 47:725–731
Wilson ID, Goetz FC (1964) Selective hypoaldosteronism after prolonged heparin administration. A case report, with postmortem findings. Am J Med 36:635–640
Wilson JX, Aguilera G, Catt KJ (1984) Inhibitory actions of calmodulin antagonists on steroidogenesis in zona glomerulosa cells. Endocrinology 115:1357–1363
Wilson M, Morganti AO, Zervoudakis I, Letcher RL, Romney BM, von Oeyon P, Papera S, Sea-

ley JE, Laragh JH (1980) Blood pressure, the renin-aldosterone system and sex steroids throughout normal pregnancy. Am J Med 68:97-104

Wilson TA, Kaiser DL, Carey RM (1983a) Dopaminergic inhibition of aldosterone secretion in man is independent of the autonomic nervous system. J Clin Endocrinol Metab 57:200-203

Wilson TA, Kaiser DL, Peach MJ, Wright EM, Carey RM (1983b) Possible mechanism of action of metoclopramide-induced aldosterone secretion: in vivo and in vitro studies in the sheep. Endocrinology 113:887-892

Winkel CA, Milewich L, Parker CR Jr, Gant NF, Simpson ER, MacDonald PC (1980a) Conversion of plasma progesterone to deoxycorticosterone in men, non-pregnant and pregnant women, and adrenalectomized subjects. Evidence for steroid 21-hydroxylase activity in non-adrenal tissue. J Clin Invest 66:803-812

Winkel CA, Simpson ER, Milewich L, MacDonald PC (1980b) Deoxycorticosterone biosynthesis in human kidney: potential for formation of a potent mineralocorticosteroid in its site of action. Proc Natl Acad Sci USA 77:7069-7073

Winkel CA, Casey ML, Simpson ER, MacDonald PC (1981) Deoxycorticosterone biosynthesis from progesterone in kidney tissue of the human fetus. J Clin Endocrinol Metab 53:10-15

Winkel CA. Casey ML, Worley RJ, Madden JD, MacDonald PC (1983) Extraadrenal steroid 21-hydroxylase activity in a woman with congenital adrenal hyperplasia due to steroid 21-hydroxylase deficiency. J Clin Endocrinol Metab 56:104-107

Winquist RJ, Faison EP, Waldman SA, Schwartz K, Murad F, Rapoport RM (1984) Atrial natriuretic factor elicits an endothelium-independent relaxation and activates particulate guanylate cyclase in vascular smooth muscle. Proc Natl Acad Sci USA 81:7661-7664

Wintour EM, Knobil E, Scoggins BA, Skinner SL, Coghlan JP (1974) The renin-aldosterone systems in the pregnant rhesus monkey *(Macaca mulatta).* Clin Exp Pharmacol Physiol 1:167-170

Wintour EM, Brown EH, Denton DA, Hardy KJ, McDougall JG, Oddie CJ, Whipp GT (1975) The ontogeny and regulation of corticosteroid secretion by the ovine foetal adrenal. In vitro and in vivo studies. Acta Endocrinol 79:301-316

Wintour EM, Blair-West JR, Brown EH, Coghlan JP, Denton DA, Nelson J, Oddie CJ, Scoggins BA, Whipp GT, Wright RD (1976) The effect of pregnancy on mineralo- and gluco-corticoid secretion in the sheep. Clin Exp Pharmacol Physiol 3:331-342

Wintour EM, Brown EH, Denton DA, Hardy KJ, McDougall JG, Robinson PM, Rowe EJ, Whipp GT (1977) In vitro and in vivo adrenal cortical steroid production by fetal sheep: effect of angiotensin II, sodium deficiency, ACTH. In: Conti C (ed) Research on steroids, vol 7. Elsevier, Amsterdam, pp 475-485

Wintour EM, Coghlan JP, Oddie CJ, Scoggins BA, Walters WAW (1978) A sequential study of adrenocorticosteroid level in human pregnancy. Clin Exp Pharmacol Physiol 5:399-403

Wintour EM, Barnes A, Cahill F, Hardy KJ, Horacek I, Scoggins BA (1979) Potassium:aldosterone relationships in pregnant ewes and chronically cannulated ovine fetuses. Pediatr Res 13: 265-267

Wintour EM, Coghlan JP, Hardy KJ, Lingwood BE, Rayner M, Scoggins BA (1980) Placental transfer of aldosterone in the sheep. J Endocrinol 86:305-310

Wisgerhof M (1984) In vitro responses of aldosterone-producing adenomas to angiotensin II and ACTH after treatment with spironolactone before adrenalectomy. J Lab Clin Med 104:160-165

Witzgall H, Hassan-Ali S (1981) A simultaneous radioimmunoassay for aldosterone and its precursors: human plasma levels following the inhibition of converting enzyme, before and after blockade of prostaglandin synthesis. J Clin Chem Clin Biochem 19:387-394

Witzgall H, von Werder K, Weber PC (1983) Mineralocorticoid and prolactin response to the dopamine antagonist metoclopramide in patients with primary aldosteronism. J Steroid Biochem 19:1671-1676

Wolfe LK, Gordon RD, Island DP, Liddle GW (1966) An analysis of factors determining the circadian pattern of aldosterone excretion. J Clin Endocrinol Metab 11:1261-1266

Womersley RA, Darragh JH (1955) Potassium and sodium restriction in the normal human. J Clin Invest 34:456-461

Woodcock EA, Johnston CI (1984) Inhibition of adenylate cyclase in rat adrenal glomerulosa cells by angiotensin II. Endocrinology 115:337-341

Woods TJC, McCaa RE, Bower JD, McCaa CS (1974) Aldosterone response to angiotensin II in patients with terminal renal failure and after bilateral nephrectomy. Trans Am Soc Artif Intern Organs 20:154-160

Wrong O (1957) Communication to the 51st annual general meeting of the Association of Physicians of Great Britain and Ireland. Q J Med 26:586

Wrong O (1964) Hyperaldosteronism secondary to renal ischaemia. In: Baulieu EE, Robel P (eds) Aldosterone. Blackwell, Oxford, pp 376-392

Yamakado M, Franco-Saenz R, Mulrow PJ (1983) Effect of sodium deficiency on β-melanocyte-stimulating hormone stimulation of aldosterone in isolated rat adrenal cells. Endocrinology 113:2168-2172

Yamakado M, Franco-Saenz R, Mulrow PJ (1985) Sites of action of β-melanocyte stimulating hormone in aldosterone biosynthesis in the rat. Proc Soc Exp Biol Med 179:318-323

Yanagibashi K, Haniu M, Shively JE, Shen WH, Hall P (1986) The synthesis of aldosterone by the adrenal cortex. Two zones (fasciculata and glomerulosa) possess one enzyme for 11β-,18-hydroxylation, and aldehyde synthesis. J Biol Chem 261:3556-3562

Yoshimura M, Takahashi H, Takashina R, Kajita Y, Miyazaki T, Hachiya T, Ijichi H, Ochi Y (1979) Effect of prostaglandin E_1 on renin and aldosterone in hypertensive patients. Endocrinol Jpn 26:481-486

Young DB, McCaa RE (1981a) Effect of sustained hypernatremia on the renin-aldosterone system in the dog. Clin Exp Pharmacol Physiol 8:261-266

Young DB, McCaa RE (1981b) Lack of prolonged effect of antidiuretic hormone on the renin-aldosterone system in the dog. Clin Exp Pharmacol Physiol 8:267-271

Young DB, Smith MJ Jr, Jackson TE, Scott RE (1984) Multiplicative interaction between angiotensin II and K concentration in stimulation of aldosterone. Am J Physiol 247:E328-E335

Young RB, Sweat ML (1967) Steroid 17α-hydroxylation in the rat adrenal gland. Arch Biochem Biophys 121:576-579

Yu R, Anderton J, Skinner SL, Best JB (1972) Renin in anephric man. Case report with physiologic studies. Am J Med 52:707-711

Yun J, Kelly G, Bartter FC, Smith H Jr (1977) Role of prostaglandins in the control of renin secretion in the dog. Circ Res 40:459-464

Zachmann M, Völlmin JA, New MI, Curtius H-Ch, Prader A (1971) Congenital adrenal hyperplasia due to deficiency of 11β-hydroxylation of 17α-hydroxylated steroids. J Clin Endocrinol Metab 33:501-508

Zachmann M, Tassinari D, Prader A (1983) Clinical and biochemical variability of congenital adrenal hyperplasia due to 11β-hydroxylase deficiency. A study of 25 patients. J Clin Endocrinol Metab 56:222-229

Zadik Z, Kowarski AA (1980) Normal integrated concentration of aldosterone and plasma renin activity: effect of age. J Clin Endocrinol Metab 50:867-869

Zadik Z, Kahana L, Kaufman H, Benderli A, Hochberg Z (1984) Salt loss in hypertensive form of congenital adrenal hyperplasia (11β-hydroxylase deficiency). J Clin Endocrinol Metab 58: 384-387

Zager PG, Luetscher JA, Hsueh WA, Chavarri MR, Dowdy AJ (1979) Adrenocorticotropin-stimulated secretion of aldosterone and cortisol, computed from plasma and red blood cell measurements. J Clin Endocrinol Metab 48:441-450

Zager PG, Hsueh WA, Luetscher JA, Biglieri EG, Dowdy AJ (1980) Effect of des-asp^1-angiotensin II on secretion and metabolism of aldosterone. J Clin Endocrinol Metab 50:874-878

Zanella MT, Bravo EL (1982) In vitro and in vivo evidence for an indirect mechanism mediating enhanced aldosterone secretion by metoclopramide. Endocrinology 111:1620-1625

Zimmermann H, Ganong WF (1980) Pharmacological evidence that stimulation of central serotonergic pathways increases renin secretion. Neuroendocrinology 30:101-107

Zipser RD, Speckart PF, Zia PK, Edmiston WA, Lau FYK, Horton R (1976) The effect of ACTH and cortisol on aldosterone and cortisol clearance and distribution in plasma and whole blood. J Clin Endocrinol Metab 43:1101-1109

Zipser RD, Meidar V, Horton R (1980) Characteristics of aldosterone binding in human plasma. J Clin Endocrinol Metab 50:158-162

Zipser RD, Davenport MW, Martin KL, Tuck ML, Warner NE, Swinney RR, Davis CL, Horton R (1981) Hyperreninemic hypoaldosteronism in the critically ill: a new entity. J Clin Endocrinol Metab 53:867-873

Žofková I (1985) The effect of long-term verapamil treatment on the secretion of cortisol and aldosterone in subjects with normal and high blood pressure. Exp Clin Endocrinol 85:217-222

Zucker A, Gleason SD, Schneider EG (1982) Renal and endocrine response to water deprivation in dog. Am J Physiol 242:R296-R302

Subject Index

MIX
Papier aus verantwortungsvollen Quellen
Paper from responsible sources
FSC® C105338

If you have any concerns about our products,
you can contact us on
ProductSafety@springernature.com

In case Publisher is established outside the EU,
the EU authorized representative is:
Springer Nature Customer Service Center GmbH
Europaplatz 3, 69115 Heidelberg, Germany

Printed by Libri Plureos GmbH
in Hamburg, Germany